Simulation and Optimization in Process Engineering

Simulation and Optimization in Process Engineering

The Benefit of Mathematical Methods in Applications of the Chemical Industry

Edited by

Prof. Dr. Michael Bortz

Dr. Norbert Asprion

ELSEVIER

Elsevier
Radarweg 29, PO Box 211, 1000 AE Amsterdam, Netherlands
The Boulevard, Langford Lane, Kidlington, Oxford OX5 1GB, United Kingdom
50 Hampshire Street, 5th Floor, Cambridge, MA 02139, United States

ISBN: 978-0-323-85043-8

For information on all Elsevier publications
visit our website at https://www.elsevier.com/books-and-journals

Publisher: Susan Dennis
Editorial Project Manager: Judith Clarisse Punzalan
Production Project Manager: Bharatwaj Varatharajan
Cover Designer: Matthew Limbert

Typeset by STRAIVE, India

Contents

3. Thinking multicriteria—A jackknife when it comes to optimization

Norbert Asprion and Michael Bortz

4. Integrated modeling and energetic optimization of the steelmaking process in electric arc furnaces: An industrial application

Jesús D. Hernández, Luca Onofri, and Sebastian Engell

5. Solvent recovery by batch distillation—Application of multivariate sensitivity studies to high dimensional multiobjective optimization problems

Jan C. Schöneberger, Daniel Staak, and Jürgen Rarey

7. The use of digital twins to overcome low-redundancy problems in process data reconciliation

Filippo Bisotti, Andrea Galeazzi, Francesco Gallo, and Flavio Manenti

8. Real-time optimization of batch processes via optimizing feedback control

Dominique Bonvin, Grégory François, Gianluca Rizzi, and Michael Amrhein

Contributors

Numbers in parenthesis indicate the pages on which the authors' contributions begin.

Heiner Ackermann (339), Fraunhofer ITWM, Kaiserslautern, Germany

Simen Akkermans (243), BioTeC+, Department of Chemical Engineering, Technology Campus Gent, KU Leuven, Ghent, Belgium

Satyajith Amaran (365), The Dow Chemical Company, Midland, MI, United States

Michael Amrhein (201), Online Control, Lausanne, Switzerland

Norbert Asprion (57), BASF SE, Ludwigshafen, Germany

Tilman Barz (273), Center for Energy, AIT Austrian Institute of Technology GmbH, Vienna, Austria; KIWI-biolab, Department of Biotechnology, Technische Universität Berlin, Berlin, Germany

Satyajeet Bhonsale (243), BioTeC+, Department of Chemical Engineering, Technology Campus Gent, KU Leuven, Ghent, Belgium

Lorenz T. Biegler (33), Department of Chemical Engineering, Carnegie Mellon University, Pittsburgh, PA, United States

Filippo Bisotti (161), Politecnico di Milano, CMIC Dept. "Giulio Natta", Centre for Sustainable Process Engineering Research (SuPER), Milano, Italy

Dominique Bonvin (201), Laboratoire d'Automatique, EPFL, Lausanne, Switzerland

Michael Bortz (57,143), Fraunhofer Institute for Industrial Mathematics (ITWM), Kaiserslautern, Germany

Mariano Nicolas Cruz Bournazou (273), KIWI-biolab, Department of Biotechnology, Technische Universität Berlin, Berlin, Germany; DataHow AG, Dübendorf, Switzerland

Luisa C. Brée (225), Process Systems Engineering (AVT.SVT), RWTH Aachen University, Aachen, Germany

Andreas Bulan (225), Covestro Deutschland AG, Leverkusen, Germany

Jorge Diaz (321), BASF SE, Ludwigshafen, Germany

Sebastian Engell (77), Process Dynamics and Operations Group, Department of Biochemical and Chemical Engineering, TU Dortmund University, Dortmund, Germany

Gregor Foltin (321), Fraunhofer Institute for Industrial Mathematics (ITWM), Kaiserslautern, Germany

Grégory François (201), HES-SO Valais-Wallis, Sion, Switzerland

Andrea Galeazzi (161), Politecnico di Milano, CMIC Dept. "Giulio Natta", Centre for Sustainable Process Engineering Research (SuPER), Milano, Italy

Francesco Gallo (161), Itelyum Regeneration s.r.l, Lodi, Italy

Federico Galvanin (273), Department of Chemical Engineering, University College London (UCL), London, United Kingdom

Jochen Gattermayer (321), BASF SE, Ludwigshafen, Germany

Thomas Gerlach (143), Bayer AG, Building E 41, Leverkusen, Germany

Joachim Gross (1), Institute of Thermodynamics and Thermal Process Engineering, University of Stuttgart, Stuttgart, Germany

Ignacio E. Grossmann (365), Carnegie Mellon University, Pittsburgh, PA, United States

David Hajnal (321), BASF SE, Ludwigshafen, Germany

Raoul Heese (143), Fraunhofer Institute for Industrial Mathematics (ITWM), Kaiserslautern, Germany

Melanie Heidgen (321), Fraunhofer Institute for Industrial Mathematics (ITWM), Kaiserslautern, Germany

Michael Helmling (339), Fraunhofer ITWM, Kaiserslautern, Germany

Jesús D. Hernández (77), Process Dynamics and Operations Group, Department of Biochemical and Chemical Engineering, TU Dortmund University, Dortmund, Germany; Acciai Speciali Terni (AST), Terni, Italy

Christoph Herwig (273), Institute of Chemical, Environmental and Bioscience Engineering, Technische Universität Wien, Vienna, Austria

Stefan Hoeser (339), BASF Italia S.p.A., Pontecchio Marconi, Italy

Shachit S. Iyer (365), The Dow Chemical Company, Midland, MI, United States

Julian Kager (273), Competence Center CHASE GmbH, Linz, Austria

Thomas Kochenburger (143), Bayer AG, Building E 41, Leverkusen, Germany

Stefan Lehner (321), BASF SE, Ludwigshafen, Germany

Neele Leithäuser (339), Fraunhofer ITWM, Kaiserslautern, Germany

Filip Logist (243), BioTeC+, Department of Chemical Engineering, Technology Campus Gent, KU Leuven, Ghent, Belgium

Jonas Mairhofer (1), BASF SE, Ludwigshafen, Germany

Flavio Manenti (161), Politecnico di Milano, CMIC Dept. "Giulio Natta", Centre for Sustainable Process Engineering Research (SuPER), Milano, Italy

David C. Miller (33), National Energy Technology Laboratory, Pittsburgh, PA, United States

Alexander Mitsos (225), Process Systems Engineering (AVT.SVT), RWTH Aachen University, Aachen, Germany; JARA-ENERGY, Aachen, Germany; Institute of Energy and Climate Research: Energy Systems Engineering (IEK-10), Forschungszentrum Jülich GmbH, Jülich, Germany

Jan Mohring (143), Fraunhofer Institute for Industrial Mathematics (ITWM), Kaiserslautern, Germany

Peter Neubauer (273), KIWI-biolab, Department of Biotechnology, Technische Universität Berlin, Berlin, Germany

Philippe Nimmegeers (243), BioTeC+, Department of Chemical Engineering, Technology Campus Gent, KU Leuven, Ghent, Belgium

Chinedu O. Okoli (33), National Energy Technology Laboratory, Pittsburgh, PA, United States

Luca Onofri (77), Acciai Speciali Terni (AST), Terni, Italy

Hector D. Perez (365), Carnegie Mellon University, Pittsburgh, PA, United States

Karen Perrey (225), Covestro Deutschland AG, Leverkusen, Germany

Jürgen Rarey (101), Rareytec Co., Ltd, Nakhon Ratchasima, Thailand

Gianluca Rizzi (201), Online Control, Lausanne, Switzerland

Kosan Roh (225), Process Systems Engineering (AVT.SVT), RWTH Aachen University, Aachen, Germany; Department of Chemical Engineering and Applied Chemistry, Chungnam National University, Daejeon, Republic of Korea

Miguel A. Romero-Valle (339), BASF SE, Ludwigshafen, Germany

Jochen Schmid (143), Fraunhofer Institute for Industrial Mathematics (ITWM), Kaiserslautern, Germany

Jan C. Schöneberger (101), Chemstations Europe GmbH, Berlin, Germany

Hergen Schultze (321), BASF SE, Ludwigshafen, Germany

Daniel Staak (101), Lonza AG, Visp, Switzerland

Ioanna Stamati (243), BioTeC+, Department of Chemical Engineering, Technology Campus Gent, KU Leuven, Ghent, Belgium

Philipp Süss (321), Fraunhofer Institute for Industrial Mathematics (ITWM), Kaiserslautern, Germany

Dries Telen (243), BioTeC+, Department of Chemical Engineering, Technology Campus Gent, KU Leuven, Ghent, Belgium

Carlos Tellaeche (339), BASF SE, Ludwigshafen, Germany

Jan F.M. Van Impe (243), BioTeC+, Department of Chemical Engineering, Technology Campus Gent, KU Leuven, Ghent, Belgium

John M. Wassick (365), The Dow Chemical Company, Midland, MI, United States

Christian Weiß (339), Fraunhofer ITWM, Kaiserslautern, Germany

Jarosław Wlazło (143), Fraunhofer Institute for Industrial Mathematics (ITWM), Kaiserslautern, Germany

Preface

Knowledge- and data-based modeling, in conjunction with setting up new technical solutions, form the basis of progress driven by engineers. A high degree of domain expertise is necessary to arrive at new insights and ideas that have a substantial practical impact.

As opposed to highly specialized engineers, mathematicians aim to find context-free structural relations, prove the existence of solutions, and design algorithms to identify them. The more general mathematical statements are, the higher their considered value.

The idea for this book is rooted in the conviction that substantial benefits can arise when these two worlds—the domain-driven engineering world and the method-driven mathematical world—come into contact. These benefits can be twofold as explained in the following.

On the one hand, from a scientific point of view, mathematical understanding of models and their structures can generate new solutions for engineering problems. The way how a model is formulated can have a significant impact on whether and how existing solutions can be revealed algorithmically. Furthermore, real-world problems posed by engineers can challenge mathematical methods and go well beyond what are considered as examples in mathematical textbooks.

On the other hand, from a practical point of view, significant progress can result from this pairing. If the application of mathematical methods leads to more reliable, more versatile, and more applicable model predictions, less trial and error is necessary, and the time needed to arrive at improvements—of whatever kind these may be—can be reduced dramatically. Eventually, this is the very promise of model-based simulation and optimization: that predictions can be made for highly complex, nonlinear systems on how to achieve high utility at not more than the necessary costs.

This book highlights applications from chemical engineering that show how mathematical advances in simulation and optimization lead to substantial impact in the real world. The order of the chapters follows the different scales involved in chemical engineering, ranging from substance properties over unit and flowsheet models, both in the steady state and dynamically, up to organizational questions when it comes to scheduling and production planning.

Throughout the book, the reader will discover how mathematical structure hidden in models, in simulation and optimization tasks, is exploited to arrive at new insights. We hope that this book will be an inspiration for both engineers and mathematicians challenged by real-world problems.

Michael Bortz and Norbert Asprion
Kaiserslautern and Ludwigshafen

Chapter 1

Prediction and correlation of physical properties including transport and interfacial properties with the PC-SAFT equation of state

Jonas Mairhofer[a] and Joachim Gross[b]
[a]BASF SE, Ludwigshafen, Germany, [b]Institute of Thermodynamics and Thermal Process Engineering, University of Stuttgart, Stuttgart, Germany

The reliability of a process simulation, to a large extent, is determined by the fitness of the applied physical property model. Adequate models are required for many different physical properties: the simulation of a distillation column using the equilibrium stage approach may require a model for activity coefficients, vapor pressure, enthalpy, and possibly vapor phase fugacity coefficients. Rate-based process models, in addition, need transport properties such as viscosity, thermal conductivity, or diffusion coefficients as input. In order to determine volume flows from mass or molar flows, for example for equipment sizing, the density of a stream needs to be known. Furthermore, entropy plays an important role in the simulation of compressors or pumps, etc. High demands are placed on the physical property models: they have to accurately correlate available experimental data, but also show robustness in extrapolating over large ranges of temperature and pressure. They have to be able to model the thermodynamic behavior of species and mixtures exhibiting complex molecular interactions. At the same time, they cannot be arbitrarily complex in order to ensure reliable numerical solutions and to avoid prohibitively long run times of a process simulation.

Several alternatives exist for providing the required physical properties. One of the most widely used approaches in a process simulation is in applying an activity coefficient model, such as the *nonrandom two-liquid model* (NRTL) [1] for modeling the liquid phase nonideality in combination with dedicated, usually empirical correlations for the pure-component vapor pressures, liquid

Simulation and Optimization in Process Engineering. https://doi.org/10.1016/B978-0-323-85043-8.00002-7

and vapor phase heat capacities, liquid density and further required properties. The gas phase is most often approximated as an ideal gas. This approach shows several drawbacks: it is not suited for high-pressure applications because the pressure dependence of the activity coefficients is usually neglected, and the ideal gas approximation will not be justified. The nonideality of the gas phase then also has to be addressed in calculating the vapor enthalpy. Furthermore, special treatment is necessary for components with a critical temperature lower than the simulation temperature.

An equation of state model complemented with expressions for ideal gas heat capacities of all pure species can provide all static thermodynamic properties. Furthermore, models for transport coefficients, such as viscosity, thermal conductivity, or diffusion coefficients, exist which build on outputs of an equation of state. However, simple cubic equations of state such as the Peng-Robinson [2] or Soave-Redlich-Kwong [3] equations of state can only be applied to simple molecules and in general, do not produce results that are accurate enough for all required properties in most real-world processes simulation applications. The need for more accurate thermodynamic models applicable also to molecules showing complex molecular interactions such as hydrogen-bonding or polar interactions led to the development of more sophisticated, physically-based equations of state such as the *statistical associating fluid theory* (SAFT) developed by Chapman et al. [4,5] and Jackson et al. [6] based on the work of Wertheim [7–10]. SAFT leads to an expression for the residual Helmholtz energy $A^{res}(T,\rho,x) = A(T,\rho,x) - A^{ig}(T,\rho,x)$, where A^{ig} denotes the Helmholtz energy of the ideal gas at temperature T, number density ρ and mole fractions x. The value of A^{res} is obtained as the sum of different contributions. Each contribution takes into account a specific type of molecular interaction. The different Helmholtz energy contributions are developed using perturbation theory which (under suitable conditions) allows to obtain the thermodynamic properties of a target fluid with specified interaction potential from the properties of a reference fluid with a simpler interaction potential and known properties. The advantage of SAFT is that it allows to include a contribution for highly directional attractive interactions such as hydrogen-bonding which makes it a suitable choice for modeling the properties of molecules (and their mixtures) exhibiting such complex molecular interactions. Different SAFT-type equations of state can be derived depending on the choice of reference fluid and interaction potential of the target fluid.

A detailed description of the fundamentals of SAFT can be found in the book by Solana [11]. A comprehensive list of successful applications of SAFT-type equations of state to complex systems including polar and associating molecules, polymers, ionic liquids, pharmaceuticals as well as bio-molecules can be found in review articles [12–18].

The scope of this chapter is to introduce the *perturbed-chain statistical associating fluid theory* (PC-SAFT). The equations to implement the model are given in Section 1. The determination of meaningful pure-component and binary interaction parameters is the topic of Section 2. Section 3 presents group-contribution methods for PC-SAFT which allow to predict the model parameters in the frequently encountered situation that not enough experimental data is available for a given molecule, thus preventing parameter regression. Finally, Sections 4 and 5 are concerned with correlating or predicting the transport and interfacial properties, respectively.

1 Model equations of PC-SAFT

In the case of PC-SAFT developed by Gross and Sadowski [19–21], the hard-chain fluid is used as the reference. Each chain is made up of m bonded hard-sphere segments of diameter parameter σ. Attractive interactions such as dispersive interactions or hydrogen bonding between the chains are then added as perturbations to this purely-repulsive reference fluid. The dimensionless residual Helmholtz energy $a^{res} \equiv A^{res}/NkT$ with molecule number N and Boltzmann's constant k is obtained as

$$a^{res} = a^{hc} + a^{disp} + a^{assoc} + a^{dd} + a^{qq} + a^{dq} \tag{1.1}$$

The Helmholtz energy has several contributions, namely the hard-chain reference fluid, a^{hc}, the change in Helmholtz energy due to dispersive interactions between the chains, a^{disp}, highly directional attractive interactions such as hydrogen-bonding, a^{assoc}, and dipole-dipole, quadrupole-quadrupole as well as dipole-quadrupole interactions. With polar contributions included, the model is usually referred to as *perturbed-chain polar statistical associating fluid theory* (PCP-SAFT). In the following section, the various terms of Eq. (1.1) will be presented for a mixture at temperature T, number density ρ, and mole fraction of component i, x_i. Using basic thermodynamic relationships, all other thermodynamic properties can be obtained as derivatives of a^{res}.

The Helmholtz energy of the hard-chain fluid is given by

$$a^{hc} = \overline{m}a^{hs} + \sum_i x_i(1 - m_i) \ln\left(g_{ii}^{hs}\right) \tag{1.2}$$

where a^{hs} denotes the Helmholtz energy of the hard-sphere fluid. The average segment number of the mixture is calculated as $\overline{m} = \sum_i x_i m_i$ with m_i as the number of segments on a chain of components i. We note that parameter m_i is in SAFT models relaxed to be a real-valued (rather than integer-valued) parameter.

The value of a^{hs} is obtained from the accurate equation of state for the hard-sphere fluid presented by Boublík [22] and Mansoori et al. [23]

$$a^{hs} = \frac{1}{\zeta_0}\left(\frac{3\zeta_1\zeta_2}{1-\zeta_3} + \frac{\zeta_2^3}{\zeta_3(1-\zeta_3)^2} + \ln(1-\zeta_3)\left(\frac{\zeta_2^3}{\zeta_3^2} - \zeta_0\right)\right) \tag{1.3}$$

with density measures ζ_n defined as

$$\zeta_n = \frac{\pi}{6}\rho\sum_i x_i m_i d_i^n \quad n \in \{0, 1, 2, 3\} \tag{1.4}$$

Here, d_i denotes the temperature-dependent segment diameter of component i

$$d_i(T) = \sigma_i\left[1 - 0.12\exp\left(\frac{-3\varepsilon_i}{kT}\right)\right] \tag{1.5}$$

with the dispersive interaction energy ε_i/k between segments within the chain of component i. Furthermore, the radial distribution function of the hard-sphere fluid at contact distance is calculated as

$$g_{ij}^{hs} = \frac{1}{1-\zeta_3} + \left(\frac{d_i d_j}{d_i + d_j}\right)\frac{3\zeta_2}{(1-\zeta_3)^2} + \left(\frac{d_i d_j}{d_i + d_j}\right)^2 \frac{2\zeta_2^2}{(1-\zeta_3)^3} \tag{1.6}$$

The contribution to the Helmholtz energy due to dispersive interactions between the chain molecules is developed as a perturbation to the hard-chain reference fluid using the second-order perturbation theory of Barker and Henderson [24,25] extended to chain fluids [19] and a perturbation potential of Lennard-Jones type, with

$$a^{disp} = a_1^{disp} + a_2^{disp} \tag{1.7}$$

The contributions of first and second-order are obtained as

$$a_1^{disp} = -2\pi\rho I_1(\eta, \overline{m})\sum_i\sum_j x_i x_j m_i m_j\left(\frac{\varepsilon_{ij}}{kT}\right)\sigma_{ij}^3 \tag{1.8}$$

$$a_2^{disp} = -\pi\rho\overline{m}\left(1 + Z^{hc} + \rho\frac{\partial Z^{hc}}{\partial\rho}\right)^{-1} I_2(\eta, \overline{m})\sum_i\sum_j x_i x_j m_i m_j\left(\frac{\varepsilon_{ij}}{kT}\right)^2\sigma_{ij}^3 \tag{1.9}$$

with

$$\left(1 + Z^{hc} + \rho\frac{\partial Z^{hc}}{\partial\rho}\right) = \left(1 + \overline{m}\frac{8\eta - 2\eta^2}{(1-\eta)^4} + (1-\overline{m})\frac{20\eta - 27\eta^2 + 12\eta^3 - 2\eta^4}{[(1-\eta)(2-\eta)]^2}\right) \tag{1.10}$$

and packing fraction $\eta = \zeta_3$. The values for ε_{ij} and σ_{ij} are determined from the Lorentz-Berthelot combining rules as $\varepsilon_{ij} = \sqrt{\varepsilon_i\varepsilon_j}$ and $\sigma_{ij} = 0.5(\sigma_i + \sigma_j)$. As discussed in Section 2.2, it is often necessary to introduce adjustable binary interaction parameters (BIP) for calculating ε_{ij} and σ_{ij} in order to improve the description of mixture properties. The perturbation approach requires the evaluation of integrals over the pair-correlation function of the reference fluid and the perturbation potential. In PC-SAFT, these integrals are approximated as power-series in density

$$I_1(\eta, \overline{m}) = \sum_{i=0}^{6} a_i(\overline{m})\eta^i \tag{1.11}$$

$$I_2(\eta, \overline{m}) = \sum_{i=0}^{6} b_i(\overline{m})\eta^i \tag{1.12}$$

with coefficients $a_i(\overline{m})$ and $b_i(\overline{m})$ that depends on the average segment number of the mixture. A simple but accurate [19] dependence on segment-number was taken in analogy to a chain-formation theory by Liu and Hu [26], as

$$a_i(\overline{m}) = a_{0i} + \frac{\overline{m}-1}{\overline{m}} a_{1i} + \frac{\overline{m}-1}{\overline{m}} \frac{\overline{m}-2}{\overline{m}} a_{2i} \tag{1.13}$$

$$b_i(\overline{m}) = b_{0i} + \frac{\overline{m}-1}{\overline{m}} b_{1i} + \frac{\overline{m}-1}{\overline{m}} \frac{\overline{m}-2}{\overline{m}} b_{2i} \tag{1.14}$$

The model constants a_{0i}, a_{1i}, a_{2i} as well as b_{0i}, b_{1i}, and b_{2i} were adjusted to experimental vapor pressure and PvT-data of n-alkanes. Their values can be found in the original publication [20].

In order to include highly directional attractive interactions such as hydrogen bonds, association sites are placed on the chain molecules. These sites can be of different types and only interactions between certain site types are allowed to occur. Site types can for example represent electron donors or acceptors. Interactions are then allowed between a donor and an acceptor site but not between two donor or two acceptor sites. Early classification of association-site schemes for several important chemical families can be found in the work of Huang and Radosz [27]. The final expression for the Helmholtz energy contribution due to association is

$$a^{assoc} = \sum_i x_i \sum\nolimits_{A \in \Gamma^i} \left(\ln \chi_A^i - \frac{\chi_A^i}{2} + \frac{1}{2} \right) \tag{1.15}$$

where Γ^i denotes the set of association sites located on a molecule of species i and χ_A^i is the fraction of nonbonded association sites A on molecules of type i. The value of χ_A^i has to be determined by solving the set of nonlinear equations given by

$$\chi_A^i = \left(1 + \rho \sum_j x_j \sum\nolimits_{B \in \Gamma^j} \chi_B^j \Delta_{A_i, B_j} \right)^{-1} \tag{1.16}$$

with

$$\Delta_{A_i, B_j} = \sigma_{ij}^3 g_{ij}^{hs} \kappa_{A_i, B_j}^{HB} \left(\exp\left(\frac{\varepsilon_{A_i, B_j}^{HB}}{kT} \right) - 1 \right) \tag{1.17}$$

Here, $\varepsilon_{Ai,\ Bj}^{HB}$ and $\kappa_{Ai,\ Bj}^{HB}$ denote the association strength between site A on molecules of component i and site B on molecules of component j and the dimensionless volume in which association between the two sites can occur, respectively. Their values are obtained from the following combining rules suggested by Wolbach and Sandler [28]

$$\varepsilon_{A_i, B_j}^{HB} = \frac{1}{2} \left(\varepsilon_{A_i, B_i}^{HB} + \varepsilon_{A_j, B_j}^{HB} \right) \tag{1.18}$$

and

$$\kappa_{A_i,B_j}^{HB} = \left(\sqrt{\kappa_{A_i,B_i}^{HB}\kappa_{A_j,B_j}^{HB}}\right)\sqrt{\left(\sigma_i\sigma_j\right)^3}/\sigma_{ij}^3 \tag{1.19}$$

Strictly speaking, combining rules have no sound justification for cross-associations between two (self-)associating molecules i and j (as opposed to dispersive interactions, where approximations can be made to derive the Berthelot-Lorentz combining rules). Cross-associations can be determined through quantum mechanical calculations. In many practical applications, however, the simple combining rules, Eqs. (1.18) and (1.19), can produce suitable approximations of the cross-association parameters. Clearly, binary interaction parameters may be introduced in Eqs. (1.18) and (1.19) to improve results, see Section 2.2. Efficient schemes to solve Eq. (1.16) and to determine derivatives of a^{assoc} have been developed for example by Michelsen [29], Michelsen and Hendriks [30], Tan et al. [31], and Langenbach and Enders [32].

In mixtures of self-associating molecules i with molecules j that do not self-associate which, however, provide proton donor or acceptor sites that can form cross-hydrogen bonds with molecules of type i, it is often advisable to assign a nonzero value for $\kappa_{A_i, Bj}^{HB}$ to component j [33]. Pure-component results for component j remain unchanged because the value for $\varepsilon_{A_i, Bj}^{HB}$ is set to zero. However, the model now takes the cross-association between molecules i and j into account because both, cross-association volume, $\kappa_{A_i, Bj}^{HB}$, as well as cross-association energy, $\varepsilon_{A_i, Bj}^{HB}$, are nonzero. To fine-tune mixture results after a (more or less arbitrary) value for association volume was assigned to component j, BIP may be introduced as presented in Section 2.2.

Helmholtz energy contributions for polar molecules were developed, among others, by Gross [34] for quadrupole-quadrupole interactions, Gross and Vrabec [35] for dipole-dipole interactions, and Vrabec and Gross [36] for dipole-quadrupole interactions. In all three cases, the resulting expressions are obtained from a third-order perturbation approach presented by Stell et al. [37,38] extended to the two-center Lennard-Jones fluid. The three polar Helmholtz energy contributions are similar in structure and only the expression for the dipole-dipole contribution a^{dd} is presented here. The reader is referred to the original publications for details on the quadrupole-quadrupole term, a^{qq}, and the dipole-quadrupole contribution, a^{dq}.

For dipole-dipole interactions, the Helmholtz energy is given as the result of the third-order perturbation in Padé approximation

$$a^{dd} = \frac{a_2^{dd}}{1 - a_3^{dd}/a_2^{dd}} \tag{1.20}$$

The second and third-order perturbation terms are

$$a_2^{dd} = -\pi\rho\sum_i\sum_j x_i x_j \frac{\varepsilon_i}{kT}\frac{\varepsilon_j}{kT}\frac{\sigma_i^3\sigma_j^3}{\sigma_{ij}^3}\mu_i^{*2}\mu_j^{*2}J_{2,ij}^{dd} \tag{1.21}$$

and

$$a_3^{dd} = -\frac{4\pi^2}{3}\rho^2 \sum_i \sum_j \sum_k x_i x_j x_k \frac{\varepsilon_i}{kT}\frac{\varepsilon_j}{kT}\frac{\varepsilon_k}{kT}\frac{\sigma_i^3 \sigma_j^3 \sigma_k^3}{\sigma_{ij}\sigma_{ik}\sigma_{jk}} \mu_i^{*2} \mu_j^{*2} \mu_k^{*2} J_{3,ijk}^{dd} \tag{1.22}$$

The dimensionless squared dipole moment is obtained as $\mu_i^{*2} = \mu_i^2/(m_i \varepsilon_i \sigma_i^3)$, with dipole moment of component i, μ_i. The perturbation approach requires the evaluation of integrals over the pair-correlation and three-body correlation functions of the reference fluid. These integrals are approximated as simple power functions of dimensionless packing fraction η:

$$J_{2,ij}^{dd} = \sum_{n=0}^{4} \left(a_{n,ij} + b_{n,ij} \frac{\sqrt{\epsilon_i \epsilon_j}}{kT} \right) \eta^n \tag{1.23}$$

$$J_{3,ijk}^{dd} = \sum_{n=0}^{4} c_{n,ijk} \eta^n \tag{1.24}$$

The coefficients $a_{n,ij}$, $b_{n,ij}$, and $c_{n,ijk}$ depend on chain length m as

$$a_{n,ij} = a_{0n} + \frac{m_{ij}-1}{m_{ij}} a_{1n} + \frac{m_{ij}-1}{m_{ij}} \frac{m_{ij}-2}{m_{ij}} a_{2n} \tag{1.25}$$

$$b_{n,ij} = b_{0n} + \frac{m_{ij}-1}{m_{ij}} b_{1n} + \frac{m_{ij}-1}{m_{ij}} \frac{m_{ij}-2}{m_{ij}} b_{2n} \tag{1.26}$$

and

$$c_{n,ijk} = c_{0n} + \frac{m_{ijk}-1}{m_{ijk}} c_{1n} + \frac{m_{ijk}-1}{m_{ijk}} \frac{m_{ijk}-2}{m_{ijk}} c_{2n} \tag{1.27}$$

The following combining rules are used for m_{ij} and m_{ijk}

$$m_{ij} = \left(\check{m}_i \check{m}_j \right)^{1/2} \quad \text{and} \quad m_{ijk} = \left(\check{m}_i \check{m}_j \check{m}_k \right)^{1/3} \tag{1.28}$$

$$\check{m}_i = \min(m_i, 2) \tag{1.29}$$

The model constants a_{0n}, a_{1n}, a_{2n}, b_{0n}, b_{1n}, b_{2n}, c_{0n}, c_{1n}, c_{2n} were adjusted to the results of molecular simulations of Stoll et al. [39]. Their values can be found in the original publication. In summary, a molecule of component i that is modeled as nonassociating and nonpolar is characterized by three pure-component model parameters: segment number m_i, segment diameter parameter σ_i, and dispersive energy ε_i. To include association, two more parameters are required: association strength $\varepsilon_{A_i,Bj}^{HB}$ and association volume $\kappa_{A_i,Bj}^{HB}$. The polar contributions require the value for the dipole or quadrupole moment. For both, literature values can be used.

2 Parameterization

2.1 Pure-component parameters

The goal in the adjustment of pure-component parameters is, of course, the accurate description of the properties of interest. The simplest way to achieve this is to include experimental data for the target properties in the parameter regression. If only data for other properties are available, one has to make sure, that parameters adjusted to this data allow satisfying predictions of the primary properties of interest.

When choosing the properties to include for parameter optimization, it has to be ensured that these properties indeed define all model parameters well. For example, in the PC-SAFT model for a nonassociating, nonpolar molecule, the values for enthalpy of evaporation and critical temperature do not depend on the value of the parameter σ and the Pitzer acentric factor ω is only a function of segment number m [40].

Also, limitations of equations of state such as PC-SAFT to accurately reproduce the critical region have to be kept in mind, and data in the vicinity of the critical point is usually excluded from the parameter optimization.

Most parameter sets found in the literature for PC-SAFT were adjusted to experimental data for vapor pressure, p^s, and liquid density, ρ_l. Other studies include data for speed of sound, surface tension [41,42], fraction of nonbonded association sites [43], or results of molecular simulations [41]. To address the conflict that exists between the subobjectives of describing different thermodynamic properties accurately, the parameter adjustment has been performed as a multicriteria optimization problem [42,44].

A recent study [45] on the parameter adjustment for nonassociating, nonpolar molecules, concluded that it is indeed sufficient to use combined data for p^s and ρ_l in order to obtain parameter values for m, σ, and ε that allow the accurate calculation of the target properties vapor pressure, saturated-liquid density, ρ_l^s, enthalpy of evaporation, Δh^{lv}, as well as saturated-liquid heat capacity, $c_p^{l,\ s}$, a set that is most relevant for process simulation. Reported average deviations for a database of more than a thousand nonassociating, nonpolar molecules with parameters adjusted to vapor pressure and saturated-liquid density account to only 0.97% for p^s, 0.85% for ρ_l^s, 3.04% for Δh^{lv} and 3.35% for $c_p^{l,\ s}$ [45]. The literature survey of Zhu et al. [46] evaluates the performance of several equations of state to reproduce heat capacities and speed of sound for nonassociating as well as associating molecules such as alcohols and water. The study shows that predictions of PC-SAFT with parameters adjusted to vapor pressure and liquid density for these derivative properties can still be in satisfactory agreement with experimental data. However, larger deviations have to be expected than for nonassociating molecules. Furthermore, results are sensitive to the choice of the association scheme [47,48]. The remedy in cases with larger deviations of the predicted derivative properties is to include these properties in the parameter optimization as demonstrated, e.g., in the work of Oliveira et al.

[49]. Furthermore, a considerable effort has been undertaken to reduce the number of model parameters that need to be adjusted, especially in the case of molecules modeled as associating fluids, where five model parameters need to be determined and different local and shallow minima are often observed. These approaches include the use of generalized pure-component parameters for members of the same chemical family [50], establishing correlations for the parameters of molecules of homologous series [51–53], performing sensitivity analysis and fixing the least-sensible parameter [54], determining the value of single parameters, often the association energy, by independent methods such as ab initio calculations [55–57], molecular simulation [58] or COSMO-RS [53] or establishing a link between the parameters of PC-SAFT and those of cubic EOS [59]. Special routes to determine PC-SAFT parameters have been developed for polymers, where vapor pressure data are unavailable [60,61] or for petroleum fractions with unspecified components [62]. Maybe the most important approach to alleviating the need for determining pure-component parameters is in group-contribution methods, as described below, in Section 3.

2.2 Binary interaction parameters

In mixtures where dispersive interactions are the dominant part of the attractive intermolecular potentials, often accurate results can be obtained without introducing binary interaction parameters (BIP) in the combining rules for ε_{ij}, σ_{ij}, $\varepsilon^{HB}_{Ai,\ Bj}$, and $\kappa^{HB}_{Ai,\ Bj}$. However, for many mixtures of industrial interest, BIP is necessary to improve mixture results. In this section, commonly applied BIP is presented and it is discussed how meaningful values can be obtained.

The dispersive energy between molecules of type i and molecules of type j, ε_{ij}, can be corrected with BIP k_{ij}, as

$$\varepsilon_{ij} = \sqrt{\varepsilon_i \varepsilon_j}\left(1 - k_{ij}\right) \tag{1.30}$$

A further BIP for dispersive energy, l_{ij}, which is especially useful for correlating liquid-liquid equilibria is sometimes used as a correction to the value of σ_{ij} as

$$\sigma_{ij} = 0.5\left(\sigma_i + \sigma_j\right)\left(1 - l_{ij}\right) \tag{1.31}$$

Alternatively, Tang and Gross [63] proposed an asymmetric mixing rule where a BIP acts on the double sum in the first-order perturbation term of the dispersive contribution, a_1^{disp}, Eq. (1.8).

Between two species modeled as associating molecules, additional BIP, k_{ij}^{HB} and l_{ij}^{HB}, can be introduced in the combining rules for the association energy and volume:

$$\varepsilon^{HB}_{A_i,B_j} = 0.5\left(\varepsilon^{HB}_{A_i,B_i} + \varepsilon^{HB}_{A_j,B_j}\right)\left(1 - k_{ij}^{HB}\right) \tag{1.32}$$

And

$$\kappa^{HB}_{A_i,B_j} = \left(\sqrt{\kappa^{HB}_{A_i,B_i}\kappa^{HB}_{A_j,B_j}} + l^{HB}_{ij}\right)\sqrt{\left(\sigma_i\sigma_j\right)^3}/\sigma^3_{ij} \tag{1.33}$$

If results over wide temperature ranges are relevant, temperature-dependent BIP can be introduced, for example, as

$$k_{ij}(T) = a + b\left(\frac{T/\mathrm{K} - 300}{100}\right) + c\left(\frac{T/\mathrm{K} - 300}{100}\right)^2 \tag{1.34}$$

In many cases, however, a temperature-dependent $k_{ij}(T)$ can be avoided by using two or more constant BIP of different kinds, for example, a constant k_{ij} and a constant l_{ij}. Analogous to pure-component parameters, the binary interaction parameters, adjusted for members of the same chemical family, often exhibit well-behaved trends which can be exploited in order to obtain parameters in the absence of experimental data. Fig. 1.1 illustrates this behavior for binary mixtures of methane with n-alkanes: despite some outliers, a clear trend can be observed for the optimized k_{ij} value with the chain length of the n-alkane.

Similar trends were also observed for more complex mixtures: Haarmann et al. [65] demonstrated that a correlation for BIP obtained from parameters adjusted to LLE of certain mixtures of n-alkane with water with increasing carbon-number of the alkane, can be extrapolated to further n-alkanes. For mixtures of benzyl-ethanoate with n-alkanes, it was shown that only one value for k_{ij}

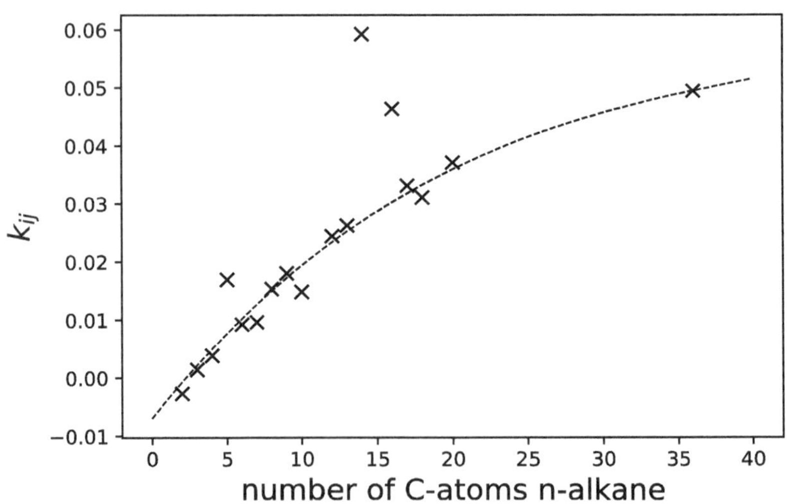

FIG. 1.1 Constant k_{ij} adjusted to binary VLE data [64] for methane with several n-alkanes as a function of carbon atoms on the n-alkane. The *dotted line* represents a cubic polynomial adjusted to the results (excluding values for tetradecane and hexadecane).

was required to obtain accurate results for excess enthalpy for all considered mixtures with *n*-alkanes [66].

In many applications, the accurate description of phase equilibria is of high priority. Thus, the BIP is most often adjusted to experimental data for binary vapor-liquid or liquid-liquid equilibria. However, in process simulation, the physical property model also needs to provide accurate results for other mixture properties, such as densities and enthalpies. Several studies [67–72] show that predicted mixture density, that is with all BIP set to zero, is sufficiently accurate in many cases. Furthermore, mixture densities obtained with BIP adjusted to phase equilibrium data were also shown to be in good agreement with experimental results [73–75].

It is furthermore desirable that the BIP of a model can be adjusted to such thermodynamic properties that are obtained by simple measurements and are prevalently available, whereas the model will be subsequently applied to another property of interest. The work of Schacht et al. [76] demonstrates that BIP adjusted to data for activity coefficients at infinite dilution, γ^{∞}, extrapolate well to describe phase equilibria over the entire composition range. Satisfactory results for excess enthalpy were obtained for mixtures of ionic liquids with alcohols [77] and solvents [78], again, using BIP adjusted to γ^{∞}. Soo [79] shows that for some mixtures, it is possible to obtain good predictions of the phase equilibrium with a single BIP adjusted to data for excess enthalpy of the system. However, systems are also presented where this approach fails. In this case, it is of course possible to adjust several BIP to phase equilibrium data as well as results for excess enthalpy. A further frequent requirement in process simulation is the satisfactory description of vapor-liquid as well as liquid-liquid equilibria using the same set of binary interaction parameters. Tumakaka et al. [16] presented convincing results for the simultaneous correlation of VLE and LLE data for the system methanol cyclohexane using a single, constant value for k_{ij}. On the other hand, the results of Grenner et al. [50] indicate that for water-alcohol systems, a constant k_{ij} is not sufficient to describe VLE and LLE well.

A case study for the binary mixture of water with 1-butanol addresses several of the issues discussed in this section. Results for this mixture are presented in Figs. 1.2 and 1.3. In order to test the capability to simultaneously correlate VLE and LLE data, BIP is simultaneously adjusted to both types of data: four temperature-independent BIP are adjusted to experimental VLE and LLE results at $p = 1$ bar only, corresponding to the black line in Fig. 1.2. VLE results are correlated very accurately and the composition of both liquid phases of the LLE is also in good agreement with the experimental data. To assess the extrapolation capability for varying temperature and pressure (not considered in the parameter optimization), the model is then applied to VLE at 2, 5, and 8 bar and to LLE at 2500 bar. Considering the large temperature and pressure range of the data, model results are very satisfactory, as seen in Fig. 1.2. Mixture density as well as excess enthalpy obtained with these BIP is presented in Fig. 1.3.

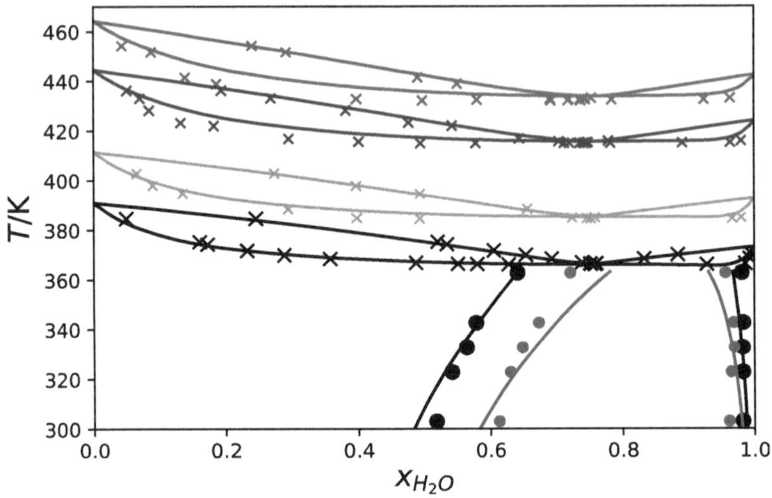

FIG. 1.2 Phase equilibria of the binary mixture of water and 1-butanol. Values for the four temperature-independent BIP, k_{ij} (−0.0223), l_{ij} [63] (−0.0102), k_{ij}^{HB} (−0.1256) as well as l_{ij}^{HB} (4.1437e-3), are adjusted to experimental data for VLE and LLE at $p = 1$ bar only *(larger black crosses and black circles)*. Results for VLE at higher pressures (*green crosses, gray in print version*: 1.98 bar, *blue crosses, dark gray in print version*: 4.92 bar, and *red crosses, gray in print version*: 7.87 bar) and LLE at 2482 bar *(magenta circles, light gray in print version)* are calculated using these BIP values. Pure-component model parameters for water and 1-butanol are taken from Stavrou [80] and Lötgering-Lin et al. [81]. Experimental data is taken from Aoki and Moriyoshi [82], Boublik [83] as well as Hessel and Geiseler [84].

The BIP only has a slight impact on mixture density and model results are in good agreement with the experimental data. Results of excess enthalpy are more sensitive toward the BIP. Deviations to experimental data are greatly reduced by applying the BIP obtained from the phase behavior at 1 bar. It is apparent that in this case, excess enthalpy data should be included in the BIP optimization in order to obtain a more accurate description of mixture enthalpy.

Various objective functions can be defined for adjusting a BIP to VLE or LLE data, where the objective function contains deviations (residuals) of the model to experimental data. Simple definitions of residuals, such as the deviation of calculated pressure to experimental phase equilibrium pressure of a bubble or dew point may be useful for some binary mixtures. For mixtures where the bubble or dew point lines are steep with pressure and possibly have a change in the sign of the slope, such a measure is not suitable. Numerically robust and fast optimization of BIP for a binary mixture is possible if two coexisting phases $'$ and $''$ are fully defined by experimental data, e.g., as (T,p,x_k',x_k'') data sets. One can then use the difference in fugacity $f_i(T,p,x_k'') - f_i(T,p,x_k')$ as a definition for residuals, noting that the difference would be zero if the model exactly reproduced the phase equilibrium. An objective function that minimizes

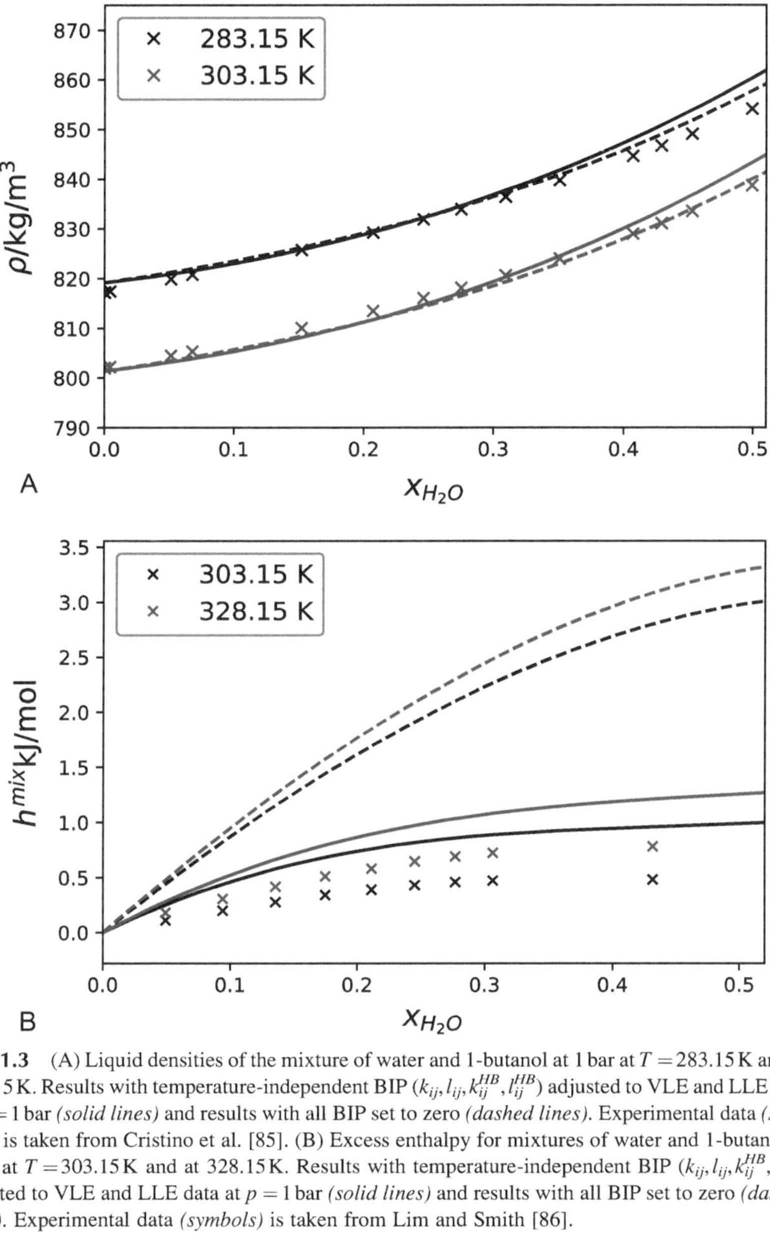

FIG. 1.3 (A) Liquid densities of the mixture of water and 1-butanol at 1 bar at $T = 283.15\,\mathrm{K}$ and at 303.15 K. Results with temperature-independent BIP ($k_{ij}, l_{ij}, k_{ij}^{HB}, l_{ij}^{HB}$) adjusted to VLE and LLE data at $p = 1$ bar *(solid lines)* and results with all BIP set to zero *(dashed lines)*. Experimental data *(symbols)* is taken from Cristino et al. [85]. (B) Excess enthalpy for mixtures of water and 1-butanol at 1 bar at $T = 303.15\,\mathrm{K}$ and at 328.15 K. Results with temperature-independent BIP ($k_{ij}, l_{ij}, k_{ij}^{HB}, l_{ij}^{HB}$) adjusted to VLE and LLE data at $p = 1$ bar *(solid lines)* and results with all BIP set to zero *(dashed lines)*. Experimental data *(symbols)* is taken from Lim and Smith [86].

residuals defined as the difference in fugacities can be strongly recommended at least for a preoptimization. Using the so-obtained BIP, a final optimization can be performed. A discussion of suitable objective functions for this final optimization can be found for example in the work of Soave et al. [87] and Höller et al. [88].

3 Group-contribution methods for PC-SAFT

It is not always possible to determine the model parameters for a pure-component or a binary mixture by adjusting the parameters to data of that specific molecule or binary pair. For many species and, even more so, for many binary pairs, such data may not exist at all or in quality or quantity that does not allow a meaningful parameter adjustment. In such circumstances, group-contribution (GC) methods often offer a suitable approach to predict the required thermodynamic properties. The basic idea of a GC approach is to characterize a given molecule by the functional groups that make up the molecule. Model parameters for the single functional groups or for the binary interaction between different types of groups are adjusted to data of molecules that include a given functional group. It is then assumed, that the properties of the molecule can approximately be determined from the contributions of all the functional groups on that molecule. For example, the model parameters for the groups —CH_3 and —CH_2— could be obtained by fitting them to the available data for all *n*-alkanes. One could then continue to adjust parameters for the —OH group to data of all alcohols using the already known parameters for the groups —CH_3 and —CH_2—. Of course, a simultaneous adjustment of all group parameters is also possible.

This procedure, however, already shows some of the challenges for the development and the limitations for the applications of a GC method: in its simplest form, a GC method assumes that the location of a functional group on a molecule is not relevant. However, the —CH_2— group surrounded only by other —CH_2— groups on a long *n*-alkane will show different interactions to other groups than the —CH_2— group located next to an OH group. The assumption that CH_2-parameters, in our example, will be transferable between molecules irrespective of the neighboring groups is imperfect. As a second example, the —OH group of primary alcohol will contribute differently to the behavior of a molecule than the —OH group on secondary alcohol. The transferability of GC contributions is particularly limited when multiple polar or hydrogen-bonding groups appear in a single molecule. In the development of a GC method, one thus has to make compromises between predictability (only one parameter set for —OH) and the introduction of different parameter sets for groups with the same chemical structure in different molecular contexts (for example allowing different parameters for primary, secondary and tertiary —OH groups). Another remedy to deal with proximity effects is the introduction of second-order groups: Second-order groups provide additional parameters for describing specific pairs of neighboring groups. One can for example define a second-order group for the neighboring pair $(CH)_{aromatic}$—OH.

The group-contribution methods for PC-SAFT can be classified as homosegmented or heterosegmented approaches: in a homosegmented approach, the chain model of a molecule is made up of identical segments and averaging rules are applied to determine molecular parameters from the group parameters.

In this case, the application of the group-contribution method is a simple preprocessing step. Once the molecular parameters have been determined, the application of PC-SAFT is identical to the non-GC case. One possible averaging rule to obtain molecular parameters (index i) from group parameters (index α) is [89]

$$\begin{aligned} m_i &= \sum_{\alpha} n_{\alpha_i} m_\alpha \\ m_i \sigma_i^3 &= \sum_{\alpha} n_{\alpha_i} m_\alpha \sigma_\alpha^3 \\ m_i \varepsilon_i &= \sum_{\alpha} n_{\alpha_i} m_\alpha \varepsilon_\alpha \end{aligned} \tag{1.35}$$

The summation runs over all types of groups and $n_{\alpha i}$ denotes how many groups of type α are present on a molecule of component i. Numerous studies have presented group-contribution parameters for homosegmented PC-SAFT, that allow to description hydrocarbon systems [90,91], asymmetric mixtures of hydrocarbons with small, volatile compounds [92–94], VLE and LLE of mixtures including polar or associating compounds [89,95–103], nitriles [104] and long-chain bi-functional molecules such as alkanoamines and alkanediols [105].

Parameters for a large set of functional groups for molecules from a broad range of chemical families were presented, for example, by Sauer et al. [106] and Emami et al. [107]. To address proximity effects, Tihic et al. [108–110] presented first- and second-order group parameters for a variety of functional groups and applied the approach also to study polymers. Burgess et al. [111] built on this work and adjusted first- and second-order group parameters for alkanes, aromatics, and cycloalkanes at high temperature and high-pressure conditions. Similarly, Vijande et al. adjusted parameters for reference and perturbing groups in their studies on nonassociating molecules [112] as well as primary alcohols and amines [113].

In a heterosegmented group-contribution approach, the segments forming the chain-molecule keep their individual properties; they can for example have different sizes. There is no averaging step to determine molecular parameters. Instead, the equation of state works directly with the various group parameters. The PC-SAFT equation of state is in this case rewritten to capture the summation over all segment types. For example, the first-order perturbation term of the dispersive contribution then reads [114]

$$a_1^{disp} = -2\pi\rho I_1(\eta, \overline{m}) \sum_i \sum_j x_i x_j \sum_\alpha \sum_\beta n_{\alpha_i} n_{\beta_j} m_\alpha m_\beta \left(\frac{\varepsilon_{\alpha\beta}}{kT}\right) \sigma_{\alpha\beta}^3 \tag{1.36}$$

Here, summations with indices i and j run over molecules, and summations with indices α and β run over group types. Values for $\varepsilon_{\alpha\beta}$ and $\sigma_{\alpha\beta}$ are obtained from the corresponding group parameters using the same combining rules as in the nongroup-contribution case, see Eqs. (1.30) and (1.31).

Building on the work of Shukla and Chapman [115] and Banaszak et al. [116], Gross et al. [114] developed a group-contribution approach for heterosegmented chain molecules to model *co*-polymers. Using this approach, Sauer et al. [106] adjusted parameters for 22 functional groups to pure-component data of molecules from 12 different chemical families, including polar and associating species. Later, binary group-group interaction parameters as well as a component-specific parameter to improve the description of molecules which are well characterized by experimental data were adjusted by Mairhofer and Gross [117]. Further applications of heterosegmented PC-SAFT are the studies of Paduszyński and Domańska [118] on pure hydrocarbons, Peters et al. [119,120] on polymer systems which includes the development of a group-contribution method to determine binary interaction parameters between polymers and solvents [121] and Haarmann et al. [122–124] on polar and associating long-chain molecules and related mixtures. To improve the description of branched alkanes, Jaber et al. [125] included second-order groups.

In general, one can assume that the accuracy of the modeling approach for given molecule declines from a component-specific description to a heterosegmented GC approach to a homosegmented GC approach, see for example Refs. [106, 118] for a comparison of the homosegmented and the heterosegmented GC method. From this, one could infer that the heterosegmented GC approach should always be preferred over the homosegmented GC approach. However, this is not necessarily the case: because in the heterosegmented GC approach the equations of the PC-SAFT model are changed, it cannot be applied to individual molecules only, every molecule, and the interaction between every binary pair of molecules has to be modeled using the heterosegmented GC approach. A homosegmented GC method, on the other hand, can be applied to selected molecules. In many applications, enough data is available to adjust the pure-component parameters for the majority of the compounds to data of that compound and only a few species require a description by group-contribution methods. In this case, it is advantageous to apply a homosegmented group-contribution approach only to these molecules and describe the remaining molecules with the more accurate, component-specific model parameters and BIP adjusted to data specific for a given binary pair.

4 Transport properties

Several alternatives exist to determine transport coefficients, such as viscosity, thermal conductivity, or diffusion coefficients based on input from an equation of states, such as friction theory [126] or free-volume theory [127]. In this section, however, the focus is on excess or residual entropy scaling. This approach allows to determine viscosity, thermal conductivity as well as self-diffusion coefficients and is strikingly simple to implement. Excess entropy scaling is based on the observation that transport coefficients, once scaled with a reference value, can rather accurately be approximated as a univariate function only of residual entropy s^{res}, where $s^{res}(T,\rho,x) = s(T,\rho,x) - s^{ig}(T,\rho,x)$ with s^{ig} as the entropy of an ideal gas. Entropy scaling was first proposed by Rosenfeld [128,129] for certain model fluids, but convincing results are also obtained for real molecules. A framework that allows investigating the theoretical foundation of excess entropy scaling is the isomorph theory of Dyre et al. [130–135]. Simple ansatz functions such as low-order polynomials can be used to correlate the logarithm of the scaled transport properties with residual entropy. The coefficients of these functions are usually adjusted to pure-component data of the considered transport coefficients (but predictive group-contribution methods exist as well) and results for mixtures are predicted using simple mixing rules [81].

Fig. 1.4 illustrates the basic idea of excess entropy scaling for methane, CO_2, and *n*-decane: experimental viscosity data covering large temperature and pressure ranges, once made dimensionless, can be correlated accurately as a function of residual entropy by a simple, cubic polynomial.

The following paragraph focuses on studies that use PC-SAFT as the model to obtain the value of residual entropy. Novak, who had already presented results for viscosity [136,137], thermal conductivity as well as self-diffusion coefficients [138] using several thermodynamic models for s^{res}, demonstrated that accurate results for viscosity of nonassociating molecules can be obtained from a predictive corresponding-state approach which uses residual entropy obtained from PC-SAFT [139,140].

Lötgering-Lin et al. [81] correlated pure-component viscosity for molecules from several chemical families and predicted mixture viscosity applying a mixing rule derived from molecular simulation results. Average deviations of around 5% are reported for mixtures of nonassociating components, while higher deviations have to be expected once associating molecules are considered. Results of similar accuracy can be obtained from the purely predictive group-contribution method of Lötgering-Lin and Gross [81,141] where residual entropy is calculated from the group-contribution PC-SAFT equation of state. Thermal conductivity was studied extensively by Hopp and Gross applying component-specific parameters to correlate pure-component results [142] but also by a predictive group-contribution method [143]. Both approaches show excellent agreement to experimental results for pure-component thermal conductivity. This includes molecules with strong hydrogen-bonding such as water and alcohol.

Hopp et al. [144] presented results for self-diffusion coefficients of pure substances from 14 different chemical families and reported an average deviation of 8% for a data set of almost 10,000 data points. In process simulation, binary diffusion coefficients are usually more relevant. Several routes exist that allow relating the binary to the self-diffusion coefficients [145,146].

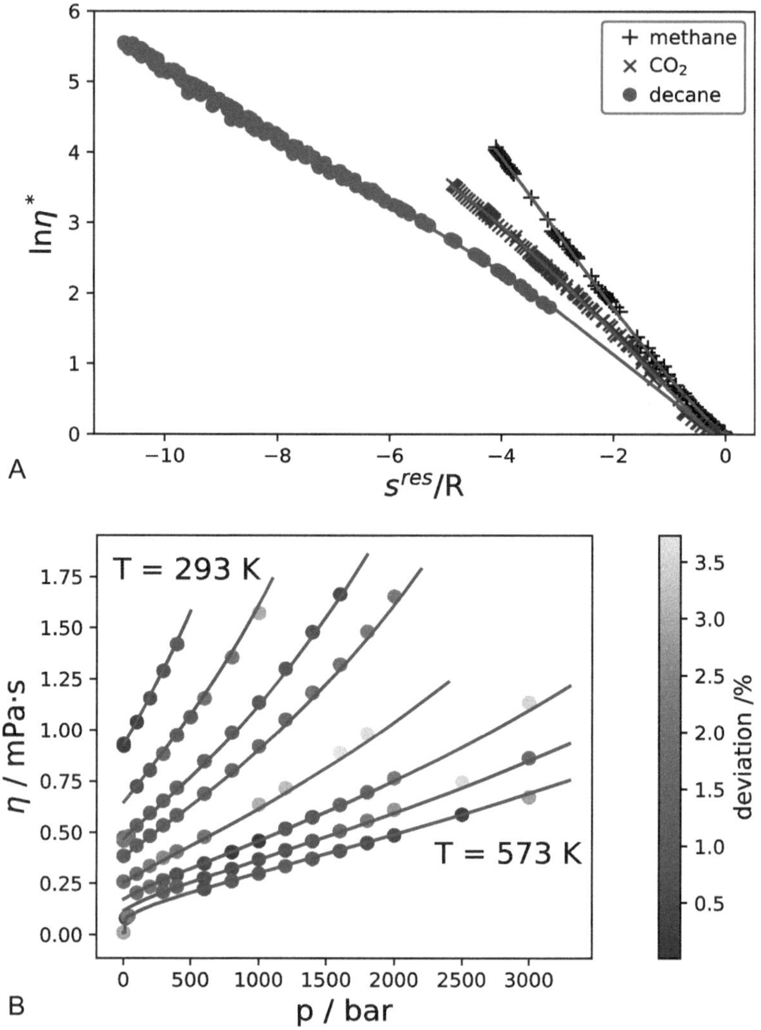

FIG. 1.4 (A) Logarithm of dimensionless shear viscosity η^* for methane, CO_2, and *n*-decane as a function of s^{res}/R. The *lines* represent cubic polynomials that correlate with the experimental data [64] *(symbols)*. The temperature and pressure ranges of the data are 91–450 K and 0.1–700 bar for methane, 219–500 K and 5.6–2600 bar for CO_2, and 289–709 K and 0.07–3000 bar for decane. (B) Viscosity of *n*-decane as a function of pressure for several isotherms. Experimental values [64] *(symbols)* and results of entropy scaling *(lines)* using the cubic polynomial from Fig. (A). The *symbol color (light gray, gray, dark gray in print version)* represents the model error for each data point.

5 Interfacial properties

In this section, density gradient theory (DGT) and classical density functional theory (DFT) will be presented briefly. The origins of DGT date back to the work of Van der Waals [147]; Cahn and Hilliard [148] later reformulated the theory. DFT was originally formulated by Hohenberg and Kohn [149] as well as Mermin [150] for the inhomogeneous electron gas. First applications to molecular systems (rather than electron fluids) are due to Ebner et al. [151,152].

Both DGT and DFT, allow to determine the density profile of every component of a mixture across the interface between two phases. The two phases can be two fluid phases, i.e., vapor-liquid or liquid-liquid equilibria, or can be a fluid phase interfacing a solid so that adsorption properties can be studied. From the density profiles across an interface, the value for surface tension and further interfacial properties such as resistivities to heat and mass transfer [153,154] can be derived. Furthermore, the density profiles allow studying phenomena such as the enrichment of certain species of a mixture at the interface. Compared to simple engineering models for surface tension, DGT and DFT do not require input from auxiliary models, they show superior predictive capability and a broader range of applicability. The DFT does not require any interface-specific parameter and can be considered predictive for interfacial properties, whereas the DGT necessitates one parameter (usually considered constant) per pure-component, whereas mixtures can often be predicted based only on the pure substances.

The starting point for a DGT or DFT calculation is a domain of volume V which contains two fluid phases in thermodynamic equilibrium. This implies that temperature T and the chemical potential of every component i, μ_i, are constant across the interface between the phases. Component densities, ρ_i, on the other hand, vary across the interface from the bulk value in phase 1, ρ_i^1, to the bulk value in phase 2, ρ_i^2. The state of equilibrium, that is, the values for μ_i, ρ_i^1, and ρ_i^2 are obtained from a preceding phase equilibrium calculation.

We shall consider a one-dimensional vapor-liquid or liquid-liquid interface, specified by values for T, V, and all μ_i. The equilibrium density profiles across the interface, $\rho_i(z)$, where z is the coordinate perpendicular to the interface, minimize the grand potential Ω that (in absence of an external field) reads

$$\Omega = \int f[\{\rho_k(z)\}] - \sum_i^N \mu_i \rho_i(z) \mathrm{d}z \tag{1.37}$$

where $f[\{\rho_k(z)\}]$ denotes the local Helmholtz energy density which is a functional of the set of density profiles of all components, $\{\rho_k(z)\}$, with the generic component index k. For a bulk phase $f/(kT) = A/(V\,kT) = a \cdot \rho$. For brevity, the dependency of f on temperature and of Ω on T, V, and all μ_i is omitted in Eq. (1.37).

The route to approximate the true value of $f[\{\rho_k(z)\}]$ for a system with spatially varying density is the central part of DGT and DFT. In DGT, the local Helmholtz energy density at position z is developed around the value of a hypothetical bulk phase with density $\rho = \sum_i \rho_i(z)$

$$f(\rho)^{DGT} = f^0(\rho) + \frac{1}{2} \sum_i \sum_j c_{ij} \nabla \rho_i \nabla \rho_j \tag{1.38}$$

The value for the hypothetical bulk phase, $f^0(\rho)$, can be calculated from any suitable equation of state. $\nabla \rho_i$ denotes the local gradient in the density profile of component i and a geometric combining rule is usually applied for calculating the value of $c_{ij} = \sqrt{c_i c_j}$ from the pure-component influence parameters c_i and c_j, respectively. Theoretically derived expressions for the influence parameter exist [155–157], however, most often, the values of the influence parameters are adjusted to experimental data for the surface tension of pure components or obtained from simple correlations [158–162]. Rehner and Gross recently proposed a predictive DGT, where the influence parameter is obtained from an expansion around the DFT formalism [163].

DFT, on the other hand, requires no interface-specific adjustable parameters. The local Helmholtz energy density is modeled as the sum of different contributions analogously to Eq. (1.1). However, now all contributions are functionals of the spatially varying density profiles which reduce to the expressions of the PC-SAFT equation of state in the bulk limit [164–166]. Once the model for $f[\{\rho_k(z)\}]$ is established, the density profiles of all components across the interface are determined by minimizing Ω with respect to the density profiles.

The applications of DGT in combination with PC-SAFT include studies on hydrocarbon mixtures [167,168], systems including polar [169,170], associating [171] or supercritical [172] components, azeotropic mixtures [173] as well as aqueous systems [174–177]. DFT based on functionals consistent with PC-SAFT was applied, for example, for determining the surface tension of binary mixtures including associating components at vapor–liquid [178] and liquid-liquid equilibrium [179] and polymer systems [180]. A comparison of PC-SAFT-based DFT and DGT for a variety of systems, including mixtures of polar and associating molecules, can be found in the study of Mairhofer and Gross [181]: for pure and mixture systems of nonassociating molecules, surface tension predicted by DFT shows comparable accuracy as results obtained from DGT with influence parameters adjusted to pure-component surface tension data. Only for pure associating molecules and some mixtures including associating compounds, the adjustable parameters of DGT led to improvements over DFT predictions. Algorithms to efficiently solve the equations arising in DFT have been identified for example by Mairhofer and Gross [182] and Stierle et al. [183]. However, the computational cost for a DFT calculation is still higher than for a DGT calculation.

To illustrate the capabilities of DFT and DGT, results for surface tensions of three refrigerant mixtures are presented in Fig. 1.5 [181]: good overall agreement to experimental data can be observed. Furthermore, only for the mixture R32 with R134a do results of the purely predictive DFT deviate noticeably more from the experimental values than results of DGT which requires a value for the influence parameter per component.

A group-contribution DFT consistent with the heterosegmented group-contribution PC-SAFT equation of state was developed by Mairhofer et al.

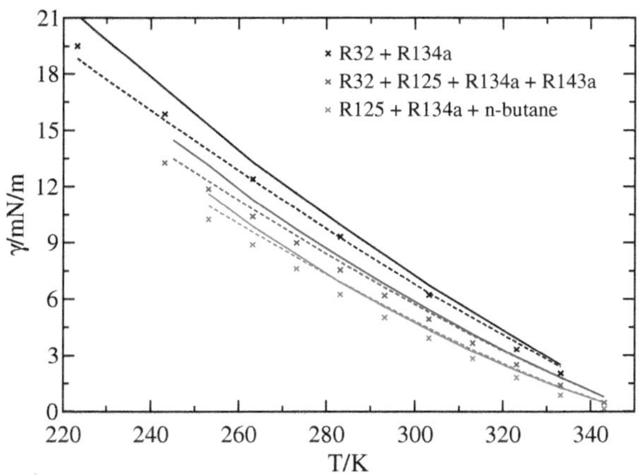

FIG. 1.5 Surface tension γ of several refrigerant mixtures. Calculated results (*solid lines*: DFT, *dashed lines*: DGT) and experimental data [184–186]. *(From J. Mairhofer, J. Gross, Modeling properties of the one-dimensional vapor-liquid interface: application of classical density functional and density gradient theory, Fluid Phase Equilib. 458 (2018) 243–252, with permission from Elsevier.)*

[117] based on interfacial SAFT [187,188] (iSAFT). Accurate surface tension predictions are obtained for biodiesel systems, among others. Haarmann et al. [189] recently presented a group-contribution DGT approach and studied the surface tension of pure long-chain compounds (fatty alcohols and methyl esters) as well as their binary mixtures with water.

References

[1] H. Renon, J.M. Prausnitz, Local compositions in thermodynamic excess functions for liquid mixtures, AIChE J. 14 (1) (1968) 135–144.

[2] D.-Y. Peng, D.B. Robinson, A new two-constant equation of state, Ind. Eng. Chem. Fundam. 15 (1) (1976) 59–64.

[3] G. Soave, Equilibrium constants from a modified Redlich-Kwong equation of state, Chem. Eng. Sci. 27 (6) (1972) 1197–1203.

[4] W.G. Chapman, G. Jackson, K.E. Gubbins, Phase equilibria of associating fluids: chain molecules with multiple bonding sites, Mol. Phys. 65 (5) (1988) 1057–1079.

[5] W.G. Chapman, K.E. Gubbins, G. Jackson, M. Radosz, SAFT: equation-of-state solution model for associating fluids, Fluid Phase Equilib. 52 (1989) 31–38.

[6] G. Jackson, W.G. Chapman, K.E. Gubbins, Phase equilibria of associating fluids: spherical molecules with multiple bonding sites, Mol. Phys. 65 (1) (1988) 1–31.

[7] M. Wertheim, Fluids with highly directional attractive forces. I. Statistical thermodynamics, J. Stat. Phys. 35 (1) (1984) 19–34.

[8] M. Wertheim, Fluids with highly directional attractive forces. II. Thermodynamic perturbation theory and integral equations, J. Stat. Phys. 35 (1–2) (1984) 35–47.

[9] M. Wertheim, Fluids with highly directional attractive forces. III. Multiple attraction sites, J. Stat. Phys. 42 (3) (1986) 459–476.

[10] M. Wertheim, Fluids with highly directional attractive forces. IV. Equilibrium polymerization, J. Stat. Phys. 42 (3) (1986) 477–492.
[11] J.R. Solana, Perturbation Theories for the Thermodynamic Properties of Fluids and Solids, CRC Press, 2013.
[12] E.A. Müller, K.E. Gubbins, Molecular-based equations of state for associating fluids: a review of SAFT and related approaches, Ind. Eng. Chem. Res. 40 (10) (2001) 2193–2211.
[13] G. Economou, Statistical associating fluid theory: a successful model for the calculation of thermodynamic and phase equilibrium properties of complex fluid mixtures, Ind. Eng. Chem. Res. 41 (5) (2002) 953–962.
[14] W.G. Chapman, S.G. Sauer, D. Ting, A. Ghosh, Phase behavior applications of SAFT based equations of state—from associating fluids to polydisperse, polar copolymers, Fluid Phase Equilib. 217 (2) (2004) 137–143.
[15] G. Sadowski, Thermodynamics of polymer systems, in: Macromolecular Symposia, vol. 206, Wiley Online Library, 2004, pp. 333–346.
[16] F. Tumakaka, J. Gross, G. Sadowski, Thermodynamic modeling of complex systems using PC-SAFT, Fluid Phase Equilib. 228 (2005) 89–98.
[17] S.P. Tan, H. Adidharma, M. Radosz, Recent advances and applications of statistical associating fluid theory, Ind. Eng. Chem. Res. 47 (21) (2008) 8063–8082.
[18] T. Spyriouni, X. Krokidis, I.G. Economou, Thermodynamics of pharmaceuticals: prediction of solubility in pure and mixed solvents with PC-SAFT, Fluid Phase Equilib. 302 (1–2) (2011) 331–337.
[19] J. Gross, G. Sadowski, Application of perturbation theory to a hard-chain reference fluid: an equation of state for square-well chains, Fluid Phase Equilib. 168 (2) (2000) 183–199.
[20] J. Gross, G. Sadowski, Perturbed-chain SAFT: an equation of state based on a perturbation theory for chain molecules, Ind. Eng. Chem. Res. 40 (4) (2001) 1244–1260.
[21] J. Gross, G. Sadowski, Application of the perturbed-chain SAFT equation of state to associating systems, Ind. Eng. Chem. Res. 41 (22) (2002) 5510–5515.
[22] T. Boublík, Hard-sphere equation of state, J. Chem. Phys. 53 (1) (1970) 471–472.
[23] G. Mansoori, N.F. Carnahan, K. Starling, T. Leland Jr., Equilibrium thermodynamic properties of the mixture of hard spheres, J. Chem. Phys. 54 (4) (1971) 1523–1525.
[24] J.A. Barker, D. Henderson, Perturbation theory and equation of state for fluids: the square-well potential, J. Chem. Phys. 47 (8) (1967) 2856–2861.
[25] J.A. Barker, D. Henderson, Perturbation theory and equation of state for fluids. II. A successful theory of liquids, J. Chem. Phys. 47 (11) (1967) 4714–4721.
[26] H. Liu, Y. Hu, Molecular thermodynamic theory for polymer systems part II. Equation of state for chain fluids, Fluid Phase Equilib. 122 (1–2) (1996) 75–97.
[27] S.H. Huang, M. Radosz, Equation of state for small, large, polydisperse, and associating molecules, Ind. Eng. Chem. Res. 29 (11) (1990) 2284–2294.
[28] J.P. Wolbach, S.I. Sandler, Using molecular orbital calculations to describe the phase behavior of cross-associating mixtures, Ind. Eng. Chem. Res. 37 (8) (1998) 2917–2928.
[29] M.L. Michelsen, Robust and efficient solution procedures for association models, Ind. Eng. Chem. Res. 45 (25) (2006) 8449–8453.
[30] M.L. Michelsen, E.M. Hendriks, Physical properties from association models, Fluid Phase Equilib. 180 (1–2) (2001) 165–174.
[31] S.P. Tan, H. Adidharma, M. Radosz, Generalized procedure for estimating the fractions of nonbonded associating molecules and their derivatives in thermodynamic perturbation theory, Ind. Eng. Chem. Res. 43 (1) (2004) 203–208.

[32] K. Langenbach, S. Enders, Cross-association of multi-component systems, Mol. Phys. 110 (11–12) (2012) 1249–1260.
[33] M. Kleiner, G. Sadowski, Modeling of polar systems using PCP-SAFT: an approach to account for induced-association interactions, J. Phys. Chem. C 111 (43) (2007) 15544–15553.
[34] J. Gross, An equation-of-state contribution for polar components: quadrupolar molecules, AIChE J. 51 (9) (2005) 2556–2568.
[35] J. Gross, J. Vrabec, An equation-of-state contribution for polar components: dipolar molecules, AIChE J. 52 (3) (2006) 1194–1204.
[36] J. Vrabec, J. Gross, Vapor-liquid equilibria simulation and an equation of state contribution for dipole-quadrupole interactions, J. Phys. Chem. B 112 (1) (2008) 51–60.
[37] G. Stell, J. Rasaiah, H. Narang, Thermodynamic perturbation theory for simple polar fluids, I, Mol. Phys. 23 (2) (1972) 393–406.
[38] G. Stell, J. Rasaiah, H. Narang, Thermodynamic perturbation theory for simple polar fluids. II, Mol. Phys. 27 (5) (1974) 1393–1414.
[39] J. Stoll, J. Vrabec, H. Hasse, Comprehensive study of the vapour–liquid equilibria of the pure two-centre Lennard–Jones plus pointdipole fluid, Fluid Phase Equilib. 209 (1) (2003) 29–53.
[40] R. Privat, E. Moine, B. Sirjean, R. Gani, J.-N. Jaubert, Application of the corresponding-state law to the parametrization of statistical associating fluid theory (SAFT)-type models: generation and use of "generalized charts", Ind. Eng. Chem. Res. 58 (21) (2019) 9127–9139.
[41] J. Mairhofer, J. Gross, Identifying pure-component parameters of an analytic equation of state using experimental surface tension or molecular simulations with a transferable force field, Ind. Eng. Chem. Res. 57 (36) (2018) 12254–12263.
[42] P. Rehner, J. Gross, Multiobjective optimization of PCP-SAFT parameters for water and alcohols using surface tension data, J. Chem. Eng. Data 65 (12) (2020) 5698–5707.
[43] G.N. Clark, A.J. Haslam, A. Galindo, G. Jackson, Developing optimal Wertheim-like models of water for use in statistical associating fluid theory (SAFT) and related approaches, Mol. Phys. 104 (22–24) (2006) 3561–3581.
[44] E. Forte, J. Burger, K. Langenbach, H. Hasse, M. Bortz, Multi-criteria optimization for parameterization of saft-type equations of state for water, AIChE J. 64 (1) (2018) 226–237.
[45] N. Ramírez-Vélez, A. Piña-Martinez, J.-N. Jaubert, R. Privat, Parameterization of SAFT models: analysis of different parameter estimation strategies and application to the development of a comprehensive database of PC-SAFT molecular parameters, J. Chem. Eng. Data 65 (12) (2020) 5920–5932.
[46] C. Zhu, X. Liu, M. He, G.M. Kontogeorgis, X. Liang, Heat capacities of fluids: the performance of various equations of state, J. Chem. Eng. Data 65 (12) (2020) 5654–5676.
[47] N.I. Diamantonis, I.G. Economou, Evaluation of statistical associating fluid theory (SAFT) and perturbed chain-saft equations of state for the calculation of thermodynamic derivative properties of fluids related to carbon capture and sequestration, Energy Fuel 25 (7) (2011) 3334–3343.
[48] C. De Villiers, A.B. Schwarz, G. Kontogeorgis, Evaluation of the PC-SAFT, SAFT and CPA equations of state in predicting derivative properties of selected non-polar and hydrogen-bonding compounds, Fluid Phase Equilib. 338 (2013) 1–15.
[49] M. Oliveira, F. Llovell, J. Coutinho, L. Vega, New procedure for enhancing the transferability of statistical associating fluid theory (SAFT) molecular parameters: the role of derivative properties, Ind. Eng. Chem. Res. 55 (37) (2016) 10011–10024.
[50] Grenner, G.M. Kontogeorgis, N. von Solms, M.L. Michelsen, Modeling phase equilibria of alkanols with the simplified PC-SAFT equation of state and generalized pure compound parameters, Fluid Phase Equilib. 258 (1) (2007) 83–94.

[51] A. Tihic, G.M. Kontogeorgis, N. von Solms, M.L. Michelsen, Applications of the simplified perturbed-chain SAFT equation of state using an extended parameter table, Fluid Phase Equilib. 248 (1) (2006) 29–43.
[52] Grenner, G.M. Kontogeorgis, N. von Solms, M.L. Michelsen, Application of PC-SAFT to glycol containing systems–PC-SAFT towards a predictive approach, Fluid Phase Equilib. 261 (1–2) (2007) 248–257.
[53] K. Albers, M. Heilig, G. Sadowski, Reducing the amount of PCP-SAFT fitting parameters. 2. Associating components, Fluid Phase Equilib. 326 (2012) 31–44.
[54] M. Fuenzalida, J. Cuevas-Valenzuela, J.R. Pérez-Correa, Improved estimation of PC-SAFT equation of state parameters using a multi-objective variable-weight cost function, Fluid Phase Equilib. 427 (2016) 308–319.
[55] N. Van Nhu, M. Singh, K. Leonhard, Quantum mechanically based estimation of perturbed-chain polar statistical associating fluid theory parameters for analyzing their physical significance and predicting properties, J. Phys. Chem. B 112 (18) (2008) 5693–5701.
[56] M. Umer, K. Albers, G. Sadowski, K. Leonhard, PC-SAFT parameters from ab initio calculations, Fluid Phase Equilib. 362 (2014) 41–50.
[57] S. Kaminski, K. Leonhard, Sepp: segment-based equation of state parameter prediction, J. Chem. Eng. Data 65 (12) (2020) 5830–5843.
[58] N. Ferrando, J.-C. de Hemptinne, P. Mougin, J.-P. Passarello, Prediction of the PC-SAFT associating parameters by molecular simulation, J. Phys. Chem. B 116 (1) (2012) 367–377.
[59] P. Hosseinifar, S. Jamshidi, Determination of perturbed-chain statistical association fluid theory parameters for pure substances, single carbon number groups, and petroleum fractions using cubic equations of state parameters, Ind. Eng. Chem. Res. 54 (45) (2015) 11448–11465.
[60] J. Gross, G. Sadowski, Modeling polymer systems using the perturbed-chain statistical associating fluid theory equation of state, Ind. Eng. Chem. Res. 41 (5) (2002) 1084–1093.
[61] A. Kouskoumvekaki, N. von Solms, T. Lindvig, M.L. Michelsen, G.M. Kontogeorgis, Novel method for estimating pure-component parameters for polymers: application to the PC-SAFT equation of state, Ind. Eng. Chem. Res. 43 (11) (2004) 2830–2838.
[62] P. Hosseinifar, M. Assareh, C. Ghotbi, Developing a new model for the determination of petroleum fraction PC-SAFT parameters to model reservoir fluids, Fluid Phase Equilib. 412 (2016) 145–157.
[63] X. Tang, J. Gross, Modeling the phase equilibria of hydrogen sulfide and carbon dioxide in mixture with hydrocarbons and water using the PCP-SAFT equation of state, Fluid Phase Equilib. 293 (1) (2010) 11–21.
[64] Dortmund Data Bank, 2015. www.ddbst.com.
[65] N. Haarmann, S. Enders, G. Sadowski, Modeling binary mixtures of n-alkanes and water using PC-SAFT, Fluid Phase Equilib. 470 (2018) 203–211.
[66] N. Haarmann, R. Siewert, A.A. Samarov, S.P. Verevkin, C. Held, G. Sadowski, Thermodynamic properties of systems comprising esters: experimental data and modeling with PC-SAFT and SAFT-γ Mie, Ind. Eng. Chem. Res. 58 (16) (2019) 6841–6849.
[67] G. Watson, T. Lafitte, C.K. Zéberg-Mikkelsen, A. Baylaucq, D. Bessieres, C. Boned, Volumetric and derivative properties under pressure for the system 1-propanol + toluene: a discussion of PC-SAFT and SAFT-VR, Fluid Phase Equilib. 247 (1–2) (2006) 121–134.
[68] M. Hashemi, M. Moosavi, A. Omrani, A.A. Rostami, Non-ideal behavior of ethanol + amines mixtures, modeling using the Peng-Robinson and PC-SAFT equation of state, J. Mol. Liq. 256 (2018) 445–455.
[69] A. Bamgbade, Y. Wu, W.A. Burgess, D. Tapriyal, I.K. Gamwo, H.O. Baled, R.M. Enick, M. A. McHugh, High-temperature, high-pressure volumetric properties of propane, squalane,

and their mixtures: measurement and PC-SAFT modeling, Ind. Eng. Chem. Res. 54 (26) (2015) 6804–6811.
[70] X. Wang, W. Du, X. Wang, Volumetric properties of n-hexadecane/ethyl octanoate mixtures from 293.15 K to 363.15 K and pressures up to 60 MPa, J. Chem. Thermodyn. (2020) 106122.
[71] I. Abala, F.E.M. Alaoui, Y. Chhiti, A.S. Eddine, N.M. Rujas, F. Aguilar, Experimental density and PC-SAFT modeling of biofuel mixtures (DBE+ 1-heptanol) at temperatures from (298.15 to 393.15) K and at pressures up to 140 MPa, J. Chem. Thermodyn. 131 (2019) 269–285.
[72] S.M. García-Abarrio, M. Torcal, M.L. Haya, J.S. Urieta, A.M. Mainar, Thermophysical properties of {(±)-linalool+ propan-1-ol}: a first stage towards the development of a green process, J. Chem. Thermodyn. 43 (4) (2011) 527–536.
[73] O. Fandiño, E.R. López, L. Lugo, J. García, J. Fernández, Solubility of carbon dioxide in pentaerythritol ester oils. New data and modeling using the PC-SAFT model, J. Supercrit. Fluids 55 (1) (2010) 62–70.
[74] K. Mejbri, A. Bellagi, Modelling of the thermodynamic properties of the water-ammonia mixture by three different approaches, Int. J. Refrig. 29 (2) (2006) 211–218.
[75] W. Yan, T. Regueira, Y. Liu, E.H. Stenby, Density modeling of high-pressure mixtures using cubic and non-cubic EOS and an excess volume method, Fluid Phase Equilib. 532 (2021) 112884.
[76] S. Schacht, L. Zubeir, T.W. de Loos, J. Gross, Application of infinite dilution activity coefficients for determining binary equation of state parameters, Ind. Eng. Chem. Res. 49 (16) (2010) 7646–7653.
[77] K. Paduszyński, M. Królikowski, U. Domańska, Excess enthalpies of mixing of piperidinium ionic liquids with short-chain alcohols: measurements and PC-SAFT modeling, J. Phys. Chem. B 117 (14) (2013) 3884–3891.
[78] K. Paduszyński, E.V. Lukoshko, M. Królikowski, U. Domańska, J. Szydlowski, Thermodynamic study of binary mixtures of 1-butyl-1-methylpyrrolidinium dicyanamide ionic liquid with molecular solvents: new experimental data and modeling with PC-SAFT equation of state, J. Phys. Chem. B 119 (2) (2015) 543–551.
[79] C.-B. Soo, Experimental Thermodynamic Measurements of Biofuel-Related Associating Compounds and Modeling Using the PC-SAFT Equation of State (PhD thesis), École Nationale Supérieure des Mines de Paris, 2011.
[80] M. Stavrou, Integrated Process and Solvent Design for CO2 Capture Using Continuous Molecular Targeting-Computer Aided Molecular Design (CoMT-CAMD) (PhD thesis), TU Delft, 2017.
[81] O. Lötgering-Lin, M. Fischer, M. Hopp, J. Gross, Pure substance and mixture viscosities based on entropy scaling and an analytic equation of state, Ind. Eng. Chem. Res. 57 (11) (2018) 4095–4114.
[82] Y. Aoki, T. Moriyoshi, Mutual solubility of n-butanol+ water under high pressures, J. Chem. Thermodyn. 10 (12) (1978) 1173–1179.
[83] T. Boublik, Gleichgewicht Flüssigkeit-Dampf xx. Bestimmung des Gleichgewichts Flüssigkeit-Dampf in Systemen, deren Komponenten in flüssiger Phase beschränkt mischbar sind, Collect. Czechoslov. Chem. Commun. 25 (1) (1960) 285–288.
[84] D. Hessel, G. Geiseler, Über die Druckabhängigkeit des heteroazeotropen Systems n-Butanol/Wasser, Z. Phys. Chem. 229 (1) (1965) 199–209.
[85] F. Cristino, L.C. Nobre, F.E. Bioucas, A.F. Santos, C.A.N. de Castro, I.M. Lampreia, Volumetric and sound speed study of aqueous 1-butanol liquid mixtures at different temperatures, J. Chem. Thermodyn. 134 (2019) 127–135.

[86] K.-H. Lim, D.H. Smith, Excess enthalpies (HE) for 2-butoxyethanol/water and determination of liquid-liquid equilibrium phase boundaries from fits to he of Redlich-Kister, Pade, and critical-scaling equations, J. Chem. Eng. Jpn. 34 (4) (2001) 513–522.

[87] G. Soave, S. Gamba, L.A. Pellegrini, SRK equation of state: predicting binary interaction parameters of hydrocarbons and related compounds, Fluid Phase Equilib. 299 (2) (2010) 285–293.

[88] J. Höller, P. Bickert, P. Schwartz, M. von Kurnatowski, J. Kerber, N. Künzle, H.M. Lorenz, N. Asprion, S. Blagov, M. Bortz, Parameter estimation strategies in thermodynamics, ChemEngineering 3 (2) (2019) 56.

[89] J. Vijande, M. Pineiro, D. Bessieres, H. Saint-Guirons, J. Legido, Description of PVT behaviour of hydrofluoroethers using the PC-SAFT EOS, Phys. Chem. Chem. Phys. 6 (4) (2004) 766–770.

[90] D. NguyenHuynh, A modified group-contribution PC-SAFT equation of state for prediction of phase equilibria, Fluid Phase Equilib. 430 (2016) 33–46.

[91] D. NguyenHuynh, D. NguyenHuynh, Application of the modified group-contribution perturbed-chain SAFT to branched alkanes, n-olefins and their mixtures, Fluid Phase Equilib. 434 (2017) 176–192.

[92] L. Thi, S. Tamouza, J.-P. Passarello, P. Tobaly, J.-C. de Hemptinne, Modeling phase equilibrium of H_2 + n-alkane and CO_2 + n-alkane binary mixtures using a group contribution statistical association fluid theory equation of state (GC-SAFT-EOS) with a k_{ij} group contribution method, Ind. Eng. Chem. Res. 45 (20) (2006) 6803–6810.

[93] D. Nguyen-Huynh, J.-P. Passarello, P. Tobaly, J.-C. de Hemptinne, Modeling phase equilibria of asymmetric mixtures using a group-contribution SAFT (GC-SAFT) with a k_{ij} correlation method based on London's theory. 1. Application to co_2 + n-alkane, methane+ n-alkane, and ethane+ n-alkane systems, Ind. Eng. Chem. Res. 47 (22) (2008) 8847–8858.

[94] D. Nguyen-Huynh, T. Tran, S. Tamouza, J.-P. Passarello, P. Tobaly, J.-C. De Hemptinne, Modeling phase equilibria of asymmetric mixtures using a group-contribution SAFT (GC-SAFT) with a k_{ij} correlation method based on London's theory. 2. Application to binary mixtures containing aromatic hydrocarbons, n-alkanes, CO_2, N_2, and H_2S, Ind. Eng. Chem. Res. 47 (22) (2008) 8859–8868.

[95] M. Mourah, D. NguyenHuynh, J. Passarello, J. De Hemptinne, P. Tobaly, Modelling LLE and VLE of methanol+ n-alkane series using gc-PC-SAFT with a group contribution kij, Fluid Phase Equilib. 298 (1) (2010) 154–168.

[96] D. Nguyen-Huynh, J.-C. de Hemptinne, R. Lugo, J.-P. Passarello, P. Tobaly, Modeling liquid–liquid and liquid–vapor equilibria of binary systems containing water with an alkane, an aromatic hydrocarbon, an alcohol or a gas (methane, ethane, CO_2 or H_2S), using group contribution polar perturbed-chain statistical associating fluid theory, Ind. Eng. Chem. Res. 50 (12) (2011) 7467–7483.

[97] J. Rozmus, J.-C. de Hemptinne, P. Mougin, Application of GC-PPC-SAFT-EoS to amine mixtures with a predictive approach, Fluid Phase Equilib. 303 (1) (2011) 15–30.

[98] D. NguyenHuynh, J.-P. Passarello, J.-C. de Hemptinne, P. Tobaly, Extension of polar GC-SAFT to systems containing some oxygenated compounds: application to ethers, aldehydes and ketones, Fluid Phase Equilib. 307 (2) (2011) 142–159.

[99] L. Grandjean, J.-C. de Hemptinne, R. Lugo, Application of GC-PPC-SAFT-EoS to ammonia and its mixtures, Fluid Phase Equilib. 367 (2014) 159–172.

[100] T.T.X. Nguyen, D. NguyenHuynh, Predicting the phase equilibria of esters/alcohols mixtures and biodiesel density from its fatty acid composition using the modified group-contribution PC-SAFT, Fluid Phase Equilib. 472 (2018) 128–146.

[101] D. NguyenHuynh, S.T. Tran, C.T. Mai, Predicting the phase behavior of alcohols, aromatic alcohols, and their mixtures using the modified group-contribution perturbed-chain statistical associating fluid theory, Ind. Eng. Chem. Res. 58 (36) (2019) 16963–16977.

[102] D. NguyenHuynh, C.T. Mai, Application of the modified group contribution PC-SAFT to carboxylic acids and their mixtures, Ind. Eng. Chem. Res. 58 (20) (2019) 8923–8934.

[103] D. NguyenHuynh, C.T. Mai, S.T. Tran, X.T. Nguyen, O. Baudouin, Modelling of phase behavior of ammonia and its mixtures using the mg-SAFT, Fluid Phase Equilib. 523 (2020) 112689.

[104] D. NguyenHuynh, T.T.X. Nguyen, Application of the modified group-contribution PC-SAFT to nitrile and their mixtures, Fluid Phase Equilib. 450 (2017) 112–125.

[105] J. Rozmus, J.-C. de Hemptinne, N. Ferrando, P. Mougin, Long chain multifunctional molecules with GC-PPC-SAFT: limits of data and model, Fluid Phase Equilib. 329 (2012) 78–85.

[106] E. Sauer, M. Stavrou, J. Gross, Comparison between a homo-and a heterosegmented group contribution approach based on the perturbed-chain polar statistical associating fluid theory equation of state, Ind. Eng. Chem. Res. 53 (38) (2014) 14854–14864.

[107] S. Emami, A. Vahid, J.R. Elliott Jr., F. Feyzi, Group contribution prediction of vapor pressure with statistical associating fluid theory, perturbed-chain statistical associating fluid theory, and Elliott–Suresh–Donohue equations of state, Ind. Eng. Chem. Res. 47 (21) (2008) 8401–8411.

[108] A. Tihic, G.M. Kontogeorgis, N. Von Solms, M.L. Michelsen, L. Constantinou, A predictive group-contribution simplified PC-SAFT equation of state: application to polymer systems, Ind. Eng. Chem. Res. 47 (15) (2008) 5092–5101.

[109] A. Tihic, N. von Solms, M.L. Michelsen, G.M. Kontogeorgis, L. Constantinou, Application of sPC-SAFT and group contribution sPC-SAFT to polymer systems—capabilities and limitations, Fluid Phase Equilib. 281 (1) (2009) 70–77.

[110] A. Tihic, N. von Solms, M.L. Michelsen, G.M. Kontogeorgis, L. Constantinou, Analysis and applications of a group contribution sPC-SAFT equation of state, Fluid Phase Equilib. 281 (1) (2009) 60–69.

[111] W.A. Burgess, D. Tapriyal, I.K. Gamwo, Y. Wu, M.A. McHugh, R.M. Enick, New group-contribution parameters for the calculation of PC-SAFT parameters for use at pressures to 276 MPa and temperatures to 533 K, Ind. Eng. Chem. Res. 53 (6) (2014) 2520–2528.

[112] J. Vijande, M.M. Pineiro, J.L. Legido, D. Bessieres, Group-contribution method for the molecular parameters of the PC-SAFT equation of state taking into account the proximity effect. Application to nonassociated compounds, Ind. Eng. Chem. Res. 49 (19) (2010) 9394–9406.

[113] J. Vijande, M.M. Pineiro, J.L. Legido, Group-contribution method with proximity effect for PC-SAFT molecular parameters. 2. Application to association parameters: primary alcohols and amines, Ind. Eng. Chem. Res. 53 (2) (2014) 909–919.

[114] J. Gross, O. Spuhl, F. Tumakaka, G. Sadowski, Modeling copolymer systems using the perturbed-chain SAFT equation of state, Ind. Eng. Chem. Res. 42 (6) (2003) 1266–1274.

[115] K.P. Shukla, W.G. Chapman, SAFT equation of state for fluid mixtures of hard chain copolymers, Mol. Phys. 91 (6) (1997) 1075–1082.

[116] M. Banaszak, C. Chen, M. Radosz, Copolymer SAFT equation of state. Thermodynamic perturbation theory extended to heterobonded chains, Macromolecules 29 (20) (1996) 6481–6486.

[117] J. Mairhofer, B. Xiao, J. Gross, A classical density functional theory for vapor-liquid interfaces consistent with the heterosegmented group-contribution perturbed-chain polar statistical associating fluid theory, Fluid Phase Equilib. 472 (2018) 117–127.

[118] K. Paduszyński, U. Domańska, Heterosegmented perturbed-chain statistical associating fluid theory as a robust and accurate tool for modeling of various alkanes. 1. Pure fluids, Ind. Eng. Chem. Res. 51 (39) (2012) 12967–12983.
[119] T. Peters, F.S. Laube, G. Sadowski, Development of a group contribution method for polymers within the PC-SAFT model, Fluid Phase Equilib. 324 (2012) 70–79.
[120] T. Peters, M. Herhut, G. Sadowski, Extension of the PC-SAFT based group contribution method for polymers to aromatic, oxygen-and silicon-based polymers, Fluid Phase Equilib. 339 (2013) 89–104.
[121] F.T. Peters, F.S. Laube, G. Sadowski, PC-SAFT based group contribution method for binary interaction parameters of polymer/solvent systems, Fluid Phase Equilib. 358 (2013) 137–150.
[122] N. Haarmann, S. Enders, G. Sadowski, Heterosegmental modeling of long-chain molecules and related mixtures using PC-SAFT: 1. Polar compounds, Ind. Eng. Chem. Res. 58 (7) (2018) 2551–2574.
[123] N. Haarmann, A. Sosa, J. Ortega, G. Sadowski, Measurement and prediction of excess properties of binary mixtures methyl decanoate + an even-numbered n-alkane (C6–C16) at 298.15 K, J. Chem. Eng. Data 64 (6) (2019) 2816–2825.
[124] N. Haarmann, S. Enders, G. Sadowski, Heterosegmental modeling of long-chain molecules and related mixtures using PC-SAFT: 2. Associating compounds, Ind. Eng. Chem. Res. 58 (11) (2019) 4625–4643.
[125] M. Jaber, W. Babe, E. Sauer, J. Gross, R. Lugo, J. de Hemptinne, An improved group contribution method for PC-SAFT applied to branched alkanes: data analysis and parameterization, Fluid Phase Equilib. 473 (2018) 183–191.
[126] S.E. Quiñones-Cisneros, C.K. Zéberg-Mikkelsen, E.H. Stenby, The friction theory (f-theory) for viscosity modeling, Fluid Phase Equilib. 169 (2) (2000) 249–276.
[127] A. Allal, C. Boned, A. Baylaucq, Free-volume viscosity model for fluids in the dense and gaseous states, Phys. Rev. E 64 (1) (2001) 011203.
[128] Y. Rosenfeld, Relation between the transport coefficients and the internal entropy of simple systems, Phys. Rev. A 15 (6) (1977) 2545.
[129] Y. Rosenfeld, A quasi-universal scaling law for atomic transport in simple fluids, J. Phys. Condens. Matter 11 (28) (1999) 5415.
[130] N.P. Bailey, U.R. Pedersen, N. Gnan, T.B. Schrøder, J.C. Dyre, Pressure-energy correlations in liquids. I. Results from computer simulations, J. Chem. Phys. 129 (18) (2008) 184507.
[131] N.P. Bailey, U.R. Pedersen, N. Gnan, T.B. Schrøder, J.C. Dyre, Pressure-energy correlations in liquids. II. Analysis and consequences, J. Chem. Phys. 129 (18) (2008) 184508.
[132] T.B. Schrøder, N.P. Bailey, U.R. Pedersen, N. Gnan, J.C. Dyre, Pressure-energy correlations in liquids. III. Statistical mechanics and thermodynamics of liquids with hidden scale invariance, J. Chem. Phys. 131 (23) (2009) 234503.
[133] N. Gnan, T.B. Schrøder, U.R. Pedersen, N.P. Bailey, J.C. Dyre, Pressure-energy correlations in liquids. IV. "Isomorphs" in liquid phase diagrams, J. Chem. Phys. 131 (23) (2009) 234504.
[134] T.B. Schrøder, N. Gnan, U.R. Pedersen, N.P. Bailey, J.C. Dyre, Pressure-energy correlations in liquids. V. Isomorphs in generalized Lennard-Jones systems, J. Chem. Phys. 134 (16) (2011) 164505.
[135] J.C. Dyre, Isomorphs, hidden scale invariance, and quasiuniversality, Phys. Rev. E 88 (4) (2013) 042139.
[136] L.T. Novak, Fluid viscosity-residual entropy correlation, Int. J. Chem. React. Eng. 9 (1) (2011).
[137] L.T. Novak, Predictive corresponding-states viscosity model for the entire fluid region: n-alkanes, Ind. Eng. Chem. Res. 52 (20) (2013) 6841–6847.

[138] L. Novak, Self-diffusion coefficient and viscosity in fluids, Int. J. Chem. React. Eng. 9 (1) (2011).
[139] L.T. Novak, Predicting natural gas viscosity with a mixture viscosity model for the entire fluid region, Ind. Eng. Chem. Res. 52 (45) (2013) 16014–16018.
[140] L.T. Novak, Predicting fluid viscosity of nonassociating molecules, Ind. Eng. Chem. Res. 54 (21) (2015) 5830–5835.
[141] O. Lötgering-Lin, J. Gross, Group contribution method for viscosities based on entropy scaling using the perturbed-chain polar statistical associating fluid theory, Ind. Eng. Chem. Res. 54 (32) (2015) 7942–7952.
[142] M. Hopp, J. Gross, Thermal conductivity of real substances from excess entropy scaling using PCP-SAFT, Ind. Eng. Chem. Res. 56 (15) (2017) 4527–4538.
[143] M. Hopp, J. Gross, Thermal conductivity from entropy scaling: a group-contribution method, Ind. Eng. Chem. Res. 58 (44) (2019) 20441–20449.
[144] M. Hopp, J. Mele, J. Gross, Self-diffusion coefficients from entropy scaling using the PCP-SAFT equation of state, Ind. Eng. Chem. Res. 57 (38) (2018) 12942–12950.
[145] X. Liu, T.J. Vlugt, A. Bardow, Predictive Darken equation for Maxwell-Stefan diffusivities in multicomponent mixtures, Ind. Eng. Chem. Res. 50 (17) (2011) 10350–10358.
[146] E. Poling, J.M. Prausnitz, J.P. O'Connell, Properties of Gases and Liquids, McGraw-Hill Education, 2001.
[147] J. Van der Waals, Thermodynamische Theorie der Kapillarität unter Voraussetzung stetiger Dichteänderung, Z. Phys. Chem. 13 (1) (1894) 657–725.
[148] J.W. Cahn, J.E. Hilliard, Free energy of a nonuniform system. I. Interfacial free energy, J. Chem. Phys. 28 (2) (1958) 258–267.
[149] P. Hohenberg, W. Kohn, Inhomogeneous electron gas, Phys. Rev. 136 (3B) (1964) B864.
[150] N.D. Mermin, Thermal properties of the inhomogeneous electron gas, Phys. Rev. 137 (5A) (1965) A1441.
[151] C. Ebner, W. Saam, D. Stroud, Density-functional theory of simple classical fluids. I. Surfaces, Phys. Rev. A 14 (6) (1976) 2264.
[152] W. Saam, C. Ebner, Density-functional theory of classical systems, Phys. Rev. A 15 (6) (1977) 2566.
[153] E. Johannessen, J. Gross, D. Bedeaux, Nonequilibrium thermodynamics of interfaces using classical density functional theory, J. Chem. Phys. 129 (18) (2008) 184703.
[154] C. Klink, C. Waibel, J. Gross, Analysis of interfacial transport resistivities of pure components and mixtures based on density functional theory, Ind. Eng. Chem. Res. 54 (45) (2015) 11483–11492.
[155] V. Bongiorno, L. Scriven, H. Davis, Molecular theory of fluid interfaces, J. Colloid Interface Sci. 57 (3) (1976) 462–475.
[156] A.J.M. Yang, P.D. Fleming, J.H. Gibbs, A molecular theory of interfacial phenomena in multicomponent systems, J. Chem. Phys. 64 (1976) 3732–3747.
[157] R. Evans, The nature of the liquid-vapour interface and other topics in the statistical mechanics of non-uniform, classical fluids, Adv. Phys. 28 (2) (1979) 143–200.
[158] B. Breure, C. Peters, Modeling of the surface tension of pure components and mixtures using the density gradient theory combined with a theoretically derived influence parameter correlation, Fluid Phase Equilib. 334 (2012) 189–196.
[159] P. Cornelisse, C. Peters, J. de Swaan Arons, On the fundamentals of the gradient theory of Van der Waals, J. Chem. Phys. 106 (23) (1997) 9820–9834.
[160] C. Miqueu, B. Mendiboure, A. Graciaa, J. Lachaise, Modelling of the surface tension of pure components with the gradient theory of fluid interfaces: a simple and accurate expression for the influence parameters, Fluid Phase Equilib. 207 (1–2) (2003) 225–246.

[161] H. Lin, Y.-Y. Duan, Q. Min, Gradient theory modeling of surface tension for pure fluids and binary mixtures, Fluid Phase Equilib. 254 (1–2) (2007) 75–90.

[162] M. Garrido, A. Mejía, M.M. Pineiro, F.J. Blas, E.A. Müller, Interfacial tensions of industrial fluids from a molecular-based square gradient theory, AIChE J. 62 (5) (2016) 1781–1794.

[163] P. Rehner, J. Gross, Predictive density gradient theory based on nonlocal density functional theory, Phys. Rev. E 98 (6) (2018) 063312.

[164] J. Gross, A density functional theory for vapor-liquid interfaces using the PCP-SAFT equation of state, J. Chem. Phys. 131 (20) (2009) 204705.

[165] C. Klink, J. Gross, A density functional theory for vapor-liquid interfaces of mixtures using the perturbed-chain polar statistical associating fluid theory equation of state, Ind. Eng. Chem. Res. 53 (14) (2014) 6169–6178.

[166] E. Sauer, J. Gross, Classical density functional theory for liquid-fluid interfaces and confined systems: a functional for the perturbed-chain polar statistical associating fluid theory equation of state, Ind. Eng. Chem. Res. 56 (14) (2017) 4119–4135.

[167] D. Fu, Y. Wei, Investigation of vapor-liquid surface tension for carbon dioxide and hydrocarbon mixtures by perturbed-chain statistical associating fluid theory combined with density-gradient theory, Ind. Eng. Chem. Res. 47 (13) (2008) 4490–4495.

[168] O.N. Amézquita, S. Enders, P.T. Jaeger, R. Eggers, Measurement and prediction of interfacial tension of binary mixtures, Ind. Eng. Chem. Res. 49 (2) (2010) 592–601.

[169] O.N. Amézquita, S. Enders, Prediction of interfacial tension of binary mixtures, in: Computer Aided Chemical Engineering, vol. 28, Elsevier, 2010, pp. 85–90.

[170] V. Vinš, B. Planková, J. Hruby, Surface tension of binary mixtures including polar components modeled by the density gradient theory combined with the PC-SAFT equation of state, Int. J. Thermophys. 34 (5) (2013) 792–812.

[171] S. Becker, S. Werth, M. Horsch, K. Langenbach, H. Hasse, Interfacial tension and adsorption in the binary system ethanol and carbon dioxide: experiments, molecular simulation and density gradient theory, Fluid Phase Equilib. 427 (2016) 476–487.

[172] O.N. Amézquita, S. Enders, P. Jaeger, R. Eggers, Interfacial properties of mixtures containing supercritical gases, J. Supercrit. Fluids 55 (2) (2010) 724–734.

[173] E. Schäfer, G. Sadowski, S. Enders, Interfacial tension of binary mixtures exhibiting azeotropic behavior: measurement and modeling with PCP-SAFT combined with density gradient theory, Fluid Phase Equilib. 362 (2014) 151–162.

[174] X.-S. Li, J.-M. Liu, D. Fu, Investigation of interfacial tensions for carbon dioxide aqueous solutions by perturbed-chain statistical associating fluid theory combined with density-gradient theory, Ind. Eng. Chem. Res. 47 (22) (2008) 8911–8917.

[175] O.N. Amézquita, D. van Putten, S. Enders, Phase equilibrium and interfacial properties of water + CO_2 mixtures, Fluid Phase Equilib. 332 (2012) 40–47.

[176] O.N. Amézquita, S. Enders, Phase equilibrium and interfacial properties of water + methane mixtures, Fluid Phase Equilib. 407 (2016) 143–151.

[177] X. Mu, S. Xi, F.O. Alpak, W.G. Chapman, Modified density gradient theory for surfactant molecules applied to oil/water interfaces, Ind. Eng. Chem. Res. 57 (22) (2018) 7643–7654.

[178] L.C. Vergara, G.M. Kontogeorgis, X. Liang, On the study of the vapor-liquid interface of associating fluids with classical density functional theory, Fluid Phase Equilib. 522 (2020) 112744.

[179] C. Klink, B. Planková, J. Gross, Density functional theory for liquid-liquid interfaces of mixtures using the perturbed-chain polar statistical associating fluid theory equation of state, Ind. Eng. Chem. Res. 54 (16) (2015) 4633–4642.

[180] X. Xu, D.E. Cristancho, S. Costeux, Z.-G. Wang, Density-functional theory for polymer-carbon dioxide mixtures: a perturbed-chain SAFT approach, J. Chem. Phys. 137 (5) (2012) 054902.

[181] J. Mairhofer, J. Gross, Modeling properties of the one-dimensional vapor-liquid interface: application of classical density functional and density gradient theory, Fluid Phase Equilib. 458 (2018) 243–252.

[182] J. Mairhofer, J. Gross, Numerical aspects of classical density functional theory for one-dimensional vapor-liquid interfaces, Fluid Phase Equilib. 444 (2017) 1–12.

[183] R. Stierle, E. Sauer, J. Eller, M. Theiss, P. Rehner, P. Ackermann, J. Gross, Guide to efficient solution of PC-SAFT classical density functional theory in various coordinate systems using fast Fourier and similar transforms, Fluid Phase Equilib. 504 (2020) 112306.

[184] R. Heide, The surface tension of HFC refrigerants and mixtures, Int. J. Refrig. 20 (7) (1997) 496–503.

[185] A. Heller, M. Rausch, F. Flohr, A. Leipertz, A. Fröba, Thermophysical properties of the refrigerant mixtures R417a and R417b from dynamic light scattering (DLS), Int. J. Thermophys. 33 (3) (2012) 396–411.

[186] P. Fröba, C. Botero, H. Kremer, A. Leipertz, Thermophysical properties of a quaternary refrigerant mixture: comparison of dynamic light scattering measurements with a simple prediction method, Int. J. Thermophys. 28 (3) (2007) 743–757.

[187] S. Tripathi, W.G. Chapman, Microstructure of inhomogeneous polyatomic mixtures from a density functional formalism for atomic mixtures, J. Chem. Phys. 122 (9) (2005) 094506.

[188] S. Jain, A. Dominik, W.G. Chapman, Modified interfacial statistical associating fluid theory: a perturbation density functional theory for inhomogeneous complex fluids, J. Chem. Phys. 127 (24) (2007) 244904.

[189] N. Haarmann, A. Reinhardt, A. Danzer, G. Sadowski, S. Enders, Modeling of interfacial tensions of long-chain molecules and related mixtures using perturbed chain-statistical associating fluid theory and the density gradient theory, J. Chem. Eng. Data 65 (3) (2019) 1005–1018.

Chapter 2

Don't search—Solve! Process optimization modeling with IDAES

Lorenz T. Biegler[a], David C. Miller[b], and Chinedu O. Okoli[b]
[a]*Department of Chemical Engineering, Carnegie Mellon University, Pittsburgh, PA, United States,*
[b]*National Energy Technology Laboratory, Pittsburgh, PA, United States*

1 Introduction

Today's energy and chemical processes are challenged by highly competitive market forces and tight regulatory and operating constraints, compounded by external pressures and challenges due to the accommodation and integration of alternate energy sources. Addressing these challenges requires future designs to support highly flexible operations that can exploit dynamic demand response. More generally, the design of next-generation chemical and energy processes will require greater environmental sustainability, higher performance, reduced cost, tolerance to uncertainty, and dynamic interactions. In particular, technological advances such as process intensification, electrification, and alternative separation techniques are driving the need for rapid and rigorous design strategies that extend beyond the current state-of-the-art, which often relies on evolutionary and heuristic approaches. Instead, they must incorporate complex decision-making strategies based on rigorous large-scale optimization tools to ensure adequate understanding of the nonintuitive interactions of potential new technologies and novel operating paradigms.

Rigorous optimization strategies have found widespread use in chemical engineering applications, especially in the engineering of process systems. Problems in this domain often have many alternative solutions with complex economic and performance interactions, and it is often intractable to identify the optimal solution through intuitive decision-making. System economics often indicate that finding the optimum solution translates into large savings, as well as a large economic penalty for sticking to suboptimal solutions. Therefore, optimization has become a major technology that is essential for new developments to remain competitive. Moreover, recent advances in large-scale

Simulation and Optimization in Process Engineering. https://doi.org/10.1016/B978-0-323-85043-8.00005-2

TABLE 2.1 Mathematical programming in process systems engineering.

	LP	MILP	QP	NLP	MINLP
Process Design and Synthesis					
Heat Exchangers	X	X		X	X
Mass Exchangers	X	X		X	X
Separations		X		X	X
Reactors	X			X	X
Flowsheeting				X	X
Process Operations					
Scheduling/Planning	X	X			X
Supply Chain	X	X			X
Real-time Optimization	X		X	X	
Process Control					
Model Predictive Control	X		X		
Nonlinear MPC			X	X	

optimization algorithms and modeling strategies are key enablers to address these challenges for both steady-state and dynamic systems.

Table 2.1 lists several chemical engineering areas that have been addressed in an extensive body of literature (e.g., summarized in Ref. [1]) for specific classes of optimization problems. These include applications in process synthesis, design operations, and control, solved as linear programs (LP), quadratic programs (QP), mixed-integer linear programs (MILP), nonlinear programs (NLP), and mixed-integer nonlinear programs (MINLP). When optimization models are represented in algebraic form, all discrete/continuous optimization problems can be written as MINLPs. In the absence of discrete decision variables, the MINLP problem reduces to a nonlinear program. This study focuses on NLP problems and briefly describes methods that locate local solutions efficiently. While this approach might first appear as a restricted form of optimization, NLPs appear in many large-scale engineering models, and they also form important components of strategies for deterministic optimization strategies for both discrete and continuous decisions.

In addition, the creation and deployment of nonlinear models for chemical processes are essential for decision-making in design, operations, and control. Engineering practice is aided by a wealth of rigorous simulation models that are trusted to predict design and operating scenarios for new processes and products. Moreover, systematic approaches for decision-making demand accurate computer models and computer-aided engineering platforms to drive and demonstrate effective decision-making. Consequently, the quality of these decisions is strongly influenced by the efficiency of optimization strategies as well as their interaction with detailed simulation models to form integrated *optimization models.*

1.1 Optimization evolution from systematic search to direct solution

Engineering practice often views the optimization exercise as a glorified case study, where a search strategy is wrapped around the process model and is directed to provide process simulation solutions until the "best" answer is found. No doubt such approaches are effective and lead to significant and consistent improvements, even when used by less experienced engineers. On the other hand, such approaches can be very expensive to execute for large problems, and often it is not clear whether all options have been explored within the allocated computational budget.

Fig. 2.1 illustrates the interaction of process models with the nonlinear optimization problem given by:

$$Min\, f(x), s.t. c(x) = 0, g(x) \leq 0 \tag{2.1}$$

which contains an objective function $f(x)$ with variables x in some domain X, represented by inequality constraints $g(x) \leq 0$, and the algebraic (or differential-algebraic (DAE)) model equations $c(z, z', u, p, t) = 0$. Included among the variables x are the state variables z, decision variables p, and profiles $u(t)$ (where t represents some independent variable, space, or time). The left graphic shows a typical, nested optimization scheme, where the simulation model is called repeatedly by the search strategy. Recent advances for the nested strategy, including derivative-free optimization (DFO), have led to superior algorithms with strong convergence properties [2]. However, search strategies still remain expensive for problems with more than a few hundred degrees of freedom and are often prohibitive for practical applications.

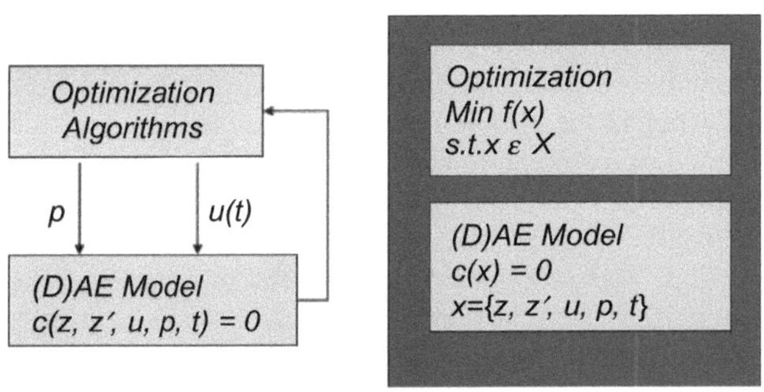

FIG. 2.1 Nested and simultaneous strategies for process optimization.

Optimization studies must move beyond the concept of *searching for improved solutions through repeated, time-consuming simulations.* Instead, *solving directly for conditions that are satisfied by the optimum point* can be done with the effort of a single simulation. For nonlinear optimization, these optimality conditions are the well-known Karush-Kuhn-Tucker (KKT) conditions, which consist of systems of the following equations and inequalities:

$$\begin{aligned} &\nabla f(x) + \nabla c(x)\lambda + \nabla g(x)\nu = 0 \\ &c(x) = 0, 0 \geq g(x) \perp \nu \geq 0 \end{aligned} \tag{2.2}$$

with KKT multipliers (λ, ν), and complementarity constraints (i.e., $\nu_i = 0$ or $g_i = 0$ (or both)).Conditions (2.2) lead to a straightforward extension of the simulation model. Efficient and robust large-scale algorithms and software are now available that quickly find local optimal solutions of process systems, with many thousands of variables and equations. Moreover, these optimization strategies *ensure that the best (local) solutions have been found,* provide efficient estimations of which neighboring solutions can be assessed, and lead to *sensitivity of the optimal solution* to exogenous inputs. Finally, the optimization strategy must be aided by optimization modeling platforms so that it can be embedded within the work process and can deal with model variations and extensions. Unfortunately, these capabilities are not often found in current engineering practice.

The right graphic of Fig. 2.1 represents a solution strategy where the optimality conditions are solved simultaneously with the simulation model. These conditions for the problem in Fig. 2.1 stem from the KKT conditions of nonlinear programming [3,4], represented by a set of equations and inequalities (2.2). Solving these optimality conditions requires about the same complexity as solving nonlinear equations and, within an effective modeling framework, scales efficiently to very large problems. In addition, second-order conditions, which confirm sufficiency and (local) uniqueness of the optimum, are easily checked at stationary points, when second derivatives are available. Finally, additional regularity conditions and constraint qualifications can be enforced through careful modeling of the optimization problem, along with reliable algorithmic implementations.

Engineering practice has started to embrace these developments, and there is tremendous potential toward improving both the work process and work product. The most efficient NLP tools now handle millions of variables and constraints with modest computational effort. As a result, complex nonlinear process models can be optimized fast enough for online purposes. Moreover, advances in modeling approaches, algorithms, software, and hardware allow the rapid creation, realization, and deployment of large-scale optimization models that outperform current search approaches. The next section discusses key concepts that lead to fast and robust nonlinear optimization strategies. Section 3 extends these concepts to the formulation and solution of dynamic optimization strategies. Section 4 then provides an overview of the Institute for the Design of Advanced Energy Systems (IDAES) Integrated Platform , an optimization modeling framework tailored to process engineering. A comprehensive case study on a novel carbon capture process demonstrates the capabilities of this platform in Section 5, and Section 6 concludes the chapter and outlines ongoing developments.

2 Solution algorithms and optimization models

Fig. 2.2 shows the family of optimization algorithms that can be applied to nonlinear optimization problems, as well as the requirements from the optimization model. NLP solvers can be classified as follows.

Derivative-free optimization algorithms (e.g., genetic algorithms, simulated annealing, particle swarm, tabu search, pattern searches [2]) require no derivatives and are primarily designed for unconstrained problems. While they are easy to apply, especially with sequential approaches, they become very expensive for optimization problems with more than about 100 variables and are not suited for large NLPs. For Black Box models that provide only input-output information, these search methods, require more than the equivalent of 100 simulation time equivalents (STEs), even with just a few degrees of freedom.

Active set strategies such as Successive Quadratic Programming (SQP) and reduced gradient methods (e.g., GRG, CONOPT [1]) allow simultaneous minimization and constraint satisfaction. They have superlinear convergence properties and work well for moderately sized NLPs (>100 decision variables).

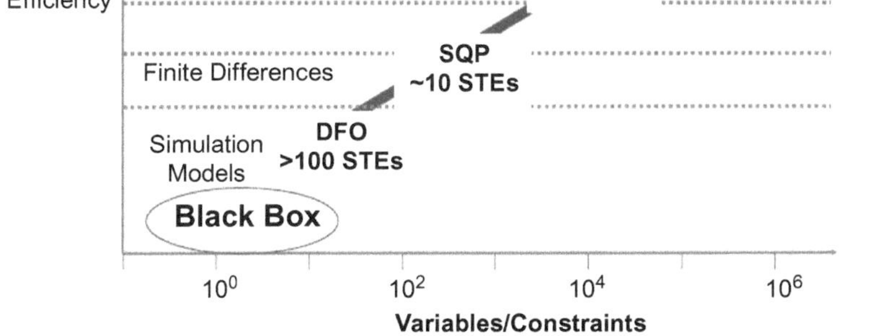

FIG. 2.2 Hierarchy of optimization models and solvers.

Optimization models that provide exact first and second derivatives, as well as structural information to the NLP solver, ensure that NLPs can be solved as cheaply as a single simulation problem. To appreciate this leap in computational efficiency, we consider the application of Newton-based solvers to the KKT conditions (2.2). First, the NLP (2.1) can be expressed equivalently as a sequence of barrier problems:

$$Min\, f(x) - \mu_l \sum_i \ln(s_i), s.t. c(x) = 0, g(x) + s = 0 \tag{2.3}$$

with the barrier parameter $\mu_l > 0$ and $\lim_{l \to \infty} \mu_l \to 0$, and KKT conditions given by:

$$\begin{aligned} r_1 &= \nabla f(x) + \nabla c(x)\lambda + \nabla g(x)v = 0 \\ r_2 &= Sv - \mu_l e = 0 \\ r_3 &= c(x) = 0 \\ r_4 &= g(x) + s = 0 \end{aligned} \tag{2.4}$$

where $e = [1,\ 1,\ \ldots 1]^T$ and $S = diag(s)$, $V = diag(\nu)$. Solution of Eq. (2.4) through a sequence of Newton's steps leads to solving the following symmetric system of linear equations at each iteration k:

$$\begin{bmatrix} W(x^k, \lambda^k, v^k) & & \nabla c(x^k) & \nabla g(x^k) \\ & (S^k)^{-1} V^k & & I \\ \nabla c(x^k)^T & & & \\ \nabla g(x^k)^T & I & & \end{bmatrix} \begin{bmatrix} \Delta x \\ \Delta s \\ \Delta \lambda \\ \Delta v \end{bmatrix}$$

$$= - \begin{bmatrix} r_1(x^k, s^k, \lambda^k, v^k) \\ (S^k)^{-1} r_2(s^k, v^k, \mu_l) \\ r_3(x^k) \\ r_4(x^k s^k) \end{bmatrix} \tag{2.5}$$

with the gradient matrices ∇c, ∇g, and W as the Hessian of the Lagrange function. It is well-known that such derivative-based solvers have fast convergence properties and scale to large problems by exploiting sparsity and structure of Eq. (2.5). Less well appreciated among practitioners are advances in the globalization of Newton-based solvers, through strong descent properties as well as line search and trust region methods [5,6] that allow convergence from distant initial points. These advances have been realized in the development of barrier NLP methods [5,7], which convert inequality-constrained optimization problems to a sequence of equation sets. Moreover, sparsity-preserving Newton-based methods (e.g., IPOPT, KNITRO, LOQO [5,8]) with exact Hessians W solve the optimality conditions directly, allow full exploitation of the problem structure, have quadratic convergence, and can handle very large NLPs (>1,000,000 variables). As a result, these NLP algorithms are the engines that drive the large-scale optimization strategies in modeling platforms such as IDAES.

3 Advanced optimization for differential-algebraic applications

We now consider the incorporation of process models that include ordinary and partial differential equations, where the equation-oriented optimization formulations are often essential for real-world process systems. Fig. 2.1 provides a clear illustration of both nested and simultaneous strategies for the optimization of differential equation systems. The simultaneous approach with equation-oriented (or glass box) models in Fig. 2.2 leads to larger NLPs because the entire process model converges simultaneously with the optimization problem. These optimization models are no longer nested, avoid repeated solutions, and exploit sparse algorithms that scale almost linearly with problem size. This also leads to much lower computational complexity for dynamic optimization algorithms.

With the incorporation of differential equation models, additional features should be embedded in order to preserve stable, consistent, and accurate problem formulations for PDEs, DAEs, and ODEs [6,9]. This requires these equations to be fully discretized in time and/or space and added to the optimization model. As can be seen from Fig. 2.1, the simultaneous approach, therefore, leads to much larger NLPs than with the nested approach, but generally leads to much faster performance, as described below.

Optimization methods can incorporate dynamic models in a number of ways. To form this system, the differential-algebraic equation (DAE) model, $c(z, z', u, p, t) = 0$, which can be written in the semiexplicit form:

$$z_d' = F(z, u, p, t), G(z, u, p, t) = 0, z_d = z_{d,0} \tag{2.6}$$

where $z(t) = (z_d(t), z_a(t))$ are the (differential and algebraic) state variables. The DAE (2.6) must be represented at successive evolutions in time or space (e.g., from t_k to $t_k + 1$) through an appropriate, discretization strategy of high accuracy (high order) and numerical stability. General DAE solvers can be classified into two types [9].

- *Single-step discretizations* (e.g., *Runge-Kutta (R-K) methods).* High order implicit R-K methods are widely used and have A-stable properties, which are useful for stiff systems. Moreover, as self-starting methods, they do not require backward interpolation and easily handle discontinuous input profiles accurately.
- *Linear Multistep discretizations* (e.g., *Adams and Backward Difference Formula (BDF) methods).* Implemented as predictor-corrector methods of high order, LMS must be initialized by single-step methods, and only BDF methods are reliable for stiff systems. To solve initial value problems (IVPs), LMS methods are generally faster than R-K methods, but because they require backward interpolation, discontinuous input profiles will require frequent restarting.

The choice of discretization methods strongly impacts the solution strategy for Eq. (2.6). The sequential (or nested) approach with an embedded initial value (IV) DAE solver is easy to implement and leads to a smaller optimization problem. However, with many decision variables, calculation of gradients through direct or adjoint sensitivity becomes expensive, and convergence noise from the IV solver also propagates to the gradients; second derivatives are rarely calculated due to the increased computational cost. Moreover, dynamic systems with open-loop instabilities may lead to the failure of the IV solver.

In contrast, a full, R-K type discretization of Eq. (2.2) (e.g., orthogonal collocation on finite elements) leads to a high order representation of the DAE and a large, sparse optimization problem (2.1). With this approach, first, and even second, derivatives for Eq. (2.1) can be evaluated through automatic differentiation and are much cheaper to calculate. The fully discretized formulation of the NLP (2.1) leads to a simultaneous strategy for the solution of the dynamic system along with the optimization problem. Moreover, such an approach has the same dichotomy property as boundary value solvers, which can handle open-loop instabilities reliably and efficiently. On the other hand, the simultaneous (or full discretization) approach requires an efficient, large-scale NLP solver for efficient performance.

3.1 Complexity of dynamic optimization strategies

For dynamic optimization, we compare two approaches. The sequential (or nested) approach embeds DAE solvers within the optimization strategy and integrates the DAEs over the entire time (or space) domain. On the other hand, the simultaneous approach applies a collocation discretization over finite elements and leads to a fully algebraic system that is directly incorporated into the optimization model. This simultaneous approach depends on efficient NLP strategies that exploit sparse linear algebra and low complexity in handling inequality constraints. These advantages can be realized by a full-space NLP formulation with exact first and second derivatives and a barrier method to handle inequality constraints [1].

With the aid of Newton-based barrier solvers, we note that while simultaneous collocation approaches lead to larger NLP formulations, the effort to solve them still remains quite reasonable. Table 2.2 lists the (order of magnitude) complexity of the major algorithmic steps for dynamic optimization using the sequential (nested) and simultaneous (collocation) strategies. While a detailed comparison of these methods is often problem-dependent, distinguishing features for each strategy allows some insight into the computational effort for each approach. Table 2.2 considers each step in nested and simultaneous approaches and presents their computational complexity in terms of the number of discretization steps N, the number of state variables n_z, and the number of control variables n_u. Detailed derivations of all of the computational costs are presented in [1]. These show that nested approaches require DAE

TABLE 2.2 Computational complexity per NLP iteration (with n_z state variables, n_u control variables, N time steps, $\alpha \in [2, 3]$, $\beta \in [1, 2]$).

Algorithmic steps	Sequential	Simultaneous
Integration	$n_z^\beta N$	–
Sensitivity	$n_z n_u N^2$	$(n_z + n_u)N$
Hessian Evaluation	$n_z(n_u N)^2$	$(n_z + n_u)N$
Decomposition	–	–
Factorization/Backsolve	$(n_u N)^\alpha$	$((n_z + n_u)N)^\beta$
Total without Hessian	$(n_u N)^\alpha + n_z n_u N^2$	$((n_z + n_u)N)^\beta$
Total with Hessian	$(n_u N)^\alpha + n_z(n_u N)^2 N$	$((n_z + n_u)N)^\beta$

integrators, sensitivity, and matrix factorization that scale to second to third order in N and n_u. On the other hand, complexity for the simultaneous approach is determined from the solution of nonlinear (KKT) equations through a sparse Newton method. As a result, Table 2.2 shows that as the number of degrees of freedom $n_u N$ increases, one sees an almost *linear scaling* with the simultaneous approach, while nested approaches have *quadratic and cubic complexity*. Consequently, the larger simultaneous formulations can solve at a much lower cost as problem sizes increase. A demonstration of this approach will be presented in Section 5.

4 The IDAES optimization modeling software platform

Many process simulators were designed decades ago, before the development of modern optimization algorithms. Consequently, they apply nested approaches to optimization and take limited advantage of the advances of Newton-based solvers and their interactions with optimization models. Moreover, nonsmooth, legacy simulation models, with inexact or approximate derivatives are not suitable for gradient-based optimization and lead to poor performance even with the most recent NLP solvers. This often leads to a frustrating optimization exercise within the work process.

In contrast, large-scale NLPs are often constructed and solved on widely used optimization modeling platforms, including GAMS [10], AMPL [11], and AIMMS [12]. These tools provide Algebraic Modeling Languages (AMLs), seamless interfaces to large-scale algorithms, automatic differentiation, as well as a focus on equation-based development of well-posed optimization models. On the other hand, these commercial tools deal with relatively closed, declarative models, generally discourage integration with external models, and define their own proprietary syntax for representing optimization problems. Consequently, they have limited application to detailed process simulation models. To overcome the gap between these platforms, there has been a recent expansion to include more detailed engineering models on python-based platforms like Pyomo (www.pyomo.org [13]). This open-source optimization modeling platform has recently undergone major re-structuring and extension to include LP, MIP, NLP, and MINLP solvers, along with automatic differentiation. Moreover, Pyomo allows the construction of meta-algorithms for optimization frameworks along with the incorporation of complex engineering models.

Pyomo has particular relevance to current research and advances in the IDAES project (www.idaes.org [14]), a US Department of Energy sponsored program to develop an open-source optimization framework for process engineering applications, envisioned as the next-generation of process engineering tools. As an equation-oriented, optimization-based integrated process modeling platform, IDAES enables rigorous analysis of multiscale, dynamic processes and operating scenarios to improve efficiency of existing systems and to develop next-generation energy systems. IDAES has seen thousands of downloads and an active, growing global user community from multiple industries. Since its release in March 2019, IDAES has demonstrated significant impacts through its unique capabilities. IDAES is the only fully equation-oriented computational platform with integrated support for steady-state design, optimization, dynamic operations, data reconciliation, parameter estimation, and uncertainty quantification of complex energy and chemical processes. IDAES was conceived to combine capabilities of state-of-the-art process simulation packages and general AMLs to address emerging challenges and opportunities in the energy and process sectors that cannot be addressed by current tools. These include the need to design and optimize processes that are more tightly coupled with external systems and incorporate new technologies with inherently dynamic operations. Finally, because IDAES is built on the Python programming language, it can be integrated and extended in complex ways to support new and existing user applications. Moreover, it leverages the existing Python ecosystem for data analysis, visualization, and application development.

Currently, IDAES consists of multiple components shown in Fig. 2.3. The core of the IDAES platform is an open and customizable process model library, built on the Pyomo AML. The core models can be combined to capture the hierarchical structure of subprocesses and complex multidimensional unit models that can consist of mass and energy balances, thermodynamics, fluid flow, heat and mass transfer, reaction kinetics, and other physical and chemical phenomena. These core capabilities include equation-based steady-state and dynamic modeling, and a hierarchical approach to support the development of custom unit models for innovative processes. In addition, the process model library can be used to support conceptual design via a superstructure formulation, process design, optimization, and integration, and dynamic optimization and control. As illustrated in Fig. 2.3, IDAES leverages the following principal features.

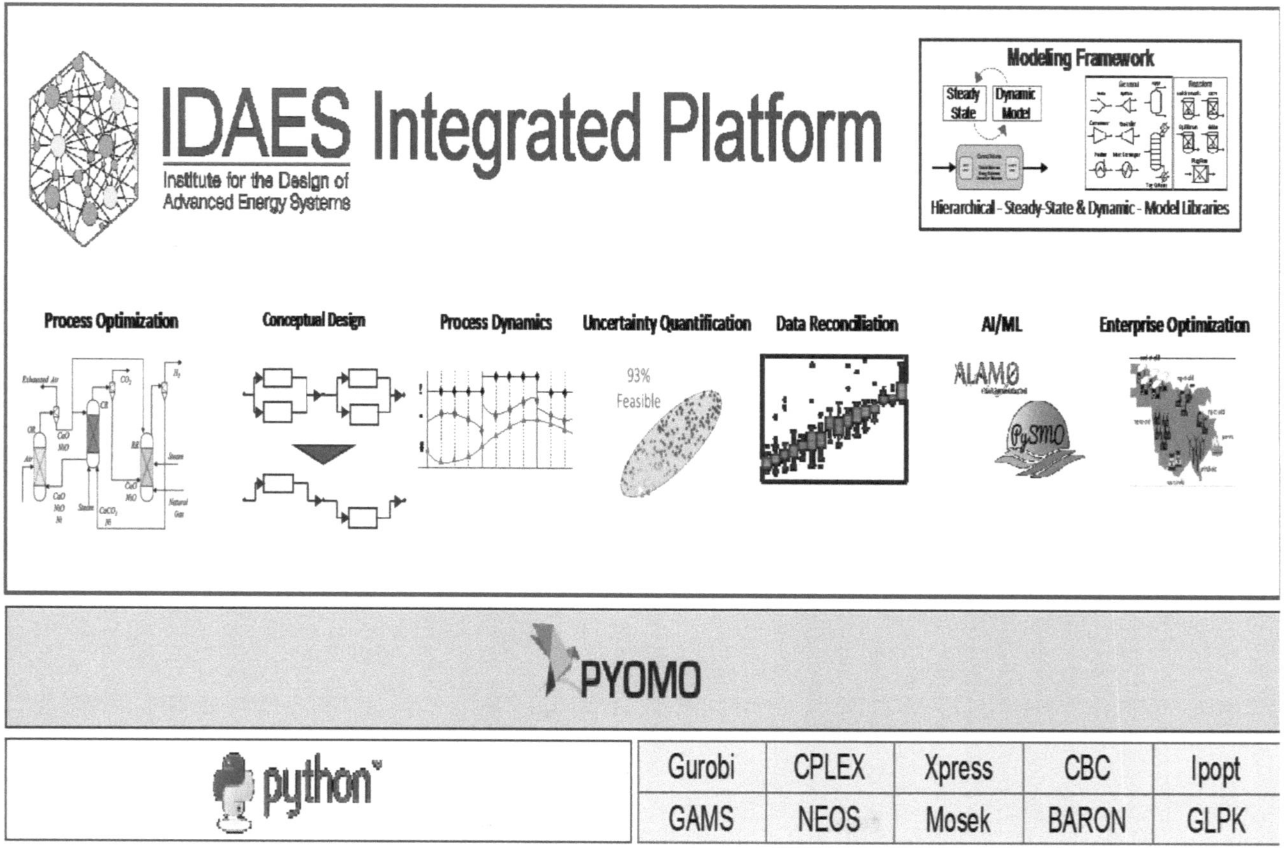

FIG. 2.3 The major components and connection of IDAES to support conceptual design, process design and optimization, process integration, dynamic optimization, and control while incorporating multiscale surrogate models.

1. *Pyomo AML:* For IDAES, this provides an extensible algebraic modeling environment that provides interfaces to dozens of optimization solvers, both open-source and commercial, and supports the execution of solvers both locally and in distributed environments. Moreover, Pyomo has extensions for advanced process modeling, including block-oriented modeling, explicit representation of network models, stochastic programming, and dynamic systems containing (ordinary and partial) differential-algebraic equations, and Generalized Disjunctive Programming (GDP). These features enable modeling flexibility in nonstandard forms not easily handled by other modeling frameworks.
2. *Hierarchical Equation-Oriented Process Models:* While AMLs readily support large-scale optimization, considerable work is required to construct process models, which are often not easily reused. In contrast, IDAES extends Pyomo's block-based hierarchical modeling constructs to create a model library for common process unit operations, where every equation is completely open and extensible, as well as a framework for rapid development of customized unit operations. The model library contains an internally consistent suite of interoperable models at different levels of complexity and accuracy for both steady-state and dynamic analysis. This flexibility enables models to be created and used for novel processes and configurations.
3. *Process Design and Optimization*: IDAES enables large-scale, system-wide optimization using rigorous models of process equipment and thermophysical properties. Complex interactions between multiple process systems can be explored simultaneously without simplifying assumptions that can risk compromising the results. Methods for enabling optimization over broad ranges of operation have also been developed that can handle nonsmooth features such as vapor/liquid or sub/supercritical phase transitions as well as vanishing phases. The process design and optimization capabilities developed within IDAES are applicable to broad applications areas, e.g., energy, chemicals, petrochemicals, pharmaceuticals, where tight integration is desired.
4. *Dynamics and Control:* This IDAES feature builds on Pyomo's native support for explicit specification of ordinary and partial differential equations, and the automated transformation, of DAE models into large-scale algebraic systems, through Pyomo.DAE [15]. The process model library includes both steady-state and dynamic versions of common unit models, which can toggle between steady-state and pressure-driven dynamic analysis with minimal changes to their flowsheet implementation. Moreover, IDAES includes advanced capabilities for state estimation and control, including moving horizon estimation and nonlinear model predictive control [16], which can be directly deployed on process flowsheets.

5. *Process Synthesis and Conceptual Design:* IDAES incorporates Generalized Disjunctive Programming (GDP), an approach that combines algebraic process models with disjunctions and logic statements that lead to specialized MINLP models. Moreover, Pyomo's support for GDP leads to strategies that operate directly on the logical problem structure and NLP subproblems that correspond to fixed flowsheet structures, thus providing increased robustness for MINLP optimization.
6. *Multiscale Surrogate Modeling and Optimization*: IDAES has developed fast *surrogate* models that preexecute complex models in a coarse-grained manner and then calculate approximations of the actual values, very quickly, during optimization. To avoid skipping over optimal solutions with this coarse-grained mesh, IDAES uses a "trust region" methodology to guarantee a solution at critical points of the complex model [17]. This surrogate modeling approach provides complete accuracy of the full model at far greater efficiency, thus allowing the solution of complex models that operate at multiple scales.
7. *Machine Learning and Parameter Estimation:* IDAES incorporates several integrated optimization-based machine learning tools for modeling thermophysical properties and reaction kinetics. These include (a) ALAMO, which builds algebraic models for an optimization framework, (b) Python-based Surrogate Modeling Objects (PySMO) package, with state-of-the-art methods for sampling, machine learning, parameter estimation, and surrogate-assisted optimization, (c) RIPE, which learns reaction networks from process data, and (d) HELMET, which learns a Helmholtz equation of state from thermodynamic data. Also, in IDAES, all models and submodels are on a common platform with access to all model parameters, for formulation and solution of simultaneous multiscale parameter estimation problems.

5 Carbon capture optimization case study

To demonstrate the unique capabilities of IDAES, we present an optimization study for a novel Chemical Looping Combustion (CLC) process for CO_2 capture. CLC is a transformative process for the combustion of fossil fuels that produces a sequestration-ready, CO_2-rich exhaust stream [18]. Current research into CLC has focused on a wide variety of areas including the choice and design of reactors, development, and testing of oxygen carriers (OCs), utilization of different types of fuels, as well as operability and process improvement [19,20]. The performance of a CLC system is dependent on the OC redox reactions and the contact between the OC and the fuel. Studies investigating the use of moving bed (MB) reactors as fuel reactors for CLC applications have emphasized the advantages of operating in the counter-current MB regime [21]. Due to the counter-current flow of the gas and solid, near-complete conversion of the fuel can be obtained, together with very high conversion of the OC, which leads to a low solids circulation rate. The MB [22] process is modeled as a rigorous steady-state, first-principles model which captures both micro-, and macro-scale phenomena present in the reactors. Here we demonstrate the performance of the optimization capabilities in IDAES through minimization of total annualized cost (TAC) of the MB process.

As sketched in Fig. 2.4, IDAES uses a modular structure that allows interchangeable "property packages" to be connected to reactor models [23]. First, a flowsheet is created to describe the process, as shown in Fig. 2.5. Property package models are then attached to provide kinetic and thermophysical properties. Models of the individual unit operations, including compressors and heat exchangers from the standard model library, are added to the flowsheet and connected together to form a representation of the process. An appropriate objective function, flowsheet constraints, and optimization variables are then specified, and the model is solved. The solution procedure includes substeps for model discretization and initialization. Furthermore, custom model initialization routines can be integrated into the solve step by leveraging the flexibility of Pyomo and Python [23].

As shown in Fig. 2.5, the Fuel (FR) and Air (AR) Reactors employ counter-current contacting of gas and solids. Here, the methane fuel feed (128.3 mol/s, with 97/5% CH_4, 2.5% CO_2) is reacted with an iron-based (Fe_2O_3/Fe_3O_4),

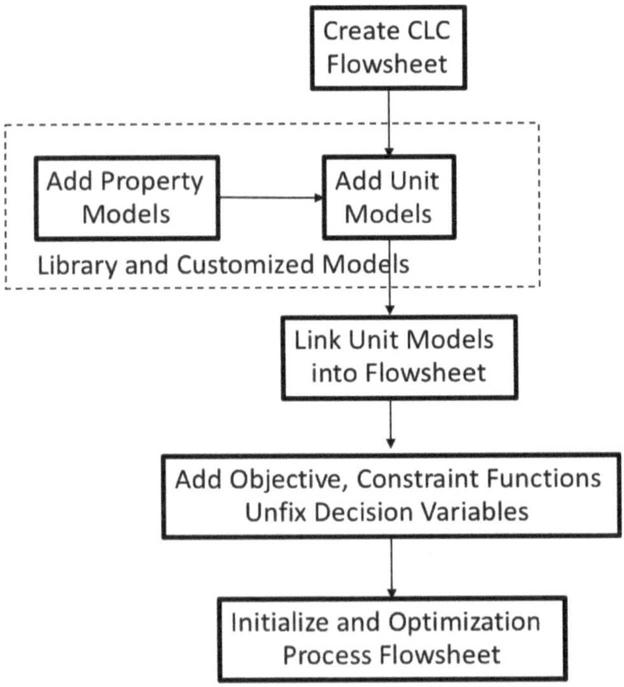

FIG. 2.4 Workflow for CLC process flowsheet optimization setup in IDAES.

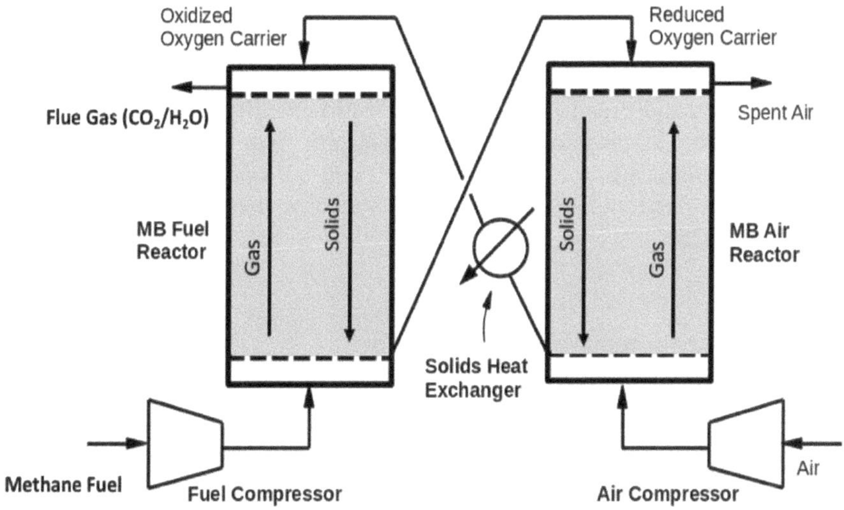

FIG. 2.5 Process flow diagram of the MB CLC system.

Al_2O_3 supported Oxygen Carrier (OC) to produce CO_2, H_2O, and the reduced form of the OC as shown below [24]:

$$\text{Fuel Reactor (OC Reduction)}: \quad CH_4 + 12\,Fe_2O_3 \rightarrow 8\,Fe_3O_4 + CO_2 + 2H_2O$$

The reduced OC from the FR is then regenerated in the AR with air according to

$$\text{Air Reactor (OC Oxidation)}: \quad 4\,Fe_3O_4 + O_2 \rightarrow 6\,Fe_2O_3$$

and is recycled back to the FR to complete the loop. Kinetic parameters for the oxidation and reduction of the iron-based OCs are obtained from Abad et al. [24]. As seen in Fig. 2.5, ancillary equipment considered in the process are compressors for the fuel and air feeds and a solids heat exchanger to cool the OC stream leaving the AR, before it is recycled back to the FR.

The model used for the MB reactors consists of two sets of differential equations representing mass, energy, and momentum balances to represent the counter-current flow of the gas and solid streams. Assuming negligible axial dispersion, thermal conduction, radial concentration, and temperature gradients, two mass balances are constructed: one for the gas and one for the solid phase. Similarly, for adiabatic operation, two energy balance equations are considered for each phase. Energy and mass are exchanged between solid and gas phases (by reaction) while moving in opposite directions. The solid phase is assumed to be moving at a constant velocity, while the gas-phase velocity through the reactor bed is computed using the Ergun equation. In both reactors, gas-phase velocity is constrained to be less than the minimum fluidization velocity to ensure particle flow in the moving-bed transport regime.

Here the minimum fluidization velocity is computed by setting the drag force from the Ergun equation equal to the gravitational force on the particle bed. Additional correlations between dimensionless groups such as the Nusselt, Reynolds, and Prandtl numbers are used to calculate heat transfer coefficients. Other assumptions include constant bed porosity, negligible entrainment and attrition of solid particles, and uniform temperature within particles. Full details of the model are provided in Ostace et al. [22] and in Okoli et al. [25].

5.1 Optimization problem formulation

Using the process model described above, the optimization problem is then formulated with the following components. The objective function in this study is the minimization of total annualized cost (*TAC*) while achieving 99.9% combustion of the methane fuel. The TAC effectively captures the trade-off between the capital cost items, determined by equipment capacities, and operating cost items, including compressor electricity use, which plays a key role in designing CLC processes. More specifically, the *TAC* is a function of the total equipment cost (*TEC*), the total operating cost (*TOC*), the overall installation factor *f*, and an annualization factor, ϕ.The TAC is given as:

$$TAC = \phi \cdot f \cdot TEC + TOC \quad (2.7)$$

The *TEC* (Eq. 2.8) is a function of the equipment cost of the CLC reactors ($Cost_{react}$), compressors ($Cost_{compr}$), OC inventory ($Cost_{inv}$) (assumed to be 1300 $/tonne), and solids heat exchanger ($Cost_{HX}$). Additional capital costs relate to equipment installation, indirect costs, and working capital, and are described in detail in Okoli et al. [25].

$$TEC = \left(\sum_i \left(Cost^i_{react} + Cost^i_{compr} + Cost^i_{inv} \right) + Cost_{HX} \right), i \in \{AR, FR\} \quad (2.8)$$

The *TOC* shown in Eq. (2.9) is a function of the total electricity required by the compressors in the process:

$$TOC = CU_{elect} \cdot \sum_i W^i, i \in \{AR, FR\} \quad (2.9)$$

All cost parameters are listed in Okoli et al. [25]. The constraints for the optimization problem include at least 99.9% combustion of the methane fuel, a superficial velocity limit, and decision variable bounds. The decision variables for the MB CLC process design and operation are given in Table 2.3.

TABLE 2.3 Total set of decision variables considered in the case studies.

Decision variables (*AR, FR*)	Lower bound	Upper bound
Design:		
Reactor bed diameter, D_b^i (m)	0.1	25
Reactor bed height, L_b^i(m)	0.1	25
Operating:		
OC flow rate, $M_{OC\_in}^{FR}$ (kg/s)	0	10,000
Gas flow rate, $F_{gas\_in}^{AR}$ (mol/s)	0	10,000
Gas inlet pressure, $P_{gas\_in}^{i}$ (kPa)	100	2000
Gas inlet temperature, $T_{gas\_in}^{i}$ (K)	298.15	1000

5.2 Problem initialization and implementation

To find a feasible starting point for the optimization, the following heuristics were used to ensure the superficial velocity (v_g) constraints are maintained, the OC circulation rate is maintained to transfer sufficient OC for high fuel conversion (and to maintain the heat balance), and the OC inventory in the reactors is sufficient to provide suitable good conversion of reacting gases [24]. The heuristics-based design is obtained from the following steps:

1. Select gas feed pressures ($P_{gas\_in}^{i}$) and temperatures ($T_{gas\_in}^{i}$) for the AR and FR. These values are assumed to be close to atmospheric conditions for the case studies in this work.
2. Size reactor areas so that $v_g < v_{mf}$ is maintained for the MB reactors.
3. Adjust the OC flowrate into the FR based on the stoichiometric amount required to fully convert (100% conversion) the fuel fed into the FR.
4. The flowrate of air into the AR ($F_{gas\_in}^{AR}$) is assumed to be in 20% excess of the stoichiometric amount needed for complete combustion of methane in the fuel feed.
5. The reactor heights are estimated to allow sufficient residence time in the beds for complete fuel conversion in the FR, and sufficient OC regeneration in the AR [26].

The MB-based CLC flowsheet models are implemented in IDAES [14,23]. The flowsheet can be solved both as a simulation (square system) and as an optimization problem. Before solving the flowsheet (either for simulation or for optimization), the MB model equations are automatically discretized with an orthogonal collocation scheme using Pyomo.DAE [15]. The discretization of the MB model is done over 40 finite elements with 3 collocation points.

The number of finite elements was determined and validated by conducting a grid independence study, while the number of collocation points uses the default for Pyomo.DAE. For the optimization studies, the resulting MB CLC discretized system consists of 20,789 variables and 20,781 equations. Ipopt [8] was used for solving both the simulation and optimization problems.

Given the nonlinearity and complexity of the CLC flowsheet models, a systematic initialization is required, which is performed with the following steps.

- Step 1. The FR is initialized and solved using an automated hierarchical approach in which submodels of the FR are solved in order of complexity, i.e., the thermophysical and kinetic submodels are solved, the hydrodynamic correlations and equations are solved next, followed by the mass balance equations, and finally the energy balance equations.
- Step 2. The solid inlet stream conditions of the AR are set equal to the solid outlet stream conditions of the FR, and the AR is solved separately, using the approach in Step 1.
- Step 3. The loop is closed by connecting the solid inlet stream of the FR to the solid outlet stream of the AR, and the CLC loop is solved.

The results for the heuristic design are shown in Table 2.4. The superficial velocities in both reactors are well below the minimum fluidization limit, and the operating conditions are within the bounds of the decision variables. While this constitutes a feasible design, we note that the compression costs are quite expensive, due to high gas inlet pressure and flowrates. Moreover, the OC inventories in the reactors are large, in order to ensure total conversion of the methane fuel as well as full regeneration of the oxygen carrier in the air reactor. These heuristic-based results along with Steps 1–3 of the initialization are used as the starting point for optimizing the MB process. An additional step is used to initialize the optimization problem.

- Step 4. The objective function and process constraints are activated, the decision variables are freed, and the flowsheet optimization model is solved using steady-state simulation results as a starting point.

IDAES enables the optimization of the MB CLC system in a computationally efficient manner, with solution times of 3–4 CPU min. Moreover, the models and optimization methodology can be leveraged to rapidly screen multiple different scenarios with different fuels and OCs. As seen in Table 2.4, the optimal design shows significant differences from the heuristic design. In particular, both the FR and AR are somewhat reduced in capacity, along with reduced OC circulation between the AR and FR. Note that the reduced air flowrate and the reduction in air pressure to near-ambient lead to a very large reduction in the air compressor capital and operating cost. On the other hand, the optimized pressure in the FR is significantly higher than in the AR. Also, the solids heat exchanger is the largest contributor to TEC, contributing 35%, while the inventory and reactors contribute 32% and 28%, respectively. The TOC only

TABLE 2.4 Cost breakdown of the heuristic and optimal designs for the MB CLC process.

	Heuristic design			Optimal design		
Variables:	*Fuel reactor*	*Air reactor*		*Fuel reactor*	*Air reactor*	
Gas inlet flow (mol/s)	128	1587		128	1199	
Gas inlet pressure (kPa)	156	156		250	102	
Gas inlet Temp. (K)	298.15	298.15		293	293	
OC circ. rate (kg/s)	591	583.4		549	540	
Reactor length (m)	5	5		4.6	2.39	
Reactor diameter (m)	8	12		4.58	13	
Compressor Work (kW)	449	5926		390	56	
Costs:			Totals			Totals
Reactor Cost (K/MW_{th}$)	5.73	9.30	15.03	3.14	10.62	13.76
Inventory Cost (K/MW_{th}$)	11.77	17.21	28.98	1.90	13.67	15.58
Compressor Cost (K/MW_{th}$)	1.60	5.23	6.82	1.51	0.61	2.12
Solids_HX cost ($K/MWth)	–	18.20	18.20	–	–	17.07
TEC (K/MW_{th}$)	–	–	69.03	–	–	48.53
TOC (K/MW_{th}$-yr)	2.43	32.00	34.43	2.23	0.32	2.55
TAC (K/MW_{th}$-yr)			**83.86**			**37.29**

contributes 7% to the TAC, with electricity for the fuel compressor the major contributor to the TOC. This is due to the higher operating pressure in the FR. On the other hand, the TAC for the optimal design is only 44% of the TAC for the heuristic design. In particular, the compressor costs and TEC in the optimized process are significantly less than for the heuristic design. This clearly demonstrates the effectiveness of the equation-based optimization strategy in IDAES.

6 Conclusions and future perspectives

Nonlinear optimization is essential for decision-making tasks in modeling and analysis, process design, and operations. These tasks can be improved significantly through major advances in NLP algorithms, optimization modeling platforms, and large-scale software and hardware. As with the development of detailed process models, engineers must take ownership of nonlinear optimization by mastering the capabilities of large-scale algorithms, re-examining optimization frameworks with tailored engineering models, and incorporating these advances within their work process. As evidenced by recent research successes, these lead to major opportunities in the process industries.

To deal with these advances, IDAES represents a paradigm shift for improving the design and operation of complex, interacting technologies and systems. IDAES is a unique fully equation-oriented platform with integrated support for steady-state design, optimization, dynamic operations, data reconciliation, parameter estimation, and uncertainty quantification of complex energy and chemical processes. In addition, it supports the process modeling lifecycle, from conceptual design to dynamic optimization and control. IDAES enables users to efficiently explore complex design spaces to discover the lowest cost, and most environmentally responsible solutions. This extensible, open platform empowers users to create models of novel processes and rapidly develop custom analyses, workflows, and end-user applications.

IDAES has been applied to numerous energy-based optimization case studies. In particular, IDAES has demonstrated an impressive impact on new and existing processes through specific use cases that have increased operational flexibility of a power plant by 44% [27] and decreased operating costs for a CO_2 capture system by 15%–18% [28]. As a result, IDAES has an active, growing user community from multiple industries, including power generation and distribution, petrochemical manufacturing, pharmaceuticals, and consumer products.

Future work in nonlinear optimization will lead to even faster solution strategies with an exploration of parallel decompositions that exploit problem structure at multiple levels, including linear algebra. For instance, the problem structure of multiscenario and dynamic optimization problems can be exploited by reordering the linear system (2.5) to the Bordered Block Diagonal (BBD) or Almost Block Diagonal (ABD) structures illustrated in Fig. 2.6. These can be solved by parallel Schur decomposition [29,30] or Cyclic Reduction strategies [31]. Moreover, these strategies are also facilitated by powerful software frameworks such as PyNumero, based on Pyomo/Python [32] and Plasmo.jl, based on JuMP/Julia [33].

Through the development of powerful meta-algorithms, optimization platforms, such as IDAES will be able to tackle even larger process system models at multiple time and length scales [17]. Moreover, this optimization framework currently allows extension into exciting new energy applications, including:

- Design and optimization of tightly coupled energy systems incorporating fossil, nuclear, and renewable energy technologies;
- Design and optimization of natural gas-based solid-oxide fuel cell systems;
- Developing highly dynamic, flexible carbon capture systems;

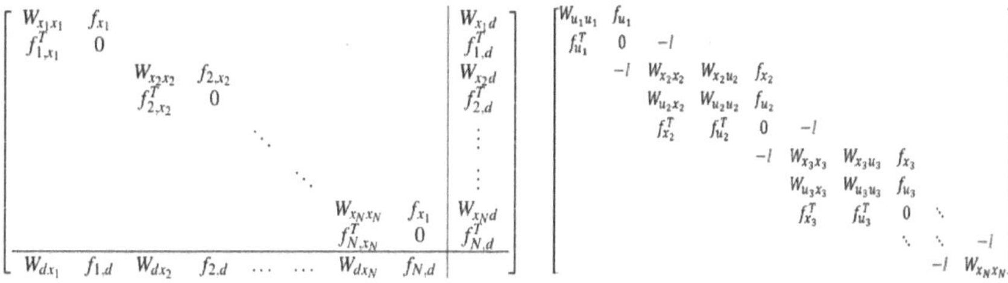

$$\begin{aligned} \min \quad & \textstyle\sum_{k=1}^{N} \varphi(x_k, d) \\ \text{s.t.} \quad & f(x_k, d) = 0 \\ & x_k \in X, d \in D \end{aligned}$$

$$\begin{aligned} \min \quad & \textstyle\sum_{k=1}^{N} \varphi(x_k, u_k) \\ \text{s.t.} \quad & x_{k+1} = f(x_k, u_k) \\ & x_k \in X, u_k \in U \end{aligned}$$

FIG. 2.6 Structured KKT matrices for multiscenario and dynamic optimization problems that allow parallel decomposition.

- Improved efficiency and reliability of power plants;
- Modeling and optimization of water treatment and desalination systems.

These activities and applications expand the scope of optimal decision-making in process engineering to a broad spectrum of time and length scales, which range from molecular synthesis for material design, all the way to integrated, distributed national energy networks. These advanced, comprehensive decision-making capabilities represent the next-generation of Computer-Aided Process Engineering.

Acknowledgments

This work was conducted as part of the Institute for the Design of Advanced Energy Systems (IDAES) with support through the Simulation-Based Engineering, Crosscutting Research Program within the U.S. Department of Energy's Office of Fossil Energy and Carbon Management.

Disclaimer

This project was funded by the United States Department of Energy, National Energy Technology Laboratory, in part, through a site support contract. Neither the United States Government nor any agency thereof, nor any of their employees, nor the support contractor, nor any of their employees, makes any warranty, express or implied, or assumes any legal liability or responsibility for the accuracy, completeness, or usefulness of any information, apparatus, product, or process disclosed, or represents that its use would not infringe privately owned rights. Reference herein to any specific commercial product, process, or service by trade name, trademark, manufacturer, or otherwise does not necessarily constitute or imply its endorsement, recommendation, or favoring by the United States Government or any agency thereof. The views and opinions of authors expressed herein do not necessarily state or reflect those of the United States Government or any agency thereof.

References

[1] L.T. Biegler, Nonlinear Programming: Concepts, Algorithms and Applications to Chemical Processes, SIAM, Philadelphia, PA, 2010.
[2] A. Conn, K. Scheinberg, L. Vicente, Introduction to Derivative Free Optimization, SIAM, Philadelphia, 2009.
[3] R. Fletcher, Practical Methods of Optimization, Wiley, Chichester, 1987.
[4] J. Nocedal, S. Wright, Numerical Optimization, second ed., Springer, New York, NY, 2006.
[5] A. Conn, N. Gould, P. Toint, Trust-Region Methods, MPS-SIAM Series on Optimization, SIAM (Society for Industrial and Applied Mathematics), 2000. Philadelphia, PA.
[6] L.T. Biegler, O. Ghattas, M. Heinkenschloss, B. van Bloemen Waanders (Eds.), Large-Scale PDE-Constrained Optimization, Lecture Notes in Computational Science and Engineering, No. 30, Springer Verlag, Berlin, 2003.
[7] R. Byrd, J. Nocedal, R. Waltz, Knitro: an integrated package for nonlinear optimization, in: G. Di Pillo, M. Roma (Eds.), Large-Scale Nonlinear Optimization, Springer, 2006, pp. 35–60.
[8] A. Wächter, L.T. Biegler, On the implementation of an interior-point filter line-search algorithm for large-scale nonlinear programming, Math. Program. 106 (2006) 25–57.
[9] U.M. Ascher, L.R. Petzold, Computer Methods for Ordinary Differential Equations and Differential-Algebraic Equations, SIAM, Philadelphia, PA, 1998.
[10] A. Brooke, D. Kendrick, A. Meeraus, R. Raman, GAMS—A User's Guide, 1998. http://www.gams.com.
[11] R. Fourer, D.M. Gay, B.W. Kernighan, AMPL: A Modeling Language for Mathematical Programming, Duxbury Press/Brooks/Cole Publishing Company, 2002.
[12] AIMMS, Advanced Interactive Multidimensional Modeling System, http://aimms.com/; https://en.wikipedia.org/wiki/AIMMS, 2014.
[13] W.E. Hart, C.D. Laird, J.-P. Watson, D.L. Woodruff, G.A. Hackebeil, B.L. Nicholson, J.D. Siirola, Pyomo—Optimization Modeling in Python, second ed., Springer Science & Business Media, 2017.
[14] D.C. Miller, J.D. Siirola, D.A. Agarwal, A.P. Burgard, A. Lee, J.C. Eslick, B. Nicholson, C. Laird, L.T. Biegler, D. Bhattacharyya, N.V. Sahinidis, I.E. Grossmann, C.E. Gounaris, D. Gunter, Next generation multi-scale process systems engineering framework, in: M.R. Eden, M. Ierapetritou, G.P. Towler (Eds.), 13th International Symposium on Process Systems Engineering (PSE 2018), San Diego, CA, 2018, pp. 2209–2214.
[15] B. Nicholson, J.D. Siirola, J.-P. Watson, V.M. Zavala, L.T. Biegler, pyomo.dae: a modeling and automatic discretization framework for optimization with differential and algebraic equations, Math. Program. Comput. 10 (2018) 187–223.
[16] D.M. Thierry, L.T. Biegler, Dynamic real-time optimization for a CO2 capture process, AIChE J. 65 (7) (2019) 1–11.
[17] J.P. Eason, L.T. Biegler, A trust region filter method for glass box/black box optimization, AIChE J. 62 (2016) 3124–3136.
[18] L.-S. Fan, Chemical Looping Partial Oxidation: Gasification, Reforming, and Chemical Syntheses, Cambridge University Press, 2017.
[19] J. Adánez, A. Abad, T. Mendiara, P. Gayán, L.F. de Diego, F. García-Labiano, Chemical looping combustion of solid fuels, Prog. Energy Combust. Sci. 65 (2018) 6–66.
[20] J. Li, H. Zhang, Z. Gao, J. Fu, W. Ao, J. Dai, CO2 capture with chemical looping combustion of gaseous fuels: an overview, Energy Fuel 31 (2017) 3475–3524.
[21] L. Zeng, A. Tong, M. Kathe, S. Bayham, L.-S. Fan, Iron oxide looping for natural gas conversion in a countercurrent moving bed reactor, Appl. Energy 157 (2015) 338–347.

[22] A. Ostace, A. Lee, C.O. Okoli, A.P. Burgard, D.C. Miller, D. Bhattacharyya, Mathematical modeling of a moving-bed reactor for chemical looping combustion of methane, in: M.R. Eden, M. Ierapetritou, G.P. Towler (Eds.), 13th International Symposium on Process Systems Engineering (PSE 2018), Computer-Aided Chemical Engineering, San Diego, CA, 2018, pp. 325–330.

[23] A. Lee, J.H. Ghouse, J.C. Eslick, C.D. Laird, J.D. Siirola, M.A. Zamarripa, D. Gunter, J.H. Shinn, A.W. Dowling, D. Bhattacharyya, L.T. Biegler, A.P. Burgard, D.C. Miller, The IDAES process modeling framework and model library—flexibility for process simulation, optimization and control, J. Adv. Manuf. Process. (2021) e10095.

[24] A. Abad, J. Adánez, F. García-Labiano, L.F. de Diego, P. Gayán, J. Celaya, Mapping of the range of operational conditions for Cu-, Fe-, and Ni-based oxygen carriers in chemical-looping combustion, Chem. Eng. Sci. 62 (2007) 533–549.

[25] C.O. Okoli, A. Ostace, S. Nadgouda, A. Lee, A. Tong, A.P. Burgard, D. Bhattacharyya, D.C. Miller, A framework for the optimization of chemical looping combustion processes, Powder Technol. (2020), https://doi.org/10.1016/j.powtec.2019.04.035.

[26] D. Sridhar, A. Tong, H. Kim, L. Zeng, F. Li, L.-S. Fan, Syngas chemical looping process: design and construction of a 25 kW subpilot unit, Energy Fuel 26 (2012) 2292–2302.

[27] J.C. Eslick, M.A. Zamarripa, J. Ma, M. Wang, I. Bhattacharya, B. Rychener, P. Pinkston, D. Bhattacharyya, S.E. Zitney, A.P. Burgard, D.C. Miller, Predictive modeling of a subcritical pulverized-coal power plant for optimization: parameter estimation, validation, and application, Appl. Energy (2021) (submitted).

[28] P. Akula, J. Eslick, D. Bhattacharyya, D.C. Miller, Model development, validation, and optimization of an MEA-based post-combustion CO2 capture process under part-load and variable capture operations, Ind. Eng. Chem. Res. 60 (14) (2021) 5176–5193.

[29] V.M. Zavala, C.D. Laird, L.T. Biegler, Interior-point decomposition approaches for parallel solution of large-scale nonlinear parameter estimation problems, Chem. Eng. Sci. 63 (2008) 4834–4845.

[30] N. Chiang, C.G. Petra, V.M. Zavala, Structured nonconvex optimization of large-scale energy systems using PIPS-NLP, in: 2014 Power Systems Computation Conference, 2014, pp. 1–7.

[31] W. Wan, J.P. Eason, B. Nicholson, L.T. Biegler, Parallel cyclic reduction decomposition for dynamic optimization problems, Comput. Chem. Eng. 120 (2019) 54–69.

[32] C.D. Laird, B.L. Nicholson, J.S. Rodriguez, Parallel Numerical Algorithms for Structured Problems With PyNumero and Pyomo, 2019. Tech. Rep. Sandia National Lab. (SNL-NM), Albuquerque, NM (United States).

[33] J. Jalving, Y. Cao, V.M. Zavala, Graph-based modeling and simulation of complex systems, Comput. Chem. Eng. 125 (2019) 134–154.

Chapter 3

Thinking multicriteria—A jackknife when it comes to optimization

Norbert Asprion[a] and Michael Bortz[b]
[a]*BASF SE, Ludwigshafen, Germany,* [b]*Fraunhofer Institute for Industrial Mathematics (ITWM), Kaiserslautern, Germany*

1 Introduction

In chemical engineering, model-based optimization carries the promise of unveiling process designs significantly better than those that have been obtained before by experience-driven improvements. However, due to conflicting objectives, the notion of "better" is by no means unambiguous: Typically, a compromise must be made between quality and cost functions. In this regard, optimal solutions are for example those that, given a threshold on the expenditures one is willing to spend, achieve the highest quality. Thus, not a single optimal solution is sought for, but the set of all such best compromises, the Pareto set.

After a short account on multicriteria optimization according to [1], this contribution highlights its benefits, especially when carried out with adaptive scalarization schemes, in industrially relevant applications, among which are process design, model adjustment, and design of experiments. The article concludes with Pareto navigation, a powerful concept that allows the user to explore the Pareto set interactively.

1.1 Short account on multicriteria optimization

A multicriteria optimization (MCO) problem is given by

$$\min_{x \in X} (f_1(x); f_2(x); \ldots; f_k(x)), \tag{3.1}$$

where X is the feasible set. In general, there is no solution that is optimal with respect to all objectives at once. Instead, the concept of dominance is introduced: A solution x dominates a solution y if $f_i(x) \leq f_i(y)$ for all components i and $f_j(x) < f_j(y)$ for at least one component j, cf. Fig. 3.1.

Simulation and Optimization in Process Engineering. https://doi.org/10.1016/B978-0-323-85043-8.00012-X

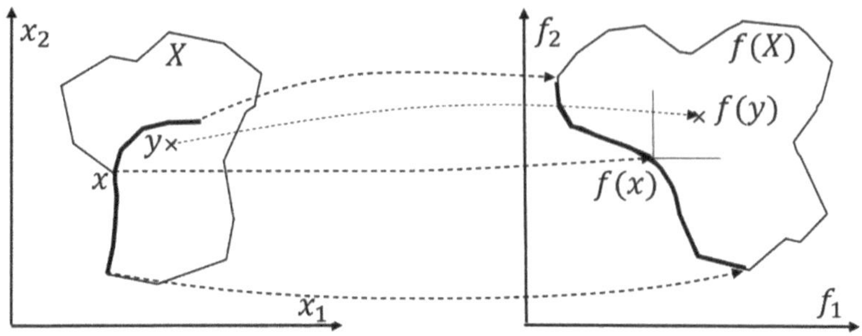

FIG. 3.1 The Pareto set. Exemplarily, a solution $f(x)$ with its dominated points are shown, among those the solution $f(y)$.

The interesting solutions are those which are not dominated by any other solution. The set of these solutions is called the Pareto set. One can consider these solutions as best compromises: Given a Pareto optimal solution, it is not possible to improve one objective, without worsening another or leaving the feasible set X.

In practice, computing the complete Pareto set is generally neither possible nor desirable. Rather, one is looking for a finite number of Pareto points as an approximation for the whole Pareto set. These are obtained by scalarization schemes, which means that the multicriteria problem (3.1) is reformulated as a standard optimization problem by defining a real-valued objective function by using the original objectives $f_{1,\ldots,k}$. The most prominent scalarization is the weighted sum, $g_{ws}(x) = \sum_{i=1}^{k} w_i f_i(x)$. It can be shown that if the original MCO problem is convex, then any Pareto optimal point is a solution to the weighted sum problem

$$\min_{x \in X} g_{ws}(x), \tag{3.2}$$

for appropriately chosen weights. However, for nonconvex MCO problems, this need not be the case, cf. below.

Another approach is the ε-constraint method. Here the new optimization problem has the following shape

$$\min_{x \in X} f_i(x) \text{s.t.} f_j(x) \leq \varepsilon_j \text{ for } j \neq i, \tag{3.3}$$

To obtain a variety of points on the Pareto frontier several values for ε_j for each objective f_j are used. All Pareto-optimal solutions can be obtained this way, irrespective of convexity assumptions of the optimization problem.

We mention a third scalarization scheme, the so-called Pascoletti-Serafini scalarization. For a reference point $r \in \mathbb{R}^k$ and a direction $d \in \mathbb{R}^k$, the problem to be solved reads:

$$\min_{x \in X; \lambda} \lambda \text{ s.t.} r_i + \lambda d_i - f_i(x) \geq 0. \tag{3.4}$$

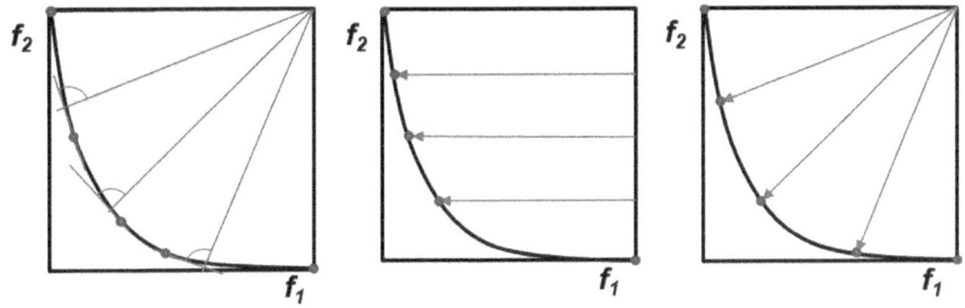

FIG. 3.2 Scalarization schemes: Weighted sum, epsilon-constraint, Pascoletti-Serafini.

Each point even on a nonconvex Pareto frontier can be obtained by the appropriate selection of the direction and the reference point. There are a number of further scalarization schemes, e.g., Refs. [2,3]. In the following, only Eqs. (3.2)–(3.4) will be used. Solving each of the scalarized problems (3.2)–(3.4) yields one Pareto point, cf. Fig. 3.2.

By varying the respective scalarization parameter, as many Pareto points as desired can be obtained. Instead of changing the scalarization parameters uniformly, adaptive scalarization schemes are employed, see, e.g., Ref. [4]. It should be noted that, as shown in Ref. [5], a uniform distribution of weights does not necessarily lead to a uniform distribution of Pareto points.

To arrive at an adaptive scalarization for the weighted sum, we can use the fact that there is a tangential hyperplane at each Pareto optimal point such that all other Pareto optimal points are on the same halfspace [6]. If a set of Pareto optimal points is already calculated, the intersection of all corresponding halfspaces is an approximation of the Pareto set containing all Pareto optimal points. The convex combination of the Pareto optimal points is an inner approximation. The location of the largest distance between both approximations yields a weight for obtaining the next Pareto optimal solution which improves both approximations. This is known as the sandwiching procedure [6–8]. A value for the largest distance is set as termination criterion and constitutes a measure for the approximation quality of the full Pareto set.

If the multicriteria optimization problem is nonconvex, the Pareto optimal solutions can be approximated by hyperboxes [4,9], and a maximally allowed value for the size of the hyperboxes is used as a termination criterion. This approach has the disadvantage that it becomes very slow if the number of objectives increases.

Generally, the structure of the optimization problem is not known in advance. One then starts with the sandwiching procedure until termination. A triangulation of the determined Pareto-optimal solution produces facets. For each of the facets, a convexity check follows. For those where the convexity checks failed the computationally more expensive hyperboxing procedure is applied [4,9].

If the multicriteria optimization problem is mixed-integer such that it can be decomposed into a continuous part and discrete variables, one can approximate the continuous part for each value of the enumeration variable and end up with a set of approximations, the Pareto patches.

The set-up of a multicriteria optimization problem as defined in Eq. (3.1), which consists in defining the objectives and optimization variables including their lower and upper bounds, has been implemented in the BASF inhouse flowsheet simulator Chemasim for steady-state operation and its dynamic counterpart Chemadis, along with the above mentioned adaptive scalarization schemes and enumeration options to generate Pareto patches if desired. The resulting optimization problems are solved with a mixed-integer sequential quadratic programming method [10]. Thus, a database of Pareto points is obtained, including the complete set of process variables for each such point. By interactive controls including projections, restrictors, and selectors, the Pareto set approximation consisting of the Pareto points and linear interpolation can be navigated by the user [4,11]. Thus, quick wins - flat regions of the Pareto set that are nearly parallel to the axes where one objective can be improved tremendously with only a minor tradeoff in the other objectives can be distinguished from regions where more substantial tradeoffs occur.

In the following, results of applying the multicriteria optimization in Chemasim to use-cases stemming from process design and model improvement are reviewed.

2 Process design

Chemical process design is a multicriteria optimization problem: Designs are sought for which lead to high-quality products while at the same time are supposed to save resources, energy, and costs. The technical challenge resides in the simulation itself: It consists of a set of nonlinear equations for the process variables x_p, which for stationary processes we summarize as $F(x_p)=0$. Since $\dim F \leq \dim x_p$, this system is underdetermined. One way to handle this is by including this complete set of equations into the MCO problem (3.1), yielding

$$\min_{x_p} \left(f_1(x_p); \ldots; f_k(x_p)\right) \text{ s.t.} F(x_p) = 0, g(x_p) \leq 0, x_{p,l} \leq x_p \leq x_{p,u}, \tag{3.5}$$

where $x_{p,l}$, $x_{p,u}$ are lower and upper bounds of the process variables and we have included additional inequality constraints $g(x_p) \leq 0$. This simultaneous approach has been successfully applied for example by Biegler's group by using their large-scale optimization solver IpOpt [12–15].

However, typical flowsheet simulators work differently: First, the system $F(x_p)=0$ is complemented by additional equalities, called specifications, $h(x_p)=x_s$, such that $\dim h + \dim F = x_p$. This system of equations and specifications constitutes the simulation model. The MCO problem is then formulated as

$$\begin{aligned} \min_{x_s} \left(f_1(x_p); \ldots; f_k(x_p)\right) \text{ s.t. } F(x_p) = 0, h(x_p) = x_s, g(x_p x_s) \leq 0, \\ x_{p,l} \leq x_p \leq x_{p,u}, x_{s,l} \leq x_s \leq x_{s,u}. \end{aligned} \tag{3.6}$$

Formally, this problem can be decoupled: An optimization solver is given the problem

$$\min_{x_s}(f_1;\ldots;f_k)\ \text{s.t.}\ g(x_p x_s)\leq 0, x_{s,l}\leq x_s\leq x_{s,u}, \tag{3.7}$$

whereas an equation solver takes care of $F(x_p)=0$, $h(x_p)=0$, $x_{p,l}\leq x_p\leq x_{p,u}$. This means that in iteration ν, for a given set of specification values $x_s^{(\nu)}$, a Newton solver seeks a solution $x_p^{(\nu)}$ for the simulation model $F(x_p^{(\nu)})=0$, $x_{p,l}\leq x_p^{(\nu)}\leq x_{p,u}$, and passes values for the objective and inequality constraint functions and their derivatives with respect to x_s to an optimization solver. This optimizer—typically an SQP method—determines a search direction, and performs a line search. For each step of the line search, a set of new specification values is generated and passed back to the Newton solver which seeks a new solution. This is repeated until the line search converges to new specification values $x_s^{(\nu+1)}$, yielding new values for the process variables $x_p^{(\nu+1)}$ determined by the Newton solver.

The advantage of this sequential approach is that the size of the optimization problem is much smaller than for the simultaneous approach. The disadvantage lies in the fact that generally, it is not clear whether the system $F(x_p^{(\nu)})=0$, $h(x_p^{(\nu)})=x_s^{(\nu)}$, $g(x_p^{(\nu)},x_s^{(\nu)})\leq 0$ is solvable at all. It may happen that no solution exists at all because the equality constraints are not passed to the optimizer. Thus, the sequential approach requires a careful choice of the specification functions and their values. In this context, let us mention that recently, the idea came up to employ machine learning techniques for exploring feasible sets for different choices of the functions g in order to provide some insight and assistance into appropriate choices of the g and the bounds $x_{s,l}$, $x_{s,u}$ [16].

In the following, results that have been obtained by following this sequential procedure within an implementation of the MCO approach in Chemasim/Chemadis are presented for different process-design-related use cases.

2.1 Continuous design variables

The first use case consists of a multicriteria comparison of different process designs for continuous design variables for a given flowsheet. As an illustrative example, consider a pressure-swing distillation to separate a binary mixture of Chloroform and Acetone as sketched in Fig. 3.3, originally discussed in Ref. [4]. This mixture exhibits a pressure-dependent azeotrope which is the heavy boiler of the system. The task is to maximize the mass fractions $x_{Cf}^{\mathrm{Cf}} x_{Ac}^{\mathrm{Ac}}$ of Chloroform (Cf) and Acetone (Ac) in their respective product streams *Cf*, *Ac* and to minimize the total heat duty. Table 3.1 shows the chosen specifications and their bounds for this MCO problem following the notation in Eq. (3.6).

The resulting Pareto surface is visualized in Fig. 3.3.

One can clearly distinguish convex and nonconvex parts of the Pareto surface, which have been resolved by the hybrid adaptive scalarization scheme as described in the introduction of this chapter. All three objectives are in conflict,

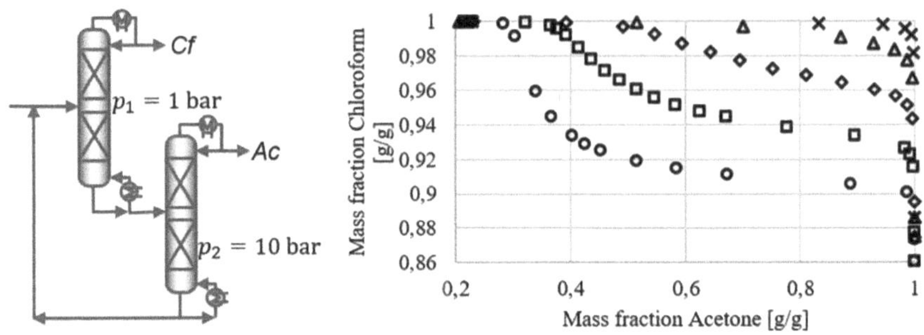

FIG. 3.3 Flowsheet (left) and Pareto surface (right) visualized by Points on the isocontours for the heating duty $\dot{Q}$ (from lower left to upper right: $\dot{Q} = 1000$ MW, 1500 MW, 2000 MW, 2500 MW, 3000 MW for the pressure swing distillation.

TABLE 3.1 Specifications and their bounds for the pressure swing distillation.

	Lower bound $x_{s,l}$	Upper bound $x_{s,u}$
Reflux ratio in first column $R_{V,1}$	1	60
Reflux ratio in second column $R_{V,2}$	1	60
Split ratio first column S_1	0.01	0.99
Split ratio second column S_2	0.01	0.99

and trade-offs can be compared transparently. The transparent comparability of alternatives with guaranteed Pareto optimality can reveal significant potential for process optimization [11,17]. This approach can also be extended to discrete design variables like the number of equilibrium stages or feed stages, as long as an adequate mixed-integer optimization solver is available. We now turn our attention to comparing flowsheet alternatives.

2.2 Discrete alternatives

In an early stage of process design, different flowsheet alternatives to solve the synthesis and separation tasks at hand are compared in order to identify the most suitable alternative. As an example, originally introduced in Ref. [18], consider the separation of 600 kmol/h of an equimolar mixture consisting of n-hexane, n-heptane, and n-octane. Four different options are compared, namely (a) a direct sequence, (b) an indirect sequence, (c) a side-draw column, and (d) a dividing wall column. The optimization variables and their ranges for these four alternatives are given in Table 3.2.

The resulting Pareto sets for the three objectives minimize total costs, maximize molar fraction of n-heptane in stream C7 and molar flow of C7 are

TABLE 3.2 Range of the optimization variables (*f*, feed; *ldw, hdw*, lower and higher end of dividing wall; *N*, number of stages; *pf*, pre-fractionator; *RR*, reflux ratio; *s*, split; *sd*, side draw) for the design of the column configurations for the separation of *n*-hexane, *n*-heptane, and *n*-octance.

	N_1	$N_{f,1}$	RR_1	N_2	$N_{f,2}$	RR_2	n/(kmol/h)	n/(kmol/h)			
Direct sequence	20–50	5–45	1–5	20–50	5–45	1–5	C8: 160–260	C7: 160–260			
Indirect sequence	20–50	5–45	1–5	20–50	5–45	1–5	C6: 160–260	C7: 160–260			
	N_1	$N_{f,1}$	RR_1	N_{sd}	n/(kmol/h)	n/(kmol/h)	Constraint				
Side-draw column	20–50	5–45	1–20	10–40	C8: 160–260	C7: 160–260	$N_{sd} > N_{f,1}$				
	N_{pf}	$N_{f,1}$	RR_1	N_1	N_{sd}	N_{ldw}	N_{hdw}	s_{ldw}	s_{hdw}	n/(kmol/h)	n/(kmol/h)
Dividing wall col.	5–41	5–45	1–20	20–50	10–40	1–40	10–50	0.2-0.9	0.1-0.9	C8: 160–260	C7: 160–260

visualized in Fig. 3.4. The objective molar flow of C7 only varies maximal 2.5% in all cases and is projected.

The four different Pareto sets belonging to the four different flowsheet alternatives make the comparison transparent: Focussing only on the lower part of the convex envelopes (neglecting the impact of the molar flow of C7) the dividing wall column leads to process designs that dominate all others. For the dominated solutions, it turns out that for not too high purity requirements, the side-draw column becomes nearly similar attractive.

2.3 The impact of uncertainties

Each process model depends on parameters ξ that for the above-sketched process optimization have been fixed, for example, thermodynamic parameters describing phase equilibrium, chemical reaction parameters, but also those related to the process itself like the feed rate or feed composition. In practice, these model parameters are not known exactly. For example, they may result from a regression of phase equilibrium models to appropriate experimental lab data. Or, in the case of feed composition, it may fluctuate around an average value. In both cases, a quantitative estimate of the confidence bounds would be possible. In other cases, practitioners know by experience that certain parameters are only known up to a certain accuracy. In the following, we assume that $\xi \in \Xi$, for a given set Ξ.

The uncertainties affect both the constraints and the objectives in the MCO formulations in Eqs. (3.5)–(3.7). Different strategies to deal with these modifications are known, and the interested reader is referred to Refs. [19–22], and references therein. In the following, for the constraints, we adopt the safe and conservative point of view that these have to be fulfilled for all $\xi \in \Xi$, i.e., $F(x_p, \xi) = 0,\ g(x_p, \xi) \leq 0 \forall \xi \in \Xi$. As far as the objective functions are concerned, this conservative position is kept, resulting in a robust reformulation of the MCO problem (3.5), i.e.,

$$\min_{x_{p,l} \leq x_p \leq x_{p,u}} \max_{\xi \in \Xi} \left(f_1\left(x_p \xi\right); \ldots; f_k\left(x_p \xi\right)\right)$$

$$\text{s.t. } F\left(x_p \xi\right) = 0, g\left(x_p \xi\right) \leq 0 \forall \xi. \tag{3.8}$$

If Ξ is a set with infinite cardinality, Eq. (3.8) leads to a semiinfinite optimization problem [23] and requires advanced solver techniques. In order to keep the problem manageable for practical purposes and tractable by the solvers at hand, two approximations are made: The first approximation consists in choosing Ξ as a set consisting of a finite number of typical scenarios representing the uncertainties, i.e., $\Xi = \{\xi_1, \ldots, \xi_u\}$. The second approximation is the replacement of the robust MCO problem (3.8) by enumerating the Pareto patches for each ξ_i while respecting all constraints, and then looking for the worst case Pareto points. This second approximation means that each decision variable is considered as adjustable once the uncertainty is actually observed,

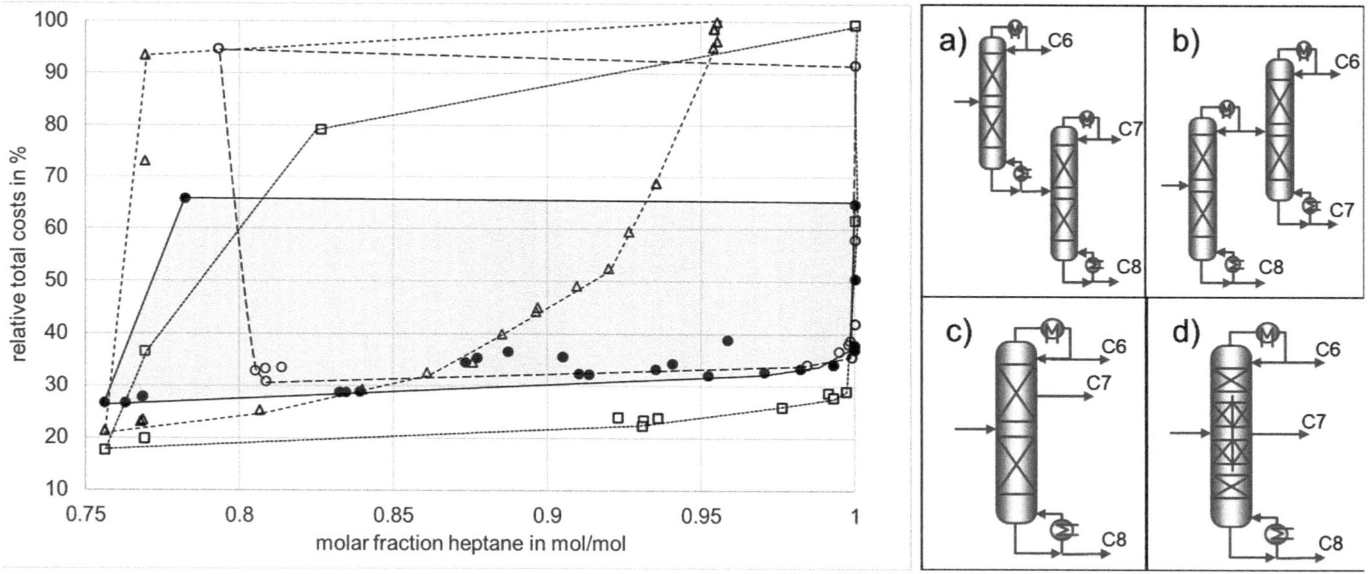

FIG. 3.4 Flowsheets and Pareto set for four different flowsheet alternatives to separate equimolar mixtures of *n*-hexane, *n*-heptane, and *n*-octane: (a) ● direct sequence of two columns, (b) ○ indirect sequence of two columns, (c) Δ side-draw column, (d) □ dividing wall column. *Lines and gray areas* only for better distinction of the convex envelopes.

that is, to be of type "wait and see". If the decision variables were of type "here and now," i.e., had to be fixed without knowing the actual realization of the uncertain variables, then the second approximation would not be valid, and the robust Pareto points had to be calculated component-wise; for a discussion of techniques to deal with "wait and see" as opposed to "here and now" variables, cf. Ref. [24]. The advantage of the way taken here is that the setting of communicating simulation and optimization solvers, cf. Eqs. (3.6), (3.7), can be directly transferred within the concept of Pareto patches.

The challenge in representing the confidence region of the parameters ξ—which typically is a set of infinite cardinality—in terms of a finite set of scenarios $\xi_{1\ \ldots,u}$ is that the choice of the ξ_i should be done such that the transformed scenarios $f_l(x,\xi_i)$ faithfully represent the transformed map $f_l(x,\Xi) := \{f_l(x,\xi) | \xi \in \Xi\}$. Different scenario-generating techniques have been studied, cf. Ref. [25] and referenced therein. For our purposes, especially interesting are screening methods, where the number of points does not increase exponentially with dim ξ. Among these are sigma-points [26] and reduced factorial designs [27]. Sigma-points can be seen as a special case of one-at-a-time designs; they are favorable to calculate an accurate estimate of the covariance matrix of $f_l(x,\Xi)$. In the following, reduced factorial designs are employed since they admit more reliable estimates of the main effects than one-at-a-time designs.

The practical benefit is illustrated for the pressure swing distillation shown in Fig. 3.3. Uncertainties are supposed to occur in the feed composition as well as in the vapor-liquid equilibrium (VLE) model, cf. Table 3.3. The uncertainty in the feed composition is described by the concentration of Chloroform x_F^{Cf} in the feed; the uncertainty in the VLE is taken into account by an uncertain limiting activity coefficient $\gamma_{\mathrm{Ac,Cf}}^{\infty}$ [28]. The set of discrete scenarios to describe these uncertainties is chosen as the four corner points of a cube resulting from factorial design in two dimensions with the min and max values given in Table 3.3.

To illustrate the effectiveness of the robust optimization scheme, the worst-case points from a sensitivity analysis are determined. That is, for each design $x_s^{(i)}$ of the ith nominal Pareto point, the points $P_{ij} := (f_1(x_s^{(i)},\xi_j),\ldots,f_k(x_s^{(i)},\xi_j))$ are calculated for the scenarios $\xi_{j=1,\ \ldots,4}$. From these points, the worst-case alternatives are retained.

TABLE 3.3 Sources of uncertainties for the pressure swing separation to separate chloroform and acetone.

	Unit	Value
x_F^{Cf}	g/g	0.86 ± 0.6
$\gamma_{\mathrm{Ac,Cf}}^{\infty}$		1 ± 0.2

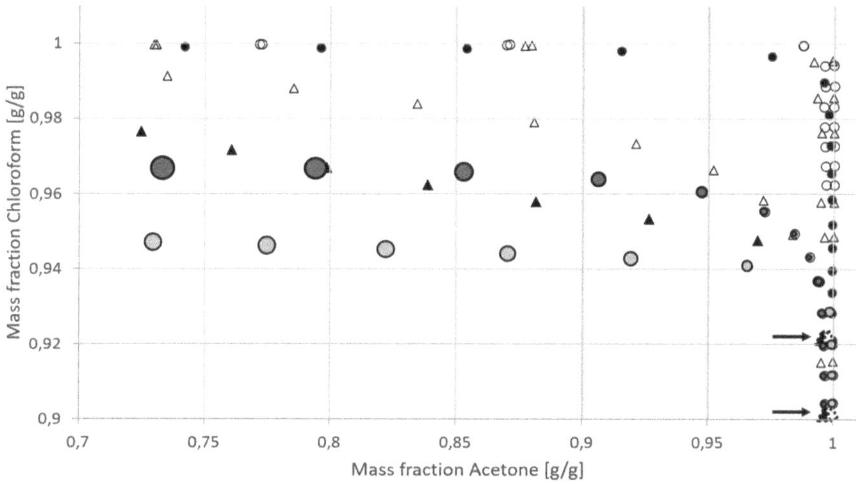

FIG. 3.5 Nominal Pareto boundary *(filled black circles)* compared with robust Pareto boundary *(filled black triangles)*. The respective scenarios from a sensitivity analysis are shown as *open circles* and *open triangles*. For two scenarios ($x_F^{Cf}=0.8, \gamma_{Ac,Cf}^{\infty}=0.8,1.2$), the violation of the constraint on the heat duty is shown by *gray bubbles* (largest bubble corresponds to a 5% violation). The robust Pareto points are obtained for ($x_F^{Cf}=0.8, \gamma_{Ac,Cf}^{\infty}=0.8$), except for large Acetone purities, where ($x_F^{Cf}=0.8, \gamma_{Ac,Cf}^{\infty}=1.2$) leads to robust Pareto points *(dented triangles, emphasized by arrows)*.

The results are shown in Fig. 3.5; in order not to overload this figure, we will limit ourselves to one isoline for the total heat duty in Fig. 3.3. That is, the MCO problem reads $\min_{R_{V,1},R_{V,2},S_1,S_2} \max_{\xi_{1,\dots,4}} \left(-x_{Ac}^{Ac}, -x_{Cf}^{Cf}\right)$ s.t. $\dot{Q} \le 3000$ MW.

Fig. 3.5 illustrates the essential results of a multicriteria optimization under uncertainties. It turns out that it is better to take account of the uncertainties while calculating the Pareto points instead of afterward: The benchmark worst cases resulting from the P_{ij} which can be seen as a sensitivity analysis of the nominal Pareto points are worse than seeking for the best worst-case scenario resulting from the Pareto patches corresponding to each scenario. The effect for both the product purities and the heat duty, the latter effect being measured by the violation of the constraints, can reach up to 5%. This is caused by a slightly different choice of the design compared to the nominal case. At the same time, demanding robustness leads to a loss of up to 5% in Chloroform purity, while keeping the heat duty at the upper limit fixed in the MCO problem.

These results illustrate that taking uncertainties into account in the optimization can lead to designs that are affected less by the uncertainties. Apart from looking for the best worst-case scenario as sketched above, valuable information from MCO consists in getting insight into the trade-offs between robustness and the other objectives. One approach to quantify these trade-offs is to replace each objective function by its average over scenarios, i.e., $E_{\Xi}[f_i(x,\Xi)]$, and furthermore to add the variances $\text{var}_{\Xi}[f_i(x,\Xi)]$ to be minimized. Originating in portfolio optimization [29], this idea has also been successfully applied in the context of chemical engineering [30]. Further ideas point toward using information about the probability distribution of the ξ if available, cf. Ref. [21] and references therein. Different concepts of MCO under uncertainties have been discussed in Ref. [31]. Still, decision support by MCO under uncertainties is an active research area of research. For example, in practice, often there is only vague information on the set Ξ, whereas often it is much clearer which size of fluctuations in the objective space is acceptable.

2.4 Extension to optimal control

The concepts for MCO of stationary processes introduced above can readily be extended to identify optimal control trajectories and recipes for dynamic processes like batch reactions and batch distillations. The solution of the corresponding optimal control problems is boiled down to the solution of a nonlinear problem. The optimization variables, in this case, are then all those parametrizing the control trajectories. The general set-up of Eqs. (3.6), (3.7) is kept. This allows to use the adaptive scalarization schemes also to solve dynamic MCO problems. It has been implemented in BASF inhouse dynamic flowsheet simulator Chemadis; results for a simple example process are summarized in the following.

In this example the model of the semibatch Williams-Otto reactor [32,33] is used with the following chemical reactions:

$$
\begin{aligned}
A+B &\rightarrow C \\
C+B &\rightarrow P+E \\
P+C &\rightarrow G
\end{aligned}
$$

Here, C is an intermediate and P and E are the target products. The last reaction is a subsequent reaction of the product P, resulting in a side-product G. The set of model equations and the following results are available in full detail in Ref. [34]. Fig. 3.6 shows a schematic of the semibatch reactor.

The objectives for the MCO are to find optimal control strategies to maximize the amount of P and to maximize the amount of E at the end of the process. Therefore, the feed rate F_B of pure B and the wall temperature T_W can be adjusted. Here, these are discretized over the time horizon of 1000 s into 40 equidistant time segments where each control variable takes constant values. Thus there are 80 optimization variables summarized in Table 3.4.

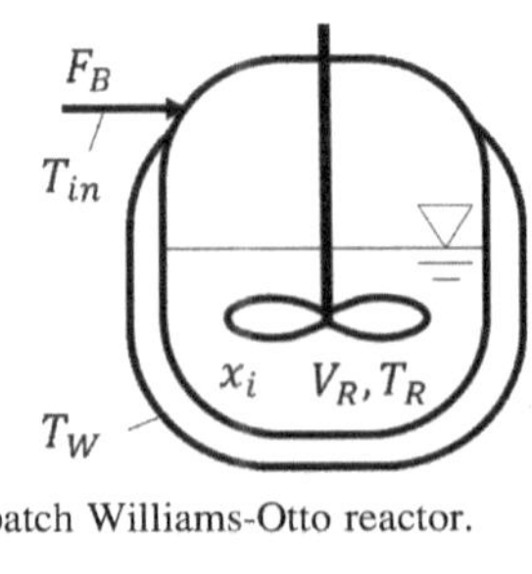

FIG. 3.6 Scheme of the semibatch Williams-Otto reactor.

TABLE 3.4 Optimization variables for the Williams-Otto process.

	Lower bound $x_{s,l}$	Upper bound $x_{s,u}$
$F_{B,1}, \ldots, F_{B,40}$/[l/s]	0	5.874
$T_{W,1}, \ldots, T_{W,40}$/°C	20	100

Additionally, three path constraints are added, namely for the reaction temperature $60\,°\mathrm{C} \leq T_R \leq 90\,°\mathrm{C}$ and for the reaction volume $V_R \leq 5\ \mathrm{m}^3$. The initial condition is chosen as $T_R(t=0) = 65$ °C; $x_A(t=0) = 1$; $V_R(t=0) = 2\ \mathrm{m}^3$.

In Fig. 3.7, the results of the multicriteria optimization are given. As far as the control variables are concerned, this view shows only values at time steps where the function has been evaluated or corresponding to the defined printing time. The underlying function is still a step function.

3 Model adjustment, model comparison and model-based design of experiments

So far, we have assumed that the model is given. In this section, the focus is on utilizing MCO in the contexts of model adjustment, model comparison, and model-based design of experiments, which all can contribute to improve the model quality and the adequate choice of a model.

Consider a set of N experiments each characterized by its D—dimensional design $x_{i=1,\ldots,N}$, lumped together in a $N \times D$ design matrix X. From the system under consideration, R responses $y_{i,j=1,\ldots,R}$ are measured, each corresponding to a model prediction $f_j(x_i, \xi)$. These responses can be single measurements or time-dependent trajectories. To keep the notation simple, we stick to single measurements here; the generalization to time-dependent trajectories is straightforward [34]. To estimate the model parameters ξ, the following multivariate nonlinear regression model is solved [35]:

$$\hat{\xi} = \arg\min_{\xi} \sum_{i=1}^{N} (f(x_i, \xi) - y_i)^T \sigma_y^{-1} (f(x_i, \xi) - y_i), \tag{3.9}$$

where σ_y is the covariance matrix of the model responses and $\hat{\xi}$ is the sought estimate. Typically, σ_y is not known a priori. Following a max-likelihood argument, an estimate $\hat{\sigma}_y$ can be obtained from $\hat{\sigma}_y = \frac{1}{N-D} \sum_{i=1}^{N} \left(f\left(x_i, \hat{\xi}\right) - y_i\right)\left(f\left(x_i, \hat{\xi}\right) - y_i\right)^T$. Then, replacing σ_y in Eq. (3.9) by $\hat{\sigma}_y$, results in a nonlinear system for $\hat{\xi}$.

Here we proceed somewhat differently: Assuming negligible correlations between the measurements means that σ_y can be considered as diagonal, so that the sum in Eq. (3.9) can be interpreted as a weighted sum of objectives $\sum_{i=1}^{N} \left(f_j(x_i, \xi) - y_{ij}\right)^2$ with weights $\sigma_{y,jj}^{-1}$. Minimizing this sum is a MCO problem that reveals whether these different objectives are conflicting—that is, whether or

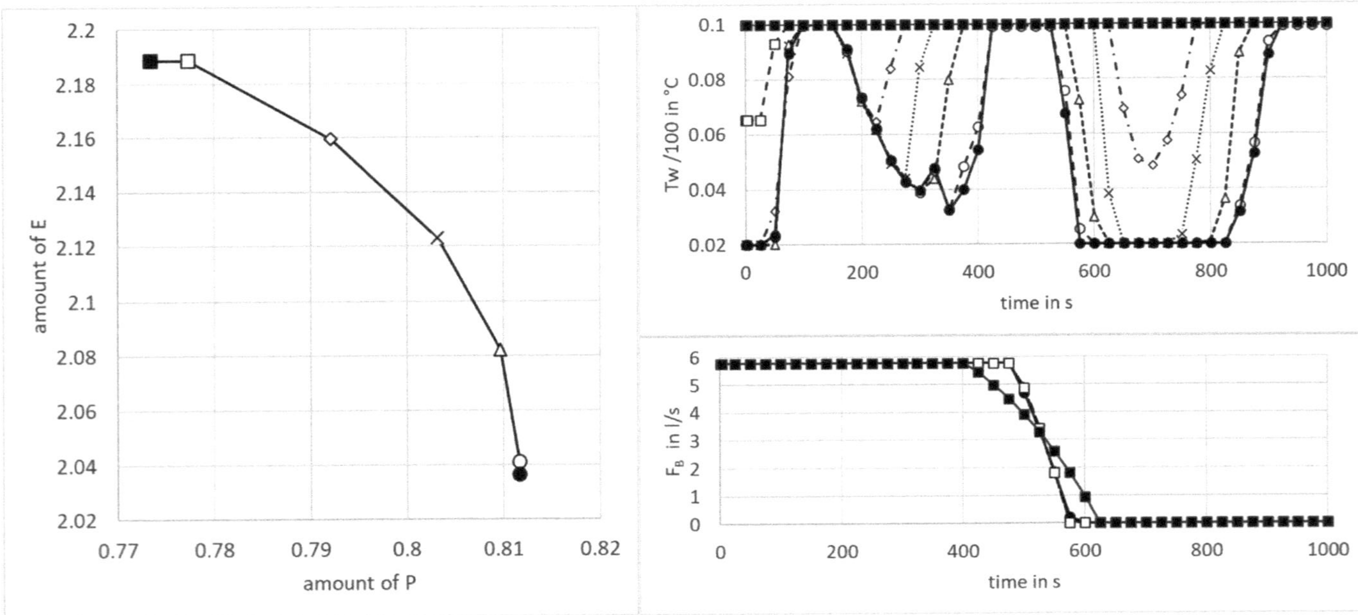

FIG. 3.7 Optimal control trajectories for the feed rate F_B (lower right panel) and wall temperature T_W (upper right panel) leading to the Pareto-optimal solutions (left panel).

not it is possible to find a $\hat{\xi}$ that minimizes all these residuals equally well. If this is the case, from an application point of view, it can make sense to consider the corresponding Pareto set as possible model choices, depending on the preference of the engineer who wants to predict certain responses more accurately than others. Furthermore, using the concept of Pareto patches, it is possible to transparently compare different models that may differ in their structures. This point of view has been successfully adopted to compare the quality of molecular models of water in Ref. [36] and in Ref. [37] for the parameterization of excess enthalpy models. Also, reconciliation tasks can benefit from the MCO approach in order to support the identification of gross errors in the residual sum of squares [38].

One step toward increasing the reliability of models is to include as informative measurement data as possible into the sum of residuals in Eq. (3.9). This is the field of optimal design of experiments (ODE), where the number and the values of the designs x_i are determined such that the largest value of the prediction errors for the functions $f_j(x, \hat{\xi})$, induced by the uncertainty in the estimate $\hat{\xi}$, are as small as possible. One can show that this point of view on ODE is equivalent to the D-criterion [39], where the determinant of the covariance matrix of $\hat{\xi}$ is minimized. In practice, however, experimental design is a MCO problem: Not only shall the experiments to be conducted lead to as small error bars for the model as possible but the experiments shall also be conducted for designs that fulfill process-related objectives, like maximizing the yield of a reaction or the concentration of product streams. Since the D-criterion tends to spread the experimental design points, these objectives are expected to be in conflict with the D-criterion. They can be straightforwardly taken into account via the ε-constraint scalarization in Eq. (3.3). How to extend the adaptive scalarization schemes to this use case is the topic of current research.

4 Decision support

The availability of Pareto points as a reliable approximation of the complete Pareto set is the prerequisite for the exploration of trade-offs. However, for more than three objectives, these can hardly be visualized. In this situation, a navigation scheme has been devised, where, via interactive controls like sliders, the user can continuously change the values of one objective toward its ideal value, while the values of the remaining objectives are equitably adjusted such that the resulting point is on an interpolating facet spanned by neighboring Pareto points. Thus, while navigating, the user has the impression of exploring a continuum. The accuracy of the interpolating facet approximating the true Pareto boundary can be estimated from the approximation accuracy, i.e., the prescribed sandwiching thickness and the size of the hyperboxes.

The navigation of the facets is done in real-time by repeatedly solving linear problems, both for the convex regions [4,6], and nonconvex regions of the Pareto set [40], wherein the latter case, a ray-tracing technique is used. Typical controls are, for example, a spider diagram and sliders with restrictors and selectors as sketched in Fig. 3.8.

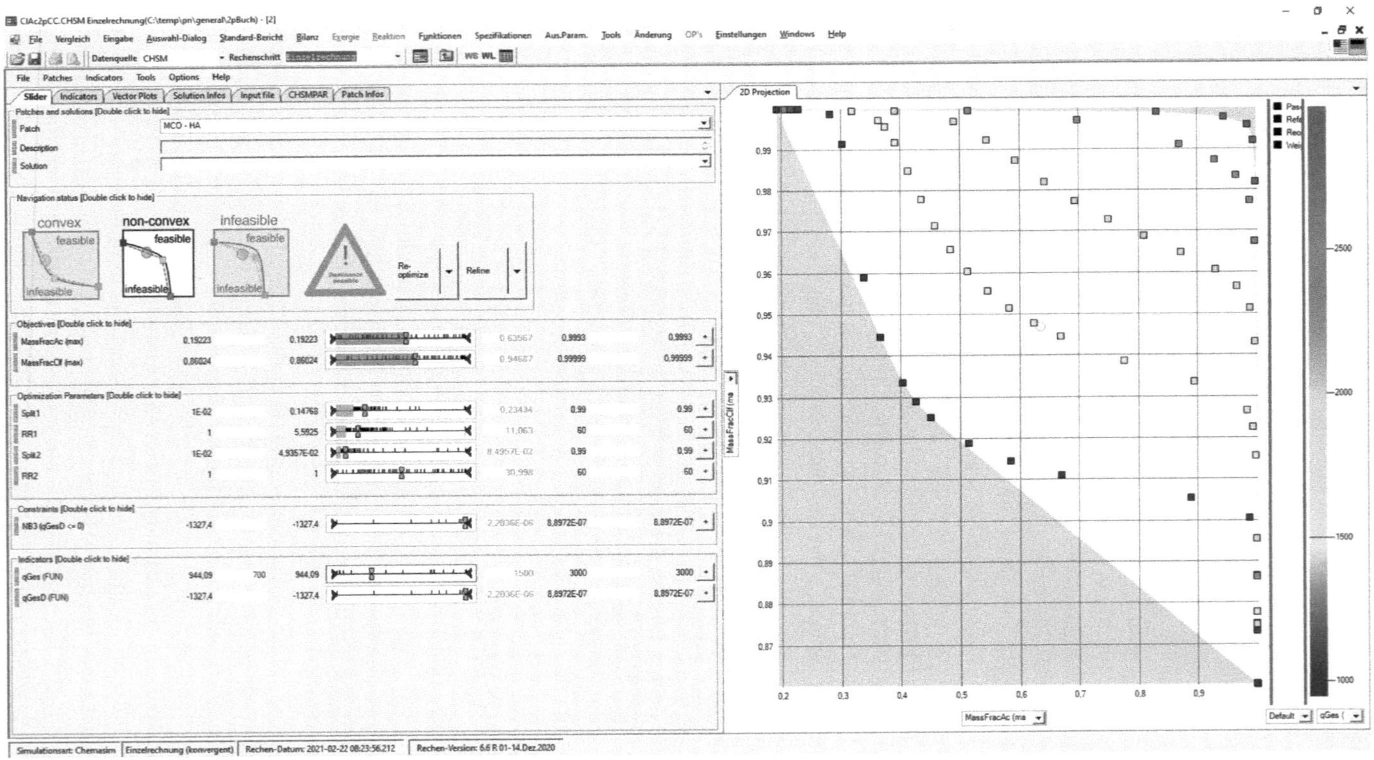

FIG. 3.8 Sliders with projection to interactively explore the tradeoffs coded by the Pareto points. Here, a nonconvex navigation situation is shown.

To summarize, we have given an overview of exemplary use cases for multicriteria optimization in chemical process engineering. All the results presented in this chapter have been obtained from an implementation of MCO algorithms in a flowsheet simulator, that is available for hundreds of process engineers every day, supporting them in comparing and thereby improving process designs.

Acknowledgments

The authors thank Roger Böttcher, Oliver Hirth, Johannes Höller, Karl-Heinz Küfer, Volker Maag, Dimitri Nowak, Jan Schwientek, Phil Süss, and Katrin Teichert for fruitful discussions and tireless coding.

References

[1] B.A. Finlayson, L.T. Biegler, I.E. Grossmann, K.-H. Küfer, M. Bortz, Multicriteria optimization, Mathematics in Chemical Engineering, Wiley Online Library, 2015.
[2] I. Hashem, D. Telen, P. Nimmegeers, F. Logist, J. Van Impe, A novel algorithm for fast representation of a Pareto front with adaptive resolution: application to multi-objective optimization of a chemical reactors, Comput. Chem. Eng. 106 (2017) 544–558.
[3] B. Schandl, K. Klamroth, M.M. Wiecek, Norm-based approximation in bicriteria programming, Comput. Optim. Appl. 20 (1) (2001) 23–42.
[4] M. Bortz, J. Burger, N. Asprion, S. Blagov, R. Böttcher, U. Nowak, A. Scheithauer, R. Welke, K.-H. Küfer, H. Hasse, Multi-criteria optimization in chemical process design and decision support by navigation on Pareto sets, Comput. Chem. Eng. 60 (01) (2014) 354–363.
[5] I. Das, J.E. Dennis, A closer look at drawbacks of minimizing weighted sums of objectives for pareto set generation in multicriteria optimization problems, Struct. Multidiscip. Optim. 14 (1) (1997) 63–69.
[6] K.H. Küfer, M. Monz, A. Scherrer, P. Süß, F. Alonso, A.S.A. Sultan, T. Bortfeld, C. Thieke, Multicriteria optimization in intensity modulated radiotherapy planning, in: P.M. Pardalos, H. E. Romeijn (Eds.), Handbook of Optimization in Medicine, Springer, 2009, pp. 123–168 (Chapter 5).
[7] K. Klamroth, J. Tind, M.M. Wiecek, Unbiased approximation in multicriteria optimization, Math. Meth. Oper. Res. 56 (2002) 413–437.
[8] J.I.S. Hernández, Multi-Objective Optimization in Mixed Integer Problems: With Application to the Beam Selection Optimization Problem in IMRT, TU Kaiserslautern, 2012 (Ph.D. thesis).
[9] K. Teichert, A Hyperboxing Pareto Approximation Method Applied to Radiofrequency Ablation Treatment Planning, TU Kaiserslautern, 2013 (Ph.D. thesis).
[10] K. Schittkowski, NLPQLP: A Fortran Implementation of a Sequential Quadratic Programming Algorithm With Distributed and Non-Monotone Line Search—User's Guide, Version 4.2, Report, Department of Computer Science, University of Bayreuth, 2009.
[11] J. Burger, N. Asprion, S. Blagov, R. Böttcher, U. Nowak, M. Bortz, R. Welke, K.-H. Küfer, H. Hasse, Multi-objective optimization and decision support in process engineering—implementation and application, Chem. Ing. Tech. 86 (7) (2014) 1065–1072.
[12] L.T. Biegler, Nonlinear Programming. Concepts, Algorithms, and Applications to Chemical Processes, MOS-SIAM Series on Optimization, Society for Industrial and Applied Mathematics and the Mathematical Optimization Society, 2010.

[13] A. Wächter, L.T. Biegler, On the implementation of an interior-point filter line-search algorithm for large-scale nonlinear programming, Math. Program. 106 (2006) 25–57, https://doi.org/10.1007/s10107-004-0559-y.
[14] A. Wächter, L.T. Biegler, Line search filter methods for nonlinear programming: motivation and global convergence, SIAM J. Optim. 16 (2005) 1–31, https://doi.org/10.1137/S1052623403426556.
[15] A. Wächter, L.T. Biegler, Line search filter methods for nonlinear programming: motivation and global convergence, SIAM J. Optim. 16 (2005) 32–48, https://doi.org/10.1137/S1052623403426556.
[16] R. Heese, M. Walczak, T. Seidel, N. Asprion, M. Bortz, Optimized data exploration applied to the simulation of a chemical process, Comput. Chem. Eng. 124 (2019) 326–342.
[17] M. von Kurnatowski, M. Bortz, A. Scherrer, A. Hoffmann, H.M. Lorenz, M. Caraucan, T. Grützner, N. Künzle, K.-H. Küfer, Multi-criteria optimization of an industrial world-scale process, Chem. Ing. Tech. 89 (2017) 1471–1478, https://doi.org/10.1002/cite.201700016.
[18] N. Asprion, Modeling, simulation, and optimization 4.0 for a distillation column, Chem. Ing. Tech. 92 (7) (2020) 879–889.
[19] D. Bertsimas, D.B. Brown, C. Caramanis, Theory and applications of robust optimization, SIAM Rev. 53 (3) (2011) 464–501.
[20] A. Ben-Tal, L. El Ghaoui, A. Nemirovski, Robust Optimization, Princeton University Press, 2009.
[21] J.R. Birge, F. Louveaux, Introduction to Stochastic Programming, second ed., Springer, 2011.
[22] A. Ben-Tal, A. Nemirovski, Robust optimization—methodology and applications, Math. Program. 92 (3) (2002) 453–480.
[23] R. Hettich, K.O. Kortanek, Semi-infinite programming—theory, methods, and applications, SIAM Rev. 35 (1993) 380–429.
[24] I. Yanıkoğlu, D. den Hertog, A survey of adjustable robust optimization, Eur. J. Oper. Res. 277 (2019) 799–813.
[25] A. Saltelli, K. Chan, E. Scott, Sensitivity Analysis, Wiley, 2009.
[26] S. Recker, P. Kühl, M. Diehl, H.-G. Bock, Sigmapoint approach for robust optimization of nonlinear dynamic systems, in: Proceedings of the 2nd International Conference on Simulation and Modeling Methodologies, Technologies and Applications, 2012, pp. 199–207.
[27] C. Douglas, Montgomery; Design and Analysis of Experiments, eigth ed., John Wiley & Sons Inc., 2012.
[28] M. Bortz, V. Maag, J. Schwientek, R. Benfer, R. Böttcher, J. Burger, E. von Harbou, N. Asprion, K.-H. Küfer, H. Hasse, Decision support by multicriteria optimization in process development: an integrated approach for robust planning and design of plant experiments, Comput. Aided Chem. Eng. 37 (2015) 2063–2068.
[29] S. Zenios, A. Consiglio, S.S. Nielsen, Mean-variance portfolio optimization, in: S. Zenios, A. Consiglio, S.S. Nielsen (Eds.), Practical Financial Optimization—A Library of GAMS Models, Wiley, 2009 (Chapter 3).
[30] M. Bortz, J. Burger, E. von Harbou, M. Klein, J. Schwientek, N. Asprion, R. Böttcher, K.-H. Küfer, H. Hasse, Efficient approach for calculating pareto boundaries under uncertainties in chemical process design, Ind. Eng. Chem. Res. 56 (2017) 12672–12681.
[31] K. Klamroth, E. Köbis, A. Schöbel, C. Tammer, A unified approach to uncertain optimization, Eur. J. Oper. Res. 260 (2017) 403–420.
[32] R. Hanneman-Tamás, W. Marquardt, How to verify optimal controls computed by direct shooting methods?—a tutorial, J. Process Control 22 (2012) 494–507.

[33] F. Logist, M. Vallerio, B. Houska, M. Diehl, J. Van Impe, Multi-objective optimal control of chemical processes using ACADO toolkit, Comput. Chem. Eng. 37 (2012) 191–199.

[34] N. Asprion, R. Böttcher, J. Schwientek, J. Höller, P. Schwartz, C. Vanaret, M. Bortz, Decision support for the development, simulation and optimization of dynamic process models, Front. Chem. Sci. Eng. 16 (2) (2022) 210–220, https://doi.org/10.1007/s11705-021-2046-x.

[35] D.M. Bates, D.G. Watts, Nonlinear Regression Analysis and Its Applications, John Wiley & Sons, Inc., 1988.

[36] A. Kulkarni, M. Bortz, K.-H. Küfer, M. Kohns, H. Hasse, Multicriteria optimization of molecular models of water using a reduced units approach, J. Chem. Theory Comput. 16 (2020) 5127–5138.

[37] E. Forte, A. Kulkarni, J. Burger, M. Bortz, K.-H. Küfer, H. Hasse, Multi-criteria optimization for parametrizing excess Gibbs energy models, Fluid Phase Equilib. 522 (2020) 112676.

[38] J. Höller, P. Bickert, P. Schwartz, M. von Kurnatowski, J. Kerber, N. Künzle, H.-M. Lorenz, N. Asprion, S. Blagov, M. Bortz, Parameter estimation strategies in thermodynamics, ChemEngineering 3 (2019) 56.

[39] V.V. Fedorov, Theory of Optimal Experiments, Elsevier, 2013.

[40] D. Nowak, K.-H. Küfer, A Ray-Tracing Technique for the Navigation on a Non-Convex Pareto Front, arXiv:2001.03634, **2020**.

Chapter 4

Integrated modeling and energetic optimization of the steelmaking process in electric arc furnaces: An industrial application

Jesús D. Hernández[a,b], Luca Onofri[b], and Sebastian Engell[a]
[a]*Process Dynamics and Operations Group, Department of Biochemical and Chemical Engineering, TU Dortmund University, Dortmund, Germany,* [b]*Acciai Speciali Terni (AST), Terni, Italy*

1 Introduction

Steel production via electric arc furnaces (EAFs) is a highly energy-intensive process that is employed for 28% of the global production of crude steel, especially stainless steel grades [1]. Electric steelmaking is the most environmentally friendly method to produce steel [2,3] but nonetheless, via the consumption of electric power, its environmental footprint can be large. Research on further improvements of the process, therefore, focuses on achieving flexible, energy-efficient, and environmentally friendly operations.

As during the operation of a steel mill, decisions on different time scales have to be made, in the optimization of the processes a hierarchical structure is proposed to compute optimal production decisions at various levels. It is presented in Fig. 4.1.

From the top to the bottom, the decisions made on each layer in Fig. 4.1 are as follows: On the first layer, production schedules are computed based on economic criteria. The schedules organize the daily production of the melt shop in groups of batches considering the time-varying and volume-dependent cost of electricity [4] and the needs for maintenance activities of the equipment, e.g., electrode replacements [5]. On the second layer, the operative setpoints for the processing of each group of batches are calculated considering the timing constraints from the scheduling layer using available process information that was gathered from previous batches. On the third layer, MPC structures can be

Simulation and Optimization in Process Engineering. https://doi.org/10.1016/B978-0-323-85043-8.00009-X

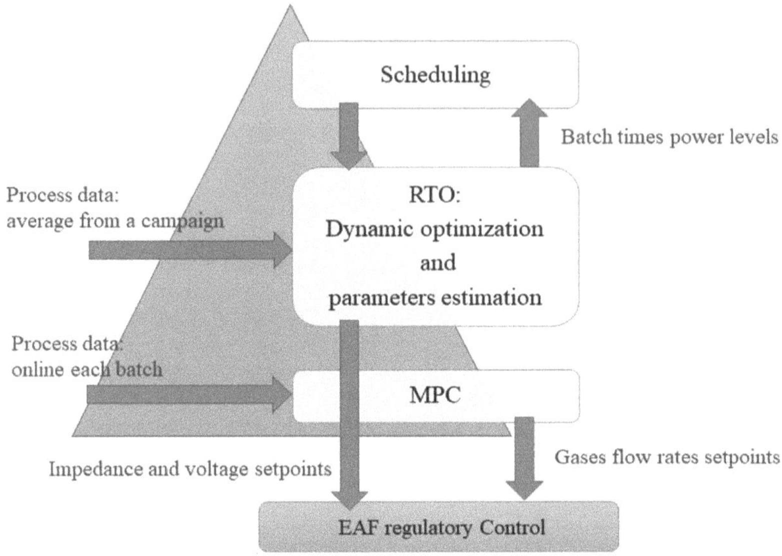

FIG. 4.1 Caption: Decision architecture for optimal operations of steelmaking meltshops.

employed with the goal of optimizing the instantaneous consumption of process gases, e.g. methane and oxygen, taking into consideration terminal constraints such as the required final temperature and chemical composition of the molten metal [6–8]. The computed process inputs are then implemented by the regulatory control layer.

State-of-the-art scheduling solutions for steelmaking plants generate optimal schedules based upon tabulated values of the batch times for given operative power levels. The flexibility of the batch times, i.e., extending or shortening the processing times of a batch is employed to adjust the operation to the hourly changing cost of electricity while still meeting the daily or weekly demands.

This article discusses in detail how dynamic optimization can be employed to compute the optimal operative setpoints for the batches that are processed in the EAF from an environmental and economical perspective, and how they can be implemented by the RTO layer between the scheduling and the MPC layers of the structure presented in Fig. 4.1. The efficacy of the presented framework is demonstrated for a real industrial case. An optimal set of operative setpoints, so-called a melting profile, was computed and implemented in one of the EAFs at Acciai Speciali Terni in Italy. The proposed dynamic optimization computed a mode of operation that, when applied to the real furnace, reduced the energy demand by an average of 4.5% over 19 batches for one type of steel and somewhat less for another type.

2 Electric arc furnace process model

A modern EAF can be viewed as a batch reactor in which metal scrap is melted, and its composition is partially or totally refined. The latent and sensible heat required for the phase transition of the metal is mainly provided by electric energy and supported by chemical energy. The EAF is a multiphase reactor that is typically modeled as a system of three to six phases [9–11]. State-of-the-art EAF models are built by deriving mass and energy balances for each modeled phase and using algebraic relationships to describe the mechanisms of heat exchange.

The level of accuracy of these models is mostly determined by the assumptions regarding the mechanisms of heat exchange. For example, models that consider convection as the main mechanism of heat transport from the arc to the other phases [9,10] tend to be less accurate than those where radiation dominates [11,12]. The predicted melting rates and the remaining amount of solid metal at specific points during a batch by some models can be over- or underestimated by as much as 50% [12].

The process model mismatch can be attributed mainly to the lack of an accurate model of the radiative heat exchange between the electric arc and the other surfaces in the furnace. Modeling the radiative exchange is a complex task because, in the EAF process, the exchange takes place in complex geometries where shadings and blockages in the lines of sight occur. For such constellations, no simple algebraic relationships can be derived. One way to improve the predictions of the radiative heat exchange is to use Monte Carlo algorithms to compute the so-called view factors for complicated geometries [13].

2.1 Hybrid EAF process model

The model assumes that the EAF is a cylindrical vessel that contains three phases: solid metal scrap, molten metal, and gaseous atmosphere. The liquid phase consists both of the molten metal and the liquid slag. The gas atmosphere is assumed to be transparent and composed only of the combustion gases from the oxy-fuel burners and the oxygen lances, see Fig. 4.2.

The trajectories of the states variables in the process mostly depend on the net amount of energy that is transferred from the inputs to the solid and the liquid phases. The energy inputs considered in the EAF model are (a) the heat irradiated by the electric arc, (b) the thermal heat from the oxy-fuel flames, (c) the heat released by the combustion of coal and the oxidation of solid and molten metal, and (d) the splashing of molten metal on the solid metal.

At the beginning of the batch, the EAF is completely filled with solid scrap. When the batch is started, the electrodes dig into the scrap creating a cavity that varies in height but that maintains a constant diameter. As the batch progresses and the electrodes reach the pool of molten metal at the bottom of the EAF, the

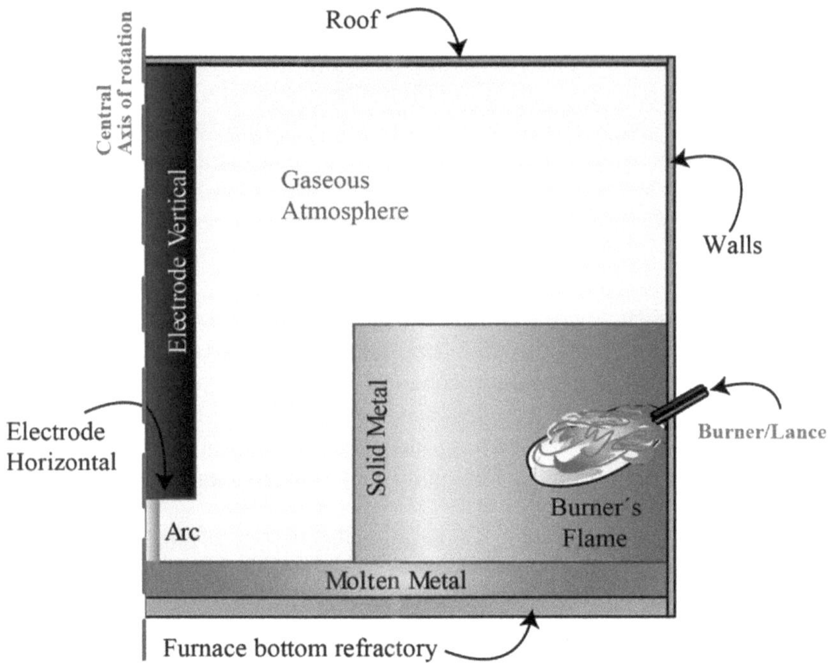

FIG. 4.2 Radial cut of the cylindrical geometry of the EAF. Modeled phases and considered surfaces in the radiative heat exchange are shown.

electrodes are locked in their position and the solid metal assumes the hollowed-cylinder geometry shown in Fig. 4.2. As the melting progresses further, the internal radius of the metal scrap increases, and its height decreases.

2.1.1 Model dynamics

The dynamics of the melting process are evaluated for the most important state variables: the masses of solid (m_{sm}) and molten metal (m_{mm}), and their temperatures T_{sm}, and T_{mm}. Their values can be computed according to Eqs. (4.1)–(4.4) that result from the energy and mass balances of the solid metal ("sm") and the molten metal ("mm").

$$\frac{dm_{sm}}{dt} = \dot{R}_B\, \zeta_{sm} - \frac{\dot{Q}_{in\_sm_m}}{\left(c_{p\_sm}\left(T_f - T_{sm}\right) + \Delta H_f\right) k_{m_{sm1,2,3}}} \tag{4.1}$$

$$\frac{dT_{sm}}{dt} = \frac{\dot{Q}_{in\_sm_h}}{c_{p_{sm}}\, m_{sm}} \tag{4.2}$$

$$\frac{dm_{mm}}{dt} = \frac{\dot{Q}_{in\_sm_m}}{\left(c_{p\_sm}\left(T_f - T_{sm}\right) + \Delta H_f\right) k_{m_{sm1,2,3}}} - \dot{R}_B\left(\zeta_{sm} + \zeta_{w/r}\right) \tag{4.3}$$

$$\frac{dT_{mm}}{dt} = -\frac{\dot{Q}_{in\_mm_h} - \left(\frac{dm_{mm}}{dt} c_{p_{mm}} \left(T_{mm} - T_f\right)\right)}{c_{p_{mm}} m_{mm}} \tag{4.4}$$

In Eq. (4.1), the variations in the thermo-physical properties of the materials are considered using the tuning parameters $k_{m_{smi}}$, where the subindex i, indicates the first, second, or third charge during a batch. They can be estimated independently from process data, in a batch-to-batch manner. The estimation procedure is discussed along with the model validation in Section 4.3. The heat capacity (c_{p_sm}), the enthalpy of fusion of the solid metal (ΔH_f), and the temperature of fusion of the metal (T_f) are assumed to be those of stainless steel, grade AISI 304. $\dot{R}_B$ represents the total amount of molten metal that is splashed by the oxygen lancing action, and ζ_{sm} and $\zeta_{w/r}$ represent the fraction of the total splashed molten metal that lands on the solid metal (sm) and the walls and the roof (w/r).

The process is driven by the heat that is absorbed by the solid metal leading to both melting ($\dot{Q}_{insm_m}$) and heating ($\dot{Q}_{insm_h}$), and the heat absorbed by the molten metal pool ($\dot{Q}_{inmm_h}$).

2.1.2 The electric arc model

The heat exchange between the arc and the rest of the surfaces is strongly dependent on the geometry of the electric arc. High-power steelmaking arcs mostly emit heat through radiation mechanisms. Their length (l_a) can be established in terms of its voltage (V_a) and impedance (Z_a) setpoints as [14].

$$l_a = \frac{V_a + 57 Z_a}{27}. \tag{4.5}$$

The active electric power of the electric arc can be computed as

$$P_a = \frac{V_a^2}{Z_a} \cos\theta. \tag{4.6}$$

The power factor ($\cos\theta$) is a measured parameter, it can be obtained from the instantaneous readings of the voltage, the current, and the active power.

2.1.3 Radiative heat exchange from the electric arc

The heat exchange between the arc and the rest of the phases in the EAF enclosure can be modeled using the DC circuit analogy. The radiative heat exchange between two gray surfaces is modeled as a series circuit composed of two voltage sources (4.7), two radiosity resistors (4.8), and one view factor resistor (4.9). In this circuit, the current flux represents the total heat that is exchanged between the two surfaces [15], see Fig. 4.3. This leads to the following equations:

$$V_{T_i} = \sigma_b T_i^4 \tag{4.7}$$

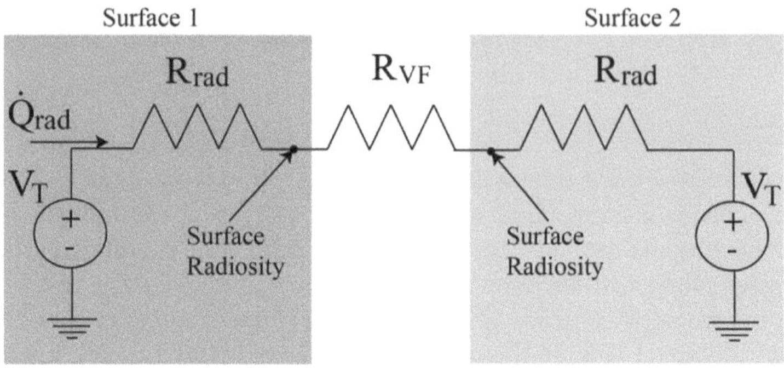

FIG. 4.3 Electrical equivalent circuit of the heat exchange between two gray surfaces.

$$R_{rad_i} = \frac{1-\varepsilon_i}{\varepsilon_i A_i} \tag{4.8}$$

$$R_{VF_{i-j}} = \frac{1}{VF_{i-j} A_i} \tag{4.9}$$

In (4.7) to (4.9), the two voltage sources (V_{Ti}) represent the potential of the surfaces to exchange heat and depend on the temperature of the surface to the fourth power (T_i^4) and on the Stefan-Boltzmann constant $\sigma_b = 5.670367 \times 10^{-8}$ - $\text{W m}^{-2}\,\text{K}^{-4}$. The two radiosity resistors (R_{radi}) describe the opposition of a given surface to absorb radiative energy and depend on both the emissivity (ε_i) and the area (A_i) of the surface. Finally, the view factor resistor (R_{VFi-j}) accounts for the fraction of radiative energy from the emitting surface that reaches the receiver surface.

For enclosures composed of many surfaces, the total energy exchange can be modeled as a network of voltage sources and resistors having as many of these branches connected as there are participating surfaces [12].

In the radiative circuit, the only resistor of unknown value is the one that is associated with the emissivity of the surface of the electric arc. For electric steelmaking arcs operating in the range of several tens of megawatts, the electrical power input equals the heat loss via radiation mechanisms at a steady state. This makes it possible to replace the branch composed by the voltage source and the unknown surface resistor of the arc with a current source of a value that is equal to the power input of the arc, see Fig. 4.4. With this simplification, the equivalent DC circuit can be solved using the node-voltage strategy for linear circuits [16]. This results in a system of linear equations of the form

$$J = A^{-1} b, \tag{4.10}$$

where each element in the matrix A is a function of various VF resistors ($R_{VF_{i-j}}$), and the vector b is composed of the voltage sources (V_{T_i}) and the radiosity resistors (R_{rad_i}). With Eq. (4.10), the radiosities of each participating surface (J_i) are

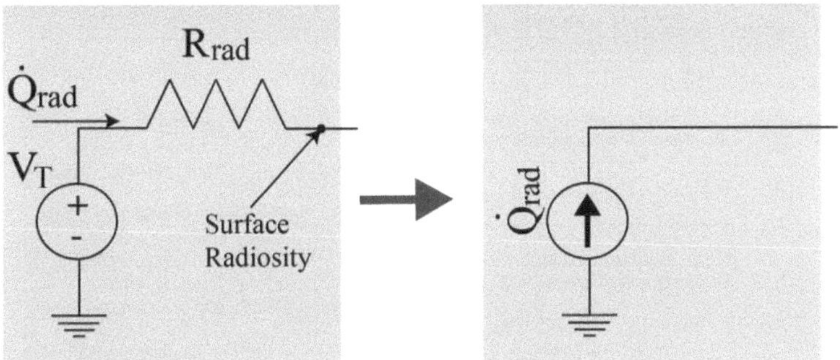

FIG. 4.4 Voltage–current circuit equivalency for the electric arc circuit.

computed. The net power that is entering or leaving every surface in the radiosity circuit can be calculated as

$$\dot{Q}_{in\_i} = \frac{J_i - V_{T_i}}{R_{rad_i}}. \tag{4.11}$$

The electrical power gains $\dot{Q}_{gain}$ and the power losses $\dot{Q}_{loss}$ of the process are defined as

$$\dot{Q}_{gain} = \dot{Q}_{in\_sm} + \dot{Q}_{in\_mm} \tag{4.12}$$

$$\dot{Q}_{loss} = \dot{Q}_{in\_roof} + \dot{Q}_{in\_wall} + \dot{Q}_{in\_elev} + \dot{Q}_{in\_eleh} \tag{4.13}$$

In Eqs. (4.12) and (4.13), $\dot{Q}_{insm}$, $\dot{Q}_{inmm}$, $\dot{Q}_{inroof}$, $\dot{Q}_{inwall}$, $\dot{Q}_{inelev}$, $\dot{Q}_{ineleh}$ represent the energy fluxes absorbed by the solid metal, molten metal, the wall, the roof, and the vertical and horizontal surfaces of the electrodes.

2.1.4 Monte Carlo calculation of the view factors

The Monte Carlo algorithm can be seen as a black-box model that for inputs x (the geometries of the pile of solid metal and the electric furnace), computes outputs y (the view factors). In short, the algorithm works by emitting a large number of bundles (energy packages) in random directions from a given surface in the enclosure. Then by computing the number of bundles that land on each of the other participating surfaces, the view factors relative to the emitting surface can be computed. The flow chart of the algorithm is presented in Fig. 4.5, for the case of the surface of the molten metal.

2.1.5 Oxy-fuel burners

The energy input from the oxy-fuel burners is estimated using a first-principles dynamic model of the oxy-fuel flame. It describes at a macro level the main physical phenomena experienced by the combustible gases as they flow through the void spaces of the metal scrap. The volume through which the gases move is

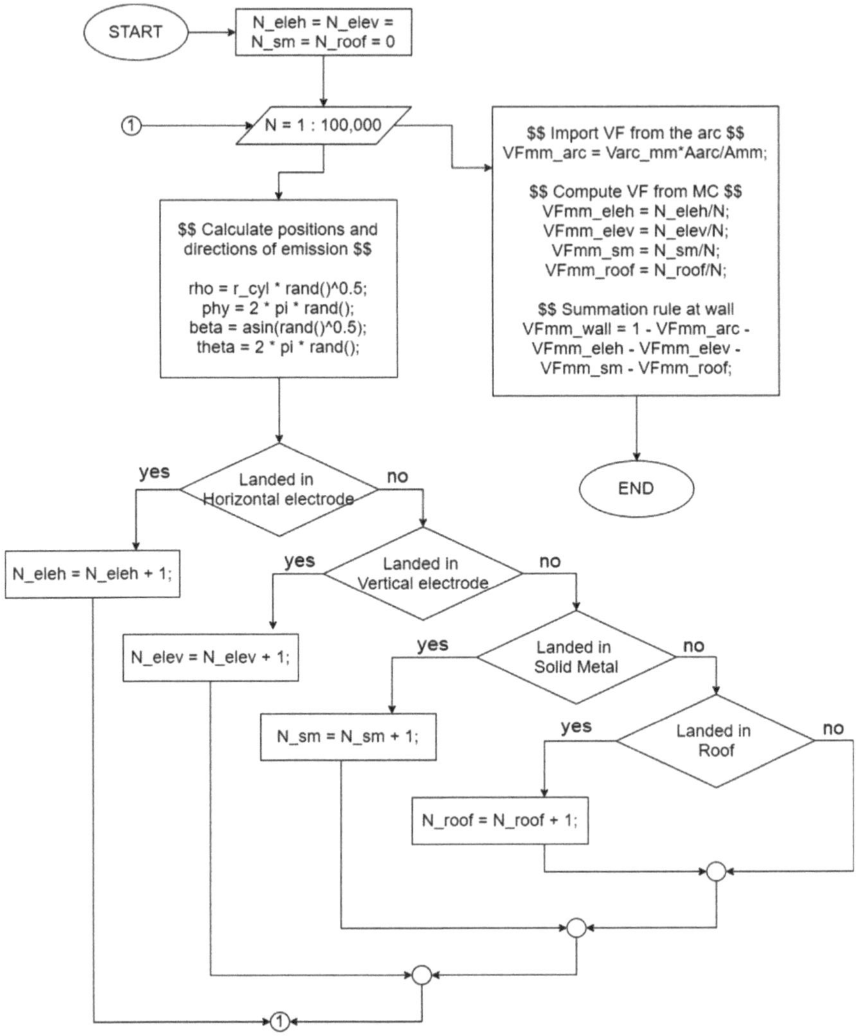

FIG. 4.5 Monte Carlo algorithm for calculating view factors for the molten metal as the emitting surface. *(Credit: T. Hay, J. Hernandez, S. Roberts, T. Echterhof, Calculation of view factors in electric arc furnace process modeling, Steel Res. Int., 92, p. 2000341, 2020.)*

first assumed to have a cylindrical shape that is then discretized. The temperature profile of the gas and the net heat exchange with the solid metal are computed using fundamental relationships of radiative and conductive heat exchange for each discretized volume [17]. By adding the energy contributions from each volume, the total energy from the flame to the metal ($\dot{Q}_{burn}$) can be estimated. The model provides dynamic efficiency curves of the heat exchange from the flame to the solid metal. They depend on the flow rates of the combustible gases, as shown in Fig. 4.6.

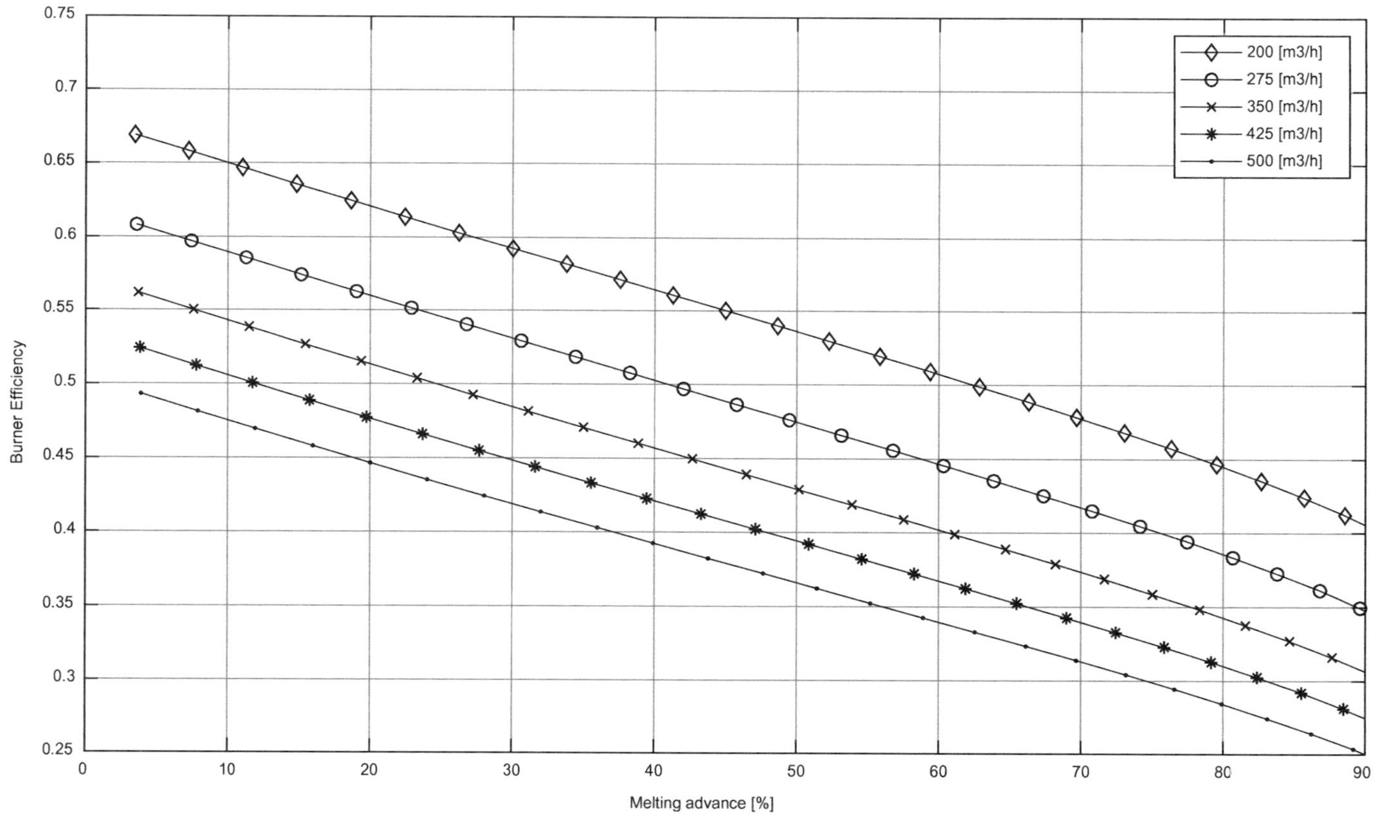

FIG. 4.6 Burner efficiency vs. batch progress for various flow rates of combustible gases. *(© 2020 Reprinted with permission of the International Federation of Automatic Control from original publication: J.D. Hernández, L. Onofri, S. Engell, Optimization of the electric efficiency of the electric steel making process, IFAC-PapersOnLine 53 (2) (2020) 11895–11900.)*

2.1.6 Combustion of coal

The energy contribution from the combustion of coal is computed assuming that its efficiency is the same as that of the oxy-fuel burner.

$$\dot{Q}_{coal} = \eta_{burn} * (\dot{m}_{coal} * 0.85) * LHV_{coal} \quad (4.14)$$

Here, the *LHV* represents the low heating value of coal.

2.1.7 Oxidation of metals

Oxidation of metals occurs both in the liquid phase and in the solid phase. The former takes place when oxygen is lanced directly into the bath of molten metal. The enthalpies of reaction of these reactions ($\Delta H_{O_2\_m}$) are known [18] and their net energy contribution to the process can be computed as

$$\dot{Q}_{LM} = \dot{F}_{O_2} \eta_{O_2} x_{O_2\_m} \Delta H_{O_2\_m}. \quad (4.15)$$

In Eq. (4.15), $\dot{F}_{O_2}$, η_{O_2}, and $x_{O_2\_m}$, represent the flow of lanced oxygen, the efficiency of the lancing, and the fraction of lanced oxygen that is consumed by the oxidation reactions of molten metal. The remaining fraction of oxygen $(1 - x_{O_2\_m})$, is consumed in the oxidation of carbon dissolved in the molten metal pool. Its energy contribution to the process is determined as in (4.15) but its enthalpy of reaction ($\Delta H_{O_2\_c}$) is much lower than that of the oxidation of the molten metal [18]. η_{O_2} and $x_{O_2\_m}$ are unknown parameters that can be estimated from process data. This is described in Section 4.3.

The oxidation of the solid metal occurs due to corrosion mechanisms. In the EAF, the solid metal reacts with the free oxygen in the atmosphere to form Iron (III) oxide. The rate at which this oxide is produced can be determined using the quadratic corrosion formula [19].

$$\frac{\Delta \dot{M}}{A_{sc} * 0.8} = \sqrt{k_p * t}. \quad (4.16)$$

Here, $\Delta \dot{M}$ [mg] represents the mass gain of the oxidized material, $k_p = 2.6 * 10^{-12}$ mg^2 cm^{-4} s^{-1} (experimental constant), t is time, and A_{sc} is the total area of the scrap in contact with the oxidizing gas [20]. The factor 0.8 accounts for the fact that the oxygen injected will expand and will occupy 80% of the void space in the scrap. The energy contribution from these reactions to the process can be approximated as the total energy released by the oxidation according to Eq. (4.16).

2.1.8 Molten metal splashing

When oxygen is lanced into the liquid pool of molten metal, large quantities of metal drops are generated. As they fly away from the impingement point, many of them land on the solid metal and solidify. The energy transfer from these drops of molten metal to the solid metal ($\dot{Q}_{sp_{sm}}$) can be estimated using Eq. (4.17). It depends on the amount of splashed metal $\dot{R}_B$ and the fraction of the metal that lands on the solid (ζ_{sm}).

$$\dot{Q}_{sp_{sm}} = \dot{R}_B\, \zeta_{sm} \left(\int_{Tref}^{T_f} Cp_{sm} \,.\, dT + \Delta H_f + \int_{T_f}^{T_{mm}} Cp_{mm} \cdot dT \right) \tag{4.17}$$

$$\frac{\dot{R}_B}{\dot{F}_{O_2}} = \frac{{N_B}^{3.2}}{\left(2.6{*}10^6 + 2{*}10^{-4}{*}{N_B}^{12}\right)^{0.2}} \tag{4.18}$$

In Eq. (4.18), the blowing number N_B has a value of 5 [21]. On the other hand, the factions of splashed metal can be computed from the geometrical configuration of the solid scrap and the EAF [22]. Here, they are computed in terms of the α_{sp} parameter, that is the dynamic ratio between the diameter of the furnace and that of the hollowed-cylinder geometry.

$$\zeta_{sm} = 0.9\left(1 - \alpha_{sp}\right) \tag{4.19}$$

The molten metal splashing occurs inside the thermodynamic boundary located at the roof and walls of the furnace but transfers energy between the molten and solid metal phases.

3 Dynamic optimization of the melting profiles

3.1 Problem statement

The dynamic optimization layer in Fig. 4.1 carries out two tasks. On the one hand, it computes the batch times for different power levels. This information is used by the top layer scheduling algorithm to generate the daily schedules. On the other hand, it computes the optimal set of operative setpoints for groups of batches.

Optimal modes of operations can be computed only if the costs of production are considered within the calculation of the batch times, providing the power level. The operation of the EAF for a given batch time should be both economically and environmentally advantageous which implies maximizing the energy efficiency of the process. As in the EAF process, 95% of the operative costs are associated with the electric power input, an economic optimization can be carried out considering only this input.

The problem of computing a set of optimal operative points for the EAF with the objective of improving its electrical energy efficiency can be formulated as the problem to minimize the radiative losses over the batch time, constrained by the power limits of the furnace. Because the radiative losses in a batch cannot be measured during operation, they must be estimated using an EAF process model.

3.2 A general formulation of the dynamic optimization problem

The dynamic optimization problem consists of the following:

- A cost function that depends on the integral of the states throughout the batch,
- Constraints related to a hybrid model (DAE+black-box model),
- Terminal constraints,
- No path constraints, and
- No inequalities other than those imposed on the control variables.

This can be formulated as:

$$
\begin{aligned}
\min \quad & \int_{t=0}^{tf} f(x_d(t), x_s(t), x_{sb}(t), u(t), p)dt \qquad (4.20) \\
\text{s.t.} \quad & \dot{\boldsymbol{x}}_d - \boldsymbol{g}_d(\boldsymbol{x}_d(\boldsymbol{t}), \boldsymbol{x}_s(\boldsymbol{t}), \boldsymbol{x}_{sb}(\boldsymbol{t}), \boldsymbol{u}(\boldsymbol{t}), \boldsymbol{p}) = \boldsymbol{0} \\
& \boldsymbol{x}_{sa} - \boldsymbol{g}_{sa}(\boldsymbol{x}_d(\boldsymbol{t}), \boldsymbol{x}_s(\boldsymbol{t}), \boldsymbol{x}_{sb}(\boldsymbol{t}), \boldsymbol{u}(\boldsymbol{t}), \boldsymbol{p}) = \boldsymbol{0} \\
& \boldsymbol{x}_{sb} - \boldsymbol{g}_{sb}(\boldsymbol{x}_d(\boldsymbol{t}), \boldsymbol{x}_s(\boldsymbol{t}), \boldsymbol{x}_{sb}(\boldsymbol{t}), \boldsymbol{u}(\boldsymbol{t}), \boldsymbol{p}) = \boldsymbol{0} \\
& x_d(t_f) - w = 0 \\
& u_{low} \le u(t) \le u_{up}
\end{aligned}
$$

Here, f represents the cost function to be minimized, g_d represent the set of differential equations, g_{sa} the set of algebraic equations, and g_{sb} the black-box model. $x_d(t)$, $x_s(t)$, $x_{sb}(t)$, p represent the dynamic states, algebraic static states, black-box states, and a vector of parameters. The terminal value of a subset of the dynamic states $x_d(t_f)$ is prescribed by the vector w. The dynamic optimization problem is solved for the control variables $u(t)$, which are constrained to a domain bounded by minimum and maximum values u_{low} and u_{up}.

3.3 Formulation of the dynamic optimization problem of the EAF process

The formulation (4.20) can represent the problem of maximizing the energy efficiency of the EAF process, subject to the predictions computed by the hybrid model and the requirement of a certain final temperature of the molten metal. From the various methods for the solution of dynamic optimization problems, the control vector parametrization approach (CVP) is employed because it can handle the hybrid model of the EAF in a straightforward manner. For a detailed description of the CVP strategy, the reader is referred to [23,24].

In CVP, the control variable $u(t)$ is discretized in N control intervals over the simulation timeline. The discretization transforms the original continuous control variable into a control vector of N components. Then, CVP employs the solution of the model by a simulator to compute the value of the objective function. Their values are passed to a numerical solver where the optimal value of each component of the control vector is computed iteratively, calling the solution by the simulator in each iteration.

Because the Monte Carlo algorithm (the black-box model) is a numerical algorithm that relies on computational loops, algorithms that generate random numbers, and logical conditions to identify the landing surface of each emitted bundle (see Fig. 4.5), the complete EAF model cannot be differentiated explicitly. Nonetheless, it is possible to solve (4.20) using efficient gradient-based numerical solvers if (a) the objective function is continuous and sufficiently smooth, (b) the solver does not need to handle the model directly, (c) the gradients are computed using a finite difference approximation.

The problem of maximizing the energy efficiency of the EAF process can then be re-written as the problem to minimize the energy losses during a batch of steel. In terms of the discretized control vector $u(n)$, it reads in a condensed form:

$$\begin{aligned} \min \quad & J(u(n), p) \\ \text{s.t.} \quad & \boldsymbol{u}_{low} \leq \boldsymbol{u} \leq \boldsymbol{u}_{up}. \end{aligned} \tag{4.21}$$

In (4.21), the objective function contains all the computations executed by the process simulator. It encapsulates all the equality constraints in (4.20): the system of dynamic and algebraic equations, the black-box model relationships of the EAF model, and the equality constraint for the required terminal state of the process. Fig. 4.7 presents a graphic representation of the evaluation of the objective function (the electrical losses) for the melting of a batch of steel in the EAF.

4 Solution using control vector parametrization

4.1 Numerical solution of the model

The hybrid EAF model was implemented and solved in MATLAB using a two-grid simulation timeline. On the coarse grid that consists of 1-min intervals, the energy contributions from each energy input are computed using linear and nonlinear solvers. The Monte Carlo algorithms are performed using the built-in random generation numbers functions and logical operators. On the fine grid, the dynamic system is solved using the solver ODE45, taking as inputs the computed energy contributions from each input. The solution of the dynamic system is done on each 1-min interval on the coarse grid and the discretization is chosen automatically by the integrator. Most of the computational load in the simulation results from the Monte Carlo calculation of view factors, which takes approximately 200ms per point. On average, the simulation of a batch of two charges and a total mass of more than 100 tons took around 7 CPU seconds. These simulations were run on a computer using an Intel® CoreTM i7-8750H processor with 16GB of RAM.

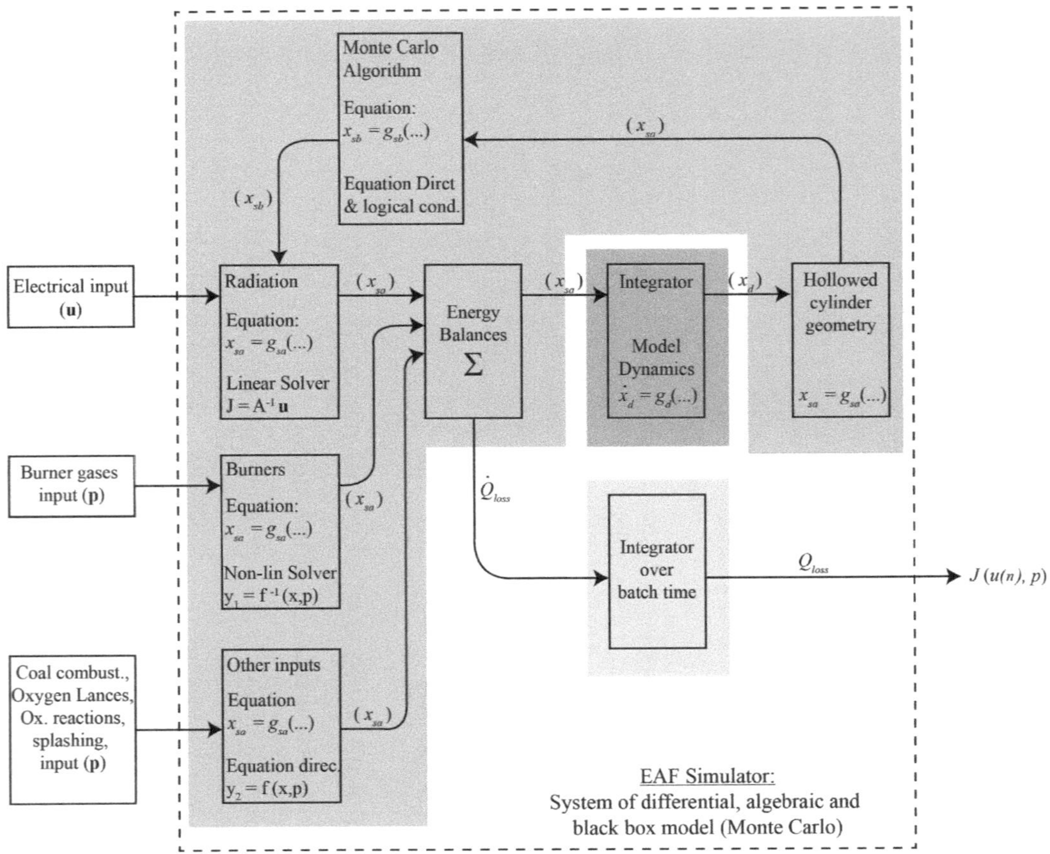

FIG. 4.7 Simulator of the EAF.

4.2 Termination conditions

For a batch of two charges, the termination conditions of the charges are as follows:

- For the first charge: the process stops if a given amount of electrical energy has been introduced to the process.
- For the second charge: the batch is terminated if no solid metal remains, and the final temperature of the molten metal has reached the desired value.

$$if\,(m_{sm}==0) \wedge \left(T_{mm} > T_{final}\right) \rightarrow Terminate\ batch \tag{4.22}$$

These two conditions are embedded in the simulator. If the termination conditions were evaluated at the discretization points of the control vector or the coarse grid of the simulator, the resulting objective function will show a nonsmooth profile which makes the optimizer terminate at fictitious local minima. Therefore, the termination conditions are evaluated on the fine grid of the ODE solver. The effect of the discretization is shown in Fig. 4.8.

4.3 Model validation and parameter estimation

While it is not possible to directly measure most of the important state variables during the batch, it is possible to obtain information about their values at specific moments in an indirect fashion. Process data that can be used for validating the model is: the boredown time of the electrodes during the first (t_{b1}) and the second melting (t_{b2}), the batch time (t_b), the final temperature of the melt ($T_{mm}|_{t=tb}$), and the amount of metal that was molten during the first ($m_{mm}|_{t=tm1}$) and the second melting stages ($m_{mm}|_{t=tm2}$).

The model parameters were adjusted to predict these measurable values well by changing the melting rate parameters $k_{m_{sm1,\,2,\,3}}$ in Eq. (4.1), and the efficiency of the oxygen lancing (η_{O_2}) and the distribution of oxygen between the oxidation reactions ($x_{O_2\_m}$) in Eq. (4.15).

The estimation was carried out by solving the minimization problem

$$\underset{x}{\text{minimize}} \quad f(x) = \sqrt{\sum_{k=1}^{K} \delta_k \left(\frac{p_{k,d} - p_{k,st}(x)}{p_{k,d}}\right)^2} \tag{4.23}$$

$$\text{subject to} \quad \begin{array}{c} \text{EAF process model} \\ x_l < x < x_u \end{array}$$

In (4.23), $p_{k,\,d}$ represents the measured data, $p_{k,\,st}$ is the approximation to the measured data computed by the model, δ_k are the weights associated to each approximated parameter, and K is the set of estimated parameters $k = \{1,2, \ldots n\}$.

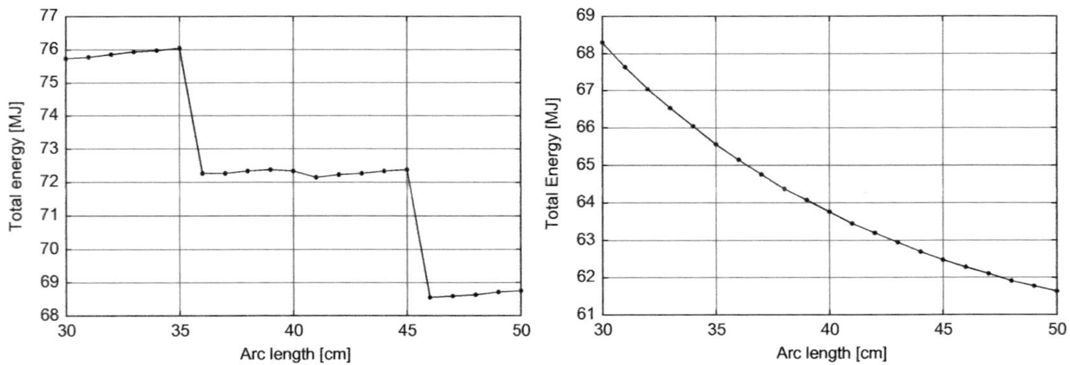

FIG. 4.8 Cost function for various arc lengths (constant throughout the batch) when the termination conditions are evaluated on the coarse grid (left), and on the fine grid (right).

Because the oxygen lancing takes place only toward the end of the melting stages and during the refining stage, its influence on the melting process is small and can be neglected. By making this simplification, the original estimation problem with four variables can be split into three simpler problems. In the first two, the melting rate parameters of the first ($k_{m_{sm1}}$) and second charge ($k_{m_{sm2}}$) are estimated. In the last problem, the oxygen lancing efficiency (η_{O_2}) and oxygen distribution ($x_{O_2\_m}$) are computed together. Because these are one- and two-dimensional problems, they can be solved efficiently in a direct fashion. Here, the cost function is computed for several values of the analyzed parameters, as they are varied from the lower to the upper bounds of the solution space.

In Table 4.1, the parameters that are adapted in the estimation problem and their optimal values are given. In Table 4.2, the normalized absolute error between the selected process data and the simulated data, using the optimal values in Table 4.1, are presented.

TABLE 4.1 Parameters estimation table.

Estimation problem	Parameter estimated x	Process data employed in the estimation $p_{k,\,d}$	Weights	Computed optimal value
Melting rate of the first melting	$k_{m_{sm1}}$	t_{d1} $m_{mm}\|_{t=tm_1}$	$\delta_{t_{d1}}=0.4$ $\delta_{m_{mm,1}}=0.6$	0.92
Melting rate of the second melting	$k_{m_{sm2}}$	t_{d2} $m_{mm}\|_{t=tm_2}$	$\delta_{t_{d2}}=0.4$ $\delta_{m_{mm,2}}=0.6$	0.9
Oxygen efficiency	η_{O_2} $x_{O_2\_m}$	t_b t_{sm_c}	$\delta_{t_{sm\_c}}=0.3$ $\delta_{t_b}=0.7$	0.72 0.32

TABLE 4.2 Errors in the predictions of the model.

Process parameter	Normalized error
t_{d1}	6.4%
$m_{mm,1}$	2.7%
t_{d2}	9.9%
$m_{mm,2}$	8.1%
t_{b_d}	0.5%

The results in Table 4.2 show that the model predicts the batch time well. From an energy perspective, this is the most important process parameter as the energy demand of the process is given by the known electrical power input multiplied by the batch time. Despite the errors in the predictions of the boredown times of the electrodes, the model can be considered sufficiently accurate because in comparison to the batch time, which typically ranges from 50 to 60 min, the boredown periods of the electrodes last only between 2 and 4 min. Although the error in the predicted mass of molten metal before the refining stage is considerable, this error is acceptable for the purposes of this study as its impact on the radiative energy streams is small.

4.4 Numerical solution of the optimization problem

The optimization problem was solved in terms of the time-varying power levels during the batch time, which was initially parametrized in 12 intervals of constant length. The discretization is equivalent to 10 times the settling time of the closed-loop time response for the impedance control of the electric arc. In practice, the impedance of the electric arc is controlled by modifying the length of the electric arc, by moving the electrodes up or down, until the set-point is achieved.

The optimization problem was solved using MATLAB's SQP solver and the gradients were estimated using finite differences.

CVP is traditionally solved by fixing the batch time and allowing the solver to iterate until the terminal conditions are satisfied. Here, because the terminal condition is embedded in the simulator, the batch time is variable, so the batch will be extended until the terminal conditions are satisfied. This methodology can be implemented for the case study of the EAF because the termination condition of the simulator, the final temperature of the molten metal, is a function that increases monotonically with respect to the electrical energy input as the processing time is extended.

A variable batch time requires a control vector of variable length. This is implemented by adding degrees of freedom to the original control vector. Sensitivity analyses around the actual operative points showed that four additional components are required to achieve the required final temperature in the case that all the components of the control vector take values at the lower bound of the solution space. Therefore, the problem is solved for a control vector with 16 elements.

4.5 Batch time constraint

As formulated in (4.21), the dynamic optimization framework will find the optimal power input trajectory for a given amount of scrap and a given final temperature. In this computation, the batch time is a free parameter. In order to obtain an optimal group of setpoints that minimizes the losses of the process for specific batch times, as required by the scheduling algorithm, a quadratic penalty term was added to the original objective function. In the EAF process, the batch time and the energy demand are inversely correlated. The higher the operative power level, the shorter the batch time.

The modified objective function is

$$J(u(n),p)+\gamma\left(t_f-t_{esp}\right)^2 \tag{4.24}$$

In Eq. (4.24), $J(u(n),p)$ is the original objective function in (4.21). t_f, t_{esp} and γ are the final time of the simulation, the specified final batch time, and the penalty weight. Here, the latter is set to the highest value of the operative electrical power of the EAF.

5 Results and discussions

5.1 Numerical case study

5.1.1 Batch simulation

In Fig. 4.9, the dynamic evolution of the energy fluxes to the metal and the energy losses during the batch, as defined in (4.12) and (4.13), are presented. The batch is operated following the standard industrial practice of operating the whole batch at a constant electrical power level. Here, the mass of solid material loaded in the EAF during each of the two charges is the same.

The main discontinuity in the model arises from the charging of the second load, which is handled within the simulator. When the first charge is terminated, the simulation is restated applying the next component of the control vector, and the system of differential equations is solved from the initial condition that

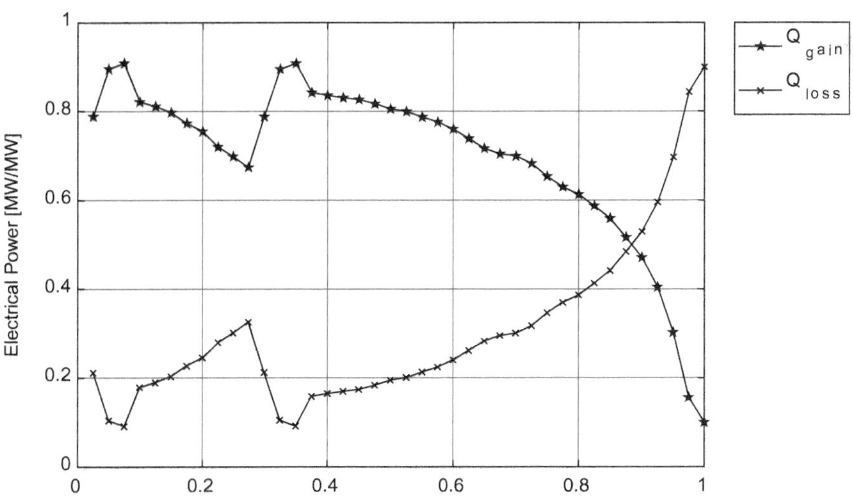

FIG. 4.9 Electrical energy gains and losses of the EAF process at operations at a constant power level. *(© 2020 Reprinted with permission of the International Federation of Automatic Control from original publication: J.D. Hernández, L. Onofri, S. Engell, Optimization of the electric efficiency of the electric steel making process, IFAC-PapersOnLine 53 (2) (2020) 11895 11900.)*

results from adding the additional solid mass to the remaining one after the first melting.

Considering that the energy fluxes are normalized w.r.t. the electric power input of the arc, Fig. 4.9 can also be understood as the dynamic efficiency curve of the process. This result suggests that the electric efficiency of the process is high at the beginning of every charge (over 80%), but it deteriorates while the batch progresses. Toward the end of the batch, the process is highly inefficient, as most of the radiative energy from the arc is irradiated toward the walls of the furnace and just a small fraction, of less than 20%, is absorbed by the solid and the molten metal phases.

5.1.2 Batch optimization

For specified batch time and molten metal temperature, the dynamic optimization (4.21), with the modified objective function (4.24) was solved 10 times, each time from a different initial guess that was generated randomly. From all initial points, the solver converged to a unique optimal point. For every run, the solver terminated after approximately 30 iterations and 600 function evaluations, which typically took between 30 and 90 min. The results of the optimization for batches of various lengths and the same final temperature are presented in Fig. 4.10.

The results in Fig. 4.10 suggest that the optimizer proposed to operate at higher power levels during the early stages of a charge, and at lower power levels toward the end of the melting stage of the batch. For all cases, the optimizer tries to reduce the operative power level at the end of the melting stage of the batch.

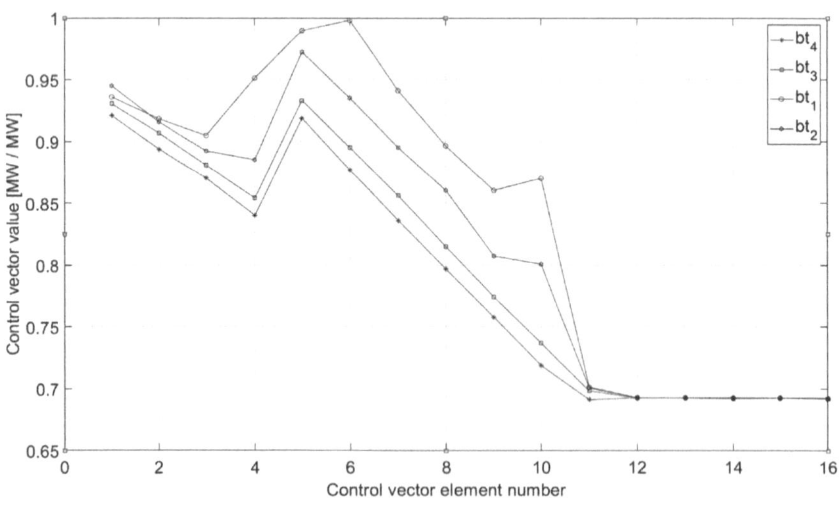

FIG. 4.10 Optimal control vector for varying batch times. b_{t1} = base batch time, $b_{t2} = b_{t1} + 2$ min, $b_{t3} = b_{t1} + 4$ min, $b_{t4} = b_{t1} + 6$ min.

5.2 Results for the real industrial process

The optimization framework was employed to compute the optimal control vector for the electric arc steelmaking process of two different grades of stainless steel. In real operations, the loading practices (number of charges, mass charged in each load, etc.) and the physical properties of the metal scrap (density, heat capacity, latent heat of fusion), are different from those of the numerical simulations described above. For each type of steel, the batch time constraint was set to the average batch time of over 300 batches. The final temperature constraint was set according to process requirements.

The computed optimal profiles for each type of steel were implemented in one of the industrial Ultra-High Power EAF (UHP-EAF) at AST and 19 test batches were run. For these batches, the average energy demand and the batch times are compared with the average data of one year of production. The results of these calculations are presented in a normalized fashion in Figs. 4.11 and 4.12.

On average, the optimal profile reduced the energy demand of steel A by 4.6% and that of steel B by 1.7%. In terms of batch time, a reduction of 4.5% was achieved for steel grade A, but no improvement was observed for steel grade B. An improvement for steel A is more prominent because the raw material used to produce this type of steel is more homogeneous than steel B. The large variations in the raw materials for steel B make it difficult to obtain consistent improvements for a large number of batches.

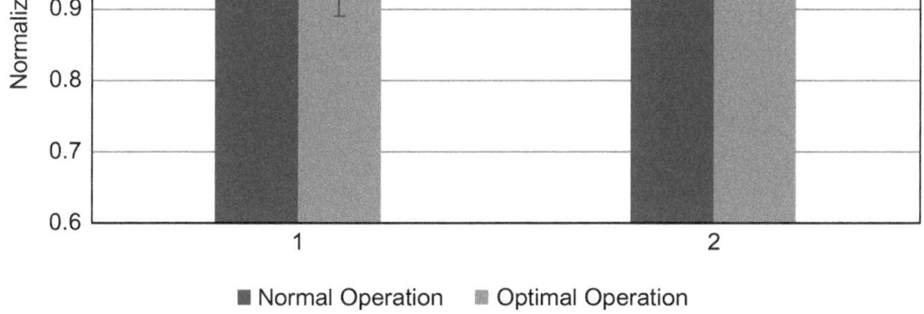

FIG. 4.11 Batch time and energy demand improvements for steel grade A. KPI 1: Batch time; KPI 2: Energy demand. *(© 2020 Reprinted with permission of the International Federation of Automatic Control from original publication: J.D. Hernández, L. Onofri, S. Engell, Optimization of the electric efficiency of the electric steel making process, IFAC-PapersOnLine 53 (2) (2020) 11895–11900.)*

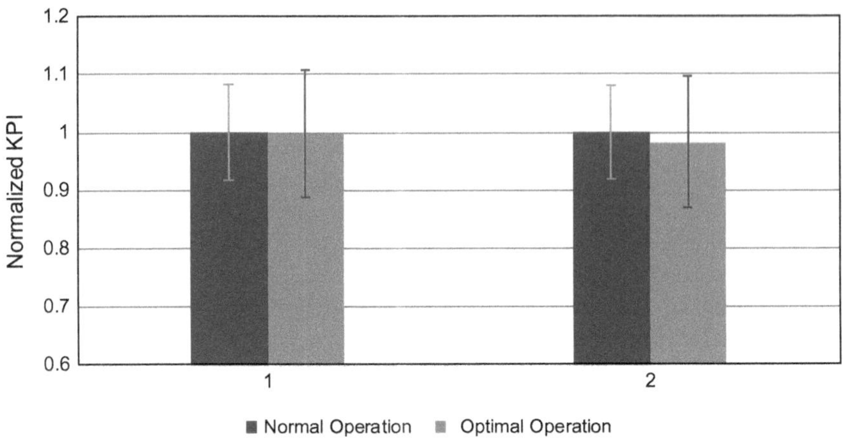

FIG. 4.12 Batch time and energy demand for steel grade B. KPI 1: Batch time; KPI 2: Energy demand. *(© 2020 Reprinted with permission of the International Federation of Automatic Control from original publication: J.D. Hernández, L. Onofri, S. Engell, Optimization of the electric efficiency of the electric steel making process, IFAC-PapersOnLine 53 (2) (2020) 11895–11900.)*

6 Conclusions

This work presented a dynamic model and an optimization framework that computes operation policies of electric arc furnaces that improve the energy performance of this steel production process. The optimization is based on a hybrid model of the EAF that includes a first-principles-based model and an on-the-fly computation of the view factors by Monte Carlo techniques. The result can be employed by state-of-the-art scheduling algorithms for optimizing the operation of the meltshop with respect to the time-varying cost of electric power. The optimization problem was solved using a control vector parameterization technique.

It was found that despite the noncontinuous model of the process, and the presence of a numerical algorithm that uses random numbers within the model, gradient-based optimization could be employed successfully to find a minimum in the nearly continuous and smooth cost surface that describes the energy losses. By extending the formulation to specified batch times, the optimizer provided power input trajectories that are dependent on the batch time and the final temperature of the molten metal which can be provided to the meltshop scheduler.

The results of the dynamic optimization provided numerical evidence that supports intuition: to improve the overall energy efficiency of the process, it should be operated at low power levels when the energy efficiency is low, and at high power levels when the efficiency is high, i.e., immediately after charging.

The optimization formulation was utilized to compute an optimal mode of operation for the steelmaking process in an industrial-scale state-of-the-art

EAF. After the implementation of the solution at the industrial plant, 19 test batches were run and energy savings of 4.6% and 1.7%, for two different grades of steel were achieved.

During the last three decades, reductions in the energy demand of the EAF process have been achieved by installing new equipment in the furnace, i.e., the oxy-fuel burners or the implementation of heat recovery solutions from the off-gas. Steelmakers have also computed setpoints using in-house process models and the operative knowledge of their operative and engineering teams. These efforts, however, so far were not successful as the marginal improvements that they achieved were not large enough to justify a change in the operational practices. The marginal improvements obtained in a model-based optimization can be justified by the fact that until now, EAF models were based on many assumptions that resulted in large process model mismatches. Complex EAF models that provide state predictions with smaller process model mismatches (as the one presented here) are more expensive to compute and more difficult to integrate with classic numerical solvers. Employing such a detailed model within an optimization framework has become possible in recent years due to growing computational power and advances in modeling [14].

With the aim of solving the question of how to best operate the EAF as a balancing element of the electrical grid, the other energy inputs to the EAF (the burners and the oxygen lances), and an additional constraint on the energy input will be considered in future work. In such a unified optimization framework, variations in the electrical energy input can be offset by the chemical energy inputs.

References

[1] T. Hay, V.V. Visuri, M. Aula, T. Echterhof, A review of mathematical process models for the electric arc furnace process, Steel Res. Int. (2020) 2000395.

[2] R. Jing, J. Cheng, V. Gan, K. Woon, I. Lo, Comparison of greenhouse gas emission accounting methods for steel production in China, J. Clean. Prod. 83 (2014) 165–172.

[3] A. Toktarova, I. Karlsson, J. Rootzén, L. Göransson, M. Odenberger, F. Johnsson, Pathways for low-carbon transition of the steel industry—a Swedish case study, Energies 13 (2020) 3840.

[4] G. Dalle Ave, I. Harjunkoski, S. Engell, A non-uniform grid approach for scheduling considering electricity load tracking and future load prediction, Comput. Chem. Eng. 129 (2019), 106506.

[5] P. Castro, G. Dalle Ave, S. Engell, I. Grossmann, I. Harjunkoski, Industrial demand side management of a steel plant considering alternative power modes and electrode replacement, Ind. Eng. Chem. Res. 59 (30) (2020) 13642–13656.

[6] D.J. Oosthuize, I.K. Craig, P.C. Pistorius, Economic evaluation and design of an electric arc furnace controller based on economic objectives, Control. Eng. Pract. 12 (2004) 253–265.

[7] M.M. Rashid, P. Mhaskar, C.L. Swartz, Multi-rate modeling and economic model predictive control of theelectric arc furnace, J. Process Control 40 (2016) 50–61.

[8] D. Wei, I.K. Craig, M. Bauer, Multivariate economic performance assessment of an MPC controlled electric arc furnace, ISA Trans. 46 (2007) 429–436.

[9] J.G. Bekker, I.K. Craig, P.C. Pistorious, Modeling and simulation of an electric arc furnace process, ISIJ Int. 39 (1999) 23–32.
[10] R.D. MacRosty, C.L. Swartz, Dynamic modeling of an industrial electric arc furnace, Ind. Eng. Chem. Res. 44 (2005) 8067–8083.
[11] V. Logar, D. Dovzan, I. Skrjanc, Modeling and validation of an electric arc furnace: part 1, heat and mass transfer, ISIJ Int. 3 (2012) 402–412.
[12] J. Hernandez, L. Onofri, S. Engell, Detailed modelling of radiative heat transfer in electric arc furnaces using Monte Carlo techniques, in: 8th International SteelSim Conference, Toronto, Canada, 2019.
[13] T. Hay, J. Hernandez, S. Roberts, T. Echterhof, Calculation of view factors in electric arc furnace process modeling, Steel Res. Int. 92 (2) (2020) 2000341.
[14] J. Hernandez, L. Onofri, S. Engell, Numerical estimation of the geometry and temperature of an alternating current steelmaking electric arc, Steel Res. Int. 92 (3) (2020) 2000386.
[15] J. Lienhard IV, J. Lienhard V, A Heat Transfer Text Book, Phlogiston Press, Cambridge, MA, 2008.
[16] J. Nilsson, S. Riedel, Electric Circuits, nineth ed., Pearson, 2010.
[17] J. Hernandez, L. Onofri, S. Engell, Model of an electric arc furnace oxy-fuel burner for dynamic simulations and optimization purposes, in: The 18th IFAC Symposium on Control, Optimization and Automation in Mining, Mineral and Metal Processing, Stellenboch, South Africa, 2019.
[18] H. Pfeifer, M. Kirschen, Thermodynamic analysis of EAF energy efficiency and comparison with a statistical model of electric energy demand, in: 7th European Electric Steelmaking Conference, Venice, 2002.
[19] A. Movahedi Rad, S.S. Pelaseyed, M. Attarian, R. Shokrallahzadeh, Oxydation behaviour of AISI 321, AISI 316, AISI 409 stainless steels: kinetic, thermodynamic and diffusion studies, J. Mater. Res. 31 (14) (2016) 2088–2096.
[20] A.C. Sabioni, A.M. Huntz, E.C. da Luz, M. Mantel, C. Haut, Comparative study of high temperature oxidation behaviour in AISI 304 and AISI 439 stainless steels, Mater. Res. 6 (2) (2003) 179–185.
[21] M. Alam, J. Nasser, G. Brooks, Computational fluid dynamics simulation of supersonic oxygen jet behavior at steelmaking temperature, Metall. Mater. Trans. B Process Metall. Mater. Process. Sci. 41B (2010) 636–645.
[22] B. Bowman, K. Kruger, Arc Furnace Physics, Stahleisen, Dusseldorf, 2009.
[23] L.T. Biegler, Nonlinear Programing. Concepts, Algorithms and Application to Chemical Processes, Society for Industrial and Applied Mathematics, Pittsburgh, PA, USA, 2010.
[24] V.S. Vassiliadis, R.W.H. Sargent, Solution of a class of multistage dynamic optimizationproblems. 1. Problems without path constraints, 2. Problems with path constraints, Ind. Eng. Chem. Res. (1994) 2111–2122. 2123–2133.

Chapter 5

Solvent recovery by batch distillation—Application of multivariate sensitivity studies to high dimensional multiobjective optimization problems

Jan C. Schöneberger[a], Daniel Staak[b], and Jürgen Rarey[c]
[a]*Chemstations Europe GmbH, Berlin, Germany,* [b]*Lonza AG, Visp, Switzerland,* [c]*Rareytec Co., Ltd, Nakhon Ratchasima, Thailand*

1 Introduction

Solvents are extensively used in industry, mainly as raw material (primarily for paints, adhesives, and plastics), for cleaning and degreasing, and as carrier or dispersion medium [1]. These applications do not allow for recovery. But solvents are also used in extractive or azeotropic distillation, extraction, recrystallization, and as reaction media, where recovery is possible. Latter processes are usually found in the chemical and pharmaceutical industry. Solvent waste results mainly from purge streams from continuous processes or as residue in the case of batch processes. Unrecycled solvent waste, either incinerated or disposed, accounts for 757,000 metric tons of dangerous wastes per year in Germany alone [2]. Recycling can reduce the carbon footprints of solvents by more than one order of magnitude [3]. This ecological advantage seldom translates into an economic gain, as most solvents are commodity chemicals with a low market price. Solvent recovery therefore most often does not justify the construction of specialized equipment, but mostly employs available capacities of multipurpose batch distillation plants. Under these tight boundary conditions, solvent recovery needs to be optimized with respect to operation time and potential extra costs. Well-designed process schedules and batch recipes are therefore mandatory. Lonza is permanently intensifying solvent recovery efforts. Lonza's customers are more

Simulation and Optimization in Process Engineering. https://doi.org/10.1016/B978-0-323-85043-8.00003-9

and more interested in reducing the ecological footprint of their product. Solvent recovery is one of the most effective ways to accomplish this. Lonza found that different from many other fields, this objective is not necessarily in contrast to an economic benefit. With a good process design, the opposite is the case. Lonza's experience shows that solvent recovery in batch processes often can be economically advantageous.

1.1 Separation of acetone and methanol

In this work, we focus on the recovery of acetone and methanol from their mixture, which results, e.g., from the oxygenation of hydrocarbons [1], the catalytic reaction of carbon monoxide with hydrogen [4], dry distillation of wood [5], and separation processes that involve a change of solvent polarity. Acetone and methanol, like most mixtures of components of similar boiling temperatures but different polarities, form a pressure maximum azeotrope [6]. This mixture cannot be separated by ordinary distillation. Most solvent-antisolvent mixtures used in crystallization fall into this category.

1.2 Continuous separation processes

In continuous processes, separation of azeotropic mixtures is often performed by either extractive or azeotropic distillation [7]. In extractive distillation, a mostly heavy boiling solvent is fed to the column a few stages below the top. Its purpose usually is to lower the volatility of the higher boiling component in the mixture, to which it shows a higher affinity. The volatility of the other component preferably is increased. In the case of acetone and methanol, methanol is the higher boiling component. The heavy boiling solvents with a high affinity to methanol would be other hydrogen bonding species, like ethanediol or glycerol. The bottom product of the extraction column is then the mixture of methanol with the heavy entrainer, which needs to be separated in a solvent recovery column.

Azeotropic distillation, on the other hand, employs a volatile solvent. The volatile solvent increases the volatility of the usually lower boiling mixture constituent to a degree, that a new azeotrope is formed. The boiling temperature of the new azeotrope must be below that of the binary azeotrope so that this turns into a middle boiler in the ternary mixture.

The new azeotrope is usually heteroazeotropic to easily separate the entrainer from the top product in a decanter, and recycle it to the column. Azeotropic distillation is most often encountered in the drying of nonpolar solvents, or the removal of nonpolar solvents from water [7]. In the case of methanol, even saturated hydrocarbons do not form a sufficiently wide miscibility gap and only at low temperatures [8]. Therefore, the technique is usually not applied.

1.3 Batch processes for separation

Batch distillation is an efficient unit operation by means of capital investment because it allows the separation of a nonazeotropic multicomponent mixture into its pure components in a single column [9]. Compared to continuous distillation the product-specific energy input commonly is higher and energy integration is more difficult. For azeotropic mixtures, the addition of an entrainer makes either azeotropic or extractive distillation feasible. The advantage of azeotropic distillation is that the entrainer is continuously removed with the distillate. In contrast, an extractive solvent accumulates in the reboiler. A common choice for a polar entrainer is water, which is studied in this work.

2 Problem definition

An amount of three cubic meters of an equimolar mixture of acetone (1545 kg) and methanol (853 kg) is to be recovered. The process of choice is an extractive batch distillation with water as the entrainer. Brüggemann and Marquardt [10] found that water is the best nontoxic entrainer with regard to the energy demand for the separation of such a mixture. Furthermore, using water minimizes the costs for additional raw materials, as it is readily available and disposable at chemical production sites.

A multipurpose batch distillation plant available at the Lonza site in Visp, Switzerland is used for the separation. It consists of a steam-heated pot, a packed column with an intermediate feed stage for the entrainer, water- and a brine-cooled condenser, and several receiver tanks.

2.1 Product specifications and constraints

The purity of the recovered solvents is a subject of this study. First, it must be determined which purity can be reached with the existing equipment, and later the purity is used as an objective in a multiobjective and multivariate sensitivity study. The purity of the commercially available, fresh products is 99.50 wt% for acetone and 99.85 wt% for methanol. Further constraints like maximum heating rates, condenser duties, tank volumes, and column vapor load arise from the limitations of the real plant.

2.2 Description of the plant

A simplified process flow diagram of the plant is depicted in Fig. 5.1, while Fig. 5.2 shows it on a photo. Table 5.1 summarizes the design conditions and the geometrical details of the process vessels.

The separation takes place in column K001 which is filled with structured CERADUR 450Y packing. The column has two sections with a packing height of 4 m each. The entrainer is fed in between these two sections. The feed mixture

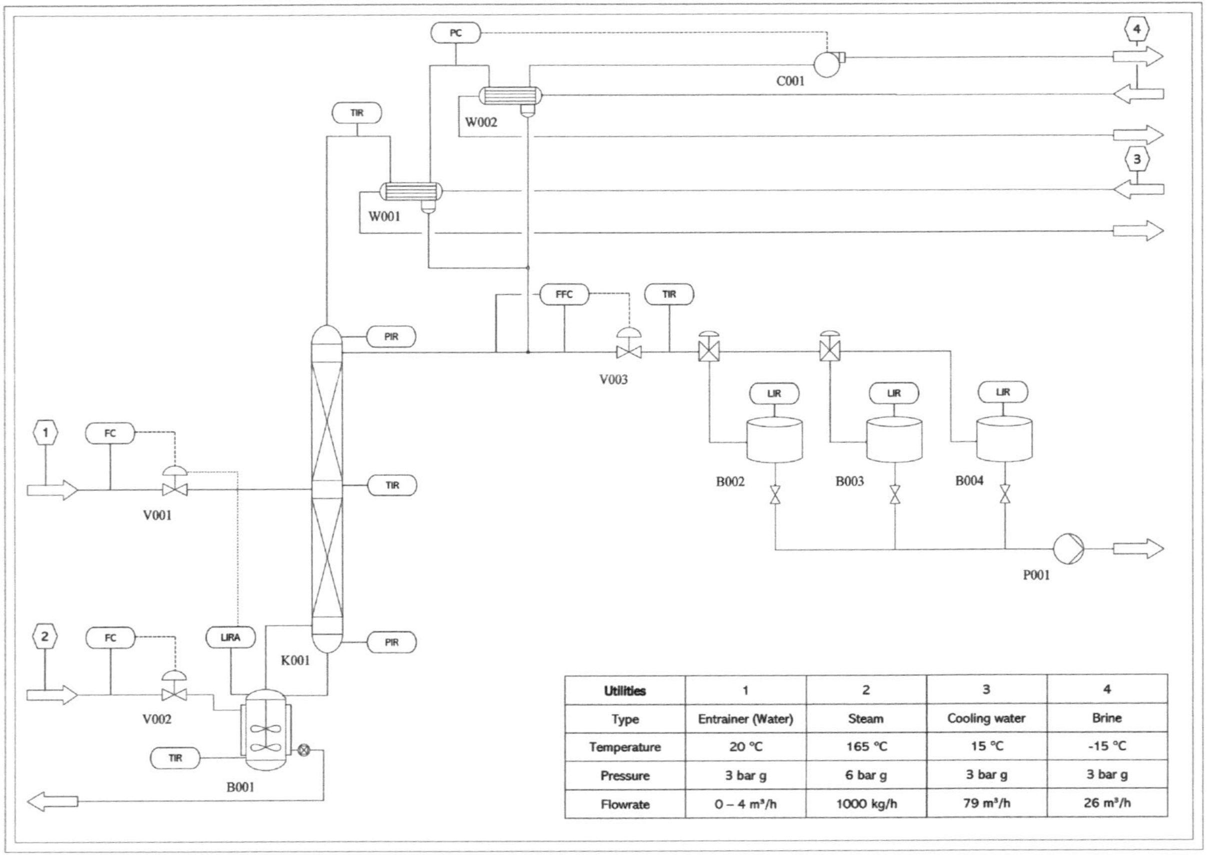

Utilities	1	2	3	4
Type	Entrainer (Water)	Steam	Cooling water	Brine
Temperature	20 °C	165 °C	15 °C	-15 °C
Pressure	3 bar g	6 bar g	3 bar g	3 bar g
Flowrate	0 – 4 m³/h	1000 kg/h	79 m³/h	26 m³/h

FIG. 5.1 Process flow diagram of the multipurpose batch distillation plant.

FIG. 5.2 The selected multipurpose batch distillation plant at Lonza site in Visp, Switzerland. *(Courtesy of Lonza AG.)*

TABLE 5.1 Summary of the process vessels.

Tag	K001	B001	B002	B003	B004
Type	Packed column	Stirred tank	Receiver tank	Receiver tank	Receiver tank
Description	Distillation column with entrainer feed	Tank for batch feed	Tank for product 1	Tank for product 2	Tank for product 3
Design pressure	−1 6 barg	−1 6 barg	−1 6 barg	−1 6 barg	−1 6 barg
Design temperature	−25 200°C	−25 200°C	−25 200°C	−25 200°C	−25 200°C
Volume	5.4 m^3	7.1 m^3	4.0 m^3	4.0 m^3	4.0 m^3
Diameter	DN800	2.0 m	1.6 m	1.6 m	1.6 m

is heated in vessel B001 with a steam jacket. The overhead vapor is condensed in the cooling water-driven heat exchanger W001. Vapor that escapes this heat exchanger is led to the guard heat exchanger W002, operated with—15°C cold brine. The condensate is split into reflux and distillate products. The first goes directly back to the column, while the latter is flow controlled by valve V003 and led to one of the receiver tanks B002 to B004. The specifications of the heat exchangers and their utilities are given in Table 5.2.

3 Literature review

A series of four articles about the feasibility of homogeneous extractive batch distillations has been published by Rodriguez-Donis, Gerbaud, and Joulia [11–14]. Their analysis is based on the topology of residue curves, for which Serafimov defined 26 classes which were nicely summarized and explained by Kiva et al. [15]. The ternary system acetone-methanol-water falls into the class 1.0–1.a. This is the most frequently occurring class in nature, as reported by Hilmen et al. [16]. The ternary diagram of the considered mixture, with two residue curves characteristic for the class 1.0–1.a, is shown in Fig. 5.3. The thermodynamic model used to calculate the curves and the azeotrope is introduced in Section 4.1.

A detailed feasibility analysis of exactly this system was performed by Lelkes et al. [17], pointing out that the mixture can only be separated with continuous feeding of the entrainer water. The authors derive that there is a minimum amount of entrainer feed required, and the number of trays in the rectifying section (trays above the entrainer feed tray) is min and max bound if a certain acetone purity must be achieved. Besides the information that the separation is generally feasible, the results are of no avail for the design of a

TABLE 5.2 Summary of the heat exchangers and their utilities.

Tag	B001	W001	W002
Type	Steam jacket	Shell and tube	Shell and tube
Heat transfer area	8–14 m^2	19.6 m^2	11.8 m^2
Estimated heat transfer coefficient	500 W/m^2K	900 W/m^2K	600 W/m^2K
Utility	Steam	Water	Brine
Inlet pressure	6 barg	3 barg	3 barg
Inlet temperature	165°C (saturated)	15°C	−15°C

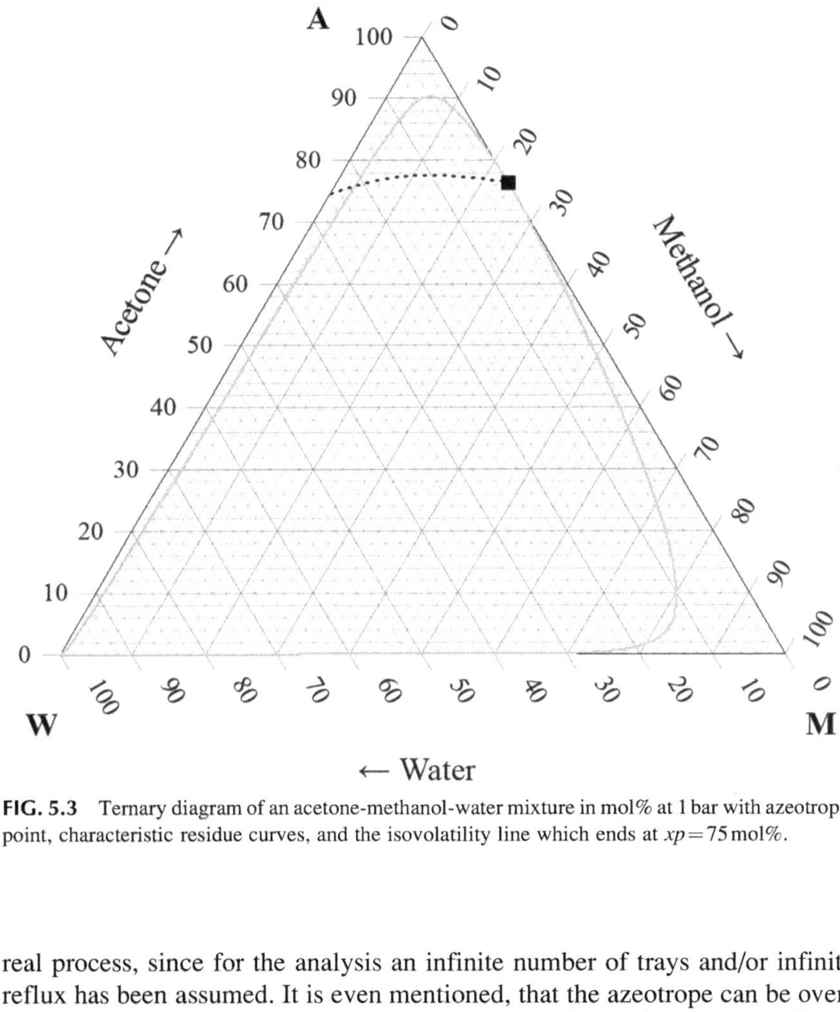

FIG. 5.3 Ternary diagram of an acetone-methanol-water mixture in mol% at 1 bar with azeotropic point, characteristic residue curves, and the isovolatility line which ends at $xp = 75$ mol%.

real process, since for the analysis an infinite number of trays and/or infinite reflux has been assumed. It is even mentioned, that the azeotrope can be overcome with an entrainer feed rate below the designated minimum. Using Eq. (5.1), this minimum value can be calculated relative to the vapor flow rate V coming from the pot. In case of an acetone-free entrainer ($z = 0$), it is a function of the acetone mol fraction xp at the point where the isovolatility line meets the binary axis (see Fig. 5.3), and the vapor mole fraction of acetone $y(xp)$ at this point.

$$\left(\frac{FE_{min}}{V}\right) = \frac{y(xp) - xp}{xp - z} \tag{5.1}$$

Using the information from Fig. 5.3 gives a minimum ratio of 0.18. However, the actually required entrainer feed rate in the real plant is around one order of magnitude higher.

Other authors analyzed the continuous extractive distillation process (using two distillation columns) with a commercial flowsheet simulator [4,18,19]. They all end up with very large columns, or a high number of theoretical plates, respectively because they set the minimum required purity for acetone to 99.5 mol%. This is above the commercially required purity of 99.5 wt%. The relative volatility between acetone and water is close to unity at such high purities. To go from 99.5 wt% (98.4 mol%) to 99.5 mol% more than 10 theoretical stages are required when using an infinite reflux ratio, as shown in Fig. 5.4.

It is impossible to achieve this purity with the given equipment, and therefore we treat the purity as an additional degree of freedom in this work. Due to the low relative volatility at high acetone concentrations and ambient pressure, the commercial-grade acetone is produced by vacuum distillation. We also consider this strategy, while considering the limitations of the given equipment.

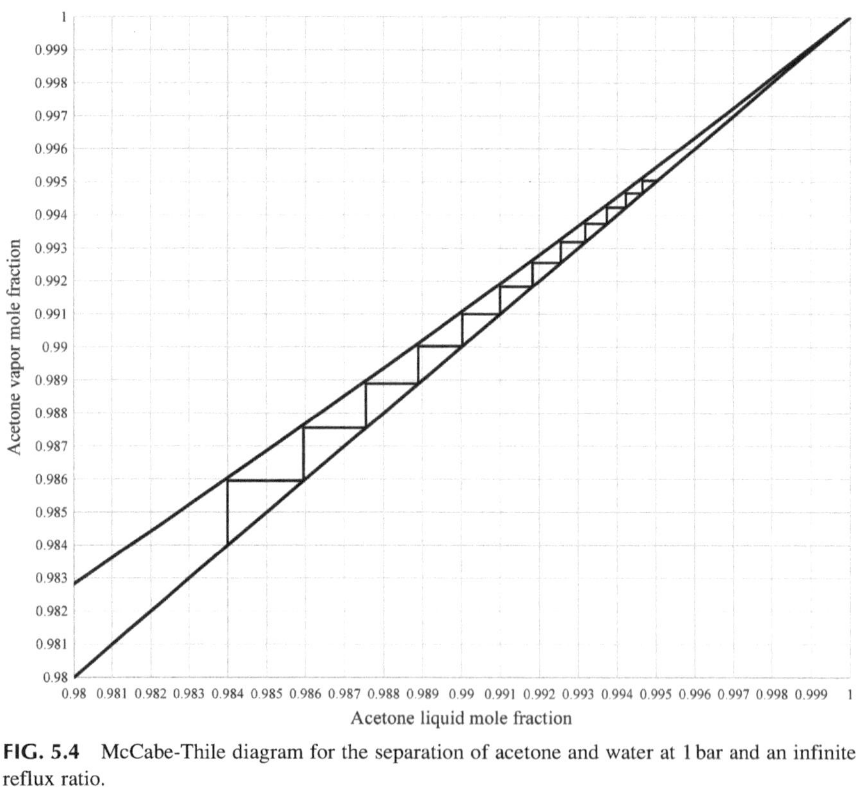

FIG. 5.4 McCabe-Thile diagram for the separation of acetone and water at 1 bar and an infinite reflux ratio.

4 Methodology

For solving the given separation problem, we must select a plant and find a feasible, ideally optimal operation strategy. This is done in four steps using heuristics and three different simulation models.

The first decision to be made is in which plant the solvent recovery is performed. Batch distillation is commonly performed in multipurpose plants. If there is enough feed to run a batch distillation full time, switching to a continuous process often is economically more attractive. Once the plant is selected, the required parameters for a flowsheet simulation are compiled. To avoid too many repetitions, the plant selection process is supported by heuristic rules.

The fully specified flowsheet simulation is then used in extensive sensitivity and optimization studies. The results are summarized in interactive graphical reports, which allow a flexible selection of constraints and a reasonable decision on the operation parameters to be applied. A validation of these operation parameters is done by running a high-fidelity process simulation of the real plant's digital twin.

4.1 Heuristics for the selection of a suitable multipurpose plant

The following points summarize best-practice heuristics to select a suitable plant for the given solvent recovery task.

(a) The batch runtime is the parameter to be minimized. Therefore, any plant should be operated at maximum heat duty or vapor load, respectively.
(b) The column must have sufficient trays or packing, to make the required separation feasible.
(c) Vacuum or higher pressure can be used to improve the separation strength.
(d) Vacuum increases the volumetric vapor load. Therefore, it can be contradictory to rule (a) and should be avoided if possible.
(e) The achievable reboiler heat duty is estimated based on the existing heat exchanger area, the available utilities (commonly steam at certain pressure levels), and the mixture boiling temperature.
(f) Vapor load and F-factor are estimated based on the achievable heat duty.
(g) The end temperature in the pot must be below the available hot utility temperature, assuring a sufficiently large temperature difference for heat transfer in the reboiler. A reduction of the available heat transfer area, due to a decreasing liquid level, must be considered.
(h) The condenser temperature should be above 30°C to allow an operation with cooling water.
(i) If (g) or (h) cannot be fulfilled, the pressure must be adjusted, considering the plants' design pressure limitations and the problems stated in point (d).

4.2 Tool for running flowsheet simulations

Flowsheet simulators are tools that solve mass and energy balances of coupled unit operations, constrained by thermodynamic equilibrium conditions. The core of each flowsheet simulator is the so-called flash algorithm, a computational strategy to solve the equilibrium separation of a feed stream into two or more phases as separate streams under given boundary conditions (isobaric, adiabatic, etc.). A specific type of flash calculation is used to find the bubble point of a liquid stream and the ratio of equilibrium vapor and liquid compositions (K-factor) for each component as required in the equilibrium condition ("E") in the MESH equations (material balances, equilibrium equations, sum equations, and heat balances). Different strategies have been developed for a safe and rapid solution of the MESH equations, e.g., various inside-out [20] and simultaneous solution methods [21].

The MESH equations are commonly used for steady-state models, but they can be expanded to dynamic models by adding hold-up terms for the material and the energy balances. By this modification, the equation system is changed from an algebraic equation system (AE) to a differential-algebraic equation system (DAE). A distillation column can be seen as a series of coupled flash units with special conditions for the reboiler or pot, the condenser, and the feed trays. The resulting DAE system cannot be solved with standard solvers, because it is very stiff [22] and it has a differential index of greater than 1 [23]. To overcome these numerical problems, Galindez and Fredenslund [24] developed a quasisteady-state method. This method was implemented in the flowsheet simulator CHEMCAD III [22]. Since then, this approach was further improved, and adopted for further unit operations beyond distillation columns, allowing a full dynamic flowsheet simulation. In this study, version 1.0.1 of the flowsheet simulator CHEMCAD NXT [25], a successor of CHEMCAD III, is used to evaluate three different models of the multipurpose plant. The first model is a screening model (SCM) with the batch distillation unit operation, which is set up and calculated very quickly. The low-fidelity model (LFM) and the high-fidelity model (HFM) also are simulating further unit operations like valves, tanks, and heat exchangers in the dynamic mode.

4.3 Algorithms for optimizing flowsheet simulations

Most flowsheet simulators have implemented optimizers. These algorithms vary design variables inside the flowsheet simulation to maximize or minimize a specific objective function, built from flowsheet variables while fulfilling certain constraints. These constraints are also based on flowsheet variables. Common algorithms are the successive quadratic programming [26], the generalized reduced gradient [27], the genetic algorithm [28], as well as the simplex downhill search [29]. The first two algorithms require derivative information, which at least in CHEMCAD can only be obtained by finite difference approaches.

This is very inaccurate, especially in combination with the quasisteady-state approach (see Section 4.2), because the step size for the difference is competing with the integration step size. In this work, the complex algorithm implemented in the flowsheet simulator is used, because it does not require gradient information. However, it still is much faster than a genetic algorithm. Box [30] developed the complex method, a modified Nelder-Mead Simplex algorithm, which can handle most kinds of constraints [31]. A fully parallelized, multiple starting point strategy is used to overcome local minima of the objective function.

Multiobjective optimizations are performed by a proprietary method implemented in the flowsheet simulator of choice. Rangaiah and Bonilla-Petriciolet [32] give a good general overview of the different methods used for chemical engineering problems.

4.4 Tool for running multivariate sensitivity studies

Up to two flowsheet variables can be varied in a sensitivity study inside the flowsheet simulator. For the variation of more variables, an external parallelization strategy is used. Simulation files with particular settings are distributed on so-called agents sitting on individual machines of a computer cluster, using the software architecture presented by Fricke and Schöneberger [33].

The results of these multivariate sensitivity studies (MSS) are stored in a database and can be exported to data tables of input and output variables. For the analysis and the visualization of the data, we use several packages implemented in the R language [34–38].

5 Set up of the flowsheet simulation

The first and most important steps in setting up a flowsheet simulation are the selection, the parametrization, and the validation of the thermodynamic models. Three flowsheet models are used in this work: a screening model, a low-fidelity model, and a high-fidelity model.

5.1 Thermodynamic models

A correct representation of the pure component and mixture behavior of the ternary mixture is critical for the numerical simulation of the process. Pure component properties like liquid vapor pressure, heat of vaporization, liquid, and vapor heat capacity, are most often satisfactorily addressed by the parameters distributed with the process simulation software [39]. The most important mixture property in the case of distillation is the relative volatility between the components, as a function of temperature or pressure and mixture composition. Excess Gibbs energy (g^E) models like Wilson, NRTL, and UNIQUAC are routinely used for mixtures of nonelectrolyte components differing in polarity. These models employ the local composition concept to predict the multicomponent behavior from binary information alone. To be able to regress data over a larger temperature range, temperature-dependent binary interaction parameters (BIPs) are used to override the implicit temperature dependence of the model equations. In this work, we use the NRTL model.

Reliable BIPs for acetone-water, and acetone-methanol were taken from Gmehling et al. [6]. In the case of methanol-water, literature data for VLE, h^E, c_P^E, SLE, and γ^∞ stored in the Dortmund Data Bank [40] were regressed simultaneously using the software RecPar [8]. All parameter values are compiled in Table 5.3. A temperature-dependent BIP expression as specified in Eq. (5.2) is used.

$$\Delta g_{ij}(T) = a_{ij} + b_{ij}T + c_{ij}T^2 \tag{5.2}$$

NRTL results and experimental data from the DDB are presented in Fig. 5.5.

5.2 Screening model

The purpose of the screening model (SCM) is a fast analysis of the impact of different design variables. The model consists of the predefined batch distillation unit operation and three receiver tanks and was implemented in CHEMCAD 7. The assumptions applied are as follows:

1. Equilibrium trays
2. Ideal control
3. No liquid hold-up on trays, in the condensers, and in pipes
4. No pressure drop
5. Constant heat duty
6. Total condenser without subcooling

5.2.1 *Number of equilibrium trays*

The column has a diameter of 800 mm and consists of two sections packed with 4 m of CERADUR 450Y each. The vendor of the packing assumes an equivalent of four theoretical plates per meter of packing [41]. This value is commonly lower when using organic substances and therefore we take the more realistic value of three. This gives 12 equilibrium trays above and below the entrainer feed. The condenser and the pot are counted as extra trays.

5.2.2 *Heat duty and molar vapor flow*

The heat duty that can be added to the pot is limited for two reasons, the heat transfer, and the column hydraulics. The transferred heat duty Q is determined by Eq. (5.3).

$$Q = U \cdot A \cdot \mathrm{d}T \tag{5.3}$$

The pot is equipped with a steam jacket, which means that the heat exchanger area A is changing with the liquid level in the pot. The heat transfer

TABLE 5.3 Binary interaction parameters for the binary mixtures of acetone-methanol-water.

a_{12}(cal/mol)	a_{21}(cal/mol)	b_{12}(cal/(mol K))	b_{21}(cal/(mol K))	c_{12}(cal/(mol K^2))	c_{21}(cal/(mol K^2))	α_{12}
Acetone (1)-water (2)						
833.97	−3146.5	0.83106	17.457	−0.0037816	−0.013622	0.5466
Acetone (1)-methanol (2)						
477.04	327.46	0.03815	−0.73193	−0.002931	0.0018051	0.7
Methanol (1)-water (2)						
687.62	−3281.75	−5.90134	21.28030	0.01204927	−0.02986956	0.6

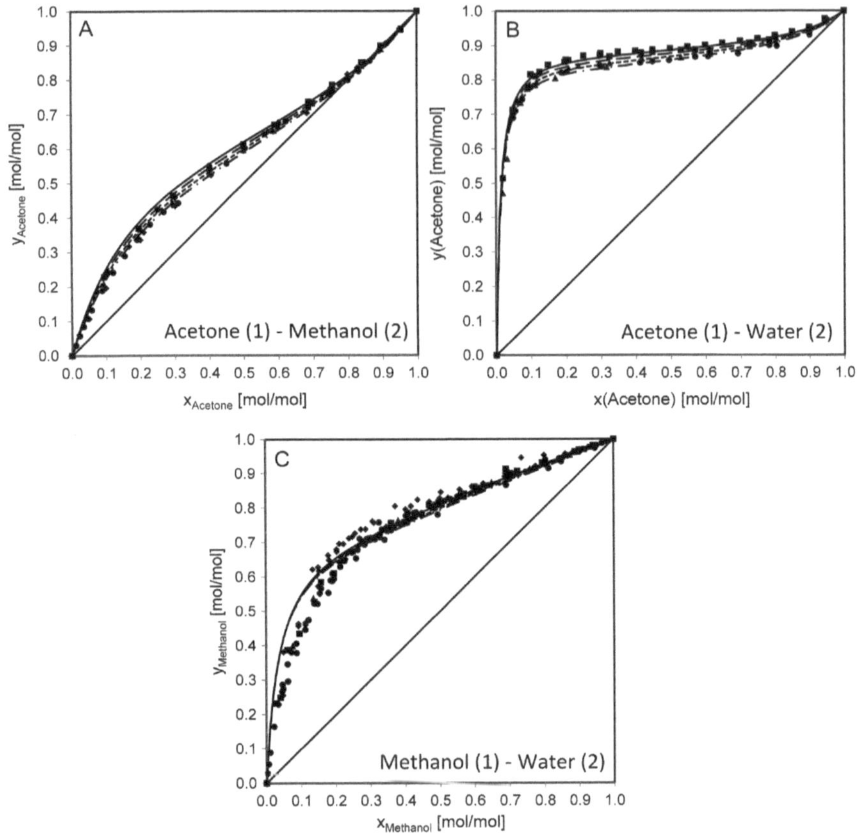

FIG. 5.5 Isothermal x-y diagrams calculated from NRTL and the binary interaction parameters given in Table 5.3 in the temperature range 308–323 K together with data taken from the Dortmund Data Bank [40].

coefficient U is assumed to be constant. Based on the equipment specifications in Table 5.2, the factor $U\,A$ ranges from 4000 to 7000 W/K. The steam temperature is constant, but the liquid temperature in the pot varies from 39°C (initial binary mixture at 0.5 bar abs) to 100°C (pure water at 1 bar abs.). Assuming a constant mean temperature difference dT of 100 K and a constant mean UA of 5000 W/K leads to a constant heat duty of 500 kW.

The heat of vaporization in the relevant pressure range is approximated to 31 MJ/kmol for acetone and 37 MJ/kmol for methanol. This gives a vapor production rate of around 53 kmol/h at the beginning of the distillation, which is reducing over time as acetone is removed. Assuming ideal gas behavior and an average vapor flow of 50 kmol/h, this leads to a volumetric flowrate of 0.73 m^3/s at 0.5 bar abs. and 40°C. The density of the vapor above the initial mixture is 0.87 kg/m^3. With that and the column diameter, the gas load factor (F-factor) can be estimated by Eq. (5.4) to $F = 1.34\sqrt{Pa}$.

$$F = w\sqrt{\rho_V} \tag{5.4}$$

This value decreases at higher pressure. However, it is well in the range where the HETP value of common packings is almost constant with regard to the F-factor [42].

5.2.3 Design variables

The SCM is defined by three operation steps. The column geometry and the heat duty added to the pot are fixed, as described above. The SCM allows the specification of the condenser pressure, stepwise constant values for the entrainer feed rate, and the reflux ratio. The first step is terminated when the pot liquid volume reaches a value of seven cubic meters. Step two ends when the methanol concentration in the distillate exceeds a specific limit. Step three ends after a specified time. This leaves the SCM with seven inputs: the condenser pressure, the entrainer feed rate, three stepwise constant reflux ratios, and two end conditions, namely the methanol quality at the end of step two and time elapsed in step three.

5.3 Low-fidelity model

The screening model is easy to set up and can be evaluated very fast. However, it is quite inflexible. It does not allow the specification of a variable entrainer feed rate or the usage of more sophisticated switch conditions. Therefore, the low-fidelity model is created in a full dynamic mode. The dynamic mode allows the usage of controller unit operations to fix simple design specifications, ramp unit operations to specify trajectories of flowsheet variables, and VBA unit operations to include complex recipes. The model assumptions are the same as for the screening model. However due to the higher flexibility, more degrees of freedom occur, and a larger variety of flowsheet variables can be used to fix them.

5.4 High-fidelity model

The high-fidelity model is developed to test the results retrieved with the low-fidelity model, for their application to the real plant. Therefore, it can be considered as an approach to build the plant's digital twin by means of its dynamic behavior. Only assumption 1 (equilibrium trays) from Section 5.2 remains, assumption 2 reduces to an ideal flow and a pressure control system.

5.4.1 Heat exchanger models

The flowsheet simulator calculates the heat duty added to the pot, by using rigorous heat transfer methods including process side heat transfer, heat

TABLE 5.4 Geometric details of the heat exchangers.

Unbaffled jacket	B001	Shell and tube	W001	W002
Jacket total volume	0.725 m^3	TEMA class	R	R
Jacket height	1.85 m	TEMA designation	AEL	AEL
Wall thickness	20 mm	Shell inner diameter	660 mm	550 mm
Wall density	8000 kg/m^3	Number of tubes	195	180
Wall heat capacity	0.4 kJ/kg/K	Tube length	2.0 m	1.3 m
Wall therm. cond.	11 W/m/K	Tube inner diameter	14 mm	14 mm
Reactor diameter	2.15 m	Tube outer diameter	16 mm	16 mm
Impeller diameter	1.90 m	Effective area	19.04 m^2	11.33 m^2
Impeller speed	1 Hz	Tube pattern	TRI60	TRI60
Heat transfer method		Tube pitch	20 mm	20 mm
- Process side	Sider-Tate [45]	Number of tube passes	2	4
- Jacket side	Lehrer [46]	Number of baffles	4	2
		Baffle cut	39%	39%

conduction, and jacket side heat transfer. The geometric details of the pot and the unbaffled jacket are summarized in Table 5.4.

The vapor coming from the top of the column is fed to a heat exchanger, which is operated with a constant cooling water flowrate of 79 m^3/h. Vapor that escapes this heat exchanger is led to a guard heat exchanger, which is operated with cold brine at −15°C with a flowrate of 26 m^3/h. Both heat exchangers are modeled with the rigorous two-phase heat transfer methods implemented in the heat exchanger simulation package of the flowsheet simulator [43,44]. Table 5.4 shows a collection of geometric details relevant for the simulation of both heat exchangers.

5.4.2 Hydraulic column model

An updated version of the Billet and Schultes [47] hydraulic column model is used to calculate the pressure drop along the packing. Raschig Pak 350Y is used instead of the real packing, as the parameters for the CERADUR 450Y packing are unavailable. The results show a quite constant pressure drop of around 5 mbar during the whole run, and for various scenarios. Therefore, we use a fixed pot pressure of 5 mbar above the condenser pressure. The effect of pressure drop on the process is quite small and we do not expect large changes of this value due to the different packing types that we used in the model.

5.4.3 Liquid hold-up model

The liquid hold-up inside the column is estimated with the approach published by Engel [48]. Using time-averaged mean values for the properties, a hold-up of 5.3 vol% is calculated. This gives 9 L per equilibrium tray.

The liquid hold-up in the piping system, connecting the condensers and the column, is set to 10 L. This is equivalent to 5 m of DN50 piping which are filled with liquid. These geometric parameters are kept constant during the simulation.

5.4.4 Flow control model

To achieve an ideal flow control, the controlled mass rate must be in the region of the control valve's rangeability. This value is commonly around 10, which means that the minimum controllable flowrate is at 10% of the maximum controllable flowrate. This limitation applies to the control valve for the distillate flow (unit V003 in Fig. 5.1). The valve is used to adjust the reflux ratio. The maximum and minimum controllable mass flow rates are 800 kg/h and 80 kg/h. Higher and lower values, which might occur from a specification of the reflux ratio, are cut. To realize an infinite reflux ratio, valve V003 is closed completely.

6 Results

We present the results in the order of increasing model complexity. The findings from each step are used to derive the settings for the next step.

6.1 Screening model

We use the screening model for multiobjective optimizations, and for a multivariate sensitivity study. Based on the results we fix the condenser pressure and determine reasonable variable ranges, e.g., for the acetone and the methanol product qualities.

6.1.1 Highest possible acetone concentration in distillate

Acetone is the main product to be recovered. Therefore, the first analysis with the screening model focuses on the purity of acetone in the distillate that can be reached with the given equipment. The model is used to find the pareto front of the concurring objectives: maximal condenser temperature versus maximal acetone concentration. Only the first step of the integration over time is performed and the distillate concentration at this point is taken. For nonoptimal combinations of entrainer flowrate and reflux ratio, this concentration is not necessarily the highest concentration that can be reached during a longer-lasting batch distillation. However, for this first analysis, we use it as an indicator for pareto optimality. The design variables, in this case, are the condenser pressure and the reflux ratio. Fig. 5.6 shows the pareto-optimal points for fixed entrainer flowrates found by the complex optimizer.

With a column operated at ambient pressure (solid line), the industrial-grade concentration of 99.5 wt% can hardly be reached. The positive effect of a higher entrainer flowrate on the acetone concentration reduces considerably and vanishes almost completely at a ratio of the entrainer feed rate to the vapor flow $FE/V = 3$.

For further analysis, a condenser pressure of 0.5 bar (dashed line) is selected. This leads to a distillate temperature of 37°C, which is more than 20 K above the available cooling water temperature and in accordance with the heuristic rules given in Section 4.1. With these settings, a maximum acetone peak concentration of 99.9 wt% can be achieved under the ideal conditions of the screening model.

The acetone product concentration is also in competition with the amount of product that can be harvested. The resulting pareto front changes from a convex shape in the region of thermodynamic limitations, to a concave shape in the range of mass balance limitations, see Fig. 5.7A. For this study, the first operation step was simulated, and the decision variables were the constant entrainer feed rate (FE) and the constant reflux ratio (RD1).

Under these conditions, the pareto front detaches from the mass balance limitation above 99.6 wt% of acetone in the distillate (Fig. 5.7B). Higher concentrations can only be reached with a considerable loss of acetone, or a reduced product amount, respectively.

6.1.2 *Impact of the design variables*

With a fixed condenser pressure, six design variables remain. The entrainer feed rate (FE) in step one, the reflux ratios in steps one to three (RD1–RD3), the methanol quality at the end of step two (zM), and the time elapsed in step three (t3) are varied in a multivariate sensitivity study (MSS). A factorial design with five intermediate points (three in case of FE) is applied, which results in 84,035 flowsheets.

In Fig. 5.8, the correlation matrix is plotted. Two findings are highlighted, marked with solid and dashed boxes. An increased entrainer feed rate (FE) has a positive impact on the acetone quality in tank B002, but a negative impact on

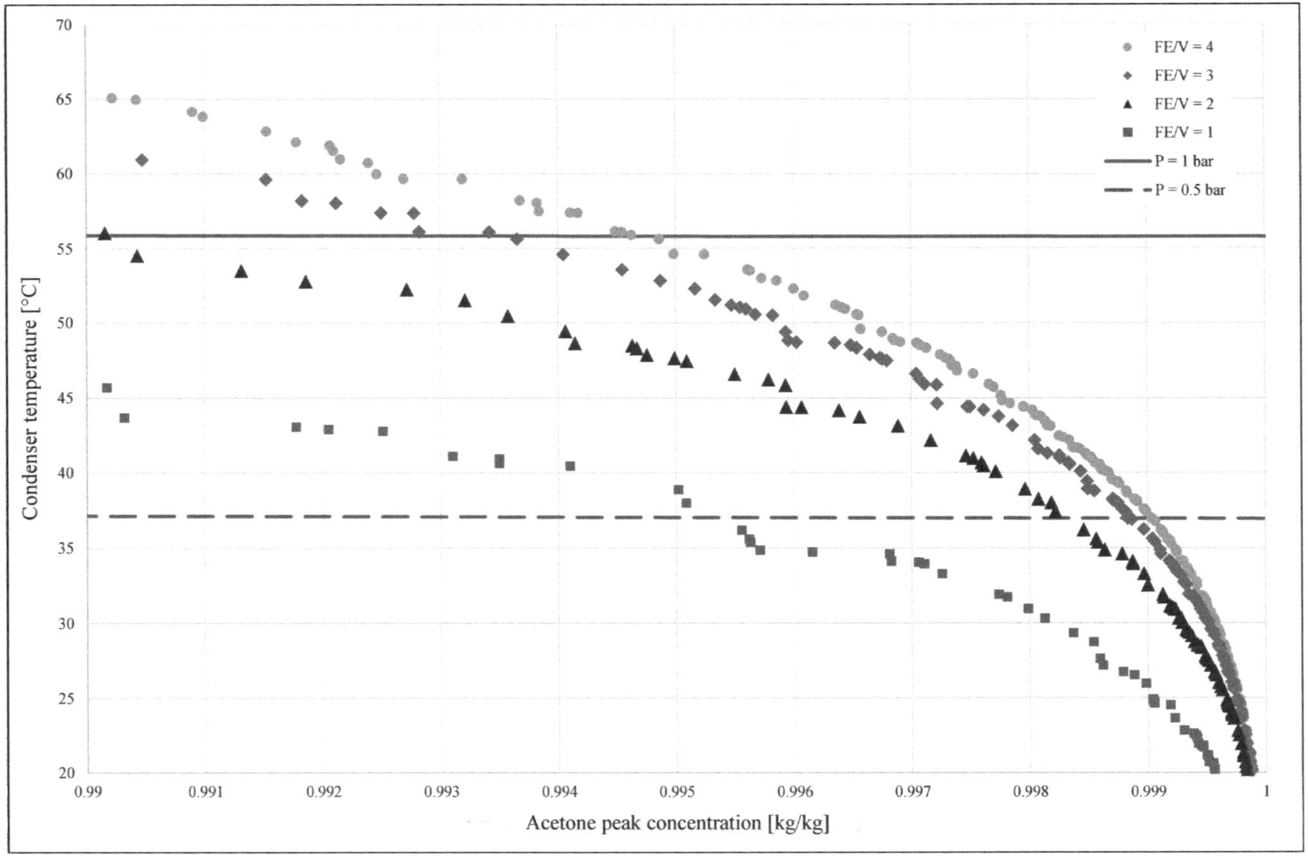

FIG. 5.6 Pareto-optimal points for the acetone peak concentrations versus the condenser temperature for different entrainer feed rates, assuming a vapor flow of $V = 50\,\text{kmol/h}$.

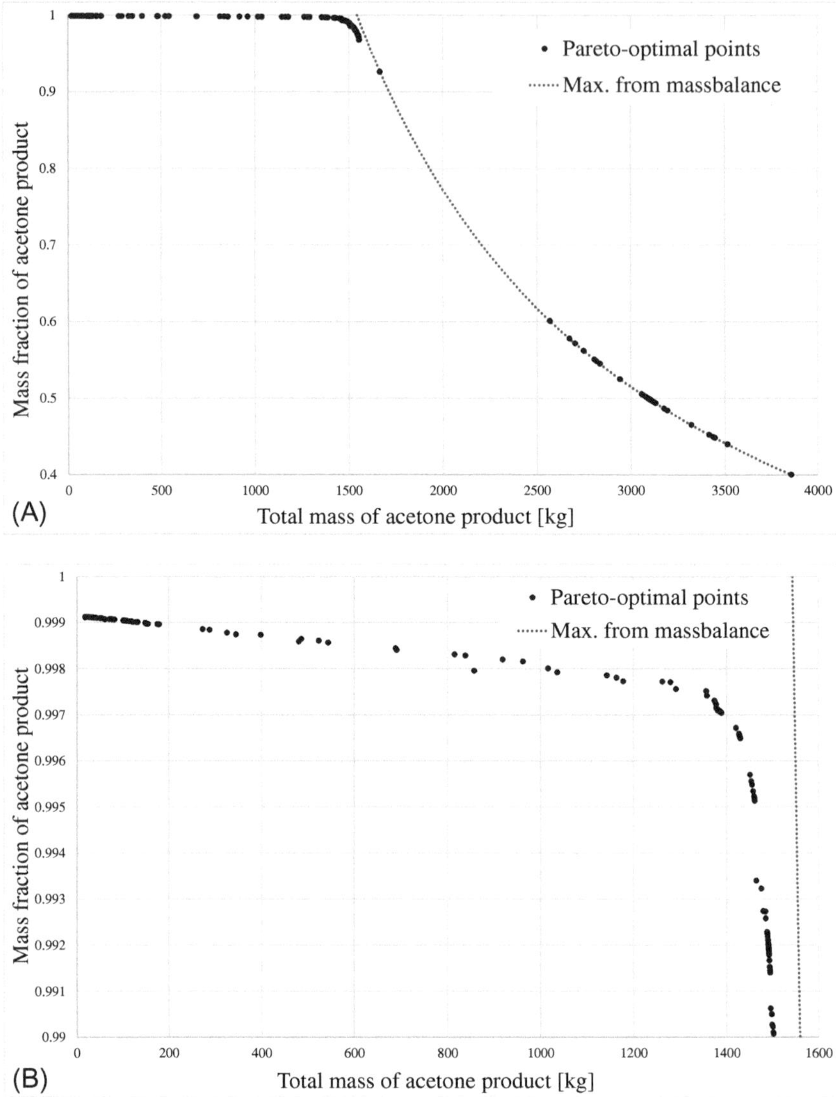

FIG. 5.7 Pareto-optimal points for the acetone concentration versus the acetone product mass. Full view (A) and zoom (B).

the mass collected in this tank. The same is valid for the reflux ratio during step one (RD1). This leads to the competing objectives in step one, which we discussed earlier. The dashed boxes show that the acetone quality in tank B002 has a positive impact on the mass collected in tank B003, and the methanol quality in this tank. This suggests that the recovery of high-quality acetone is not in competition with the methanol product.

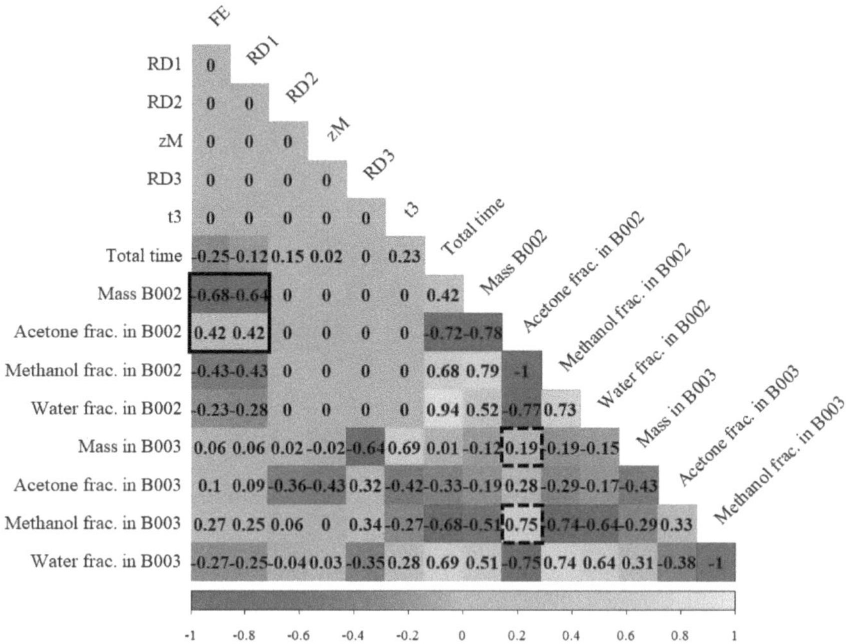

FIG. 5.8 Correlation matrix of input and output variables based of a factorial design with 84,035 points.

Another result of the MSS is that a reasonable range of the design variables was selected. Fig. 5.9 shows the result space for the acetone and the methanol product covered by the input variables.

A video of the interactive report is available at Video 5.1 in the online version at https://doi.org/10.1016/B978-0-323-85043-8.00003-9 where individual cuts of the result space are created by moving sliders and selecting points.

6.2 Low-fidelity model

With the low-fidelity model, we develop a recipe for the operation of the multipurpose batch plant. We analyze and optimize each of the four operation steps individually. Finally, the full recipe is executed, and the simulation results are discussed.

6.2.1 Acetone recovery

The first task is the recovery of acetone. Above the minimum feasible entrainer flowrate, there are up to two solutions for the reflux ratio, when the distillate concentration of acetone is fixed. This is depicted in Fig. 5.10, where the acetone concentration is plotted over the reflux ratio for different entrainer feed

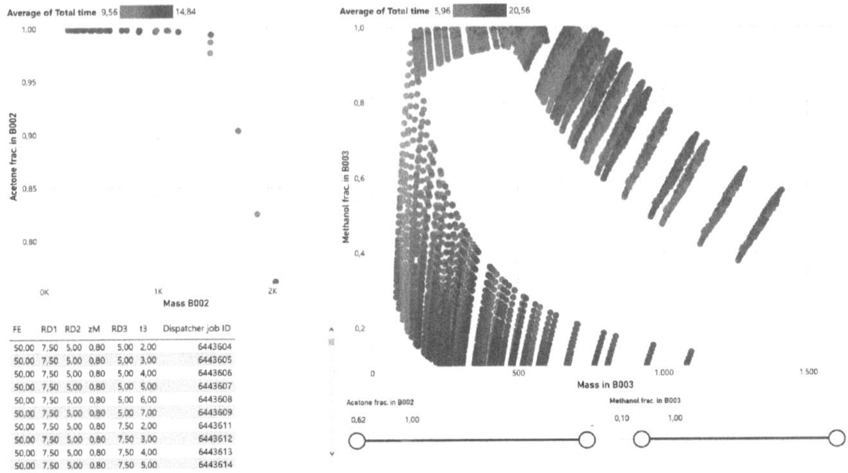

FIG. 5.9 Interactive visualization of the MSS results, currently showing operation time, acetone, and methanol products.

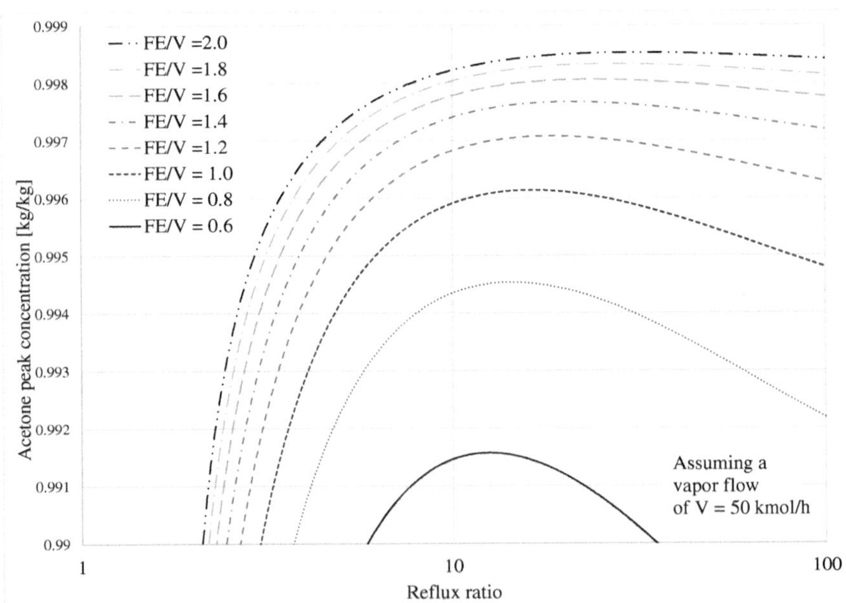

FIG. 5.10 Acetone peak concentration as a function of the reflux ratio and the entrainer feed rate FE.

rates. For this graph, only one time step was calculated, and the condenser pressure was set to 0.5 bar abs.

With a fixed distillate concentration and a fixed entrainer feed rate, no solution, one solution, or two solutions exist for the reflux ratio. In the latter case, one of the solutions is at high reflux ratios and therefore is not economically feasible. However, the existence of either none or two solutions makes the convergence of the equation system, with a fixed distillate concentration, extremely difficult or even impossible. With initial values that are on the right side of the maximum, the solver converges to very high reflux ratios. With initial values too far on the left side of the maximum, there is no solution at all. To make things worse, the maxima of the curves in Fig. 5.10 change with time, as the acetone concentration in the pot is decreased and the water concentration is increased.

The two design variables in this operation step are the entrainer feed rate and the reflux ratio. An increased entrainer feed rate widens the operation window for reflux ratios on the left side of the maximum, which means that lower reflux ratios can be used. On the other hand, a high entrainer feed rate means that the pot is filled up faster with water. When the maximum level in the pot is reached, the first operation step ends because the water feed must be switched off to avoid an overflow. From that point on no more high purity acetone can be recovered.

An attractive operation point for a maximum acetone recovery is at the point where a maximum of the curves in Fig. 5.10 hits the desired acetone concentration. At this point, the minimum reflux ratio meets the minimum entrainer feed rate. But due to the numerical problems stated above, this point cannot be found for every time step.

In the next step, we use the low-fidelity model to analyze individual steady states with different pot compositions. Fig. 5.11 shows the required reflux ratio regarding the entrainer feed rate, for a fixed acetone concentration of 99.5 wt%. The curve parameter is the elapsed operation time, using a fixed entrainer feed rate of 50 kmol/h. After 170 min, the minimum feasible entrainer feed rate approaches this value, which leads to rapidly increasing reflux ratios.

By applying a linear ramping of the entrainer feed rate, more than 99% of the acetone can be recovered from the solvent mixture at a concentration of 99.5 wt %. Fig. 5.12 shows the trajectories of the remaining acetone mass together with the resulting reflux ratio, the entrainer feed rate, and the pot liquid volume.

The plot suggests that a higher than proportional increase of the entrainer feed rate after around 150 min is required to maintain a reasonable reflux ratio or distillate rate, respectively. At this point in time, more than 90% of the acetone has been recovered.

Fig. 5.13 shows an alternative solution calculated with the screening model, using a stepwise increase of the entrainer feed rate. The rate is increased automatically, whenever the reflux ratio reaches a certain bound. We set these bounds to 15, 25, 60, and 100 while we increase the entrainer feed rate ratio two times by 0.5 and then by 1.

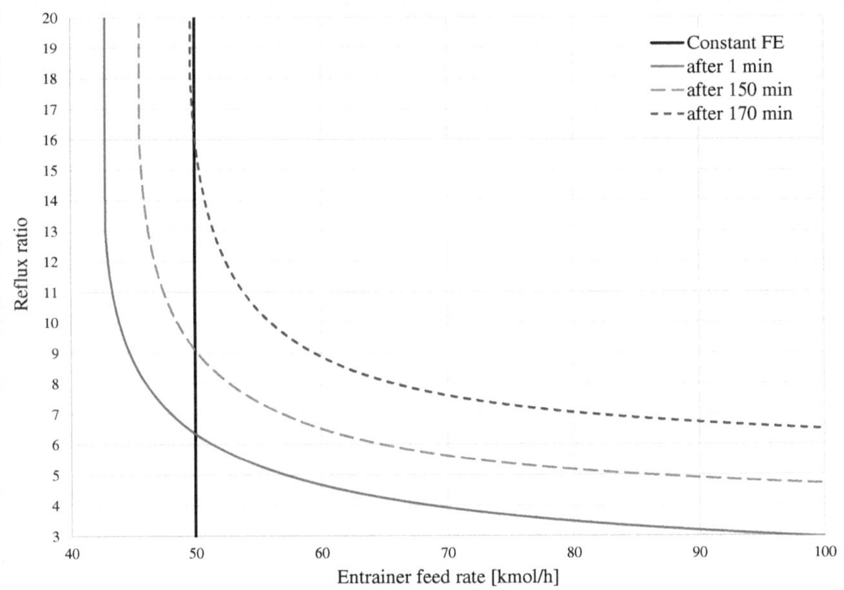

FIG. 5.11 Required reflux ratio RD to get an acetone concentration of 99.5 wt% as a function of the entrainer feed rate FE and the elapsed time of operation step 1 using a constant feed rate of 50 kmol/h.

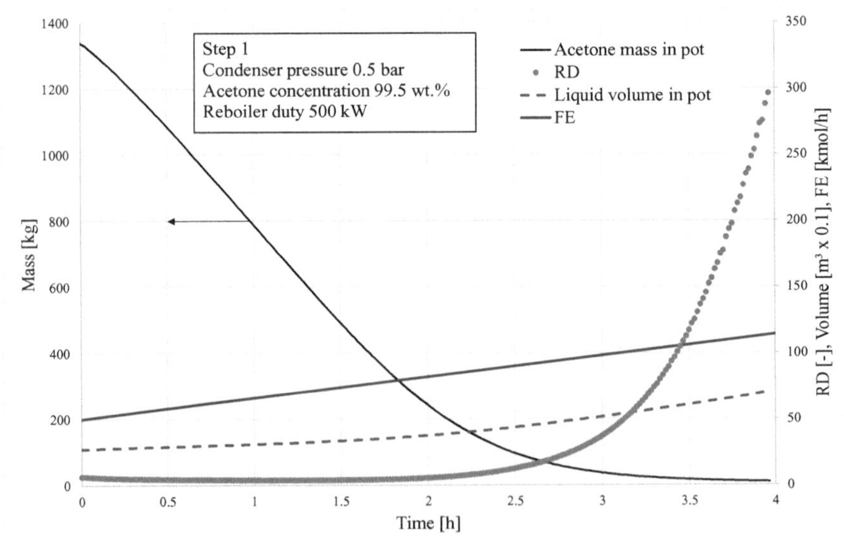

FIG. 5.12 Trajectories of the acetone mass in the pot, the reflux ratio (RD), the entrainer feed rate (FE), and the liquid volume in the pot during step one using a linear ramping of the entrainer feed rate.

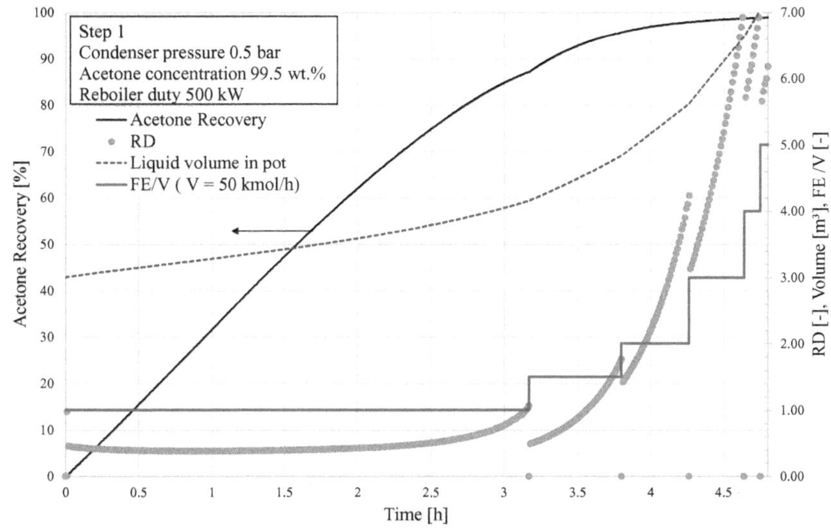

FIG. 5.13 Trajectories of the acetone recovery, the reflux ratio RD, the entrainer feed rate FE, and the liquid volume in the pot during step one using a stepwise increasement of FE.

None of the strategies was optimized, but with the applied settings, the latter strategy requires higher reflux ratios and a longer operation time to reach the same grade of acetone recovery. Therefore, we prefer a ramped entrainer feed rate, which also reduces the number of design variables.

Both cases show that costs, by means of time and energy, increase disproportionally high for a recovery of the last 10% of acetone in the pot. However, the remaining acetone is seriously impacting the efficiency of the methanol recovery.

6.2.2 Methanol recovery

The first operation step (acetone removal) ends when the level of the pot reaches its maximum value, and the water feed is switched off. Without entrainer feed, the maximum concentration of acetone in the distillate is limited by the binary azeotrope of acetone and methanol. This results in a concentration of 90 wt% of acetone at a condenser pressure of 0.5 bar. Acetone still is the light boiling component in the ternary mixture, and thus it must be removed before high purity methanol can be recovered. All remaining acetone from the first step must be purged or it will end in the methanol product. Therefore, the maximization of the acetone product and the maximization of the methanol product are no competing objectives. But even at high refluxes, which lead to the azeotropic concentration in the distillate, every kilogram of acetone removed comes with at least 0.1 kg of methanol. This gets worse when the acetone concentration in the pot drops, and the azeotropic concentration in the distillate cannot be maintained anymore with reasonable refluxes. The loss of methanol and the time required to reduce the acetone content in the pot to the desired amount are the competing objectives in this case.

Fig. 5.14 shows the pareto fronts for these objectives starting with 18 kg of acetone (1% remaining), 846 kg of methanol (99% remaining), and 5764 kg of water in the pot. This is the final composition after step one as per Fig. 5.12. The curve parameter is the mass of acetone in the pot at the end of step two.

The relatively high operation times for this second operation step already indicate that only a very small distillate flow is drawn. As the initial point for the succeeding analysis, we use a reflux ratio of 100 until the acetone mass in the pot is reduced to 0.1 kg. This occurs after 142 min and causes a methanol loss of 5%. This is the point where the upper curve slope in Fig. 5.14 changes significantly.

Operation step three (recovery of methanol) is straightforward. It is a batch distillation of a binary nonazeotropic mixture. The design variables in this step are the reflux ratio and the endpoint of the distillation. The concurring objectives are the maximization of the recovered methanol mass, the methanol purity, and the minimization of the elapsed time.

Fig. 5.15 shows the pareto front between the recovered mass and the elapsed time, with the methanol product purity as curve parameter. It is almost impossible to distinguish between the curves for different product purities. This happens because very little acetone is left from step two to contaminate the product. More notable is the difference in the required reflux ratios, which are also depicted in Fig. 5.15.

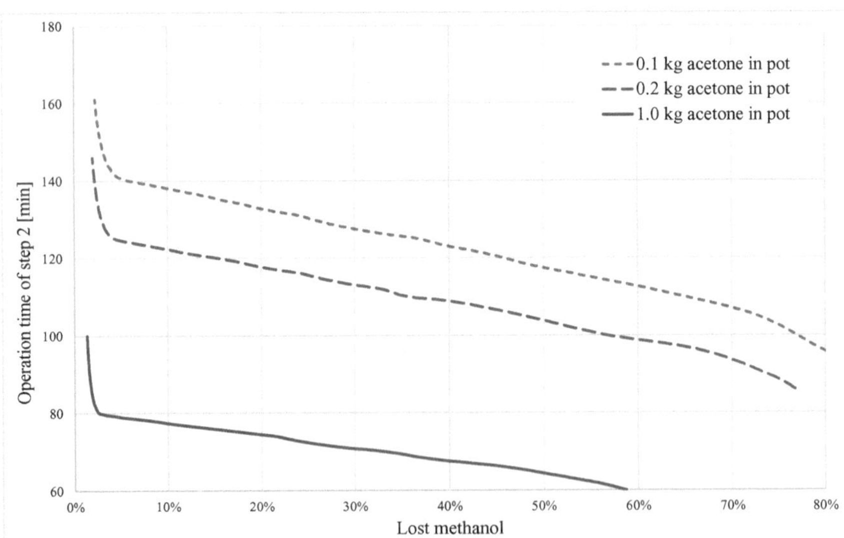

FIG. 5.14 Pareto-optimal points for the time elapsed during step two versus the loss of methanol for different amounts of acetone residue in the pot.

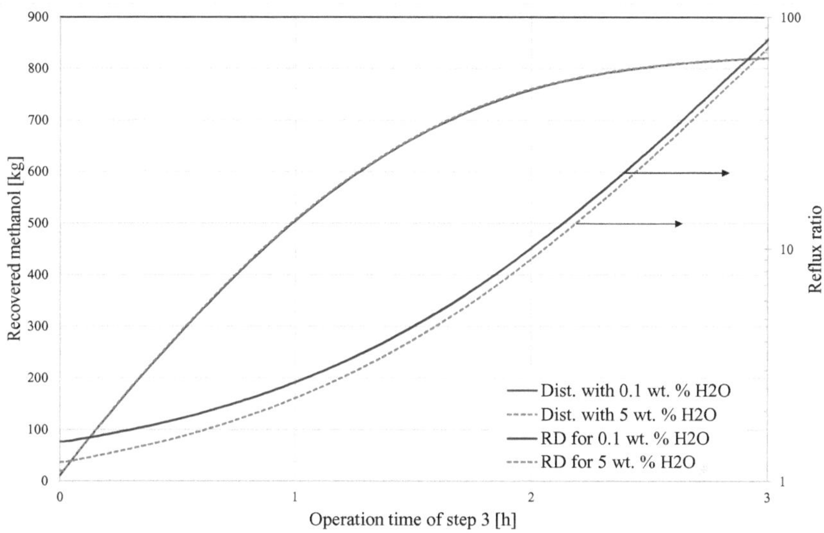

FIG. 5.15 Mass of the methanol product as a function of the time elapsed during step three and the methanol quality.

For the recovery of 90% (765 kg), around 2 h of operation are required. From that point on, the distillation time for more methanol recovery increases significantly. Therefore, not recovering all methanol seems to be an economically attractive option. In this ideal scenario, we run step three for 180 min and recover 95% of the initially available methanol.

6.2.3 Water purification

At the end of operation step three, water remains in the pot which is contaminated with methanol. There are three options for this residual: it can be purified and used for the next batch, sent to a wastewater treatment plant, or collected and added to the pot of the next batch. The wastewater treatment option can be attractive since the remaining methanol can be used by the bacteria of the biological plant. This means, that less additional methanol must be fed.

However, here we focus on the first option, which means that we close the material cycle as far as possible.

The methanol concentration in the pot is 0.15 wt% at the beginning of step four. The relation between the operation time of step four and the amount of waste that is generated is almost linear, until the curves approach the vertical line that represents the existing amount of waste, see Fig. 5.16. We varied the reflux ratio from 1 to 100 for this analysis. This parameter does not affect much the operation time required to achieve certain water purity. Here we decided to use a reflux ratio of 100 until a residue of 10 wt ppm methanol in the pot is reached. This gives an operation time of 2.1 h.

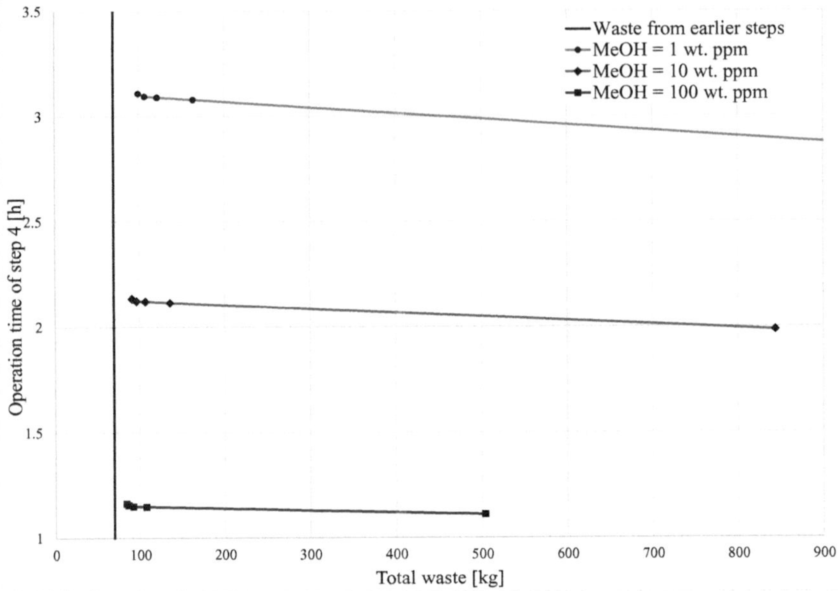

FIG. 5.16 Operation time as a function of the waste remaining from the overall process with the final methanol concentration in the pot as curve parameter.

6.2.4 *Consecutive simulation of all steps*

Table 5.5, summarizes all the operation steps for one batch run, with the selected design variables to achieve: 99% acetone recovery at quality of 99.5 wt%, 95% methanol recovery at quality of 99.85 wt%, and water recovery

TABLE 5.5 Recipe for the recovery of acetone, methanol, and water under ideal conditions.

Step	Task	Design variables	End condition
0	Establishing total reflux profile	RD = ∞	Elapsed time = 1 h
1	Recovery of acetone	$z_{D,Acetone} = 99.5$ wt% FE = 50 kmol/h + 20 kmol/h² × tE	Liquid volume in pot > 7 m³
2	Purging of acetone	RD = 100	Acetone in pot < 0.1 kg
3	Recovery of methanol	$z_{D,Water} = 0.1$ wt%	95% MeOH recovery
4	Purification of water	RD = 100	10 wtppm MeOH

of 97% with contamination of 10 wt ppm. The low-fidelity model starts with a column at total reflux. Operation step 0 was added to reflect that this state takes some time to be established in the real plant.

The total run time is slightly more than 12 h (733 min), with 221 min for step one, 139 min for step two, 180 min for step three, and 133 min for step four. This results in energy consumption of 6108 kWh or a steam consumption of 10.7 tons. By assigning the energy consumption of steps 0 and 1 to the acetone recovery, 2 and 3 to the methanol recovery, and 4 to the water recovery, the specific energy consumption for each product can be calculated. The results are summarized in Table 5.6.

In Fig. 5.17 some trajectories of the batch run specified in Table 5.5 are plotted.

During step 1 and step 3, a constant distillate quality is specified. This leads to variable reflux ratios, which become very high at the end of the steps due to the specified end conditions. Step 3 shows the typical behavior of an increasing reflux ratio with time. In step 1, the reflux ratio is decreasing before it starts an exponential increment. This counterintuitive behavior is a result of the existence of a maximum feasible reflux ratio as described in Section 6.2.1: A higher reflux ratio does not necessarily lead to a higher distillate quality. In Fig. 5.18, the column profiles are plotted at the end of step zero (A), at the beginning of step one (B), and at the end of step one (C). The profile jumps immediately after the entrainer feed is started or stopped because the low-fidelity model ignores the tray and the condenser hold-ups.

TABLE 5.6 **Products, plant time, and energy use when executing the recipe from Table 5.5.**

Product	Amount	Plant time	Energy input	Steam input
Acetone 99.5 wt%	1550 kg	38%	5.44 MJ/kg	4.1 t
	27 kmol		314 MJ/kmol	
Methanol 99.89 wt%	821 kg	44%	11.66 MJ/kg	4.6t
	26 kmol		373 MJ/kmol	
Water MeOH <10 wt ppm	5745 kg	18%	0.69 MJ/kg	1.9t
	319 kmol		13 MJ/kmol	
Waste (42/45/13) wt%	91 kg			
Total		**100%**		**10.6 t**

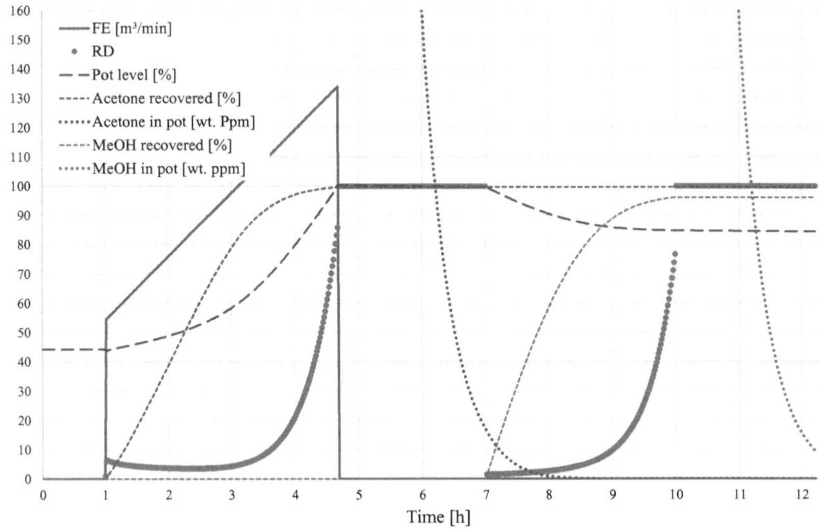

FIG. 5.17 Trajectories of decision and state variables while executing the recipe from Table 5.5.

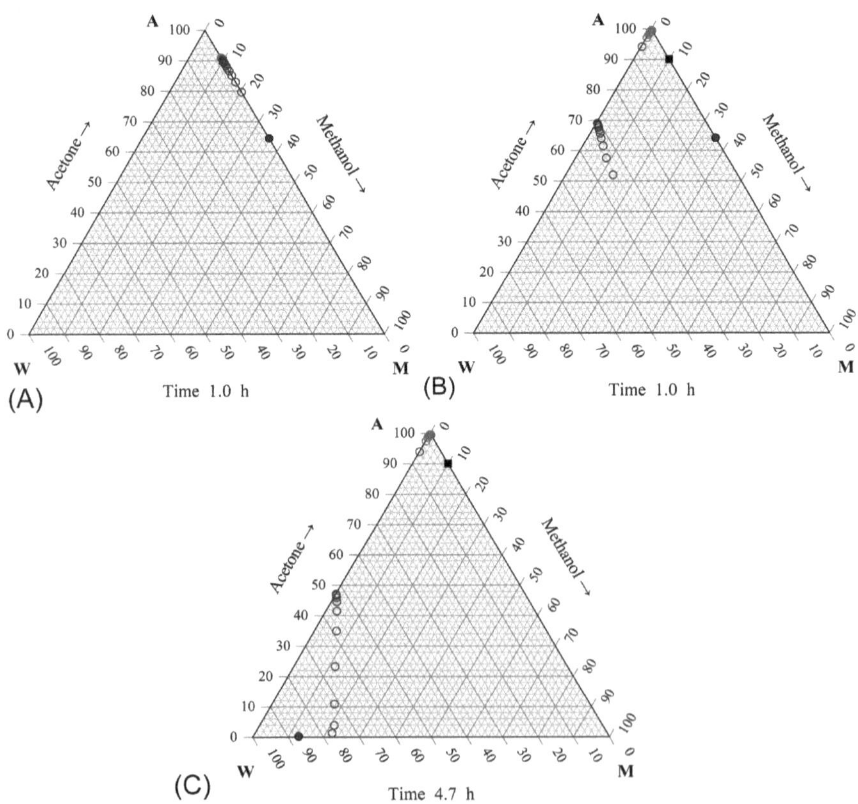

FIG. 5.18 Column profiles in wt% at characteristic time points of operation step 1 using the LFM and the recipe from Table 5.5. *Solid circles* show the distillate and the pot composition.

6.3 High-fidelity model

In the first step, we apply the trajectories and switch times found with the low-fidelity model directly to the high-fidelity model. This leads to a loss in product amount and product quality, due to the liquid hold-ups and the nonconstant heat duties in the pot and in the condenser. Based on these results, we discuss trajectory and switch point tuning strategies.

6.3.1 Applying the low-fidelity model results

Maintaining a fixed distillate concentration is almost impossible in a real plant. Commonly, temperature control is used instead. However, the temperature range from the entrainer feed tray to the distillate is in the range of 1 K, and therefore out of the range of accuracy of industrial temperature measurements. This is depicted in Fig. 5.19, which shows the boiling curve of acetone and water at condenser pressure and for different compositions in the column. Tray 12 refers to the section above the entrainer feed.

A direct jump to new distillate concentrations, as performed with the low-fidelity model, is impossible due to the liquid hold-up in the condensers and their piping system. Therefore, we use the calculated reflux ratios and switch times to specify the high-fidelity model. In Fig. 5.20, the distillate compositions calculated with the high-fidelity model (HFM) are compared with those from

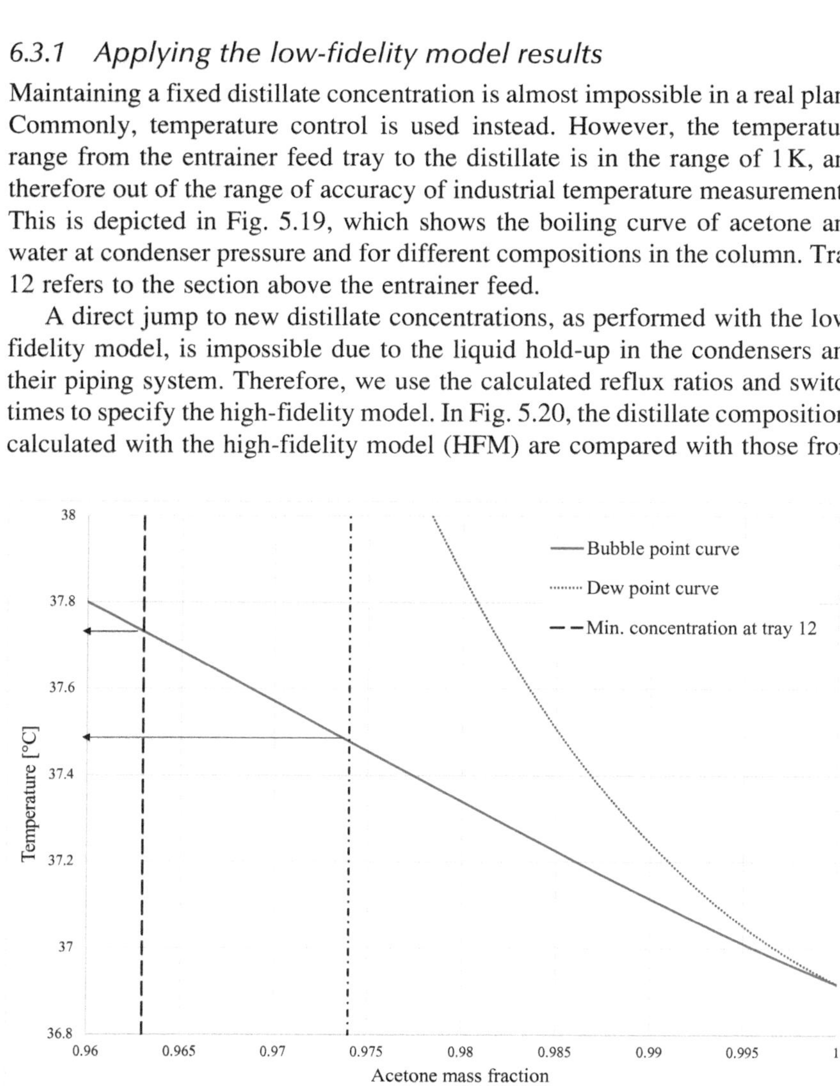

FIG. 5.19 Dew and boiling point temperatures of the binary mixture acetone-water at condenser pressure.

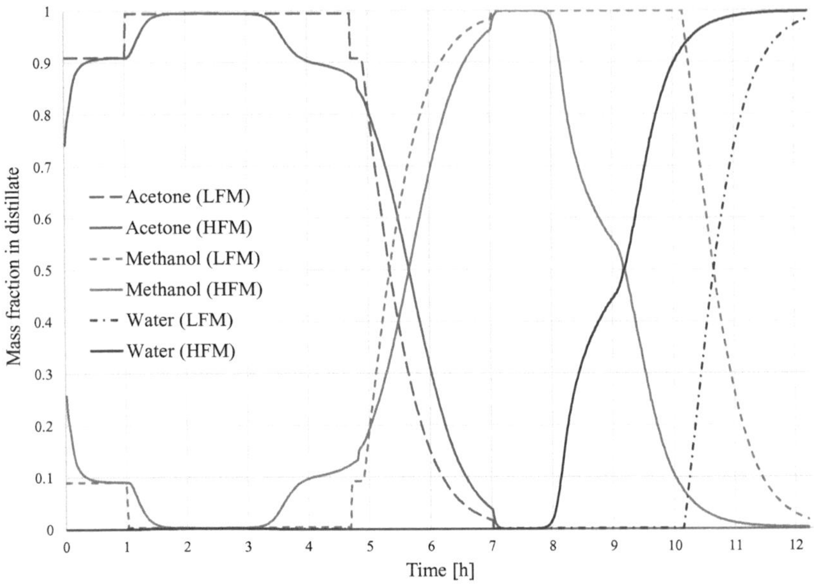

FIG. 5.20 Comparison of the distillate compositions over time calculated with the low-fidelity model (LFM) and the high-fidelity model (HFM) using the recipe from Table 5.5.

the low-fidelity model (LFM). The liquid hold-ups cause a delay in the concentration profiles and it takes a longer time until the desired concentration is reached. Later, the reflux ratios specified for step one become too high and the methanol concentration in the distillate is increasing. The poor acetone recovery during step one affects the methanol recovery and the water purification. It leads to a loss in quality for acetone and methanol but makes water purification easier. In Table 5.7, the products predicted with the low-fidelity model and the high-fidelity model are compared.

The deviations of the model predictions depend mainly on the liquid hold-up in the condensers and the piping from the condensers back to the column. These volumes must be kept as small as possible. With the volumes assumed in this case study (Section 5.4.3), the real product qualities might be acceptable for company internal usage. However, in the next section, we discuss approaches to return them to their original specification.

Further deviations are a result of the detailed heat exchanger and control valve modeling. The heat duty in the pot changes with the pot temperature (change in driving temperature difference) and the liquid level (change in available heat transfer area). The rigorous modeling of the condensers leads to a stronger subcooling of the distillate. In Fig. 5.21, the ideal values used for the low-fidelity model are compared to the rigorously calculated values.

TABLE 5.7 Comparison of the product amounts and qualities calculated with the low-fidelity model (LFM) and the high-fidelity model (HFM) using the recipe from Table 5.5.

		Low-fidelity model	High-fidelity model
Acetone product	Amount	1550kg	1538kg
	Acetone	99.5wt%	97.9wt%
	Methanol	0.3wt%	1.9wt%
	Water	0.2wt%	0.2wt%
Methanol product	Amount	821 kg	829kg
	Acetone	137wtppm	0.2wt%
	Methanol	99.9wt%	87.8wt%
	Water	0.1 wt%	12.0wt%
Water product	Amount	5745kg	5527kg
	Acetone	Traces	Traces
	Methanol	10wtppm	2wtppm
	Water	100wt%	100wt%

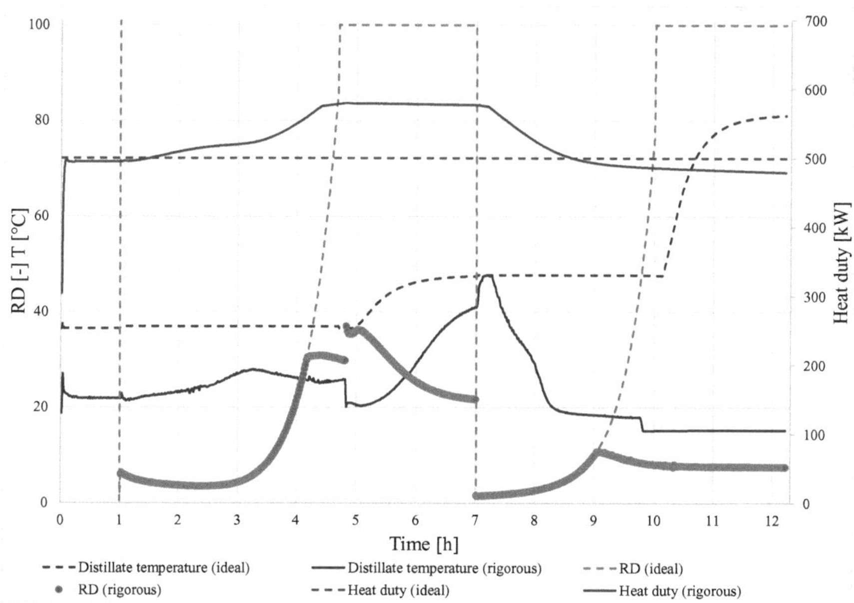

FIG. 5.21 Comparison of ideal and rigorously calculated distillate temperature, reflux ratio, and heat duty.

In the real plant, the high reflux ratios that were found to be suitable for the recovery process cannot be applied because the minimum controllable flowrate for the distillate flow is 80 kg/h.

In Fig. 5.22, the column profiles are plotted some minutes after the beginning (A), in the middle (B), and at the end of step one (C). In contrast to Fig. 5.18, the profile does not jump immediately. Therefore, it takes some time until it touches the binary axis. It also detaches earlier from this axis because a too high reflux ratio is used.

6.3.2 Trajectory and switch point tuning

The basis for tuning trajectories and switch points are the curves in Fig. 5.20. The time reserved for step 0 is sufficient and the total reflux profiles can be reached.

It takes around 30 min after the feeding of the entrainer until the desired distillate concentration is reached. One strategy is to postpone the product drawing by a delay time and send the distillate to the waste tank instead. The distillate

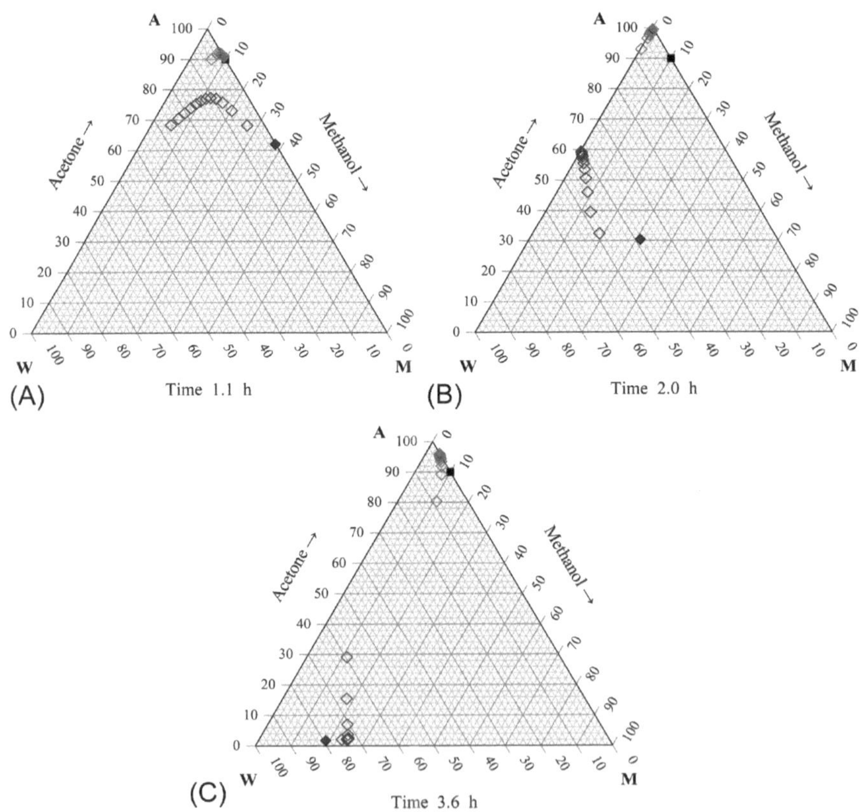

FIG. 5.22 Column profiles in wt% at characteristic time points of operation step 1 using the HFM and the recipe from Table 5.5. *Solid squares* show the distillate and the pot composition.

concentration of acetone also drops earlier than in the ideal case, because at some point too high reflux ratios are used and the distillate is contaminated with methanol. Unfortunately, the control of the distillate concentration or a related temperature in the column is not possible, as discussed above. Therefore, a trajectory for the reflux ratio must be specified. During step one, both too high and too low values lead to a reduced acetone concentration in the distillate, while an increased entrainer feed rate always increases this concentration.

In the following, we keep the trajectory for the reflux ratio as it is and only adjust the entrainer feed rate. In further optimization, a better trajectory might be found, e.g., by solving a minimum time problem for step one [49]. But such optimization with the high-fidelity model is computationally expensive, and it only gives a very plant-specific result. Consequently, it should not be performed before the high-fidelity model was validated with real plant data, and hold-ups, heat transfer, and mass transfer parameters have been adjusted.

As for tuning parameters for adjusting operation step one, we selected the initial entrainer feed rate (FE0), its slope (FEinc), and the delay time before the product is drawn (tD). A three-dimensional sensitivity study is performed to find a good setting. We reduce the plottable design variables to two by fixing the delay time to adjust the acetone product tank quality to 99.5 wt%. Fig. 5.23 shows the acetone product mass using a quadratic interpolation. The parameters that give the maximum amount of product (1266 kg) are FE0 = 50 kmol/h, FEinc = 50 kmol/h^2, and tD = 17 min.

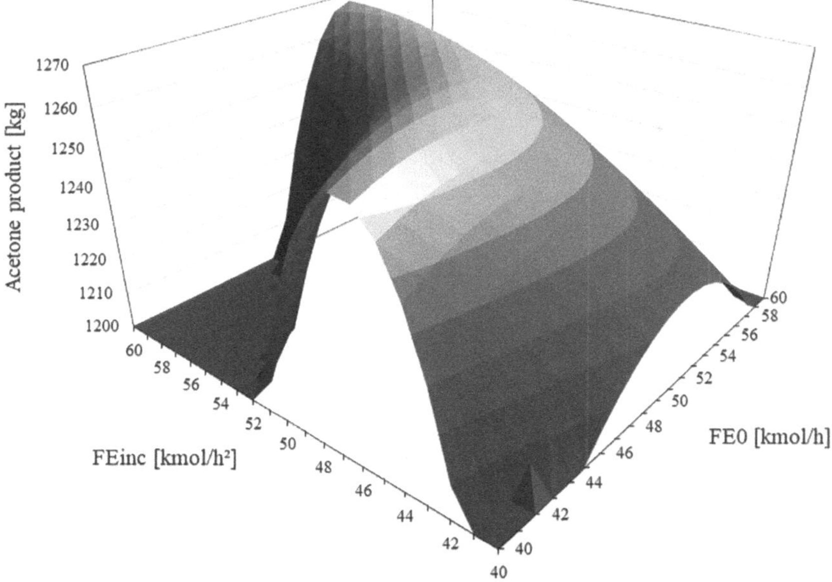

FIG. 5.23 Plot of the acetone product mass as a function of FE0 and FEinc during operation step one assuring 99.5 wt.%.

Due to the increased entrainer feed rate, operation step one ends earlier than in the ideal case. This has the positive effect that the reflux ratios do not become as high as before. High reflux ratios often are a result of mathematical optimization under ideal conditions. In praxis reflux ratios higher than 20 are rarely applied. In the high-fidelity model, the reflux ratio is limited indirectly by the minimum flowrate of the drawn distillate. This leads to reflux ratios of around 40 when the vapor flow from the column is high and to ratios of 20 when this flow is low, see Fig. 5.21.

Time and reflux ratio specified for the acetone purge during step two remain unchanged. However, they cannot be realized as mentioned above and the minimum controllable distillate flow is applied instead. The trajectory of the reflux ratio for methanol recovery is also maintained, but the product drawing is initiated and ended based on the distillate temperature. It starts at a temperature of 47.4°C and ends at a temperature of 49°C. This relates to acetone contamination of 1 wt% in the beginning and water contamination of 6 wt% at the end of operation step three. High reflux ratios are avoided again.

Due to the reduced operation times of step one and step three, the operation time for all these steps is reduced to 8 h. An additional hour is added to operation step four to account for the higher amount of methanol that remains in the pot after step three. In this allotted hour additional methanol is recovered. The new recipe is summarized in Table 5.8 and the trajectories for the decision variables and some state variables are plotted in Fig. 5.24.

The new end condition for the methanol recovery allows considerable product harvesting during step 4. The quality achieved there is still high enough, and methanol stored in the column, the condensers, and the piping system is recovered.

By applying the recipe from Table 5.8, 80% of both products, acetone, and methanol, can be recovered at industrial-grade quality.

TABLE 5.8 Recipe derived with the high-fidelity model.

Step	Task	Design variables	End condition
0	Establishing total reflux profile	RD = 1,000,000	Elapsed time = 1 h
1	Recovery of acetone	RD as in Fig. 5.17 FE = 50 kmol/h + 50 kmol/h^2 × tE	Liquid volume in pot >7 m^3
2	Purging of acetone	RD = 100	Elapsed time = 2.3 h
3	Recovery of methanol	RD as in Fig. 5.17	47.4°C < T < 49°C
4	Purification of water	RD = 100	Elapsed time = 3 h

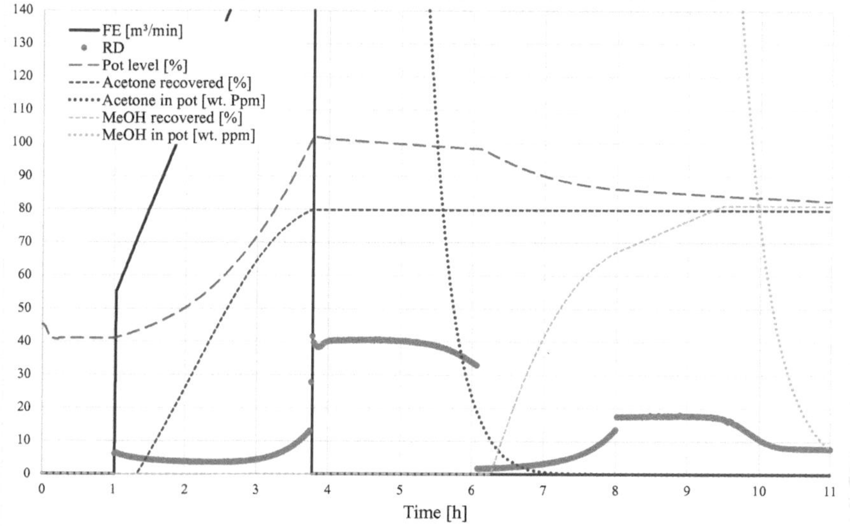

FIG. 5.24 Trajectories of decision and state variables while executing the recipe from Table 5.8.

In Fig. 5.25, the column profiles are plotted some minutes after the beginning (A), in the middle (B), and at the end of step one (C). In contrast to the old recipe, the profile stays connected to the binary axis until the end of step one. This is the effect of the increased entrainer feed rate.

Four videos available as Videos 5.2–5.5 in the online version at https://doi.org/10.1016/B978-0-323-85043-8.00003-9 show the differences in the dynamic column profiles of the low-fidelity model, the high-fidelity model, and the different recipes.

6.4 Economic evaluation

With the results from the last section, we perform a short economic evaluation of the solvent recovery process. We estimate the steam costs based on a steam price of 4 € per metric ton. This is in accordance with prices for low-pressure steam obtained by rigorous calculation as recommended by the US Department of Energy [50]. The market prices for a metric ton of product are set to 900 € for acetone [51] and to 400 € for methanol [52]. The resulting steam costs and product revenues are summarized in Table 5.9.

Revenue of 1416 € can be achieved by product sales or saved purchase costs, respectively. This must be balanced with steam costs of 41 €, and the internal costs for blocking and operating the multipurpose plant for 11 h plus time required for set up and cleaning. The highest benefit is achieved with the acetone product, while the recovery of water has no economic advantage. However,

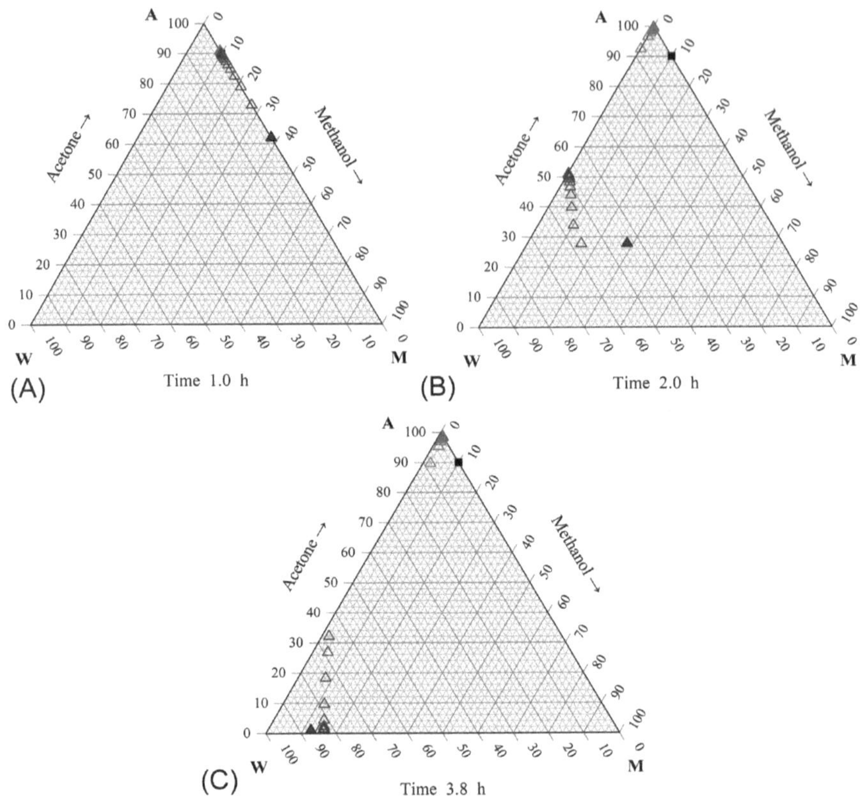

FIG. 5.25 Column profiles in wt% at characteristic time points of operation step 1 using the HFM and the recipe from Table 5.8. *Solid triangles* show the distillate and the pot composition.

the whole process of recovering acetone and methanol from a waste solvent mixture can be considered economically feasible and ecologically attractive. Fresh acetone has a carbon footprint of 1.45 kg/kg [53], fresh methanol 0.76 kg/kg [54], and low-pressure steam 0.17 kg/kg. The carbon dioxide that is not emitted, because the initial solvent mixture is not incinerated or sent to a wastewater treatment plant is 2.27 kg/kg for acetone and 1.37 kg/kg of methanol. With these values, we can calculate the avoided carbon dioxide emissions to 4457 kg for one batch.

The specific energy consumption per mole of product relates to an equivalent reflux ratio of 9.5 for the recovery of acetone (heat of vaporization is 31 MJ/kmol, the distillate is vaporized 10.5 times) and to 12.7 for the recovery of methanol (heat of vaporization is 37 MJ/kmol) in a continuous distillation plant. You et al. [4] found an optimal reflux ratio of 2.48 in an acetone recovery column with 57 equilibrium stages and of 1.14 in a methanol recovery column of 26 equilibrium stages for a similar separation problem.

TABLE 5.9 Steam costs and product revenues for a batch run based on the recipe from Table 5.8 calculated with the high-fidelity model.

Product	Amount	Plant time	Energy input	Steam input	Steam cost	Revenue
Acetone	1266 kg	34%	5.64 MJ/kg	3.5 t	14 €	1139 €
99.5 wt%	22 kmol		325 MJ/kmol			
Methanol	692 kg	52%	15.76 MJ/kg	5.3 t	21 €	277 €
99.90 wt%	22 kmol		505 MJ/kmol			
Water	5596 kg	14%	0.52 MJ/kg	1.4 t	6 €	0 €
MeOH <10 wt ppm	311 kmol		9 MJ/kmol			
Waste (57/ 31/12) wt%	605 kg					
Total		**100%**		**10.1 t**	**41 €**	**1416 €**

The maximum feed flowrate that can be processed by this plant is three cubic meters in 12 h (0.25 m^3/h), assuming a liquid transfer time of 1 h and no cleaning time. When there is more waste mixture occurring, the process must be moved to a continuous distillation plant or the methanol recovery and/or the water recovery must be omitted.

Taking 730 cycles of 12 h per year (cyclic-continuous operation of the batch plant at maximal feed rate) leads to steam costs of approximately 30,000$. We conclude that it will not justify the investment in a plant with two large columns, no matter how much energy can be saved by this measure.

7 Summary

Starting from a clear problem definition, a recipe for a real existing multipurpose batch distillation plant was derived. Thermodynamic insights into the mixture's behavior were used to find and analyze different operation strategies. These were tested with models with increasing levels of complexity, to reflect the different aspects of the real plant. A screening model was used to find the basic operating conditions, design parameter ranges, and feasible product qualities. With this model, an interactive report was created that allows an analysis of the batch distillations variable interdependency. A low-fidelity model was used to study the effects of the design variables in more detail, specifically the application of time-dependent design variables. The recipe, that was created based on the results of the low-fidelity model, was then applied to the high-fidelity model. The high-fidelity model was built to reflect the real plant's

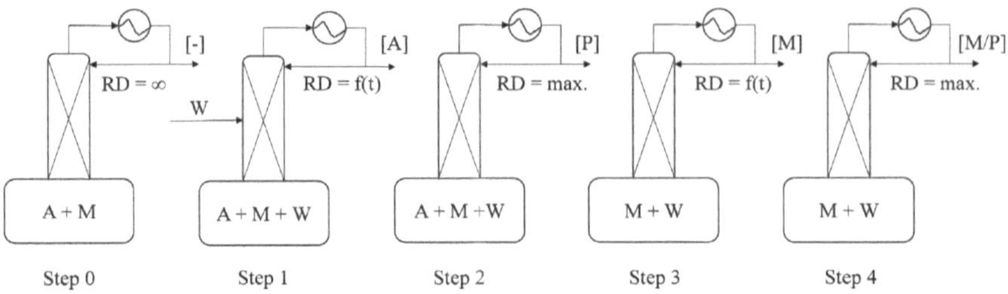

FIG. 5.26 Batch recipe for the recovery of acetone (A) and methanol (M) with water (W) as entrainer. RD indicates the reflux ratio, and the brackets show the connected product tank. [P] refers to the purge tank B004.

dynamic behavior and includes effects such as liquid hold-ups, pressure drops, and heat transfer limitations. After an adaption of the recipe to the real plant requirements, realistic time and energy consumptions were calculated, and an economic evaluation of the solvent recovery process was performed. The results show that recovery of acetone and methanol, from an equimolar binary mixture of three cubic meters, is economically feasible in the existing plant when we apply the developed recipe. Roughly 80% of both solvents are recovered with an industrial-grade quality in an 11-h lasting batch run. Around 4.5 tons of carbon dioxide are not emitted per batch run performed, or per recovery of three cubic meters of solvent waste, respectively. The overall recovery process is summarized in Fig. 5.26.

References

[1] Ullmann's Encyclopedia of Industrial Chemistry, Wiley, Chichester, 2010.

[2] Statistisches Bundesamt (Destatis), Abfallbilanz (Abfallaufkommen/-verbleib, Abfallintensität, Abfallaufkommen nach Wirtschaftszweigen), Statistisches Bundesamt, 2018. www.destatis.de.

[3] A. Azapagic, D. Amienyo, R.M. Cuéllar Franca, H.K. Jeswani, Carbon Footprints of Recycled Solvents, European Solven Recycler Group, 2013. www.esrg-online.e.

[4] X. You, I. Rodriguez-Donis, V. Gerbaud, Improved design and efficiency of the extractive distillation process for acetone–methanol with water, Ind. Eng. Chem. Res. 54 (2015) 491–501, https://doi.org/10.1021/ie503973a.

[5] M. Lewandowski, E. Milchert, Modern technology of dry distillation of wood, CHEMIK 65 (2011) 1301–1306.

[6] J. Gmehling, M. Kleiber, B. Kolbe, J. Rarey, Chemical Thermodynamics for Process Simulation, Wiley, 2019.

[7] E.-K. Hilmen, Separation of Azeotropic mixtures: Tools for Analysis and Studies on Batch Distillation Operation (PhD thesis), Norwegian University of Science and Technology (NTNU), 2000.

[8] Dortmund Data Bank Software & Separation Technology GmbH, DDB Software Package.

[9] M.W.R. Düssel, Batch-Rektifikation azeotroper Gemische in Verstärkungs- und Abtriebskolonnen, Chem. Ing. Tech. 72 (2000) 675–682.

[10] S. Brüggemann, W. Marquardt, Shortcut methods for nonideal multicomponent distillation: 3. Extractive distillation columns, AIChE J. 50 (2004) 1129–1149, https://doi.org/10.1002/aic.10108.

[11] I. Rodríguez Donis, V. Gerbaud, X. Joulia, Thermodynamic insights on the feasibility of homogeneous batch extractive distillation, 1. Azeotropic mixtures with a heavy entrainer, Ind. Eng. Chem. Res. 48 (2009) 3544–3559, https://doi.org/10.1021/ie801060n.
[12] I. Rodríguez Donis, V. Gerbaud, X. Joulia, Thermodynamic insights on the feasibility of homogeneous batch extractive distillation, 2. Low-relative-volatility binary mixtures with a heavy entrainer, Ind. Eng. Chem. Res. 48 (2009) 3560–3572, https://doi.org/10.1021/ie801061y.
[13] I. Rodríguez Donis, V. Gerbaud, X. Joulia, Thermodynamic insights on the feasibility of homogeneous batch extractive distillation. 3. Azeotropic mixtures with light entrainer, Ind. Eng. Chem. Res. 51 (2012) 4643–4660, https://doi.org/10.1021/ie201942b.
[14] I. Rodríguez Donis, V. Gerbaud, X. Joulia, Thermodynamic insights on the feasibility of homogeneous batch extractive distillation. 4. Azeotropic mixtures with intermediate boiling entrainer, Ind. Eng. Chem. Res. 51 (2012) 6489–6501, https://doi.org/10.1021/ie2019432.
[15] V.N. Kiva, E.K. Hilmen, S. Skogestad, Azeotropic phase equilibrium diagrams: a survey, Chem. Eng. Sci. 58 (2003) 1903–1953, https://doi.org/10.1016/S0009-2509(03)00018-6.
[16] E.K. Hilmen, V.N. Kiva, S. Skogestad, Topology of ternary VLE diagrams: elementary cells, AIChE J. 48 (2002) 752–759, https://doi.org/10.1002/aic.690480410.
[17] Z. Lelkes, P. Lang, B. Benadda, P. Moszkowicz, Feasibility of extractive distillation in a batch rectifier, AIChE J. 44 (1998) 810–822.
[18] W.L. Luyben, Comparison of extractive distillation and pressure-swing distillation for acetone – methanol separation, Ind. Eng. Chem. Res. 47 (2008) 2696–2707, https://doi.org/10.1021/ie701695u.
[19] I.D. Gil, D.C. Botía, P. Ortiz, O.F. Sánchez, Extractive distillation of acetone/methanol mixture using water as entrainer, Ind. Eng. Chem. Res. 48 (2009) 4858–4865, https://doi.org/10.1021/ie801637h.
[20] J.F. Boston, Inside-out algorithms for multicomponent separation process calculations, in: R. G. Squires, G.V. Reklaitis (Eds.), Computer Applications to Chemical Engineering: Process Design and Simulation Based on a Symposium Sponsored by the Division of Industrial and Engineering Chemistry at the 178th Meeting of the American Chemical Society, Washington, D.C., September 11–13, 1979, American Chemical Society, Washington, DC, 1980, pp. 135–151.
[21] L.J. Christiansen, M.L. Michelsen, A. Fredenslund, Naphtali-Sandholm distillation calculations for NGL mixtures near the critical region, Comput. Chem. Eng. 3 (1979) 535–542,- https://doi.org/10.1016/0098-1354(79)80090-3.
[22] R.H. Perry, D.W. Green, J.O. Maloney, Perry's Chemical Engineers' Handbook, seventh ed., McGraw-Hill, New York, 1997.
[23] K.E. Brenan, S.L. Campbell, L.R. Petzold (Eds.), Theory of DAE's, in: Numerical Solution of Initial-Value Problems in Differential-Algebraic Equations, Society for Industrial and Applied Mathematics (SIAM, 3600 Market Street, Floor 6, Philadelphia, PA 19104), Philadelphia, PA, 1996, pp. 15–39 (Chapter 2).
[24] H. Galindez, A. Fredenslund, Simulation of multicomponent batch distillation processes, Comput. Chem. Eng. 12 (1988) 281–288, https://doi.org/10.1016/0098-1354(88)85039-7.
[25] Chemstations, CHEMCAD NXT, 2021.
[26] T.F. Edgar, D.M. Himmelblau, L.S. Lasdon, Optimization of Chemical Processes, second ed., McGraw-Hill, Boston, 2001.
[27] R. Fletcher, Optimization, Academic Press, London, 1969.
[28] J.H. Holland, The Adaptation in Natural and Artificial Systems: An Introductory Analysis With Applications to Biology, Control, and Artificial Intelligence, The MIT Press, 1992. https://mitpress.mit.edu/books/adaptation-natural-and-artificial-systems.

[29] J.A. Nelder, R. Mead, A simplex method for function minimization, Comput. J. 7 (1965) 308–313.
[30] M.J. Box, A new method of constrained optimization and a comparison with other methods, Comput. J. 8 (1965) 42–52, https://doi.org/10.1093/comjnl/8.1.42.
[31] V.K. Mehta, B. Dasgupta, A constrained optimization algorithm based on the simplex search method, Eng. Optim. 44 (2012) 537–550, https://doi.org/10.1080/0305215X.2011.598520.
[32] G.P. Rangaiah, A. Bonilla-Petriciolet, Multi-Objective Optimization in Chemical Engineering: Developments and Applications, Wiley A John Wiley & Sons Ltd. Publication, Chichester, West Sussex, 2013.
[33] A. Fricke, J.C. Schöneberger, A blueprint for software architectures in process optimization, Chem. Ing. Tech. 89 (2017) 515–526, https://doi.org/10.1002/cite.201600130.
[34] R Core Team, R: A Language and Environment for Statistical Computing, 2020, Vienna https://www.R-project.org/.
[35] T. Wei, V. Simko, R Package "corrplot": Visualization of a Correlation Matrix, 2017. https://github.com/taiyun/corrplot.
[36] H. Wickham, J. Bryan, readxl: Read Excel Files, 2019. https://CRAN.R-project.org/package=readxl.
[37] M.R. Smith, Ternary: An R Package for Creating Ternary Plots, Zenodo, 2017.
[38] J. Ooms, av: Working with Audio and Video in R, 2020. https://CRAN.R-project.org/package=av.
[39] Design Institute for Physical Property Data (U.S.), DIPPR 801, Design Institute for Physical Property Research/AIChE, Houghton, MI, 2011.
[40] Dortmun Data Bank Software & Separation Technology GmbH, Dortmund Data Bank.
[41] A.W.T. Hunan Liling, Separation Technology, CERADUR—Ceramic Structured Packing. http://www.awt-septec.com/CERADUR.asp. (Accessed 3 March 2021).
[42] Sulzer Chemtech, Structured Packings for Distillation, Absorption and Reactive Distillation.
[43] VDI-Wärmeatlas, Springer Berlin Heidelberg, Berlin, Heidelberg, 2013.
[44] D.Q. Kern, Process Heat Transfer, Mac Graw Hill, Auckland, Paris (etc.), 1983.
[45] E.N. Sieder, G.E. Tate, Heat transfer and pressure drop of liquids in tubes, Ind. Eng. Chem. 28 (1936) 1429–1435, https://doi.org/10.1021/ie50324a027.
[46] I.H. Lehrer, Jacket-side Nusselt number, Ind. Eng. Chem. Process. Des. Dev. 9 (1970) 553–558, https://doi.org/10.1021/i260036a010.
[47] R. Billet, M. Schultes, Modelling of pressure drop in packed columns, Chem. Eng. Technol. 14 (1991) 89–95, https://doi.org/10.1002/ceat.270140203.
[48] V. Engel, Fluiddynamik in füllkörper- und packungskolonnen für gas/flüssig-systeme, Chem. Ing. Tech. 72 (2000) 700–703, https://doi.org/10.1002/1522-2640.
[49] U.M. Diwekar, Batch Distillation: Simulation, Optimal Design, and Control, second ed., CRC Press, Boca Raton, FL, 2012.
[50] US Department of Energy, How to Calculate the True Cost of Steam, 2003.
[51] Echemi Group, Acetone Price Analysis, https://www.echemi.com/productsInformation/pid_Rock3929-acetone.html, [Accessed 16 February 2021].
[52] M. Pérez-Fortes, J.C. Schöneberger, A. Boulamanti, E. Tzimas, Methanol synthesis using captured CO2 as raw material: techno-economic and environmental assessment, Appl. Energy 161 (2016) 718–732, https://doi.org/10.1016/j.apenergy.2015.07.067.
[53] sphera, GaBi LCA Datenbanken.
[54] M. Pérez-Fortes, E. Tzimas, Techno-Economic and Environmental Evaluation of CO2 Utilisation for Fuel Production: Synthesis of Methanol and Formic Acid, Publications Office, Luxembourg, 2016.

Chapter 6

Modeling and optimizing dynamic networks: Applications in process engineering and energy supply

Jan Mohring[a], Jochen Schmid[a], Jarosław Wlazło[a], Raoul Heese[a], Thomas Gerlach[b], Thomas Kochenburger[b], and Michael Bortz[a]

[a]*Fraunhofer Institute for Industrial Mathematics (ITWM), Kaiserslautern, Germany,* [b]*Bayer AG, Building E 41, Leverkusen, Germany*

1 Introduction

In discussions between engineers and mathematicians, one can often observe a fundamentally different approach to technical problems. While the engineer is typically more oriented toward the application at hand, the mathematician tends to focus on the structure underlying the problem. In many cases, combining the two perspectives leads to innovative solutions. In particular, the structural approach allows strategies to be transferred from one field of application to another. In this chapter, we want to tell one such success story.

It revolves around the modeling, simulation, and optimization of general dynamic networks, which consist of components and nodes connected by flows like material fluxes or electric currents, for instance. Such networks arise in many application domains—for instance:

- In energy supply (in the form of power grids or district heating networks) or
- In chemical engineering (in the form of alternating equilibrium stages in a distillation column or in the form of flowsheets of an overall chemical process).

In particular, such networks arise when components from different technical or physical areas are coupled (yielding so-called multiphysics systems). While in some applications, a stationary description of the considered technical network is sufficient, there are many applications in which transient effects are essential and which, thus, require a dynamic mathematical model. In order to obtain such

Simulation and Optimization in Process Engineering. https://doi.org/10.1016/B978-0-323-85043-8.00013-1

a model, the differential equations describing the dynamics of the network's individual components have to be coupled according to the algebraic or differential relations describing the connections between the network's components. In this manner, one arrives at a system of generally nonlinear differential-algebraic equations (DAE).

We use the custom software AD-Net in order to build—and solve—these differential-algebraic network equations in a comfortable, modular way from the individual component models. In principle, the components can arise from arbitrary application domains and they can be interconnected according to an arbitrary network topology—the only limitation of AD-Net in its current implementation is that differential-algebraic equations with partial derivatives (PDAE) are not supported. As the name suggests, AD-Net makes essential use of automatic differentiation (AD), that is, all quantities computed in AD-Net are automatically differentiated w.r.t. user-specified input parameters. In particular, this allows for an effortless solution of dynamic network optimization problems by means of a great variety of external gradient-based optimizers.

AD-Net has been developed at the Fraunhofer Institute for Industrial Mathematics since 2015 as a general-purpose C++ template library, tailored to the modeling, simulation, and optimization of dynamic technical networks. Originally, AD-Net was specialized and applied to solve optimal control problems in district heating (AD-Net District Heating) or power networks (AD-Net Power). Yet, due to the structural approach, AD-Net could be extended to batch distillation processes within a short amount of time, leading to AD-Net Distillation. We present results from all these application areas, with a focus on those achieved in batch distillation in collaboration with Bayer AG, Leverkusen.

2 AD-Net

AD-Net is a general-purpose dynamic network software, tailored to the modeling, simulation, and optimization of technical networks. It is implemented as a C++ template library, which describes the typical structure of a technical network with nodes linked by components, but has no notion of the actual application, yet. Adding the relations which describe how the dynamics of a specific component depend on its environment, AD-Net combines all equations to a large nonlinear system of differential-algebraic equations.

In order to solve the differential-algebraic system equations of a given technical network of interest, AD-Net uses a collocation method, adaptive step size control, and event detection. Additionally, automatic differentiation (forward AD) is employed [1,2]. In fact, it is employed in a twofold manner. On the one hand, the nonlinear relations between states $\boldsymbol{x}$ of different network components and time steps are linearized symbolically to get accurate derivatives of the implicit time-stepping scheme. On the other hand, states of a new time step are linearized in terms of the control parameters $\boldsymbol{p} \in \mathbb{R}^{n_p}$ by means of linearizations of previous states and the implicit function theorem. Typically, the parameters are values of a control input signal $t \mapsto u_p(t)$ at some interpolation points, e.g., of the reflux ratio controlling a batch distillation process.

In a single forward simulation, observables $\boldsymbol{y}_p$ are collected, which are certain (explicit or implicit) functions of interesting state variables like point evaluations, integrals, or zeros of state trajectories. After each forward simulation, these observables can then be composed to a cost function $J(\boldsymbol{p})$ and constraints $\boldsymbol{c}(\boldsymbol{p})$ to solve the corresponding dynamic optimization problem.

Within AD-Net, these optimization tasks are solved by means of the *sequential approach* to dynamic optimization (Section 8.6 of Ref. [3]). In this approach, the optimization and the solution of the differential-algebraic system are largely separated, so that any external gradient-based optimizer can be used, for example, Ipopt [4], Knitro [5], or the SQP or trust-region solvers from the scipy.optimize package. In each iteration, the optimizer provides new control parameters which are translated into time-dependent input signals. AD-Net then performs a single forward simulation for the given parameter $\boldsymbol{p}$ and returns the values $J(\boldsymbol{p})$, $\boldsymbol{c}(\boldsymbol{p})$ of the cost function and constraints along with their derivatives $DJ(\boldsymbol{p})$, $D\boldsymbol{c}(\boldsymbol{p})$. See Fig. 6.1 for a schematic illustration of this sequential approach. While this procedure is likely to fail for open-loop unstable systems, it has several advantages in the case of moderate systems. In contrast to the *simultaneous approach* to dynamic optimization, equality constraints related to the discretized system dynamics do not show up in the optimizer. Instead,

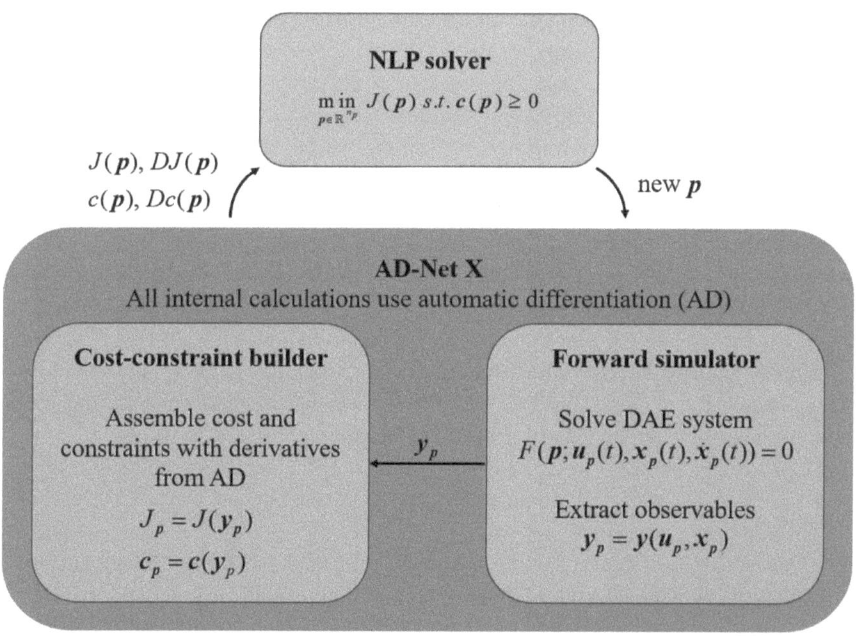

FIG. 6.1 Optimizing with AD-Net and an external NLP solver.

they are eliminated by the forward simulator, which keeps in memory only state variables of subsequent time steps. This is especially advantageous for large dynamic systems or long simulation periods. Moreover, the number of internal states may vary for different control parameters and with adaptive time-stepping.

3 Applications in energy supply

Energy is supplied via transport networks for electricity, gas, and district heating. In times of energy transition, it is no longer sufficient to simulate the behavior of a single network for monitoring purposes. In particular, the decentralized, volatile feed-in to the power grid leads to significantly more demanding optimization problems in operational management, even across sector boundaries. The aim in developing AD-Net was to provide an answer to these new challenges, i.e., to deliver a software framework that understands optimization issues directly as part of the model and in which networks from different sectors can be modeled and coupled in a uniform way.

3.1 Power transmission

AD-Net Power allows for steady-state and dynamic power system analysis. Common benchmark networks such as the Cigre Nordic 32 grid [6] can be handled at the same level of detail as, for example, with the commercial software DiGSILENT PowerFactory [7]. Model descriptions in its DiGSILENT format can be imported. In addition to the usual component models for generators, transformers, lines, etc., also the standard controllers are stored. In a small example, we now illustrate two special advantages of AD-Net-Power: other power networks can be simulated within the same framework and gradient-based optimization is directly integrated. The electric grid shown in Fig. 6.2 is connected to the gas grid via a gas turbine and an electrolyzer. The gas power plant compensates for the fluctuating feed-in of a wind farm. The goal is to start up the comparatively inefficient electrolyzer only when the gas turbine runs the risk of leaving its optimal operating range, i.e., during particular power peaks of the wind farm.

Fig. 6.2 shows the power shares of gas turbine and electrolyzer before and after optimization. Note that the additional models of a gas grid and the electrolyzer could be added easily within the scope of a master's thesis [8].

3.2 District heating

AD-Net District Heating [9] was developed together with the energy supplier Technische Werke Ludwigshafen (TWL) and the engineering office GEF Ingenieur AG. Normally, TWL can meet the customers' heating needs from a waste incineration plant. On colder days, however, supplementary firing with oil or

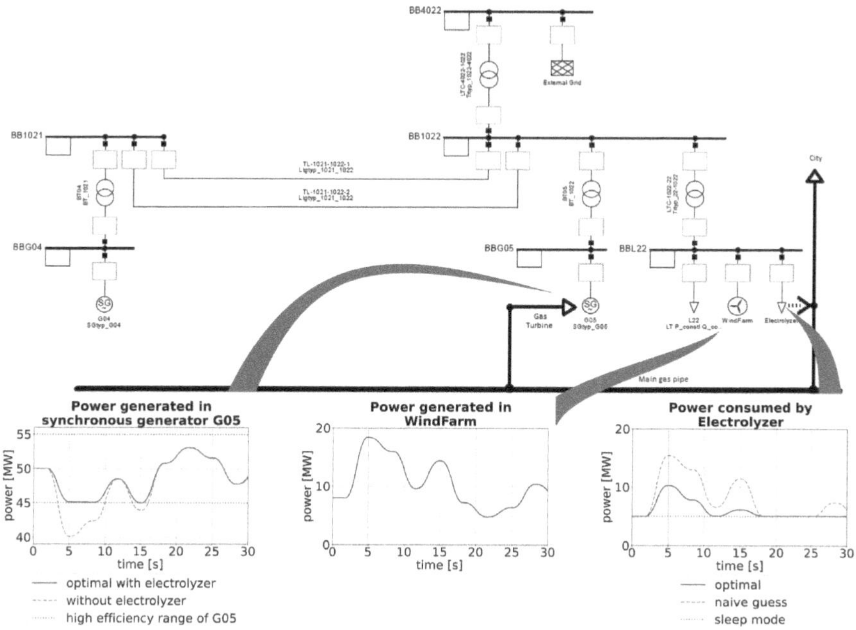

FIG. 6.2 Optimal coupling of power and gas grids.

gas is needed to satisfy the peak demand. This additional firing can be reduced by using the district heating network as a storage facility: by clever preheating, energy is already introduced at times of low heat demand, which customers can then call up additionally when demand is high. Such time-delay effects cannot be captured with conventional quasistationary simulators. With AD-Net District Heating, on the other hand, it is possible to determine optimal preheat profiles.

Fig. 6.3 shows the simulated temperature distribution in TWL's inner-city network at a time of high demand. The grid includes about 1000 consumers and 6000 pipes. With a preview length of 3 days and a temporal resolution of a quarter of an hour, the optimization problem to be solved comprises about 19 million degrees of freedom.

4 Applications in batch distillation

Batch distillation is a standard process to separate a chemical mixture into its components exploiting the components' different boiling temperatures [3,10–15]. In contrast to continuous distillation, batch distillation is used to produce seasonal or high-purity chemicals. A typical batch distillation column is sketched in Fig. 6.4.

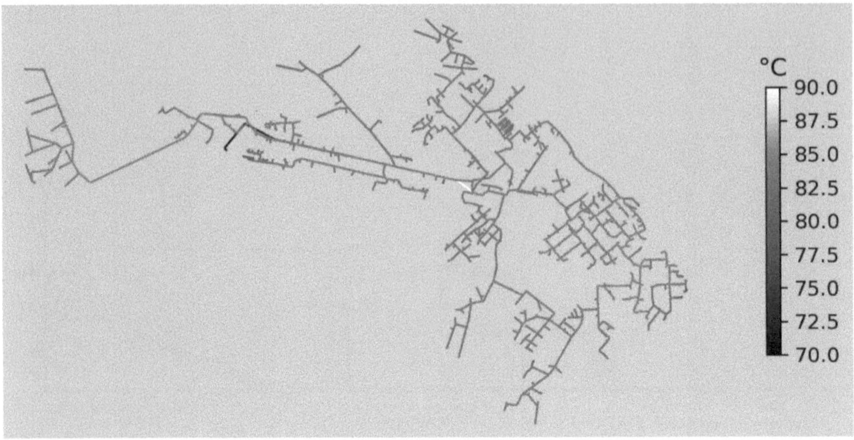

FIG. 6.3 Temperature distribution in the district heating network of TWL.

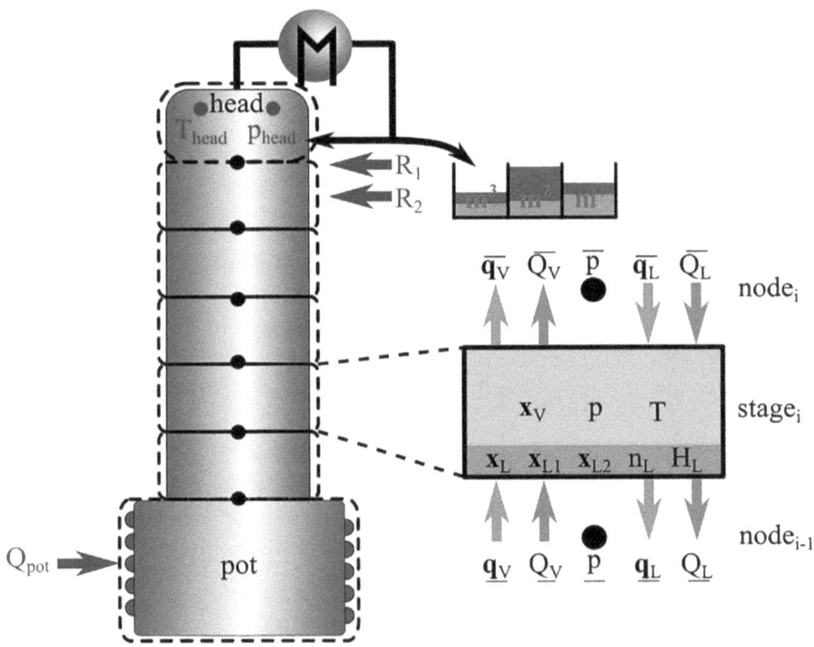

FIG. 6.4 Schematic of a batch distillation column.

The source material is given to a heated pot. Rising vapor streams through the column toward the head where it is liquefied in a condenser. Part of the liquid is fed back to the column according to a given reflux ratio. The rest is withdrawn and collected in fractions, the future products. The column is composed of stages, which are either physical compartments separated by perforated plates or just computational units filled homogeneously with packing material. The purpose is to have a higher exchange of heat and matter between vapor upstream and liquid downstream. The amount of liquid stored in a stage is called the liquid holdup. It can split into two phases [12,16], typically an organic and an inorganic one. The two phases can be fed back to the column at two different reflux ratios.

Within AD-Net Distillation, batch processes are modeled as a simple, linear network of alternating components and nodes. Pressures at stage boundaries and flows through boundaries are stored in the nodes. The other state variables are assigned to pot, head, and an arbitrary number of internal stages. Fig. 6.4 shows the states of an internal stage and its surrounding nodes as an example. In this figure, T and p denote the mean temperature and pressure in the stage, while n_L and H_L are the total moles and enthalpy of the liquid holdup, respectively. $\boldsymbol{x}_V$ and $\boldsymbol{x}_L$ are the vectors representing the mole fractions of the mixture's components in the vapor and the liquid phase, respectively. Both vectors are normalized, i.e., their components sum up to one. Also, $\boldsymbol{x}_{L1}, \boldsymbol{x}_{L2}$ denote the fractions in the two liquid phases, and $\boldsymbol{x}_{L1} + \boldsymbol{x}_{L2} = \boldsymbol{x}_L$. If there is only one liquid phase, we have $\boldsymbol{x}_{L2} = 0$. And finally, the symbol $\boldsymbol{q}$ denotes vectors of componentwise molar flows and the symbol Q stands for enthalpy flows. Underscores and bars indicate the adjacent node below and above the current stage, whereas V and L mark vapor and liquid phase, respectively. The dynamics of the pot, the internal stages, and the head are described by mass conservation, energy conservation, and thermodynamic (vapor-liquid-liquid) equilibrium equations [12,16] in conjunction with a holdup and a pressure drop relation. Spelled out, the equations for an internal stage read as follows. Similar equations hold true for the pot and the head, but they are omitted here for brevity.

$$\begin{gathered}
\frac{\mathrm{d}}{\mathrm{d}t}(n_L\boldsymbol{x}_L) = \overline{\boldsymbol{q}}_L + \underline{\boldsymbol{q}}_V - \underline{\boldsymbol{q}}_L - \overline{\boldsymbol{q}}_V \\
\underline{\boldsymbol{q}}_L = sum\left(\underline{\boldsymbol{q}}_L\right)\boldsymbol{x}_L, \quad \overline{\boldsymbol{q}}_V = sum(\overline{\boldsymbol{q}}_V)\boldsymbol{x}_V \\
(\boldsymbol{x}_{L1}, \boldsymbol{x}_{L2}, \boldsymbol{x}_V, T) = \boldsymbol{f}_{VLLE}(p, \boldsymbol{x}_L) \\
n_L = f_{holdup}\left(\underline{\boldsymbol{q}}_L\right), \quad \left(p, \underline{p}\right) = \boldsymbol{f}_{\Delta p}\left(\overline{p}, T, \underline{\boldsymbol{q}}_L, \overline{\boldsymbol{q}}_V\right) \\
\frac{\mathrm{d}}{\mathrm{d}t}H_L = \overline{Q}_L + \underline{Q}_V - \underline{Q}_L - \overline{Q}_V \\
H_L = n_L\boldsymbol{x}_L \cdot \boldsymbol{h}_L(p, T, \boldsymbol{x}_L), \quad \underline{Q}_L = \underline{\boldsymbol{q}}_L \cdot \boldsymbol{h}_L(p, T, \boldsymbol{x}_L), \quad \overline{Q}_V = \overline{\boldsymbol{q}}_V \cdot \boldsymbol{h}_V(p, T, \boldsymbol{x}_V)
\end{gathered} \tag{6.1}$$

In addition to the standard laws of mass and energy conservation, the process dynamics are described by several algebraic functions, namely $\boldsymbol{f}_{VLLE}$, $\boldsymbol{f}_{LLE}$, f_{holdup}, $\boldsymbol{f}_{\Delta p}$, $\boldsymbol{h}_L$, and $\boldsymbol{h}_V$. Specifically, $\boldsymbol{f}_{VLLE}$ provides the temperature and partial compositions at which the different phases of a mixture are in thermodynamic equilibrium, the vapor-liquid-liquid equilibrium. A similar function $\boldsymbol{f}_{LLE}$

describes the phase equilibria in the condenser. Internally, the two equilibrium functions add the isoactivity equations [17,18] and equations for partial pressures and mole balance to the global DAE system. Liquid-liquid splitting is detected by a tailor-made mechanism that does not require the global minimization of the Gibbs energy [19–21] or the solution of relaxed equilibrium conditions featuring complementarity constraints [3,22]. Also, f_{holdup} links the holdup to the liquid downflow, while $\boldsymbol{f}_{\Delta p}$ describes the pressure drop along a stage. And finally, $\boldsymbol{h}_L$ and $\boldsymbol{h}_V$ denote the molar liquid and vapor enthalpies, respectively. In addition to state variables, these algebraic functions depend also on substance properties like Antoine and NRTL parameters [12,23] or geometric quantities like the cross-section of the column.

Altogether, we allow six input signals: the reflux ratios R_1, R_2 for two liquid phases, the pressure p_{head} and temperature T_{head} in the head, the pressure drop Δp from pot to head, and the heat duty Q_{pot}. Two of the three inputs p_{head}, Δp and Q_{pot} can be freely specified by the user. The third one is then completely determined and computed by the forward simulator of AD-Net Distillation. Note that this approach models a perfect controller in an easy way. All input signals are represented as piecewise polynomial splines either in time or total distillate. Apart from the forward simulation functionality, AD-Net Distillation also features two optimization modules for batch distillation processes: one for the estimation of unknown model properties, like Antoine or NRTL parameters, and one for the optimal design of reflux ratio profiles and fractions. Since we take the sequential approach to dynamic optimization, the corresponding optimization problems take the simple form of nonlinear programs.

$$\min J(\boldsymbol{p}) \quad \text{s.t.} \quad c_i(\boldsymbol{p}) \leq 0 \text{ for all } i \in I \tag{6.2}$$

What the optimization parameters $\boldsymbol{p}$ look like in detail will be explained in Sections 4.2 and 4.3 below. In these sections, it will also be explained from which observables the objective function J and the constraints c_i are constructed.

4.1 Forward simulation

Simulating distillation processes, we face three major challenges: sudden changes, vanishing components, and splitting of the liquid phase. This requires adaptive time-stepping and, as components may vanish at a steep slope, also zero detection. In particular, when the liquid begins or ends splitting into two phases, compositions change discontinuously, which requires tricky re-initialization of the vapor-liquid-liquid equilibrium solver.

Fig. 6.5 shows sample control input signals and Fig. 6.6 depicts the corresponding simulation results for a batch distillation column with a diameter of 5 cm, height 1 m, and 11 stages. Initially, the pot is filled with 650 g butanol, 100 g methylisobutylketon (MIBK), 80 g methanol, 50 g water, 30 g acetone, 10 g toluene, and 80 g high boilers. The distillation progress is measured in

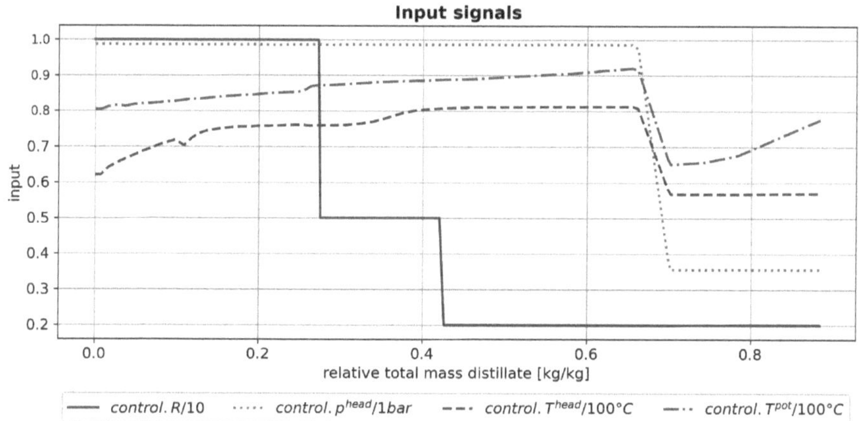

FIG. 6.5 Sample control input signals (normalized).

terms of the relative total mass distillate while compositions are plotted in mol/mol. Two liquid phases are present only between 12% and 26% of the total distillate. At these values, we observe steep gradients and instantaneous phase splitting and the reader may guess the associated numerical challenges to be overcome.

Since AD-Net consistently uses automatic differentiation, sensitivities come for free. When identifying NRTL parameters, this fact may be exploited to detect periods of an experiment where careful measurements provide maximal information. As can be seen from Fig. 6.7B, for instance, the NRTL-B parameter of the pair toluene-water may be identified most accurately in the period where the methanol content is highly curved. Therefore, experimenters should be aware of the fact that the segment of particularly informative measurements is interrupted at about 10% of the total mass distillate. Here, the derivatives of some mole fractions w.r.t. the selected NRTL parameter change sign.

4.2 Parameter identification

An important feature of AD-Net Distillation is its module on parameter estimation which allows the user to identify unknown parameters of substances or substance pairs from experimental fractionation data [24]. As an illustrative example, we consider a multicomponent system consisting of a set of known and one new component whose properties have not been well estimated, yet. Parameter estimation can also be formulated as an optimization problem of the form (6.2) and therefore fits into our framework. The unknown property parameters form the vector $\boldsymbol{p}$, and the mass compositions in the considered fractions constitute the vector of observables $\boldsymbol{y} = (\xi^{1,\,1}, \ldots, \xi^{1,\,n_C}, \ldots, \xi^{n_F,\,1}, \ldots \xi^{n_F,\,n}{}_C)$, where n_C is the number of components, $n_F$is the number of fractions, and

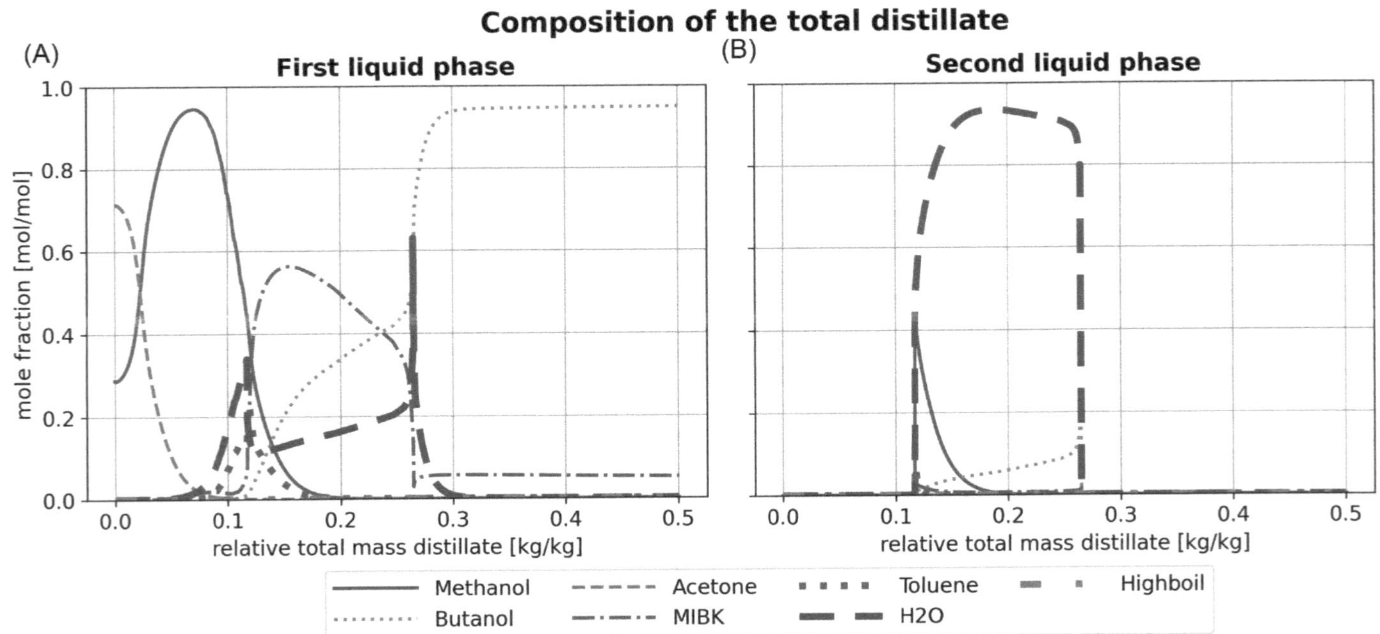

FIG. 6.6 Composition of the first (A) and second (B) liquid phase in the total distillate.

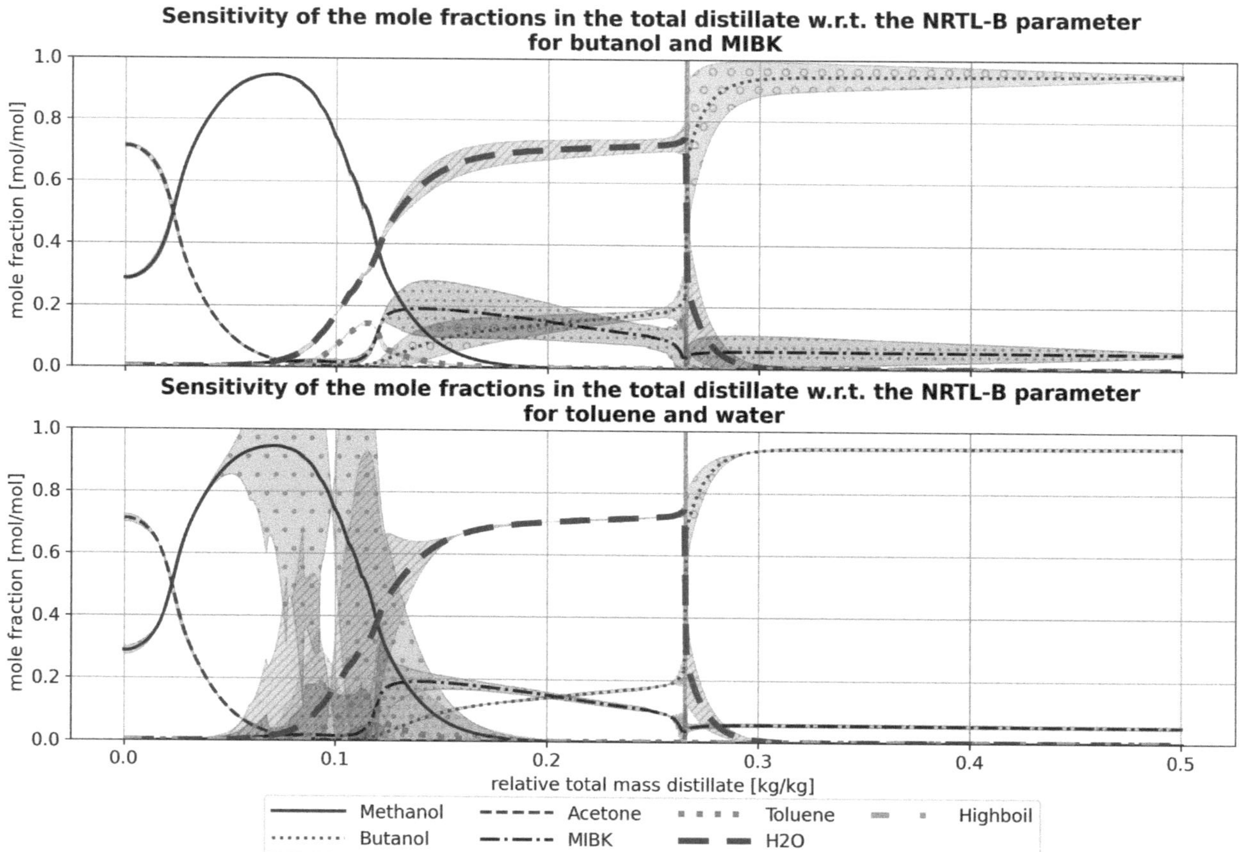

FIG. 6.7 Sensitivity of the mole fractions in the total distillate with respect to certain NRTL parameters. Sensitivities are indicated by *shaded envelopes* around the predicted mole fraction curves, that is, the upper and lower enveloping curves are given by $t \mapsto x(t,p_0) \pm \partial/\partial p\ x(t,p_0) \cdot \Delta p$, where $t \mapsto x(t,p_0)$ is the mole fraction curve for any of the substances for a nominal value p_0 of the considered parameter (here: $p_0 = 244.61$ for the figure in the top and $p_0 = 14{,}759.76$ for the figure in the bottom) and where Δp is some increment (here: $\Delta p = 1$).

$n_p = n_C n_F$ is the total number of observables. The objective function measures the deviation of simulated and measured compositions:

$$J(\boldsymbol{p}) = \sum_{i=1}^{n_p} \frac{\left(y_{\boldsymbol{p},i} - \hat{y}_i\right)^2}{\sigma_i^2} \tag{6.3}$$

In this equation, $\boldsymbol{y_p}$ are the compositions simulated for parameters $\boldsymbol{p}, \hat{\boldsymbol{y}}$ are the measured compositions, and $\boldsymbol{\sigma}$ are the associated measurement errors. We impose no constraints, so that Eq. (6.2) is an unconstrained nonlinear regression problem here. In addition to the mass compositions, the observables may also comprise pot and head temperatures at the fraction endpoints [25].

We now present the results of a real parameter identification applied to a system with six substances. Five of them are well known: *acetone*, *methanol*, *toluene*, *ETHGL*, and *water*. But the sixth component, called *unknown*, still requires some investigation. As the structural formula is known, some NRTL parameters can be determined by predictive methods in a sufficiently accurate way. However, the NRTL parameter of the pair *water-unknown* proves to be very uncertain and has to be calibrated by experiment. The measurements have been taken at Bayer AG. Here, all the unary and binary parameters for subsystems other than *water-unknown* are taken from the literature or are determined using some other predictive methods. During the real distillation process, the mass compositions of 46 distillate fractions are measured together with the head pressure and pot and condenser temperatures. We use the latter three signals and the given reflux ratios as input signals to our model. They control the distillation process, its speed, and distribution of energy.

In order to achieve the expected degree of separation, we equip our model with 10 stages. Since multiple liquid phases are observed in the experiment, we activate the liquid-liquid splitting checker. Applying the SQP solver from *scipy.optimize* with initial guess $\boldsymbol{p}_{init} = 0$, we obtain optimized property parameters $\boldsymbol{p}_{opt}$ typically after less than 25 iterations. Fig. 6.8 shows the measured and predicted fractionation of the distillate obtained for the property parameters $\boldsymbol{p}_{init}$ and $\boldsymbol{p}_{opt}$ before and after the optimization, respectively. We can observe that after the optimization, the predicted fraction compositions match the measured ones significantly better than before. Moreover, in comparison to the initial guess, the optimized property parameters $\boldsymbol{p}_{opt}$ lead to a completely different character of the investigated subsystem *water-unknown*, cf. Fig. 6.9: while the subsystem behaves like a nonazeotrope for $\boldsymbol{p}_{init}$, the true azeotropic behavior is detected for $\boldsymbol{p}_{opt}$. This underlines the robustness of our method.

4.3 Optimal control

Another important feature of AD-Net Distillation is its optimal control module. With this module, it is possible to design reflux ratio profiles and fractionation

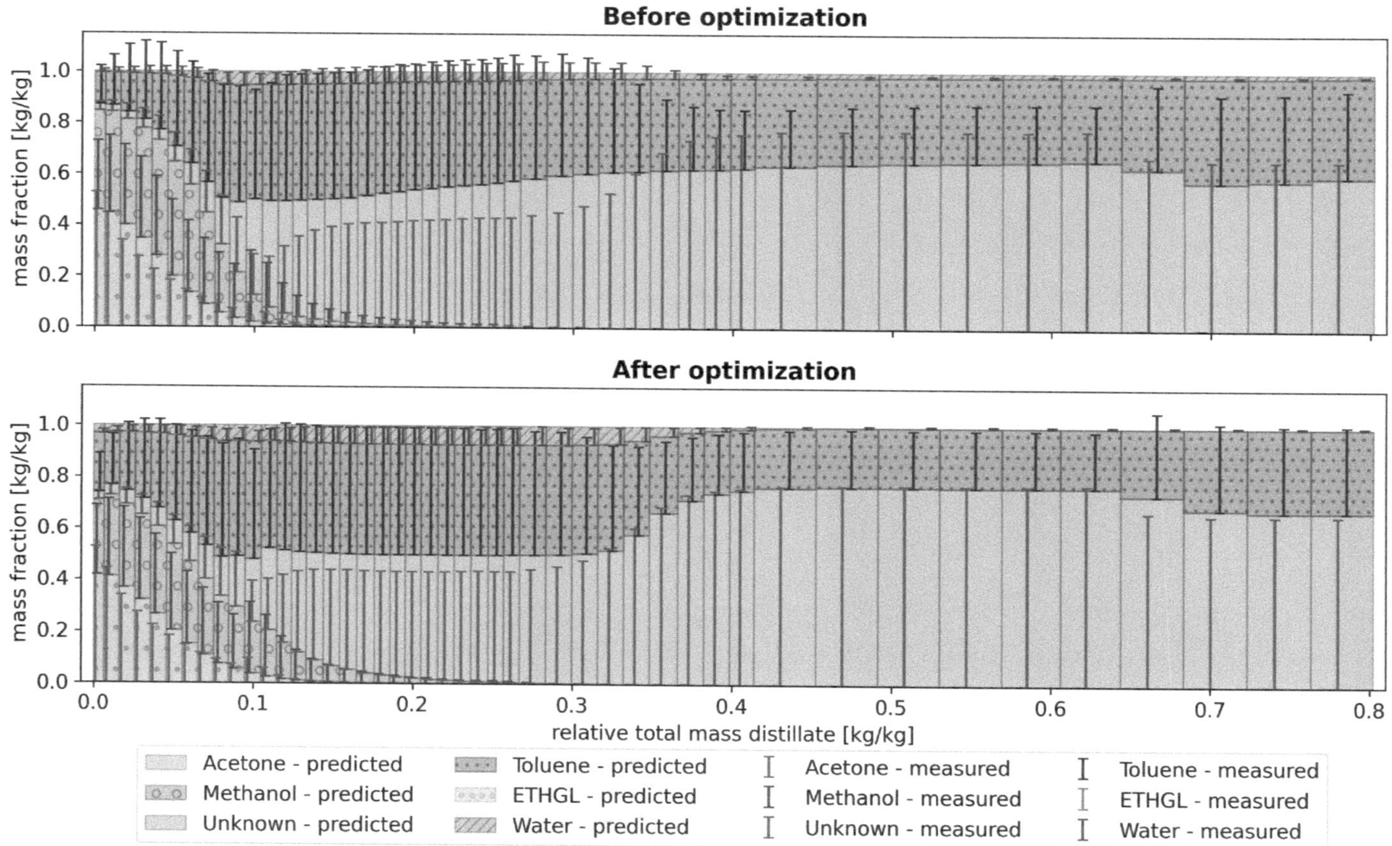

FIG. 6.8 Comparison of predicted and measured mass compositions. The predicted mass fractions of each distillate fraction are represented by *slices* stacked on top of each other. The measured mass fraction of a given component is represented by a *vertical line* anchored at the same height as the slice representing the corresponding predicted mass fraction.

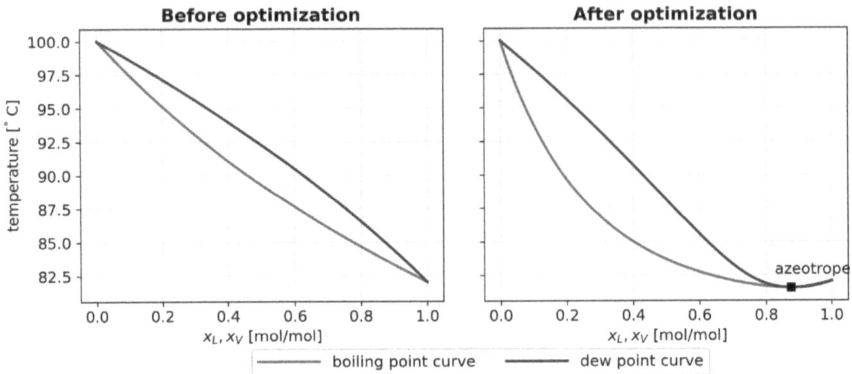

FIG. 6.9 McCabe-Thiele-diagram (Txy-diagram) for the guessed and for the optimized NRTL parameter of the pair water-unknown.

schemes in such a way that, on the one hand, certain user-specified objectives are optimally achieved and that, on the other hand, certain constraints are satisfied. Some typical examples of such objectives are the total mass of a fraction, the total distillation time, or the total heat energy used in the distillation process. Constraints are typically imposed on the purity of a certain component in some distillate fraction. After discretizing the reflux ratio profile as a piecewise constant signal with n_R internal knots, the optimal control problems described above take the form (6.2), and the respective optimization parameter is given by

$$\boldsymbol{p} = \left(m^{R,1}, \ldots, m^{R,n_R}, R_0, \ldots, R_{n_R}, m^{F,1}, \ldots, m^{F,n_F}\right),$$

where $m^{R,\,1}, \ldots, m^{R,\,n_R}$ and $R_0, \ldots, R_{n_R}$ are the internal knots and the values of the piecewise constant reflux ratio signal, respectively, and where $m^{F,\,1}, \ldots, m^{F,\,n_F}$ are the endpoints of the n_F fractions. We express the quantities $m^{R,\,k}$, $m^{F,\,k}$ in total distillate units, because in these units—as opposed to physical time units—the instant at which the apparatus runs dry (and beyond which the forward simulation makes no sense) is independent of the optimization parameter $\boldsymbol{p}$ and known in advance (namely $m^{head} := m_{init}$, the initial mass input to the apparatus). It should be pointed out that, even though the numbers n_R, n_F of reflux ratio knots and fractions can be chosen arbitrarily by the user, they are fixed and not optimized. Also, the objective function J is a weighted sum

$$J(\boldsymbol{p}) = w_1 J_1(\boldsymbol{p}) + \ldots + w_m J_m(\boldsymbol{p}) \tag{6.4}$$

with user-specified contributions $J_1, \ldots, J_m$ from the following list of possible objectives:

- the total mass $m^{F,\,k} - m^{F,\,k-1}$ of one or several fractions k
- the total distillation time t^{dist}, which is defined to be the endtime $t^{F,\,n_F}$ of the last fraction

- the total mass $m^{pot}(t^{dist}_{t^{dist}})$ in the pot at the end of the distillation
- the total heat energy $\int_0 Q^{pot}(\tau)d\tau$

Apart from the objectives, the user also has to specify weights $w_1, \ldots, w_m$ expressing the importance of the individual objectives, and to specify, for each objective, whether it should be minimized or maximized (which means that the corresponding weight has to be chosen positive or negative, respectively). And finally, the constraint functions c_i from Eq. (6.2) can be chosen from the following list of possible constraints:

- the purity $\xi_i^{F,k}$ of some component i in fraction k is above (or below) some threshold τ
- the purity $\xi_i^{pot}(t^{dist})$ of some component i in the pot at the end of the distillation is above (or below) some threshold τ
- some of the fraction endpoints are coupled to some of the reflux ratio knots, for example:

$$m^{R,1} = m^{F,1} \text{ and } m^{R,2} = m^{F,2} \tag{6.5}$$

As in Section 4.2 above, the vector $\boldsymbol{\xi}^{F,k}$ represents the mass composition of fraction k, while $\boldsymbol{\xi}^{pot}(t)$ represents the mass composition in the pot at time t. We point out that, in principle, the user can choose an arbitrary number of objectives and constraints from the lists above.

As a simple illustration of the general framework described above, we now present the optimization results for a very simple mixture, namely *(acetone, butanol, methanol)*, which is distilled in a column with *six* stages. In this example, we choose the piecewise constant reflux ratio to have just $n_R := 2$ internal knots (and, correspondingly, $n_R + 1 = 3$ possible values). We also choose to fractionate the distillate into $n_F := 3$ fractions and to allow switches in the reflux ratio values only at the endpoints $m^{F,1}$, $m^{F,2}$ of fractions 1 and 2, that is, we impose the coupling constraint (6.5) above. As objectives, we choose to maximize the mass of fraction 2 and to minimize the total distillation time, the mass in the pot at the distillation end, and the total heat energy. And finally, as purity constraints, we require that the purity of methanol in fraction 2 be at least 95% and that the purity of methanol in the pot at the distillation end be at most 1%. Initializing the trust-region optimizer from *scipy.optimize* available in AD-Net Distillation with some initial guess $\boldsymbol{p}_{init}$ (Fig. 6.10A), we obtain the reflux ratio profile and fraction endpoints from Fig. 6.10B to be optimal. With the SQP optimizer, the same optimal solution is found.

Fig. 6.11 illustrates the mass compositions of the pot and the total distillate over time, namely for the initial and the optimal values $\boldsymbol{p}_{init}$, $\boldsymbol{p}_{opt}$ of the optimization parameter (Fig. 6.11A and B). It is worth noting that even though the employed trust-region solver is an interior-point method, the initial guess need not be feasible, that is, the purity constraints need not be satisfied at the initial guess.

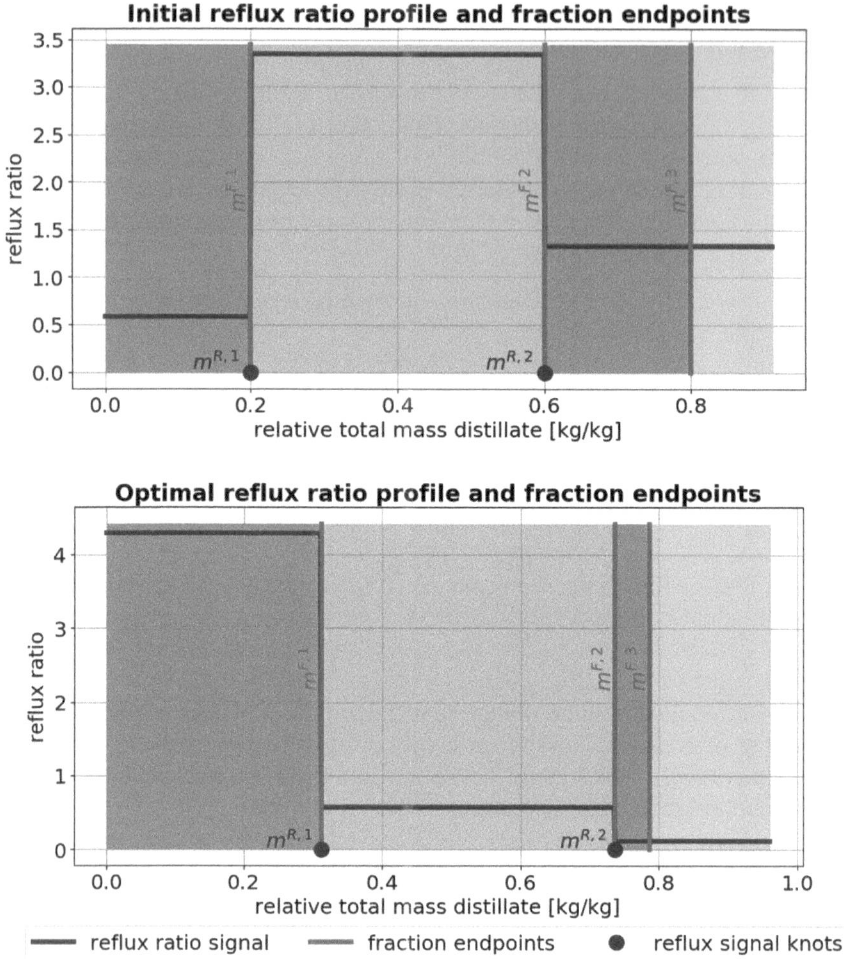

FIG. 6.10 Initial and optimal reflux ratio profile and fraction endpoints. *Vertical red lines (gray in print version)* indicate the fraction endpoints, while *blue dots (dark gray in print version)* mark the internal knots of the reflux ratio signal. The period after the last fraction or, in other words, after the distillation endtime is colored *gray*.

5 Conclusion

In this chapter, we have illustrated the power of combining the deep technical expertise of engineers with a mathematical view of recurring structures of technical networks. The template library AD-Net provides a general framework to model, simulate, and optimize dynamic technical networks. By populating this framework with application-specific relations between parameters, states, and observables, and by formulating cost functions and constraints, AD-Net can be easily extended to new technical fields. Examples are given for power transmission and district heating, while the main focus is on batch distillation. Here, we

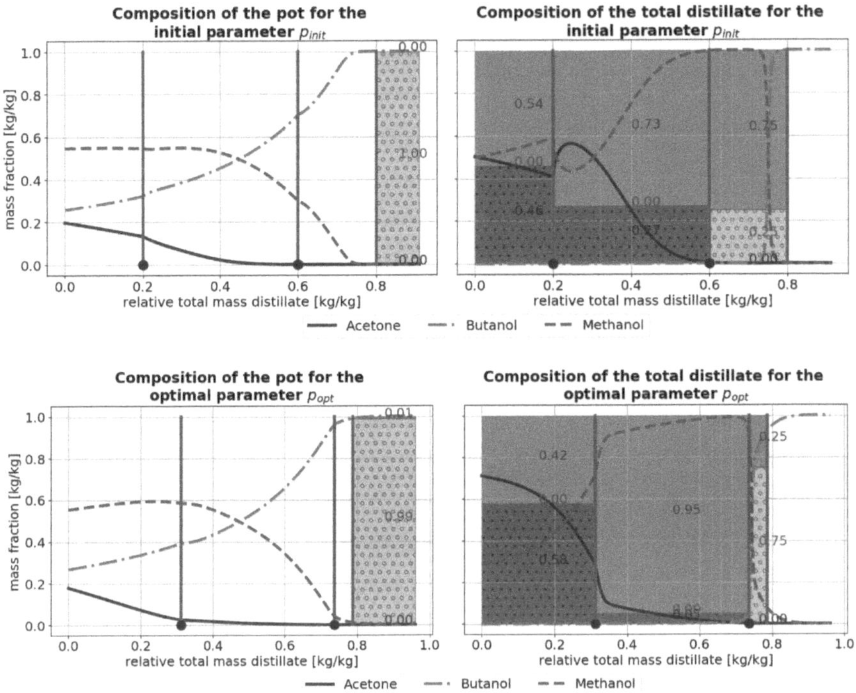

FIG. 6.11 Composition in the pot and the total distillate for the initial value of the optimization parameter (A) and the optimal value of the optimization parameter (B). As in Fig. 6.10, the fraction endpoints are indicated by *vertical red lines (gray in print version)*, while the composition of the fractions is indicated by *colored slices (gray in print version)* stacked on top of each other.

address two typical tasks—estimating the NRTL parameters of a new pair of substances and maximizing distillate fractions with a given purity.

Acknowledgment

AD-Net-Power was developed in the project MathEnergy supported by the Federal Ministry for Economic Affairs and Energy (funding code 0324019A). We thank Bayer AG for the release of the measurement data.

References

[1] M. Bücker, G. Corliss, P. Hovland, U. Naumann, B. Norris (Eds.), Automatic Differentiation. Applications, Theory, and Implementations, Springer, 2006.

[2] A. Griewank, A. Walther, Evaluating Derivatives. Principles and Techniques of Algorithmic Differentiation, second ed., Society for Industrial and Applied Mathematics, 2008.

[3] L.T. Biegler, Nonlinear Programming. Concepts, Algorithms, and Applications to Chemical Processes, Society for Industrial and Applied Mathematics, 2010.

[4] A. Wächter, L.T. Biegler, On the implementation of a primal-dual interior point filter line search algorithm for large-scale nonlinear programming, Math. Program. 106 (2006) 25–57.

[5] R.H. Byrd, J. Nocedal, R.A. Wright, Knitro: An Integrated Package for Nonlinear Optimization, Available at https://web.archive.org/web/20161012043221/https://www.artelys.com/downloads/pdf/composants-numeriques/knitro/papers/integratedpackage.pdf.
[6] S. Peyghami, P. Davari, M. Fotuhi-Firuzabad, F. Blaabjerg, Standard test systems for modern power system analysis: an overview, IEEE Ind. Electron. Mag. 13 (2019) 86–105.
[7] https://www.digsilent.de/de/powerfactory.html.
[8] M. Aehle, Simulation of Gas Pipelines via Asymptotic Expansions (Master's thesis), University of Kaiserslautern, 2020.
[9] J. Mohring, D. Linn, M. Eimer, M. Rein, N. Siedow, District heating networks—dynamic simulation and optimal operation, in: Mathematical Modeling, Simulation and Optimization for Power Engineering and Management, Springer, Cham, 2021, pp. 303–325.
[10] L.T. Biegler, I.E. Grossmann, A.W. Westerberg, Systematic Methods of Chemical Process Design, Prentice Hall, 1997.
[11] U. Diwekar, Batch Distillation. Simulation, Optimal Design, and Control, CRC Press, 2012.
[12] J. Gmehling, B. Kolbe, M. Kleiber, J. Rarey, Chemical Thermodynamics for Process Simulation, Wiley, 2012.
[13] S. Skogestad, Chemical and Energy Process Engineering, CRC Press, 2008.
[14] M.F. Doherty, J.D. Perkins, On the dynamics of distillation processes—I. The simple distillation of multicomponent non-reacting, homogeneous liquid mixtures, Chem. Eng. Sci. 33 (1978) 281–301.
[15] D.B. van Dongen, M.F. Doherty, On the dynamics of distillation processes—VI. Batch distillation, Chem. Eng. Sci. 40 (1985) 2087–2093.
[16] M.L. Michelsen, J.M. Mollerup, Thermodynamic Models. Fundamentals and Computational Aspects, second ed., Tie-Line Publications, 2007.
[17] J.-S. Wu, P.R. Bishnoi, An algorithm for three-phase equilibrium calculations, Comput. Chem. Eng. 10 (1986) 269–276.
[18] Z. Li, K.A. Mumford, K.H. Smith, J. Chen, Y. Wang, G.W. Stevens, Solution structure of isoactivity equations for liquid-liquid equilibrium calculations using the nonrandom two-liquid model, Ind. Eng. Chem. Res. 55 (2016) 2852–2859.
[19] L.E. Baker, A.C. Pierce, K.D. Luks, Gibbs energy analysis of phase equilibria, Soc. Pet. Eng. J. 22 (1982) 731–742.
[20] D.V. Nichita, S. Gomez, E. Luna, Multiphase equilibria calculation by direct minimization of Gibbs free energy with a global optimization method, Comput. Chem. Eng. 26 (2002) 1703–1724.
[21] G.A. Iglesias-Silva, A. Bonilla-Petriciolet, P.T. Eubank, J.C. Holste, K.R. Hall, An algebraic method that includes Gibbs minimization for performing phase equilibrium calculations for any number of components or phases, Fluid Phase Equilib. 210 (2003) 229–245.
[22] V. Gopal, L.T. Biegler, Smoothing methods for complementarity problems in process engineering, AICHE J. 45 (1999) 1535–1547.
[23] H. Renon, J.M. Prausnitz, Local compositions in thermodynamic excess functions for liquid mixtures, AICHE J. 14 (1986) 135–144.
[24] C.-Y. Gau, J.F. Brennecke, M.A. Stadtherr, Reliable nonlinear parameter estimation in VLE modeling, Fluid Phase Equilib. 168 (2000) 1–18.
[25] M. Bortz, R. Heese, A. Scherrer, T. Gerlach, T. Runowski, Estimating mixture properties from batch distillation using semi-rigorous and rigorous models, in: A.A. Kiss, E. Zondervan, R. Lakerveld, L. Özkan (Eds.), Proceedings of the 29th European Symposium on Computer Aided Process Engineering, 2019.

Chapter 7

The use of digital twins to overcome low-redundancy problems in process data reconciliation

Filippo Bisotti[a], Andrea Galeazzi[a], Francesco Gallo[b], and Flavio Manenti[a]
[a]*Politecnico di Milano, CMIC Dept. "Giulio Natta", Centre for Sustainable Process Engineering Research (SuPER), Milano, Italy,* [b]*Itelyum Regeneration s.r.l, Lodi, Italy*

1 Introduction

A chemical plant includes many process units interconnected by a complex network of streams. Measurements of mass flow rates, temperatures, and other variables are continuously performed [1] since process data are the base for process control, production accounting goals, as well as supply chain and online optimization [2,3]. All the measurements are affected by errors, whose effect is reflected in discrepancies in material and energy balances. Such errors would usually be expected to be random with a normal distribution that has a null expected value, but occasionally also systematic errors with an expected value other than zero can be found. These are also called bias and could occur for instance through malfunction of instruments, miscalibration, or poor sampling [4].

To obtain accurate data or at least reduce the presence of misleading observations, data reconciliation (DR) is required. DR is the process of adjusting, or reconciling, the process measurements to obtain more accurate estimate of flow rates, temperatures, etc., that are consistent with material and energy balances [5].

Reconciled values allow to obtain a more accurate evaluation of process yield as compared to the use of raw data and they also help in better identification of energy inefficiencies in processes and equipment by assessing the consumption of utilities in different process units.

Simulation and Optimization in Process Engineering. https://doi.org/10.1016/B978-0-323-85043-8.00011-8

The online optimization is used to provide setpoints to the process distributed control system (DCS) in order to maintain the process near its optimum operating conditions [6]. This optimization requires an accurate process model that usually contains some parameters which have to be estimated through plant data [7]. This is also known as model tuning which requires accurate data, indeed the use of erroneous measurements in model tuning may give rise to incorrect model parameters identification that nullifies the benefits achievable through optimization [6].

It should be emphasized that, in today's highly competitive world market, correcting even small errors can lead to significant improvements in plant performance and economy [5].

Data reconciliation can be very useful also in scheduling maintenance of process, in fact, it can be used to accurately estimate key performance parameters of process equipment [8,9] such as the heat transfer coefficient of heat exchangers [10–12] or the level of catalyst activity in reactors [6].

Unfortunately, if the measurements are adjusted to conform to the conservation laws in the presence of biases, all the adjustments are greatly affected by such biases and would not, in general, be reliable indicators of the state of the process. Thus biases, or gross errors (GE), must, whether possible, be identified and either corrected or the measurement discarded before performing the DR [4]. Gross error detection not only improves the estimation accuracy of data reconciliation procedures but is also useful in identifying instrumentation problems, which require special maintenance and correction. Incipient detection of gross errors can reduce maintenance costs and provide smoother plant operations [6].

The present chapter proves that our in-house Steady-State Data Reconciliation model implemented in Matlab properly works since it enables us to adjust industrial observations to reduce the variance of the overall experimental dataset. The developed framework enables to operate also on low-redundancy systems such as in the case of industrial plants. Moreover, our numerical tool allows recognizing outliers and which streams are affected by misleading observations. Hence, the DR applied on industrial-scale benefits the controllability and data interpretation, indeed, it is possible to improve the measurement's distribution and suggest to the industry the optimal location of the instrumentation in the field without impacting on the pre-existing control system with a reduced investment cost. In addition, the use of digital twin in the industry is going to become relevant for Dynamic Data Reconciliation (DDR) to overcome the limited number of measured process variables and its impact on data analytics. Despite a feasible solution for DR in a poor-redundant system that has been provided in the literature and the role of models have been highlighted [13], more additional improvements can be achieved in this direction and some challenges should be overcome.

2 Data reconciliation

2.1 Variable classification

Typically, in a chemical plant, due to high costs related to process instrumentation, maintenance, and data collection and storage, there is an incomplete set of instruments, thus not all the unmeasured variables can be estimated. Hence, process variables are classified as measured and unmeasured variables.

The unmeasured variables (vector $\boldsymbol{u}$) are furtherly divided into the following two groups: determinable or estimable and indeterminable or inestimable [5]:

- An unmeasured variable is determinable if it can be evaluated from the available measurements using the balance equations through inference.
- An unmeasured variable is indeterminable if it cannot be evaluated from the available measurements using the balance equations.

Similarly, also the measured variables (vector $\boldsymbol{x}$) can be divided into two groups, redundant and nonredundant:

- A measured process variable is redundant, or overdetermined if it can be also computed from the balance equations using other measured variables. These measurements are corrected by the reconciliation.
- A measured process variable is called nonredundant or just-measured if it cannot be computed from the balance equations using the rest of the measured variables. These measurements are not corrected by the reconciliation but are used only to evaluate the unmeasured variables by inference.

2.2 Steady-state data reconciliation (DR)

Every measurement is represented as a contribution of two different terms, the mean value ($\boldsymbol{x}$), and the residue ($\boldsymbol{\varepsilon}$). In the absence of gross errors and considering that the variables are directly measured, the measurement vector can be written as:

$$\boldsymbol{y}=\boldsymbol{x}+\boldsymbol{\varepsilon}, \quad \boldsymbol{y}\in\Re^{l}, \quad \boldsymbol{x}\in\Re^{g} \tag{7.1}$$

The assumptions that usually are made are:

1. The expected value of $\boldsymbol{\varepsilon}$ is the null vector, i.e., $E(\boldsymbol{\varepsilon})=\boldsymbol{0}$
2. Successive vectors of measurements are independent, i.e., $E(\varepsilon_i\varepsilon_j^T)=\boldsymbol{0}$, for $i\neq j$
3. The covariance matrix of the measurement errors is known and positive definite, i.e., $\text{Cov}(\boldsymbol{\varepsilon})=\boldsymbol{\Psi}=E(\varepsilon_i\varepsilon_i^T)$
4. The random errors $\boldsymbol{\varepsilon}$ are normally distributed, i.e., $\boldsymbol{\varepsilon}\sim N(\boldsymbol{0},\boldsymbol{\Psi})$ [5,14,15],

According to the assumptions (2) and (3), the diagonal elements of the matrix $\boldsymbol{\Psi}$, σ_{ii}^2, is the variance of the measurement i, and the off-diagonal element, σ_{ij}^2, is the covariance of measurement i and j. It is common practice to consider all the covariances null [6].

Considering this assumption, the form of the matrix $\boldsymbol{\Psi}$ is reported in Eq. (7.2).

$$\boldsymbol{\Psi} = \begin{bmatrix} \sigma_{11}^2 & 0 & 0 \\ 0 & \ddots & 0 \\ 0 & 0 & \sigma_{ll}^2 \end{bmatrix} \quad (7.2)$$

Some of the largest parts of the process variables must conform to some physical relationships arising from laws of conservation. In general, these are a set of nonlinear algebraic equations involving both measured and unmeasured variables.

The data reconciliation problem can be generally stated as a constrained weighted least-square estimation problem [16]:

$$\begin{aligned} &\min_{x,u} (\boldsymbol{y}-\boldsymbol{x})^T \boldsymbol{\psi}^{-1} (\boldsymbol{y}-\boldsymbol{x}) \\ &\text{s.t.} \\ &\varphi(\boldsymbol{x}, \boldsymbol{u}) = \boldsymbol{0} \\ &\boldsymbol{x}^L \le \boldsymbol{x} \le \boldsymbol{x}^U \\ &\boldsymbol{u}^L \le \boldsymbol{u} \le \boldsymbol{u}^U \end{aligned} \quad (7.3)$$

φ is the constraints function that is generally defined as the product of a constraint matrix (**A**) and measured variables ($\boldsymbol{x}$). Since the objective function weights each measurement correction on its precision, the reconciled value of the most accurate data is closer to the measurement, whereas the least precise one is largely corrected by the reconciliation to respect and fulfill the constraints.

Different methodologies are required for solving problem (7.3) depending on whether the constraints are a linear or a nonlinear set of algebraic equations.

When the constraints are a set of linear equations, the data reconciliation problem is linear. This is the typical feature of flow rate measurements reconciliation when the constraints are material conservation laws.

In the data reconciliation context, two scenarios can be distinguished:

- All the variables included in the process model are measured.
- Some variables included in the process model are not measured.

The latter situation is the most frequent in industrial practice due to instrumentation cost and maintenance. The key parameter in steady-state data reconciliation is the (degree of) redundancy or DOR defined as the minimum amount of process variables that can be independently determined by the remaining variables [5,13]. Hence, redundancy is estimated as

$$DOR = (n.measured\ variables + n.constraints) - n.reconciled\ variable$$

The problem stated in Eq. (7.3) can be solved only in case of positive redundancy, i.e., when that system contains an excess of (redundant) information.

2.3 Gross error detection

In practice, process data may also contain other types of errors, called Gross Errors (GE). GE is caused by nonrandom events and therefore they are systematic. In general, they are classified into two categories:

- Instrument Malfunctioning: they are related to the measurement instrument. They can be caused by miscalibration, wear or corrosion of sensor and solid deposits [6]. They can be classified in Martini et al. [17]:
  - Bias: it occurs when a sensor has an offset or bias.
  - Complete failure: it occurs when the output of the sensor provides completely abnormal and constant results.
  - Drifting: it occurs when the output of the sensor changes slowly regardless of the nature of the measured quantity.
  - Precision degradation: it occurs when the precision of the instrument decreases in time.
- Leaks in units: they are related to the inaccuracies of the process model equations, that is the model is significantly different from reality.

The presence of these nonstochastic errors invalidates the statistical basis of data reconciliation procedures. It is also impossible, for example, to prepare an adequate process model on the basis of erroneous measurements or to assess production accounting correctly [5]. For these reasons, one of the main issues and greatest research topics in the field of data reconciliation in process plants is the gross errors detection in order to avoid the aforementioned shortcomings [18].

2.4 Gross error effect and how to handle

It is intuitive that detecting the gross error before performing the reconciliation is of primary importance since they impact the objective function and constraints. A data reconciliation provided without removing GE brings misleading pieces of information concerning the real performance and state of the monitored processes. Many experts proposed several numerical techniques to identify and remove gross errors from observations. Once all the GE are detected two actions are possible [5,19]:

- Discard the measurements affected by the gross error and perform the reconciliation considering the data eliminated as unmeasured.
- Correct the measurements affected by gross error estimating their magnitude and perform the reconciliation (Fig. 7.1).

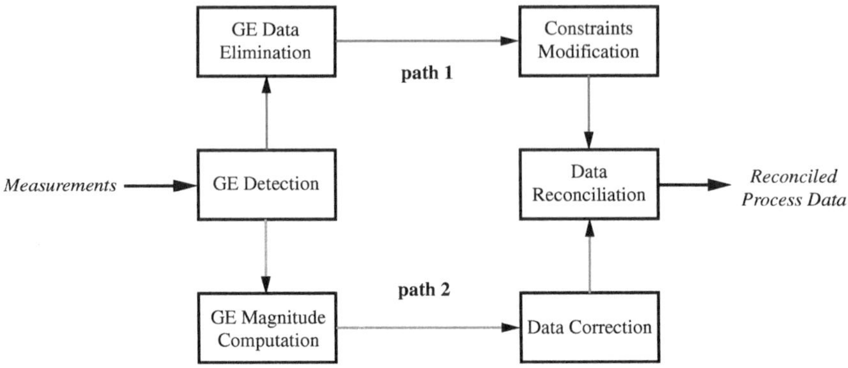

FIG. 7.1 Possible path for gross error (GE) handling in data reconciliation.

2.5 Gross error detection: Statistical methods

Any comprehensive gross error detection strategy should include the following capabilities [6]:

- Detect the presence of one or more gross errors in the data (detection problem)
- Identify the type and the location of the gross error (identification problem)
- Locate and identify multiple gross errors which may be present simultaneously in the data (multiple gross error identification problem)
- Estimate the magnitude of the gross errors (estimation problem)

The basic principles in gross error detection are derived from the detection of outliers in statistical applications [6]. The most used statistical techniques for detecting gross errors are based on hypothesis testing. In the literature advanced [20] and less complex statistical methods [21] have been proposed, the most common are as follows:

1. Global Test (GT)
2. Measurement Test (MT)
3. Constrained Test (CT)

The concept at the base of these statistical tests is that random errors are normally distributed around null values. In the case of influential observations and gross errors, this property is verified no longer. Hence, through statistical tests (i.e., fixing a confidence interval) it is possible to detect the presence of gross error. The cited statistical tests differ according to the statistics adopted and confidence interval.

2.6 GE statistical detection algorithms

To correctly identify the measurement affected by the gross error, some strategies based on the use of the above-mentioned statistic tests have been developed. The GE detection algorithms are classified as either serial or simultaneous strategies [6].

The serial algorithms identify gross error serially one by one exploiting different kinds of statistical tests. Unlike the serial algorithms, the simultaneous ones do not use an iterative approach but identify all the gross errors present in the data simultaneously. This technique allows avoiding the so-called smearing effect that leads to many mispredictions [17]. The smearing effect is also known as the contamination effect occurs when the least-squares objective function is used in DR while gross errors are present in measurements. In this scenario, the reconciliation drives the solution toward the measurement with gross errors overcorrecting the "healthy measurements." Due to this effect, the test statistics of good measurements can exceed the test criterion identifying them as measurements affected by gross error. On the other hand, the test statistics of the real gross error measurements will probably be lower than the test criterion and therefore they will be considered as good measurements.

The smearing effect is more evident when multiple gross errors are present. In this scenario, the simultaneous algorithms perform better than the serial ones. In fact, the simultaneous algorithms detect all the gross errors in a single iteration differently from the serial techniques which identify a single gross error each iteration. Therefore, in these methods, at least a gross error will be present in various iteration leading to higher probability of smearing effect. Simultaneous Estimation of Gross Errors (SEGE) proposed by Romagnoli and Sánchez [5] is an excellent prototype for this family of algorithms. In the first stage of the procedure, a set of measurements suspected to be affected by GE is constructed using the Global Test. Then, all the possible combinations of measurements present in the suspect set are considered. Gross errors magnitudes are estimated simultaneously for each combination [22].

2.7 Numerical method for low-redundant system

These methods are not based on classical statistical tests, but they exploit some robust estimators such as clever mean, clever variance, median and mad [23,24]. Different from the algorithms based on statistical tests, the detection of the gross errors does not exploit neither all streams measurements nor the plant features. These methods take advantage only of the data of a single stream (historical data and real-time data) to detect outliers (or better influential observation). Therefore, theoretically, redundancy is not a necessary condition for their application.

A point will be identified and then classified as an outlier whether its distance from the true value estimated by the robust estimator is greater than a threshold value opportunely set. This procedure is repeated for each stream. The main drawback of this kind of GE detection technique is the use of plant historical data that is usually stored on the distributed control system (DCS) and this may cause some computational inefficiencies in data processing and analysis [25].

It is worth mentioning that the historical data must refer to the same steady state. If the measurements coming from the plant and the historical data do not refer to the same steady-state, all the new measurements will be classified as gross errors. Notice that these methods do not suffer the smearing effect since the streams are analyzed one by one.

Traditionally, the procedure to detect outliers in a set of n data is the following [25]:

1. Evaluate the arithmetic mean $\overline{y}$

$$\overline{y} = \frac{\sum_{i=1}^{n} y_i}{n} \tag{7.4}$$

2. Evaluate the variance s^2

$$s^2 = \frac{\sum_{i=1}^{n} (y_i - \overline{y})^2}{n-1} \tag{7.5}$$

3. Evaluate the standardized residuals r_i

$$r_i = \frac{y_i - \overline{y}}{s} \tag{7.6}$$

4. If $r_i \geq \delta$ the data y_i is an outlier. δ is a threshold value and usually $\delta = 2.5$

Since both the arithmetic mean and the standard deviation are nonrobust estimators, they are biased even though there is only a single outlier [23]. Therefore, it is erroneous to use these estimators to detect possible outliers, hence, the importance of robust indicators raises [24].

The correct methods should use robust estimators. These methods do not neglect outliers, rather they are simply uninfluenced by them, therefore a robust estimator should always be used to check the presence of bad data.

3 Clever mean and clever variance (cm and cv)

Novel robust criterion for detecting outliers in a set of data is based on two robust estimators: the clever mean (cm) and the clever variance (cv) [25]. Considering a set of n data extracted (n very large) from the population Y with unimodal symmetrical distribution, the procedure to detect outliers involving cm and cv is the following:

1. Select a predetermined number of maximum and minimum values among the data
2. Evaluate the zeroth-order clever mean (cm_0) and the zeroth-order clever variance (cv_0)

$$cm_0 = \frac{\sum_{i=1}^{n} y_i}{n} = \overline{y} \tag{7.7}$$

$$cv_0 = \frac{\sum_{i=1}^{n} (y_i - cm_0)^2}{n-1} = s^2 \tag{7.8}$$

3. Assuming y_1^* as the first possible outlier among the data selected at point (1), remove its value and evaluate the first-order clever mean (cm_1) and the first-order clever variance (cv_1)

$$cm_1 = \frac{\sum_{i=1}^{n} y_i - y_1^*}{n-1} \tag{7.9}$$

$$cv_1 = \frac{\sum_{i=1}^{n} (y_i - cm_1)^2 - (y_1^* - cm_1)^2}{n-2} \tag{7.10}$$

4. If the following condition holds the measurement is classified as a suspicious influential observation

$$\left|cm_1 - y_1^*\right| > \delta\sqrt{cv_1} \tag{7.11}$$

where δ is a threshold value (i.e., 2.5), the data y_1^* is considered an outlier and the procedure is repeated for each y-th element (Fig. 7.2).

It is worth stressing these following points [25]:

- The value of the clever mean might remain constant while outliers are progressively removed for basically two reasons. When treating large set of

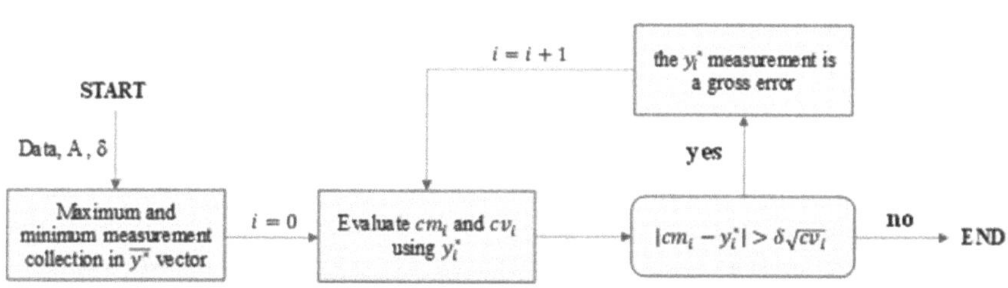

FIG. 7.2 Scheme for clever mean (cm) and clever variance (cv) method.

data, the arithmetic mean could change slightly even though an outlier is removed. Moreover, if two outliers are symmetrical with respect to the mean, their removal does not influence the clever mean. In conclusion, it is wrong to classify a point as outlier only by looking at the value of the clever mean.

- The clever variance has a monotone decreasing trend as outliers are gradually removed, while it remains the same or slightly increases when the point removed is not an outlier.

4 Median and mad

Another method to identify outliers involves other two robust estimators: *median* and *mad* [25]. The median, m_y, of a continuous distribution, is a value of the random variable such that the probability that an observed value of the distribution is less than or equal to the median is 0.5, and the probability that an observed value of the distribution is greater than or equal to the median is 0.5.

If the probability distribution of the population is symmetric, the median is a consistent estimator of the true value of the variable. Moreover, the median is very robust, its estimate of the true value is good even if almost half the experimental data are outliers.

The *mad* (median absolute deviations) is a consistent and very robust estimator of the standard deviation σ if the population is normally distributed [23]. It can be evaluated from the median through Eq. (7.12).

$$mad = 1.253313 \cdot m_y |y_i - m_y| \tag{7.12}$$

The procedure to detect outliers is as follows:

1. Evaluate the median, m_y
2. Evaluate the *mad* by mean of the definition (7.12)
3. Evaluate the standardized residuals r_i as

$$r_i = \frac{y_i - m_y}{mad} \tag{7.13}$$

4. If $r_i \geq \delta$ the data y_i is an outlier. δ is the threshold value and usually is set equal to $\delta = 2.5$ (Fig. 7.3)

This is exactly the traditional procedure based on normalized residuals but using median and *mad* instead of mean and variance. In this way, the robustness is guaranteed and increased. Nevertheless, a major drawback of this method is its high computational time due to the need of ordering the data to evaluate the median.

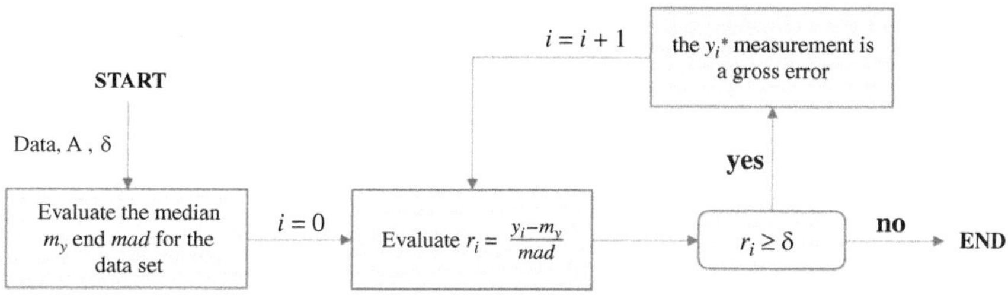

FIG. 7.3 Scheme of median and mad method.

4.1 Dynamic data reconciliation

As in the case of the steady-state data reconciliation, also the Dynamic Data Reconciliation (DDR) is formally a constrained minimization problem. The constraints are represented by the dynamic model of the system. The general formulation for the NDDR problem is reported in Eq. (7.14)

$$\min_{\hat{y}} \boldsymbol{\Phi}[\boldsymbol{y}, \hat{\boldsymbol{y}}, \boldsymbol{\sigma}]$$

$s.t.$

$$\boldsymbol{f}\left[\frac{d\hat{\boldsymbol{y}}}{dt}, \hat{\boldsymbol{y}}(t)\right] = \boldsymbol{0}$$

$$\boldsymbol{\eta}[\hat{\boldsymbol{y}}(t)] = \boldsymbol{0}$$

$$\boldsymbol{\omega}[\hat{\boldsymbol{y}}(t)] \geq \boldsymbol{0} \tag{7.14}$$

where

- $\boldsymbol{\Phi}$ is the objective function
- $\boldsymbol{y}$ are the sets of measurements at different times (input and state variables)
- $\hat{\boldsymbol{y}}(t)$ are the estimate functions
- $\boldsymbol{\sigma}$ is the measurements standard deviation vector
- $\boldsymbol{f}$ is the set of differential equation constraints
- $\boldsymbol{\eta}$ is the set of algebraic equality constraints
- $\boldsymbol{\omega}$ is the set of algebraic inequality constraints

The most used objective function is defined as:

$$\boldsymbol{\Phi}[\boldsymbol{y}, \hat{\boldsymbol{y}}, \boldsymbol{\sigma}] = \sum_{k=0}^{c} \frac{1}{2} [\hat{\boldsymbol{y}}(t_k) - \boldsymbol{y}_k]^T \boldsymbol{\Psi}^{-1} [\hat{\boldsymbol{y}}(t_k) - \boldsymbol{y}_k] \tag{7.15}$$

where

- $\hat{\boldsymbol{y}}(t_k)$ is the vector of estimate functions at time t_k
- $\boldsymbol{y}_k$ is the measurement vector at time t_k
- $\boldsymbol{\Psi}$ is the covariance matrix of the measurements

- t_0 is the initial reconciliation time
- t_c is the current time

The objective function is the sum over all the sampling times from t_0 to t_c of the weighted least squares defined exactly as for the steady-state reconciliation.

4.2 Moving time-window approach

An ideal data reconciliation scheme would use all the information (process measurements) from the startup of the process to the current time. Unfortunately, such a scheme would necessarily result in an optimization problem of ever-increasing dimension.

For the practical implementation, it can be used a moving time window to reduce the optimization problem to manageable dimensions [5,6,26].

In this technique, a history horizon $H\Delta t$ is defined from $(t_c - H\Delta t)$ to t_c, where Δt is the sampling time and t_c is the time at which the most recent measurements are available. Only data measurements within the horizon are reconciled during the dynamic data reconciliation run. Once a reconciliation run is performed, the time window shifts forward by Δt in such a way that the measurements at the new time t_c are included as shown in Fig. 7.4.

It is worth noticing that with this approach a single set of measurements at a certain time t^* is included in as much time windows as the number of points considered in a single history horizon, hence each set undergoes several reconciliations equal to the number of points considered in a single history horizon. Every time that a set of measurements is reconciled, in the next reconciliation run their measurements are replaced by the last estimates.

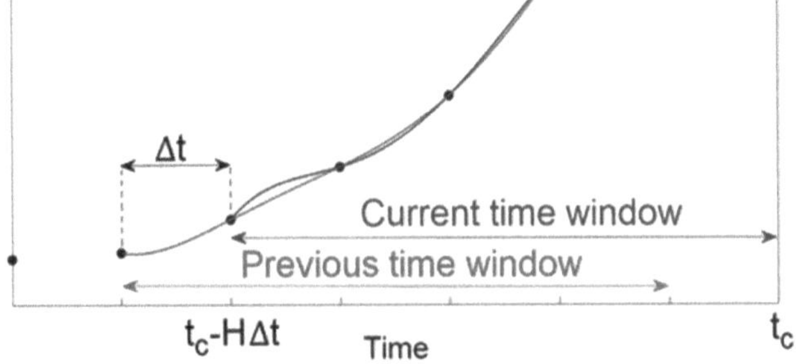

FIG. 7.4 Moving time horizon concept.

A set is considered completely reconciled when it exits from the last time window where it appears and therefore when it has already been reconciled multiple times. This approach allows reconciling a single set of data at a certain time t^* taking advantage of all the information provided from the data present in all the time windows where t^* is included, thus, all the measurements from time $(t^* - H \Delta t)$ to time $(t^* + H \Delta t)$.

4.3 Solution of DDR with orthogonal matrix

DDR reported in Eq. (7.14) is a differential/algebraic optimization problem where differential-algebraic solver and optimization algorithms are simultaneously applied. The orthogonal collocation matrix enables to partially simplify the problem since differential equations are approximated as a set of algebraic ones exploiting [27].

Hence, by mean of orthogonal matrix ($\boldsymbol{M_C}$), the original problem becomes a nonlinear algebraic optimization rewritten as:

$$\begin{aligned}
&\min_{\hat{y}} \boldsymbol{\Phi}[\boldsymbol{y}, \hat{\boldsymbol{y}}, \boldsymbol{\sigma}] \\
&s.t. \\
&\boldsymbol{M_C}\hat{\boldsymbol{y}} - \boldsymbol{f}(\hat{\boldsymbol{y}}) = \boldsymbol{0} \\
&\boldsymbol{\eta}[\hat{\boldsymbol{y}}(t)] = \boldsymbol{0} \\
&\boldsymbol{\omega}[\hat{\boldsymbol{y}}(t)] \geq \boldsymbol{0}
\end{aligned} \tag{7.16}$$

The solution of a differential equation at discrete time points can be approximated by a Lagrange interpolating polynomial. Considering, for instance, four discrete points, the polynomial is:

$$y(t) = A + Bt + Ct^2 + Dt^3 \tag{7.17}$$

In this derivation, a third-order polynomial approximates the solution at the four points in the horizon. Increasing the number of collocation points increases the corresponding polynomial order.

The following discussion demonstrates the existence of a matrix $\boldsymbol{M_c}$ that satisfies the relationship reported in Eq. (7.18).

$$\begin{bmatrix} \dot{y}_1 \\ \dot{y}_2 \\ \dot{y}_3 \end{bmatrix} = \boldsymbol{M_c} \left(\begin{bmatrix} y_1 \\ y_2 \\ y_3 \end{bmatrix} - \begin{bmatrix} y_0 \\ y_0 \\ y_0 \end{bmatrix} \right) \tag{7.18}$$

where the dot indicates the derivative terms and y_0 is a fixed initial condition or otherwise equal to the final point from the prior interval.

For initial value problems, the coefficient A in Eq. (7.17) is equal to y_0 when the initial time is arbitrarily defined as zero. To determine the matrix $\boldsymbol{M_c}$, Eq. (7.17) is differentiated and substituted in Eq. (7.18) obtaining:

$$\begin{bmatrix} B+2Ct_1+3Dt_1^2 \\ B+2Ct_2+3Dt_2^2 \\ B+2Ct_3+3Dt_3^2 \end{bmatrix} = \boldsymbol{M_c} \begin{bmatrix} Bt_1+2Ct_1^2+3Dt_1^3 \\ Bt_2+2Ct_2^2+3Dt_2^3 \\ Bt_3+2Ct_3^2+3Dt_3^3 \end{bmatrix} \tag{7.19}$$

Note that in the right-hand side term of the Eq. (7.19), the coefficient A is canceled by y_0.

Eq. (7.19) can be rearranged as:

$$\begin{bmatrix} 1 & 2t_1 & 3t_1^2 \\ 1 & 2t_2 & 3t_2^2 \\ 1 & 2t_3 & 3t_3^2 \end{bmatrix} \begin{bmatrix} B \\ C \\ D \end{bmatrix} = \boldsymbol{M_c} \begin{bmatrix} t_1 & t_1^2 & t_1^3 \\ t_2 & t_2^2 & t_2^3 \\ t_3 & t_3^2 & t_3^3 \end{bmatrix} \begin{bmatrix} B \\ C \\ D \end{bmatrix} \tag{7.20}$$

Finally, rearranging and solving the Eq. (7.20) for $\boldsymbol{M}$ it can be obtained the weighting orthogonal collocation matrix:

$$\boldsymbol{M_c} = \begin{bmatrix} 1 & 2t_1 & 3t_1^2 \\ 1 & 2t_2 & 3t_2^2 \\ 1 & 2t_3 & 3t_3^2 \end{bmatrix} \begin{bmatrix} t_1 & t_1^2 & t_1^3 \\ t_2 & t_2^2 & t_2^3 \\ t_3 & t_3^2 & t_3^3 \end{bmatrix}^{-1} \tag{7.21}$$

This matrix will be used to approximate the derivatives present in the differential equations exploiting the relationship reported in Eq. (7.18) to generate a set of algebraic equations.

To better understand the derivative approximation, in Fig. 7.5 is reported a data interpolation using a polynomial that fits the five points of the current time

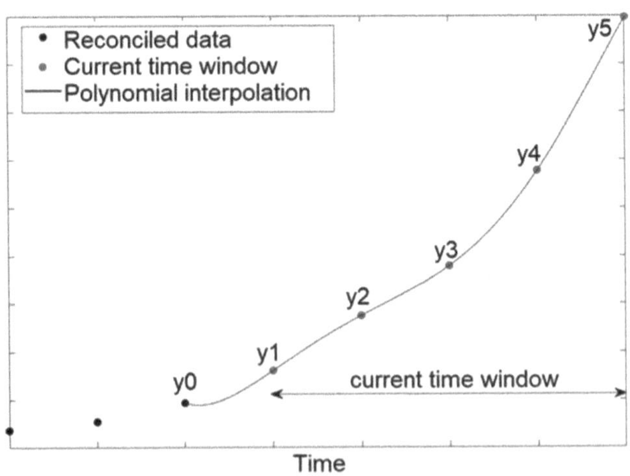

FIG. 7.5 Moving time horizon mechanism.

windows and using as initial point the last one reconciled. Finally, it is remarkable that the time-window length is a key aspect for DDR: a short time-moving window (i.e., few points within the time discretization) are in principle able to catch the main dynamic variations of process variable whereas a long time window is useful to have a full picture of the general trend of the system, but, owing to the internal workings for the moving window method, a lag is introduced. Time lag is more evident for a large time window.

4.4 Implementation and the role of digital twin

In our perspective, DR and DDR coupled with GE detection assume a central role in Process System Engineering as already emphasized in the Computer-Aided Process Engineering (CAPE) working party programmatic manifest [28] and other eminent researchers in chemical engineering community [3,29]. DR and DDR are some of the sublayers depicted in Fig. 7.6. According to this view of Process Engineering in the digital era, process data flow in a circular motion originated from the process and return to the process itself. Process variables are collected and registered in the DCS (bottom layer) meaning that raw data are at the base of the digitalization of industry. Raw data are useless without a data analytic process as raw feedstock is not economically appealing without chemical transformation. In the sudden-upper layer digital twin and process simulation are the main actors: they offer a reliable simplified picture of the real process behavior and process variable trends. Digital twin may be developed on commercial simulation software or not. For instance, in the resulting paragraph, it will be shown that it is possible to model amine washing and the simple unit just using material balance equation and proper constraint without process simulation software. Data analytics and optimization occupy the top of the proposed hierarchical structure. Data analytics enables to classify,

FIG. 7.6 Hierarchical structure of data flow in process industry.

elaborate, and handle/manipulate raw data coming from DCS. The digital twin takes part in this phase as primary support; indeed, it allows to interpret whether data are coherent and consistent with the monitored process. After data analytics, reliable data are fed to robust process optimization step that properly works only when coherent data are provided. The reverse flow (i.e., top-bottom movement of data) implies that reliable data and optimizer solutions are sent back to improve, for instance, process performances and/or process conceptual design.

In this hierarchical structure, our framework for DR, DDR, and GE detection is one of the main steps involved in the Middleware interface. Indeed, Fig. 7.7 describes the general structure of your framework for steady-state DR. Only two inputs are present: raw data from DCS (or even generated) and the digital twin (model) of the process. These two inputs correspond to the two bottom layers in Fig. 7.6. As already discussed, the digital twin may derive from a simulation flowsheet or a set of equations able to describe the main features of the process. Digital twin is a simplified model of the real unit or (section) plant; hence, simplifications and assumptions are determinants to reduce degree of complexity

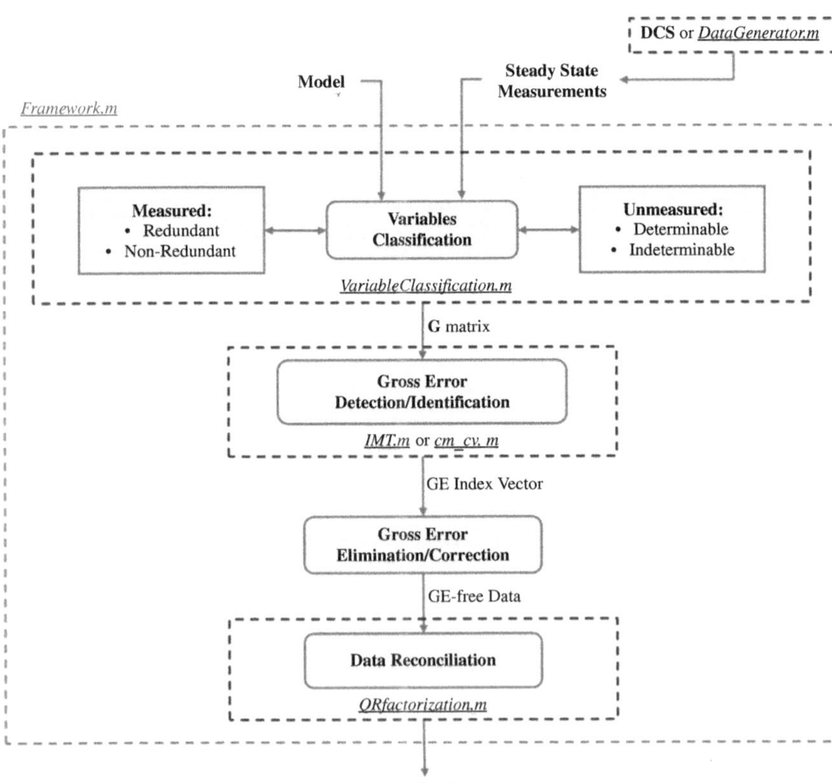

FIG. 7.7 Framework structure for DR.

without compromising accuracy as will be shown in the proposed case study. The DR framework is subdivided into inner layers each with a specific role. Inner data flow is strictly ordered, and it is not possible to skip or modify the sequence of the sublayers:

1. Data classification—Data are clustered in categories and according to digital twin and data feed, this step determines which variables (or streams) will be corrected in the DR phase
2. GE detection according to statistical or numerical algorithm—this layer acts as a filter of information: whether present, suitable algorithms accomplish the GE identification
3. GE elimination or correction—it is possible to correct or delate datasets affected by influential observations. The proper decision is done looking at the impact of datasets removal on the system redundancy
4. Data Reconciliation

The DDR framework conceptual structure is identical to the steady-state DR one. The only difference stands in the raw data that the structure manipulates: dynamic ones instead of steady-state measurements.

5 Industrial case study: Itelyum Regeneration amine washing unit

The implemented framework for the steady-state data reconciliation has been finally applied on an industrial scale. The gas sweetening (amine washing) has been chosen as the case study to validate performance and robustness also in the case of Itelyum Regeneration extracted the data from DCS of the production plant based in Pieve Fissiraga (Lodi).

5.1 Process description

The sweetening process is depicted in Fig. 7.8.

This plant section enables to process two kinds of streams: sulfur-rich flows where H_2S is removed from the hydrogen and the standard acid gas flows when hydrogen has to be removed from hydrocarbons vapors. This flexibility is guaranteed thanks to the presence of three different absorption columns: the high-pressure absorber (HP), low-pressure absorber (LP), and the regeneration unit (REG). As in common practice, the sour gas streams enter the plant through an inlet separator (D-2 and D-3) to remove entrained liquid drops.

After these separators, the gas streams F_1 and F_6 enter the bottom of the absorber columns and flow upward through the absorber in intimate countercurrent contact with the aqueous amine solution where the amine absorbs the H_2S. The sweetened gas outflowing from the top of the high-pressure column (F_{19}) passes through an additional separator D-1 while the rich amine solution leaving this column from the bottom is fed to the low-pressure column.

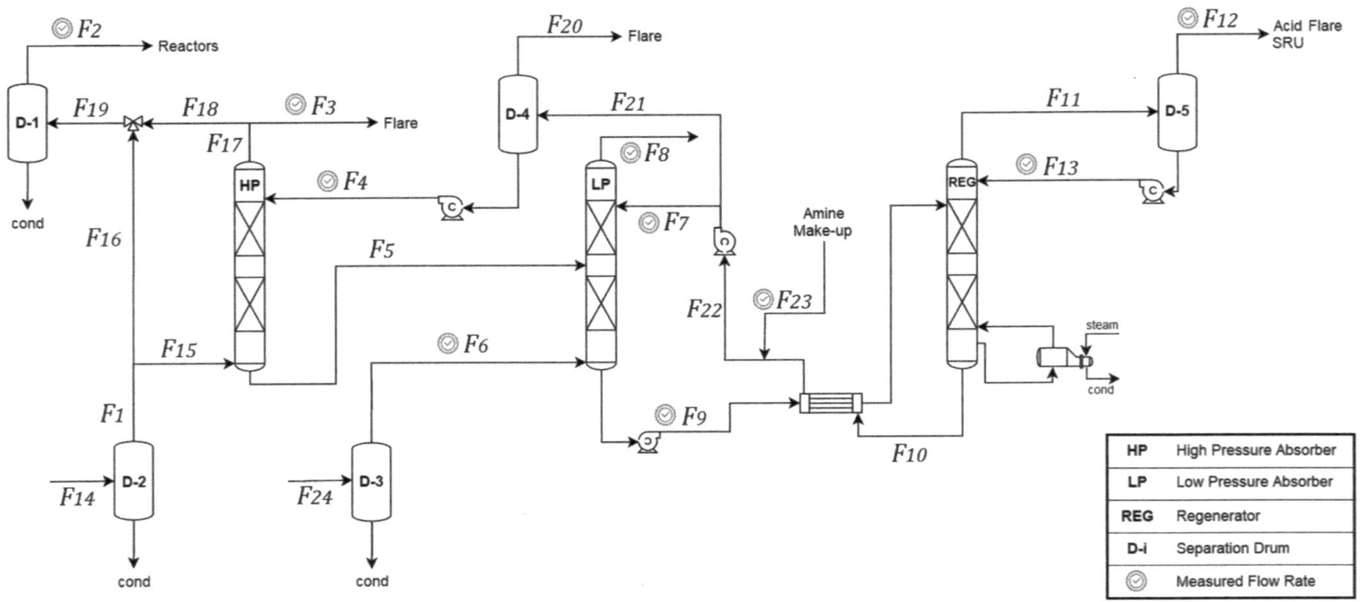

FIG. 7.8 Sweetening unit scheme at Itelyum Regeneration.

The solvent containing the absorbed H_2S that leaves the low-pressure column (F_9) is sent to the regeneration column, after being preheated by the hot regenerated solvent (F_{10}). The process-process heat exchanger reduces the energy requirements for the process [34] guaranteeing the necessary heat recovery. In the regenerator, the H_2S is stripped from the rich solution by means of the reboiler vapor. The acid gases and the vapor leave the top of the regenerator and are cooled, in this way the major portion of the vapor is condensed. The acid gases are separated in the separator D-5 and sent to the acid flare or to the Sulfur Recovery Unit (SRU). The condensed vapor is returned to the top of the column as reflux.

The main goal for this industrial case study is to apply the steady-state reconciliation framework developed in the previous sections on this plant to reconcile the mass flow rates measurements. The major drawback of this system is that, according to Fig. 7.8, only a few streams are measured (these are highlighted with a green tick). Considering also the condensates leaving from the bottom of the separators D-1, D-2, and D-3, only 10 streams out of 27 have a flow measurement instrument installed. This is a common situation in the industry due to instrumentation costs [32].

The constraints that the variables must fulfill are 13 steady-state global mass balances on the following units:

Recalling that the definition of redundancy is the difference between the amount of available information (measurements and constraints) and the minimum amount necessary for a unique determination of the independent variables, the system has a negative redundancy:

$$r = (m + l) - g = (13 + 10) - 27 = -4$$

Since positive redundancy is the necessary condition for the estimability of a system, it is not possible to reconcile all the streams of the plant reported in Fig. 7.8.

5.2 Assumptions

Some assumptions are needed to have a positive redundancy and then solve the reconciliation problem. It is remarkable that the application of these assumptions should increase the system redundancy without impacting the accuracy and reliability of reconciled variables:

(1) A black box around the high-pressure column that includes the streams F_{15}, F_{16}, F_{17}, F_{18}. In this way, the reconciliation of these unmeasured streams is neglected reducing the number of total variables by four elements.

(2) The amine make-up is neglected since in normal operating conditions no fresh solvent is needed. In fact, the amine make-up pump is turned on only occasionally, when the solvent circulating through the plant is no more effective due to the amine thermal degradation or solvent losses.

(3) The condensate streams outflowing from the separator drums D-1, D-2, and D-3 are neglected since they are considered very no-flow or negligible with respect to the others circulating in the plant.
(4) The gas stream leaving the top of the separator drum D-4 sent to the flare is neglected since in normal operating conditions it is supposed to have a very small flow rate.
(5) The stream F_{12} mass flow rate was null during the entire data sampling. This was due to the absence of sulfur in the hydro-finishing feed at the moment the data have been collected.

Notice that assumptions (1) and (2) do not introduce any error to the plant steady-state model, whereas the assumptions (3) and (4) lead to some model inaccuracies that however should not bias the final reconciliation results. Condition (5) is related to a specific operating condition of the amine washing unit during the DSC data extraction.

These assumptions lead the system to a positive redundancy condition, despite that, the redundancy is the minimum requested to perform a DR. The plant then can be schematized as reported in Fig. 7.9. This new configuration still has the same redundancy because three variables have been removed (F_{11}, F_{12}, and F_{13}) along with two measurements and their common constraint around unit D-5.

Before proceeding to the reconciliation, it is worth classifying the measured variables. Once accomplished, results allow to predict which variables will be subjected to numerical manipulation: unmeasured flow rate but will not be modified by the reconciliation since there are no balances that constraint them due to the low redundancy of the system. Variable classification is sketched in Fig. 7.10.

6 Results

In the charts, flowrates have been normalized with respect to their mean value due to Non-Disclosure Agreement. Despite the nominal flow is not reported, results allow to draw relevant conclusions concerning the potential of data reconciliation applied on real industrial processes.

6.1 Steady-state data reconciliation results discussion

Measurements have been extracted from DCS in Itelyum Regeneration s.r.l production plant based in Pieve Fissiraga (Lodi, North of Italy). Data refers to the flowrates measured in the amine washing section to purify the process stream for acid gas (in this case carbon dioxide). The observation collection covers 80 s.

Reconciliation requires the variance-covariance matrix $\boldsymbol{\Psi}$. For sack of simplicity, the covariance terms are considered null, therefore only the measurements variances are estimated. The variances have been evaluated from the

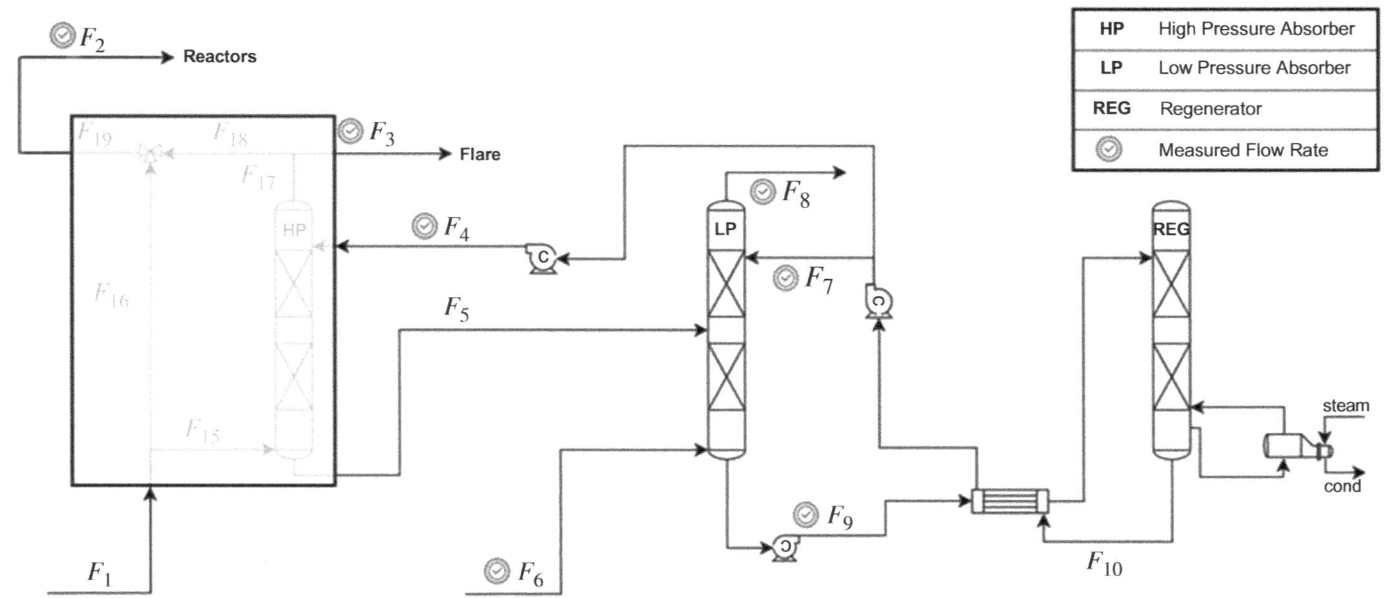

FIG. 7.9 Simplified amine washing scheme.

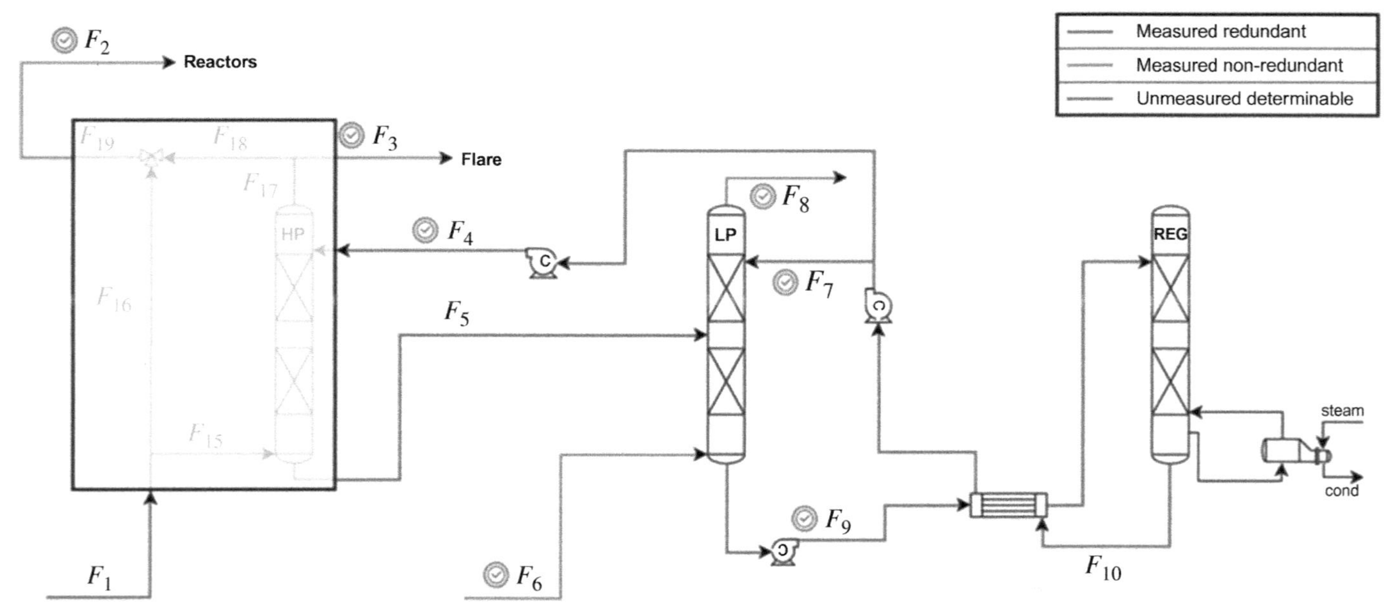

FIG. 7.10 Variable classification.

TABLE 7.1 Flowrate normalized variance.

	Normalized variance [−]
F_2	134.15
F_3	1916.79
F_4	8237.40
F_6	8.59
F_7	23.51
F_8	1.00
F_9	216.03

data. The normalized variances are reported in Table 7.1, the normalization has been done dividing all the variances by the smallest one to not disclose sensitive data.

According to the variable classification reported in Fig. 7.10, the only measurements that data reconciliation manipulate the redundant ones: F_4, F_7, and F_9. The results of the reconciliation of these streams' measurements, without considering any gross error detection algorithm, are reported in Fig. 7.11.

F_4 is highly modified by the data reconciliation algorithm. This is due to the lower accuracy of its measurement instrument with respect to the others that leads to higher oscillations of the stream observations and thus to a higher variance as confirmed in Table 7.1. In the light of F_7 and F_9 measurements accuracy, their reconciled values are closer to the measurements, hence, the reconciliation does not correct them significantly as depicted in Figs. 7.12 and 7.13.

An objective index of the reconciliation performances is the redundant measured variables variance percentage reduction ($VR_{\%}$) which estimates the impact of the data reconciliation on the reconciled stream. The expected result is a beneficial effect on the variance as reported in Table 7.2. It is confirmed that the F_4 data has been considerably corrected, while the variances for other redundant streams have been slightly reduced, leading to a negligible variation despite data reconciliation benefits also to these streams.

In Fig. 7.14 charts report the results for the nonredundant measured variables. As expected, the reconciliation does not correct streams measurements and their reconciled values are equal to the measurements since, due to the low redundancy, no constraints involving these flows are available. The unmeasured variables can be evaluated by inference using the nonredundant measurements and the reconciled values of the redundant measured variables.

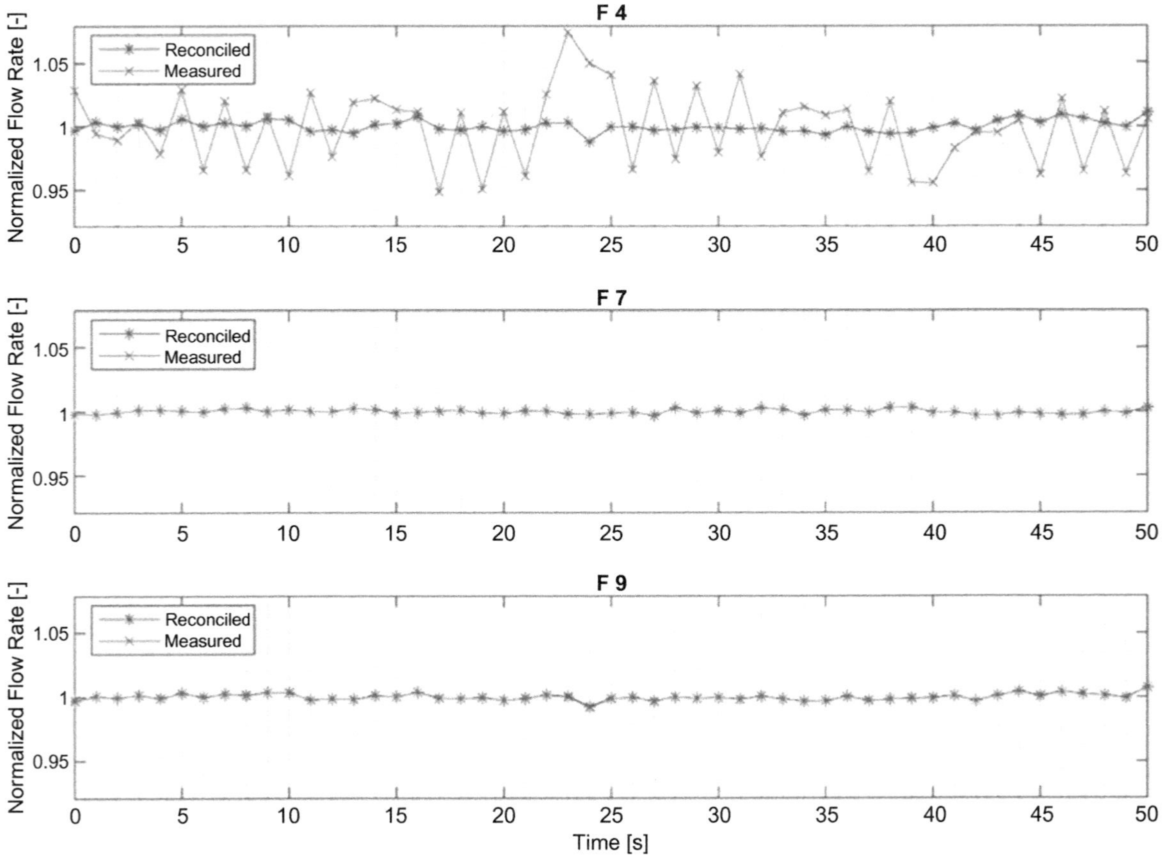

FIG. 7.11 Reconciled redundant streams.

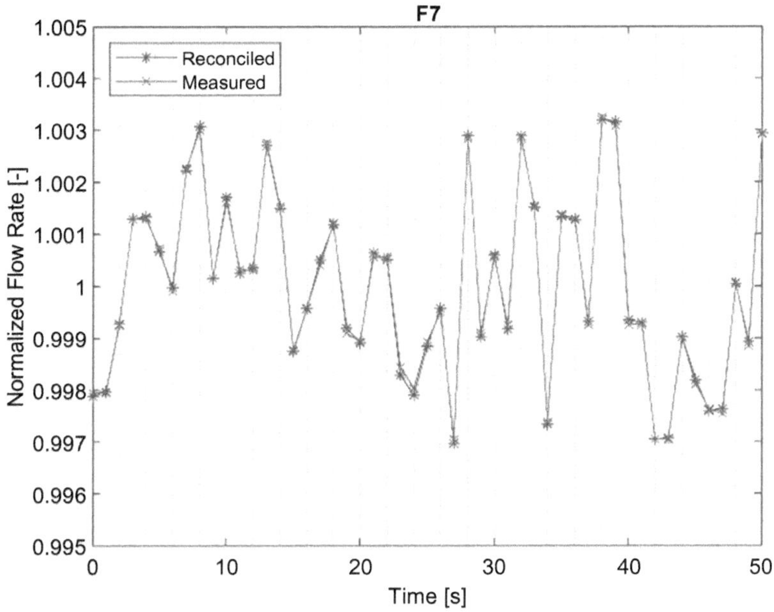

FIG. 7.12 Detail of redundant stream F_7.

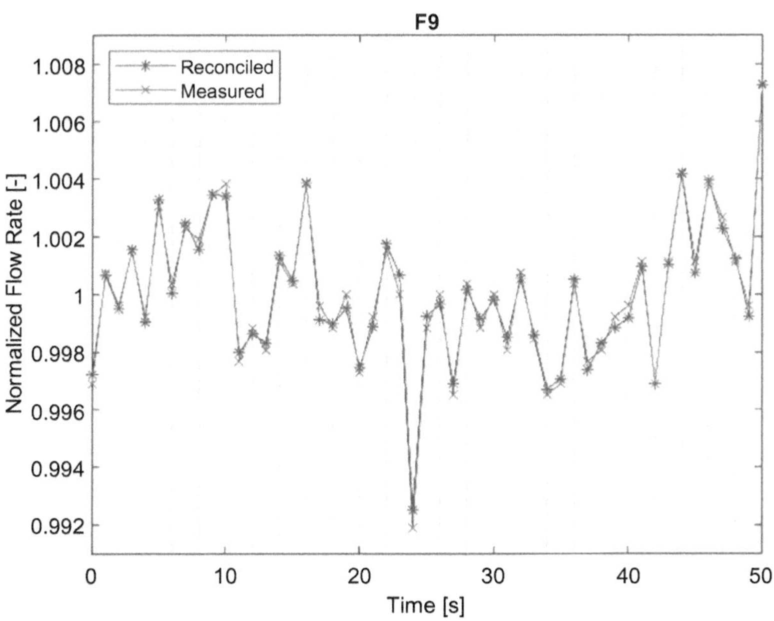

FIG. 7.13 Detail of redundant stream F_9.

TABLE 7.2 Variance percentage reduction for redundant flowrates.

	$VR_{\%}$ [%]
F_4	−97.7
F_7	−1.8
F_9	−8.4

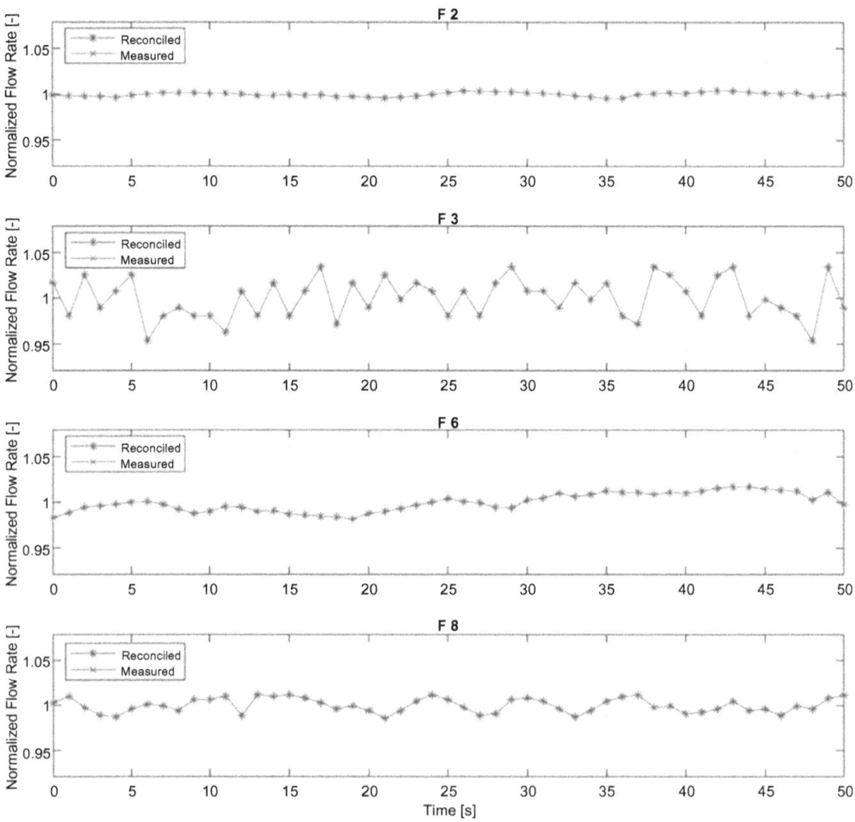

FIG. 7.14 Effect of DR on nonredundant measurements.

6.2 Gross error detection results discussion

Previous results do not consider any gross error detection algorithm. It is interesting to investigate how the results change when the complete steady-state reconciliation framework including the gross error detection function is used.

In a plant with a maximum redundancy equal to 1, hence, not all the gross error detection algorithms can be applied. For instance, the algorithm based on statistical tests (SE, SEGE, and IMT) cannot be exploited since they remove suspicious streams measurements and performs reconciliation excluding these streams to effectively verify whether the temporarily removed streams were gross errors or not. This can be done only in case that the redundancy is still positive after having removed suspicious stream measurements. In the considered industrial case study the application of the statical method is unfeasible since redundancy drops to null or even negative in case of multiple streams elimination.

In this condition, only numerical gross error detection methods such as clever mean and clever variance allow to reconcile data without impacting on the system redundancy. In fact, this algorithm analyzes individually the measurements for each stream checking for any potential suspicious maximum or minimum in the data.

The data reconciliation framework involving the *cm* and *cv* algorithm has been therefore exploited. The framework has been slightly modified to cope with the plant poor redundancy: when a GE is detected, it is deleted and replaced with the reconciled value obtained at the previous time to keep positive redundancy.

The results for the measured redundant variables are reported in Fig. 7.15 and details for streams F_7 and F_9 are in Fig. 7.16 and 7.17, respectively.

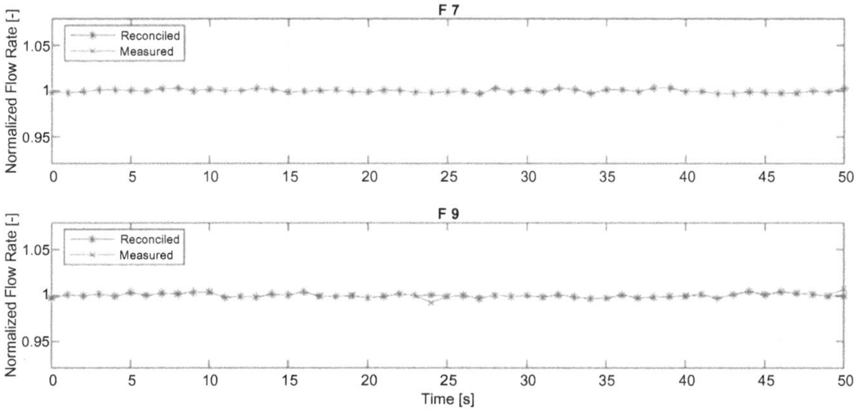

FIG. 7.15 Redundant measurements reconciliation with GE detection.

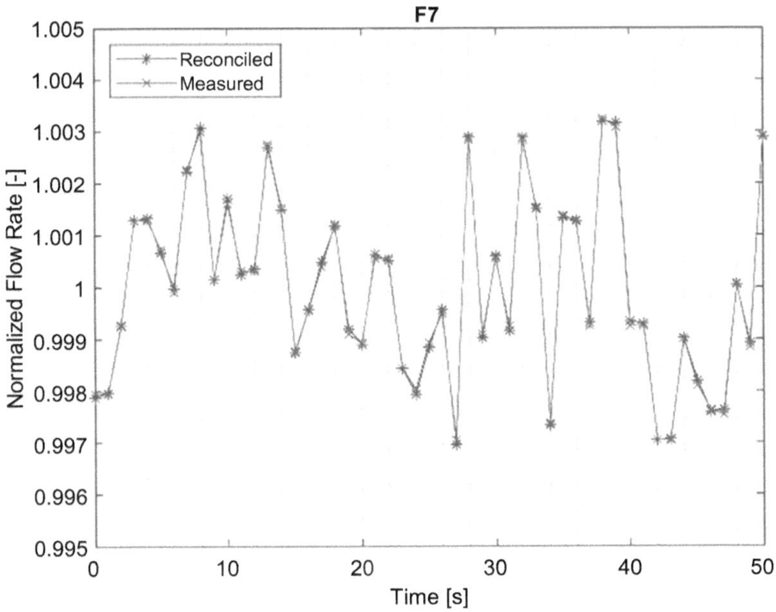

FIG. 7.16 Detail of reconciliation with GE detection for stream F_7.

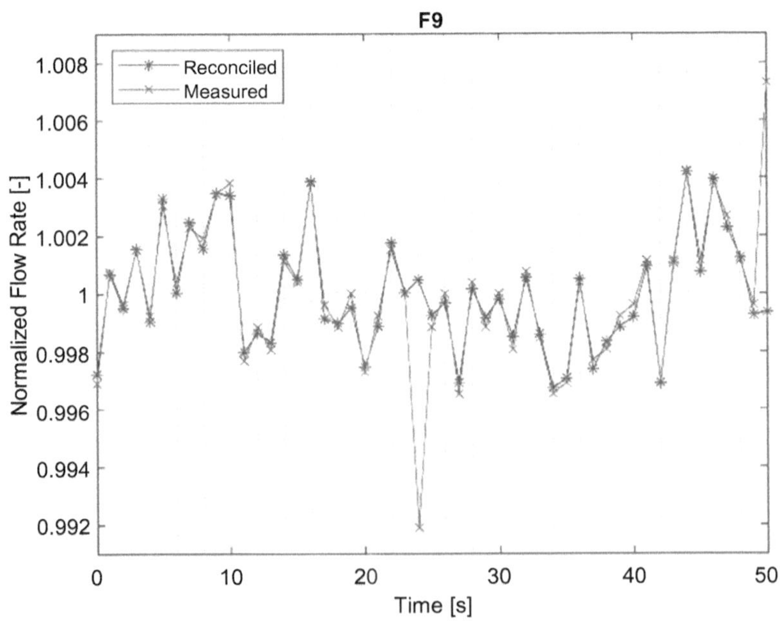

FIG. 7.17 Detail of reconciliation with GE detection for stream F_9.

TABLE 7.3 Effect of GE detection on the variance reduction for redundant flowrates.

	$VR_{\%}$ without GE detection [%]	$VR_{\%}$ with GE detection [%]
F_4	−97.7	−98.3
F_7	−1.8	−1.1
F_9	−8.4	−40.8

Three gross errors have been detected:

- Stream F_4 at time 23 [min]
- Stream F_9 at time 24 [min] and 50 [min]

Using the gross error detection and therefore eliminating the outliers, the reconciliation performance has increased, as shown in Table 7.3. Clever mean and variance algorithm (*cm* and *cv*) benefits the data reconciliation. The impact of the gross error identification is mainly visible on stream F_9 where variance is strongly reduced.

Since it is not possible to identify and then modify nonredundant streams affected by GE (or better influential observations), reconciliation does not correct those measurements. Results are the same reported in Fig. 7.14. On the other hand, since the reconciled data of the redundant measured variables, in this case, are different, also the inference of the unmeasured variables is slightly different.

6.3 Dynamic data reconciliation case study: Amine tank dynamics

DDR has been applied on the exhaust amine solution vessel which is an ancillary unit of the sweeting process (Fig. 7.18).

In this application of DDR, it will be proven the importance of digital twin to describe the dynamic evolution of the experimental observation extracted from Itelyum DCS. More in general, DDR requires more details (i.e., equipment geometry, control loop tuning parameters, etc.) and additional pieces of information which are not of interest for steady-state DR. Hence, DDR model is tuned according to available measurements, hence, conversely to steady-state DR, a general framework is not available. This reflects the importance of digital twin that, despite introduced simplification, it enables to catch the main features and qualitatively follow the dynamic variation of the main variables involved.

The measured variables are as follows:

- Tank level percentage $h^{[\%]}$
- Outlet mass flow rate F_{out}
- Valve opening percentage x_v

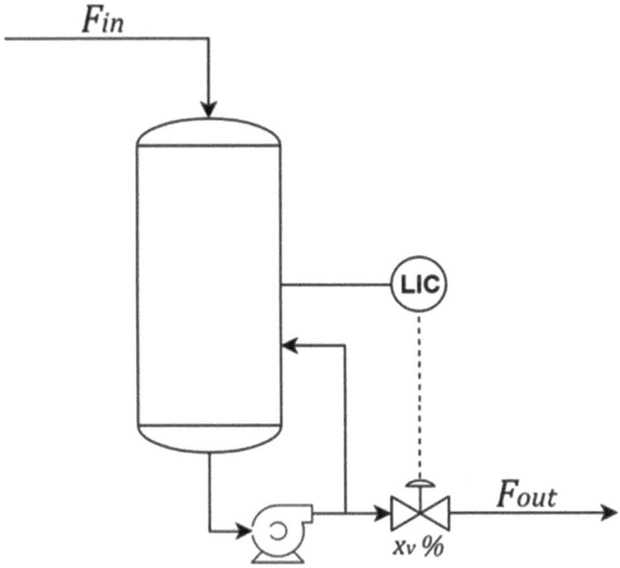

FIG. 7.18 Amine solution accumulation tank scheme.

The data cover a total time of 300 [s] with an estimated sampling time of 1 [s]. The system is poorly redundant since details on the inlet streams are missing. The inlet flowrate can be only inferenced from reconciled data for the outlet stream. To reconcile these data, constraints involving measured parameters are needed. All geometrical properties and control loop tuning are known.

The first constraint is the overall mass balance:

$$\rho A H_{tot} \frac{dh^{[\%]}}{dt} = F_{in} - F_{out} \tag{7.22}$$

The outlet flow rate represents the second constraint and it is a function of the valve position [30]:

$$\frac{F_{out}}{\rho} = f(x_v) \cdot C_v \sqrt{\frac{\Delta P}{\gamma}} \tag{7.23}$$

where:

- C_v is the valve flow coefficient that represents the gallons of fluid flowing through the valve in 1 min when the pressure drop across the valve is 1 [psi]
- ΔP is the pressure drop across the valve
- γ is the specific gravity of the fluid
- $f(x_v)$ is the flow characteristics that put into relation the valve opening to the volumetric flow rate and depends on the valve type

The third constraint is given by control law that relates the valve opening to the level as.

$$x_v(t) = x_v^0 + K_c \varepsilon(t) + \frac{K_c}{\tau_I} \int_{t^0}^{t} \varepsilon(t)\, dt \tag{7.24}$$

where:

- $\varepsilon(t)$ is the error signal defined as $\varepsilon(t) = h^{[\%]}(t) - sp^{[\%]}$ and $sp^{[\%]}$ the desired level percentage set point
- x_v^0 is the bias, that is the control signal used to maintain the valve in the desired position when the error signal is null
- K_c is the proportional gain
- τ_I is the integral time

As already mentioned, the valve characteristic flow $f(x_v)$ is strictly related to the valve. In this case, the valve is an equal percentage type, therefore its theoretical or inherent characteristic is:

$$f(x_v) = a^{x_v - 1} \tag{7.25}$$

where a is a constant depending on the valve size [31].

When a valve is installed in series with other pieces of equipment that produce a large pressure drop in the line compared to the pressure drop across the valve, for instance, a pump, the actual valve characteristics deviate from the inherent characteristics. To evaluate the installed characteristic curve a fitting of the experimental data is needed. In this work, the fitting has been done using a third-order polynomial function and it is reported in Fig. 7.19. On the y-axis is reported the volumetric outlet flow rate normalized dividing the data by the maximum one.

Nevertheless, considering that the initial data collection time, t_c^0, does not correspond to the starting time for the integration contribution, it is necessary to estimate the accumulated integral error up in the interval $[t^0; t_c^0]$:

$$C = \frac{K_c}{\tau_I} \int_{t^0}^{t_c^0} \varepsilon(t) dt \tag{7.26}$$

Since this integral term is unknown, the only possible approach to estimate C is a nonlinear fitting of the DSC data (reported as experimental data) including Eq. (7.26) in Eq. (7.24):

$$x_v(t) = C + x_v^0 + K_c \varepsilon(t) + \frac{K_c}{\tau_I} \int_{t_c^0}^{t} \varepsilon(t) dt \tag{7.27}$$

Fig. 7.20 shows a good match for refitted valve opening *(red dots, gray in print version)* and DSC data *(blue, dark gray in print version)*.

Now digital twin is complete and all the process variables of interest are suitably modeled for DDR. The length of the historical horizon is the only tuning

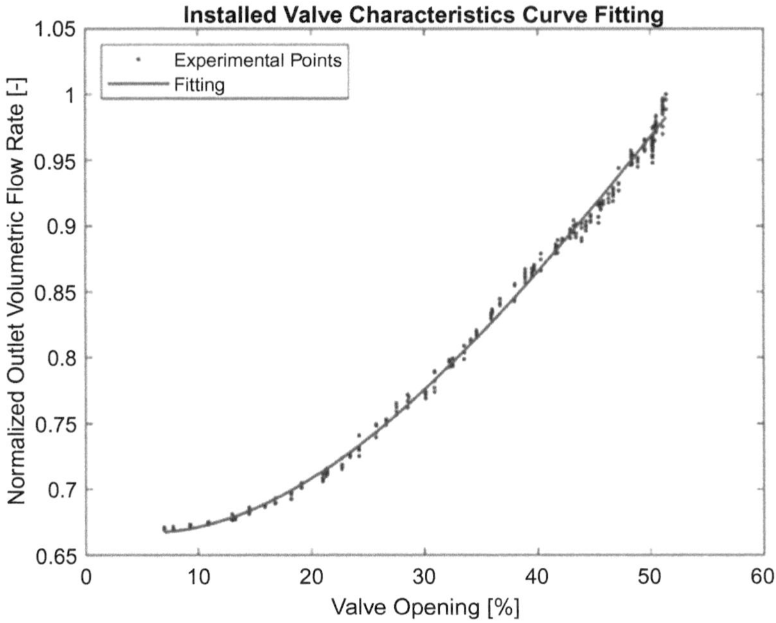

FIG. 7.19 Installed valve characteristic curve fitting.

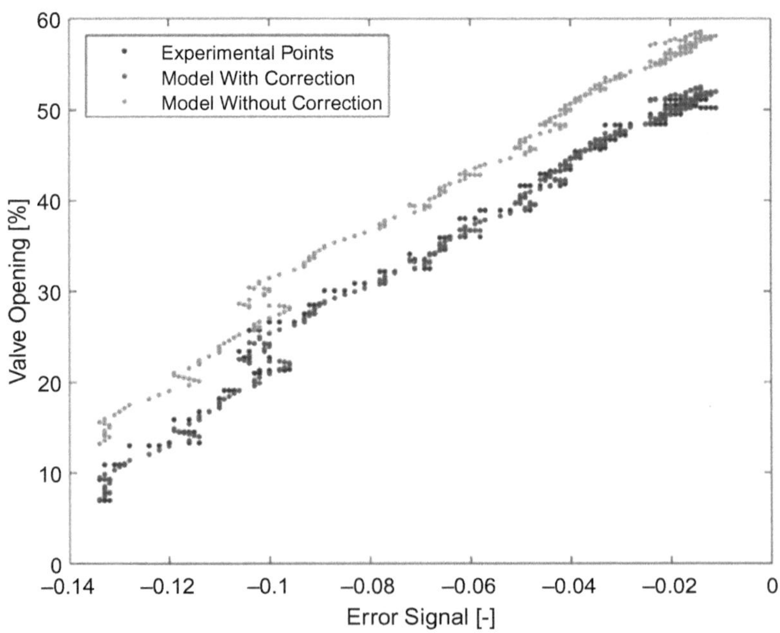

FIG. 7.20 Control law model.

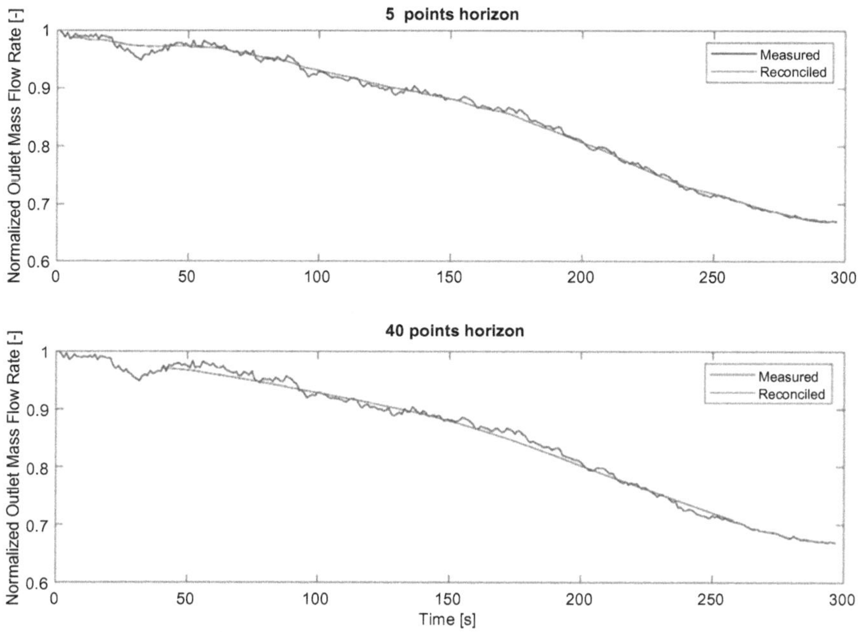

FIG. 7.21 Comparison of outlet mass DDR with different time horizon.

parameter needed to adequately tune the performance of the DDR. The optimal history horizon length represents is the best compromise between the variance reduction and the lag introduced. For this analysis, two very different history horizon lengths have been used to stress the aforementioned trade-off: 5 and 40 points time-window. The best-fitting function of the available data has proven to be a polynomial since allows fast computation and reasonable accuracy. The results of the dynamic data reconciliation are reported from Figs. 7.21 to 7.23. In the reported charts, on the y-axis the normalized flowrate is plotted, the normalization has been done by dividing the measurements and the reconciled value by the maximum measured value.

The reconciliation has significantly reduced the measurement oscillations in both cases leading to smoother trends. The reconciled values obtained using 40 points in the time-window as historical horizon are smothered and therefore in some points are do not follow the real trend of the variables as can be clearly seen in Fig. 7.21: a larger historical horizon means a greater amount of data to be constrained together (i.e., greater temporal redundancy); hence, reconciled values cannot follow the current variable trend when sharp changes occur. Moreover, since a point is considered completely reconciled when it has been processed exactly for the number of times equal to the number of points in the historical horizon, using a length of 40 points it is not possible to have the first and the last 39 data completely reconciled.

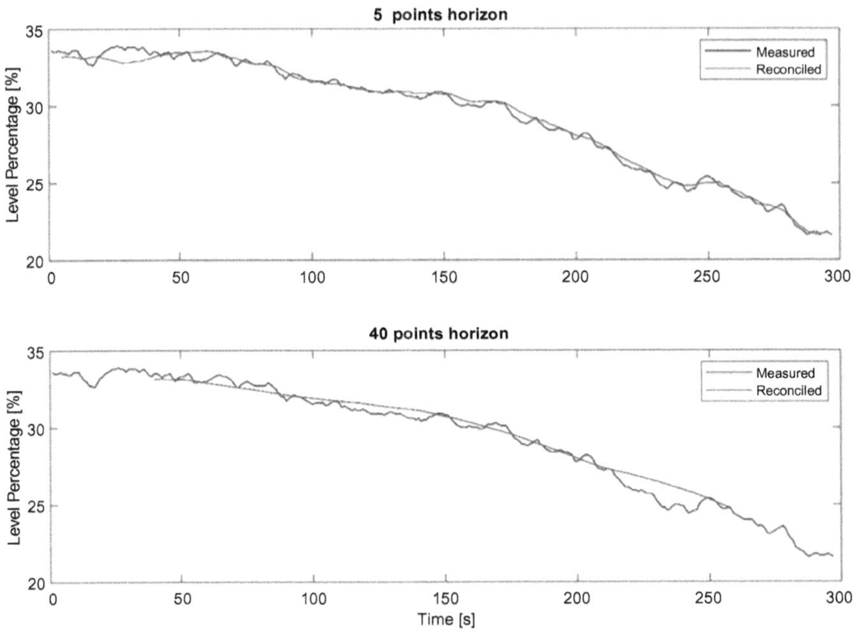

FIG. 7.22 Comparison of level percentage DDR with different time horizon.

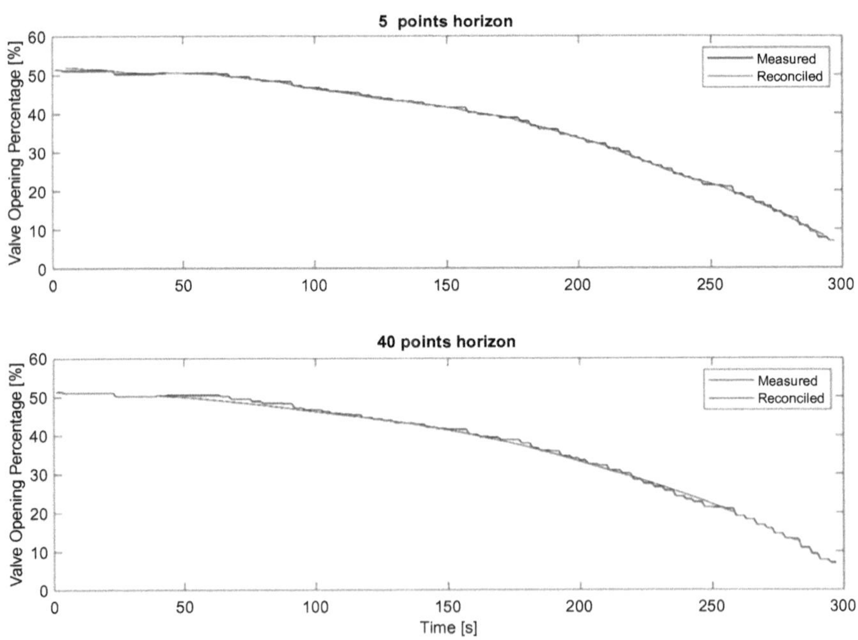

FIG. 7.23 Comparison of open valve DDR with different time horizon.

7 Conclusions

7.1 Steady-state data reconciliation

Data reconciliation is a powerful tool with a deep impact also in industrial applications. Just considering the proposed case study, it is possible to conclude that:

- The reconciliation potential cannot be fully exploited in low-redundant system; however, relevant information may be deduced. There are alternatives to positively impact the redundancy: (1) increase measurements in the system or (2) improve the distribution of instrumentation
- The reconciliation enables to correct only redundant streams measurements. Since low redundancy is a common feature in industrial practice, few streams normally may be reconciled providing few reliable data which can be reasonably used to analyze and estimate process performances
- The low accuracy (i.e., large variance) on streams measurements highly affects the data reconciliation performance: in case of few redundant streams, low accuracy prevents the reconciliation algorithm from significantly correcting more accurate measurements. This means that reconciliation properly acts only on a few streams while others are not touched. To solve this issue, a more accurate measurement instrument
- The clever mean and clever (*cm* and *cv*) variance method is suitable in a low-redundant system. Advantages lie where statistical methods fail or show some weaknesses and shortcomings: (1) it allows to detect influential observations without impacting on the global system redundancy, (2) this method solves inefficiencies and numerical efforts related to standard mean and variance as investigated in [25] and, as a consequence of point (2), (3) it requires less numerical and computational efforts since it does not require the investigation of possible single, couple and higher other data combination.

These are general considerations and conclusions on data reconciliation. However, data reconciliation allows to improve the controllability of plant and provided the process engineers with useful information on the process and the current state of the equipment and installed instrumentation. For instance, the oscillating profile for F_4 stream measurements denotes some potential issues in the valve and/or corresponding mass flowrate indicator:

1. The measurement instrument may be damaged or suffer miscalibration.
2. The control valve may be oversized. A slight change in its opening leads to high flow rate variation.
3. The controller parameters are not adequately tuned introducing strong oscillations in the manipulated variable (controller ringing). Nevertheless, in these situations the system is underdamped and the oscillations tend to decrease in time that is not the case of the F_4 measurements [32].

The more probable hypotheses are the number (1) and (3), but unfortunately, no additional data and parameters have been provided to validate one of them.

Finally, data reconciliation permits optimizing installed instrumentations distribution. Considering that installation and maintenance costs should be optimized, data reconciliation provides a useful piece of information to increase the system redundancy without heavily impact on the preexisting control system. For instance, variable classification (a prerequisite for reconciliation) suggests measuring stream F_1 in this way almost all flowrates become redundant as depicted in Fig. 7.24. The benefit is double: firstly, redundancy is increased with low investment costs, and, secondly, data reconciliation performances are improved since more streams are redundant meaning that the algorithm can handle and modify a larger measurements dataset to fulfill constraints. The letter consideration implies that more streams may be reconciled meaning that the process would gain more reliable data to be used for key performance estimates, economical evaluation, and so forth.

7.2 Dynamic data reconciliation (DDR)

In DDR it has been proven that digital twin covers a key role since only a proper model of the system enables to properly reconcile dynamic data and an effective digital twin (or model) is a relevant concern in system with poor measurements. In such conditions, even though in DDR measurements redundancy is in part meaningless but time redundancy is strictly necessary, the few available measurements are of primary importance for reliable DDR. Moreover, results have shown that moving time horizon implies a trade-off between accuracy and predictive target. Short time windows are preferable whether catching process variables fluctuation is the aim of DDR. Conversely, large time horizon prevents from following the real dynamic of the system and sharp variations since DDR tends to provide smoother trends. Therefore, large time horizon is the optimal choice to have a general tracking of process variables without giving weight to local oscillations or sudden variations. However, it is important to remark that time window length impacts the time delay in predictions and the time lag is magnified for large time horizon; hence, for real-time purposes and application, large time windows are discouraged. These different features are directly related to time redundancy: the larger is the time horizon, more are the processed data and DDR algorithm is more and more constrained to minimize the gap between experimental data and constraints fulfillment.

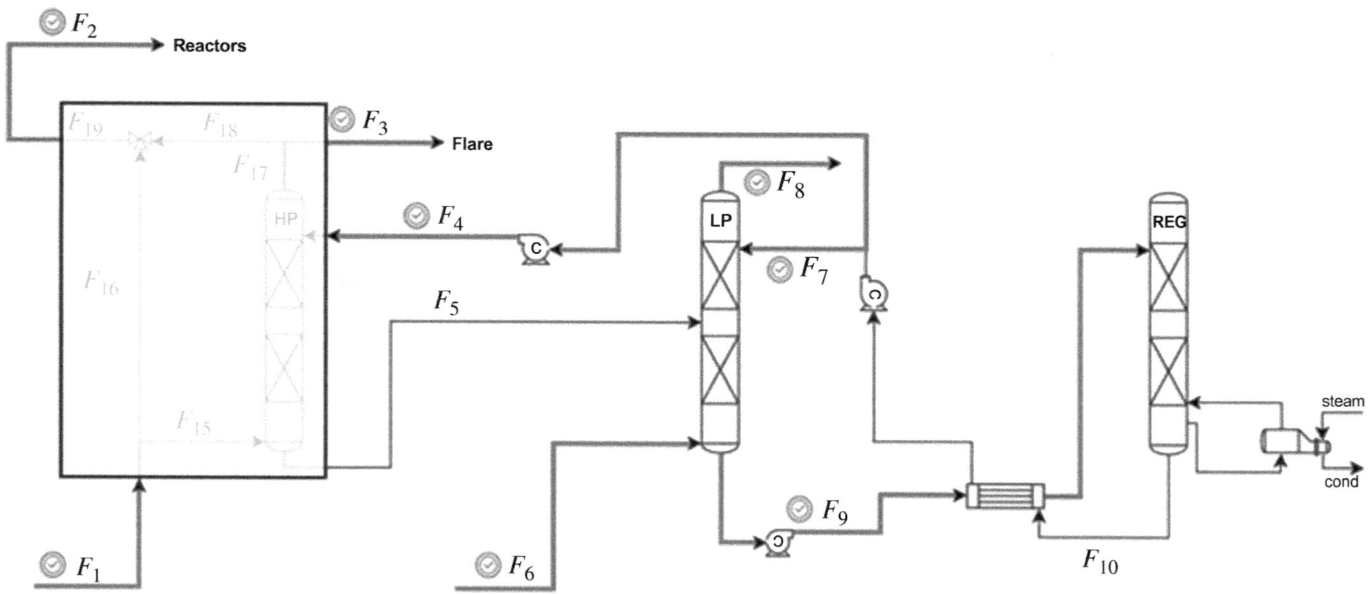

FIG. 7.24 Redundancy for different measurements distribution.

Acknowledgments

Authors gratefully acknowledge the invaluable support of M.D. Eng. Stefano Apruzzese and M.D. Eng. Davide Fratelli for their work during the M.Sc. Thesis Project in Politecnico di Milano. Also, the constant support by field and control room operators of Pieve Fissiraga (LO) exhausted-oil refinery is deeply acknowledged.

Credits: Filippo Bisotti—writing original chapter draft, conceptualization, data mining and elaboration, software and review; Andrea Galeazzi—writing original chapter draft, conceptualization, data mining and elaboration, software and review; Flavio Manenti—conceptualization, supervision and review; Francesco Gallo—data mining and elaboration, supervision and review.

References

[1] A.C. Tamhane, R.S.H. Mah, Data reconciliation and gross error detection in chemical process networks, Technometrics 27 (1985) 409–422, https://doi.org/10.1080/00401706.1985.10488080.

[2] I.E. Grossmann, Challenges in the new millennium: product discovery and design, enterprise and supply chain optimization, global life cycle assessment, Comput. Chem. Eng. 29 (2004) 29–39, https://doi.org/10.1016/j.compchemeng.2004.07.016.

[3] D.C. Miller, J.D. Siirola, D. Agarwal, A.P. Burgard, A. Lee, J.C. Eslick, B. Nicholson, C. Laird, L.T. Biegler, D. Bhattacharyya, N.V. Sahinidis, I.E. Grossmann, C.E. Gounaris, D. Gunter, Next Generation Multi-Scale Process Systems Engineering Framework, Elsevier Masson SAS, 2018, https://doi.org/10.1016/B978-0-444-64241-7.50363-3.

[4] C.M. Crowe, Data reconciliation—progress and challenges, J. Process Control 6 (1996) 89–98, https://doi.org/10.1016/0959-1524(96)00012-1.

[5] J.A. Romagnoli, M.C. Sánchez, Data Processing and Reconciliation for Chemical Process Operations, vol. 2, Academic Press, 2000.

[6] S. Narasimhan, C. Jordache, Data Reconciliation and Gross Error Detection: An Intelligent Use of Process Data, 2000, https://doi.org/10.1111/j.1749-6632.1972.tb54778.x.

[7] D.B. Özyurt, R.W. Pike, Theory and practice of simultaneous data reconciliation and gross error detection for chemical processes, Comput. Chem. Eng. 28 (2004) 381–402, https://doi.org/10.1016/j.compchemeng.2003.07.001.

[8] D. Brown, F. Maréchal, G. Heyen, J. Paris, Application of multi-objective optimisation to process measurement system design, Comput. Aided Chem. Eng. 20 (2005) 1153–1158, https://doi.org/10.1016/S1570-7946(05)80034-3.

[9] R. Hayati, M. Sadi, S. Shokri, M.A. Marvast, S.H. Boroojerdi, A.H. Abedi, Model based monitoring using integrated data validation, simulation and parameter estimation, World Acad. Sci. Eng. Technol. 50 (2011) 588–592.

[10] J. Loyola-Fuentes, R. Smith, Data reconciliation and gross error detection in crude oil pre-heat trains undergoing shell-side and tube-side fouling deposition, Energy 183 (2019) 368–384,-https://doi.org/10.1016/j.energy.2019.06.119.

[11] J. Loyola-Fuentes, R. Smith, Classification and estimation of unmeasured process variables in crude oil pre-heat trains subject to fouling deposition, Comput. Chem. Eng. 137 (2020), https://doi.org/10.1016/j.compchemeng.2020.106779.

[12] E. Masato Ishiyama, S. John Pugh, Effect of flow distribution in parallel heat exchanger networks: use of thermo-hydraulic channeling model in refinery operation, Heat Transfer Eng. 41 (2020) 189–198, https://doi.org/10.1080/01457632.2018.1522098.

[13] F. Manenti, M.G. Grottoli, S. Pierucci, Online data reconciliation with poor redundancy systems, Ind. Eng. Chem. Res. 50 (2011) 14105–14114, https://doi.org/10.1021/ie202259b.

[14] L. Zhou, J. Li, A study of gross error detection and data reconciliation in process industry, in: 2010 International Conference on E-Product E-Service and E-Entertainment, ICEEE2010, 2010, pp. 1–3, https://doi.org/10.1109/ICEEE.2010.5661554.

[15] R.W. Serth, W.A. Heenan, Gross error detection and data reconciliation in steam-metering systems, AIChE J. 32 (1986) 733–742, https://doi.org/10.1002/aic.690320503.

[16] C.A. Floudas, P.M. Pardalos, Recent Advances in Global Optimization, Princeton University Press, 2014.

[17] A. Martini, D. Coco, A. Sorce, A. Traverso, P. Levorato, Gross error detection based on serial elimination: applications to an industrial gas turbine, in: Proceedings of the ASME Turbo Expo, vol. 3A, 2014, https://doi.org/10.1115/GT2014-26746.

[18] Q. Jiang, M.J. Bagajewicz, On a strategy of serial identification with collective compensation for multiple gross error estimation in linear steady-state reconciliation, Ind. Eng. Chem. Res. 38 (1999) 2119–2128, https://doi.org/10.1021/ie980431e.

[19] J.M. Bagajewicz, Data Reconciliation and Instrumentation Upgrade. Overview and Challenges, FOCAPO 2003, 2003.

[20] I.B. Tjoa, L.T. Biegler, Simultaneous strategies for data reconciliation and gross error detection of nonlinear systems, Comput. Chem. Eng. 15 (1991) 679–690, https://doi.org/10.1016/0098-1354(91)85014-L.

[21] J. Rosenberg, R.S.H. Mah, C. Iordache, Evaluation of schemes for detecting and identifying gross errors in process data, Ind. Eng. Chem. Res. 26 (1987) 555–564, https://doi.org/10.1021/ie00063a023.

[22] M. Sánchez, J. Romagnoli, Q. Jiang, M. Bagajewicz, Simultaneous estimation of biases and leaks in process plants, Comput. Chem. Eng. 23 (1999) 841–857, https://doi.org/10.1016/S0098-1354(99)00104-0.

[23] G. Buzzi-Ferraris, F. Manenti, Interpolation and Regression Models for the Chemical Engineer, John Wiley & Sons, 2010. ISBN: 3527326529.

[24] C.E. Llanos, M.C. Sanchéz, R.A. Maronna, Robust estimators for data reconciliation, Ind. Eng. Chem. Res. 54 (2015) 5096–5105, https://doi.org/10.1021/ie504735a.

[25] G. Buzzi-Ferraris, F. Manenti, Outlier detection in large data sets, Comput. Chem. Eng. 35 (2011) 388–390, https://doi.org/10.1016/j.compchemeng.2010.11.004.

[26] M.J. Leibman, T.F. Edgar, L.S. Lasdon, Efficient data reconciliation and estimation for dynamic processes using nonlinear programming techniques, Comput. Chem. Eng. 16 (1992) 963–986, https://doi.org/10.1016/0098-1354(92)80030-D.

[27] J.D. Hedengren, R.A. Shishavan, K.M. Powell, T.F. Edgar, Nonlinear modeling, estimation and predictive control in APMonitor, Comput. Chem. Eng. 70 (2014) 133–148, https://doi.org/10.1016/j.compchemeng.2014.04.013.

[28] R. Gani, J. Bałdyga, B. Biscans, E. Brunazzi, J.C. Charpentier, E. Drioli, H. Feise, A. Furlong, K.M. Van Geem, J.C. de Hemptinne, A.J.B. ten Kate, G.M. Kontogeorgis, F. Manenti, G.B. Marin, S.S. Mansouri, P.M. Piccione, A. Povoa, M.A. Rodrigo, B. Sarup, E. Sorensen, I.A. Udugama, J.M. Woodley, A multi-layered view of chemical and biochemical engineering, Chem. Eng. Res. Des. 155 (2020) 133–145, https://doi.org/10.1016/j.cherd.2020.01.008.

[29] E.N. Pistikopoulos, A. Barbosa-Povoa, J.H. Lee, R. Misener, A. Mitsos, G.V. Reklaitis, V. Venkatasubramanian, F. You, R. Gani, Process systems engineering—the generation next? Comput. Chem. Eng. 147 (2021) 107252, https://doi.org/10.1016/j.compchemeng.2021.107252.

[30] N. Crane Co., Crane Technical Paper 410: Flow of Fluids Through Valves, Pipes and Fittings, 2009.

[31] B.W. Bequette, Master Process Control, 2002.

[32] G. Stephanopoulos, Chemical Process Control AN Introduction to Theory and Practices, Prentice-Hall, 1996, p. 716.

Chapter 8

Real-time optimization of batch processes via optimizing feedback control

Dominique Bonvin[a], Grégory François[b], Gianluca Rizzi[c], and Michael Amrhein[c]
[a]Laboratoire d'Automatique, EPFL, Lausanne, Switzerland, [b]HES-SO Valais-Wallis, Sion, Switzerland, [c]Online Control, Lausanne, Switzerland

1 Introduction

Batch processing is widely used in the manufacturing of goods and commodity products, in particular in the chemical, pharmaceutical, and food industries [1]. Batch operation differs significantly from continuous operation. While in continuous operation the process is maintained at an economically desirable operating point, in batch operation the process state evolves from an initial to a final time. Discontinuous operations can be of batch or semibatch type. In batch operations, the products to be processed are loaded in a vessel and processed without material addition or removal. This operation permits more flexibility than continuous operation by allowing adjustment of the operating conditions and the final time. Additional flexibility is available in semibatch operations, where reactants are added by adjusting the feedrate profiles, or products are removed via some outflow. We use the term batch processes to include semibatch processes.

In the chemical industry, many batch processes deal with reaction and separation operations. Reactions are central to chemical processing and can be performed in a homogeneous (single-phase) or heterogeneous (multiphase) environment. Separation processes can be of very different types, such as distillation, absorption, extraction, adsorption, chromatography, crystallization, drying, filtration, and centrifugation. The operation of batch processes follows recipes developed in the laboratory. The sequence of tasks to be carried out on each piece of equipment is predefined. The desired production volume is then achieved by repeating the processing steps on a predetermined schedule.

Simulation and Optimization in Process Engineering. https://doi.org/10.1016/B978-0-323-85043-8.00008-8

Optimization is the method of choice for improving process performance, while enforcing operational constraints [2]. However, one of the strengths of optimization, namely, its inherent mathematical rigor, can also be perceived as a weakness since engineers might sometimes find it difficult to obtain an appropriate mathematical formulation to solve their practical problems. Furthermore, even when process models are available, the presence of plant-model mismatch and process disturbances makes the direct use of model-based optimal inputs hazardous.

In the last 30 years, the field of real-time optimization (RTO) has emerged to help overcome the aforementioned modeling difficulties. RTO integrates process measurements into the optimization framework. This way, process optimization does not rely exclusively on a (possibly inaccurate) process model but also on process information stemming from measurements. The first widely available RTO approach was the two-step approach that adapts the model parameters on the basis of differences between predicted and measured outputs and uses the updated process model to recompute the optimal inputs [3]. However, in the presence of *structural* plant-model mismatch, this method is very unlikely to drive the plant to optimality [4]. Hence, alternatives to the two-step approach have been developed. For example, in modifier adaptation, correction for uncertainty is implemented via modifier terms that are added to the cost and constraint functions of the optimization problem [5]. As the modifiers include information on the differences between the predicted and the plant necessary conditions of optimality (NCO), this approach is prone to reach plant optimality upon convergence. Yet, another field has emerged, for which numerical optimization is *not* used online. With the so-called self-optimizing approaches [6–8], the optimization problem is recast as a control problem that uses measurements to enforce certain optimality features for the plant.

The optimization of batch processes falls naturally in the category of *dynamic optimization*, that is, with infinite-dimensional time profiles as decision variables. However, the introduction of input parameterization can transform the dynamic optimization problem into a static optimization problem with a countable number of decision variables [9], a so-called nonlinear program (NLP). This chapter describes a feedback-based strategy for the RTO of batch processes.

The chapter is organized as follows. Section 2 discusses the specificities of batch processes, and the models that can be used, while Section 3 formulates the dynamic optimization problem and reformulates it as a static optimization problem that can be solved on a batch-to-batch manner. The two main steps, namely, (i) the offline determination of an appropriate feedback structure and (ii) the implementation of online optimizing feedback control are discussed in Section 4. Section 5 illustrates the approach via the example of a batch distillation column, while Section 6 concludes the presentation.

2 Representation of batch processes

The operation of batch processes requires considerable attention, in particular regarding the coordination in time of various processing tasks such as charging, heating, reacting, separation, cooling, and the determination of optimal temperature and feeding profiles. The main features that distinguish batch from continuous processes are given next.

2.1 Distinguishing features

The two main distinguishing features of batch processes are [10]:

- *No steady-state operating point.* In batch processes, chemical and physical transformations proceed from an initial state to a very different final state. Hence, there does not exist a steady-state operating point for which the operating conditions can be optimized. Instead, the decision variables are infinite-dimensional time profiles.
- *Repetitive nature.* Batch processing is characterized by the frequent repetition of batch runs. Hence, it is appealing to use the results from previous runs to optimize the operation of subsequent ones. This has generated the industrially relevant topics of run-to-run control and, of interest here, run-to-run or batch-to-batch optimization.

In addition to these two main features, a number of issues tend to complicate the operation of batch processes, in particular their nonlinear behavior, the fact that the models are often poor for covering a large operating range, the presence of both path and terminal constraints, the presence of disturbances, the irreversible nature of batch processes and the fact that the potential of corrective action decreases with batch duration. These issues are discussed elsewhere [11].

2.2 Mathematical models

Model-based optimization techniques rely on appropriate mathematical representations. Modeling aspects of batch processes are addressed in Ref. [12]. The models used can be of three main types, namely, data-driven (black-box) models, knowledge-driven (white-box) models, and hybrid (gray-box) models [11]. Next, we will briefly introduce the formulation of knowledge-driven models that will be used in the context of this chapter.

A mechanistic state-space representation based on energy and material balances is the preferred approach for modeling batch processes. The rate expressions describe the effect that temperature and concentrations have on the various rates. The model relates the independent variables (inputs and disturbances) to the states (concentrations, temperature, and total mass) and the corresponding outputs. The run-time dynamics of the kth batch can be written as follows [11, 13]:

$$\dot{\mathbf{x}}_k(t) = \mathbf{F}(\mathbf{x}_k(t), \mathbf{u}_k(t)), \quad \mathbf{x}_k(0) = \mathbf{x}_{0,k}, \tag{8.1}$$

$$\mathbf{y}_k(t) = \mathbf{H}(\mathbf{x}_k(t), \mathbf{u}_k(t)), \tag{8.2}$$

$$\mathbf{z}_k = \mathbf{Z}(\mathbf{x}_k[0, t_f], \mathbf{u}_k[0, t_f]), \tag{8.3}$$

where $t \in [0, t_f]$ denotes the run time, $k = 1,2,\ldots$ the run or batch index, $\mathbf{x}$ is the n-dimensional state vector, and $\mathbf{u}$ is the m-dimensional input vector. There are two types of measured outputs, namely, the p-dimensional vector of run-time outputs $\mathbf{y}(t)$ that are available online within a batch and the q-dimensional vector of run-end outputs $\mathbf{z}$ that are available at the final time t_f. Note that $\mathbf{z}$ is not limited to quantities measured at the final time but can include quantities inferred from the profiles $\mathbf{x}_k[0, t_f]$ and $\mathbf{u}_k[0, t_f]$, such as the maximal temperature reached during the batch.

In addition to the run-time dynamics specific to a given run, one has the possibility of updating the initial conditions and the input profiles on a batch-to-batch manner as follows[a]:

$$\mathbf{x}_{0,k} = \mathbf{I}(\mathbf{x}_{k-1}[0, t_f], \mathbf{u}_{k-1}[0, t_f]), \quad \mathbf{x}_0[0, t_f] = \mathbf{x}_{init}[0, t_f], \tag{8.4}$$

$$\mathbf{u}_k[0, t_f] = \mathbf{K}(\mathbf{x}_{k-1}[0, t_f], \mathbf{u}_{k-1}[0, t_f]), \quad \mathbf{u}_0[0, t_f] = \mathbf{u}_{init}[0, t_f], \tag{8.5}$$

where $\mathbf{x}_{init}[0, t_f]$ and $\mathbf{u}_{init}[0, t_f]$ represent the initial state and input profiles, respectively.

2.3 Static view of a batch process

Due to the absence of a steady state, most signals in a batch process evolve with time. Consequently, the decision variables (here the input profiles $\mathbf{u}[0, t_f]$ for a given batch run) are of infinite dimension. However, it is possible to represent a batch process as a *static map* between a finite number of input parameters (used to define the input profiles before the batch starts) and the run-end outputs (representing the outcome at batch end). The key element is the parameterization of the input profiles as

$$\mathbf{u}[0, t_f] = \mathcal{U}(\boldsymbol{\pi}, t), \tag{8.6}$$

with the input parameters $\boldsymbol{\pi}$. This parameterization is achieved by dividing the input profiles into time intervals and using a polynomial approximation (of which the simplest form is a constant value) within each interval. The parameter

a. The initial conditions $\mathbf{x}_{0,k}$ and the input profiles $\mathbf{u}_k[0, t_f]$ can be updated on the basis of *several* prior batches; the relationships proposed here consider only the previous batch for simplicity of notation. Furthermore, although the batch time can differ from batch to batch, it is chosen to be the same for all batches in this formulation.

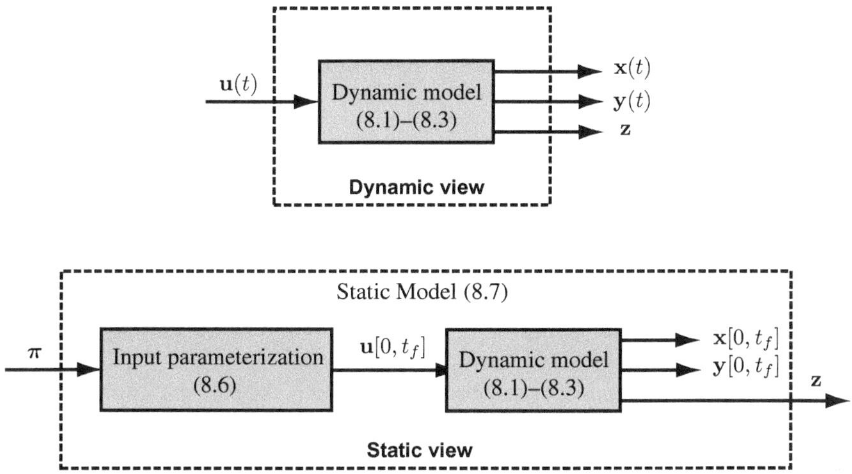

FIG. 8.1 Two different views of the same batch run. *Dynamic view*: The dynamic model describes the behavior of the states $\mathbf{x}(t)$, the run-time outputs $\mathbf{y}(t)$ and the run-end outputs $\mathbf{z}$ as functions of the inputs $\mathbf{u}(t)$. *Static view*: The static model $\mathbf{z} = \mathbf{M}(\boldsymbol{\pi})$ explicits the relationship between the input parameters $\boldsymbol{\pi}$ and the run-end outputs $\mathbf{z}$.

vector $\boldsymbol{\pi}$ can also include switching times between intervals [9]. With such a parameterization, the batch process can be seen as a static map between the finite set of input parameters $\boldsymbol{\pi}$ and the run-end outputs $\mathbf{z}$:

$$\mathbf{z} = \mathbf{M}(\boldsymbol{\pi}). \tag{8.7}$$

This static model indicates that once the input parameters $\boldsymbol{\pi}$ have been specified, it is possible to compute the run-end outputs $\mathbf{z}$. For this, one needs to generate the input profiles $\mathbf{u}[0, t_f]$ using Eq. (8.6), integrate Eq. (8.1), and generate $\mathbf{z}$ via Eq. (8.3). Alternatively, with an experimental setup, one can generate the input profiles $\mathbf{u}[0, t_f]$, apply them to the batch process, and collect information regarding the run-end outputs $\mathbf{z}$. Although the static model (8.7) looks simple, one has to keep in mind that the batch dynamics are in fact implicit in this relationship between input parameters (before the run) and run-end outputs (after the run). These two different ways of looking at the operation of a given batch are illustrated in Fig. 8.1. Very naturally, the "dynamic view" will be relevant for online within-batch operations, while the "static view" will be more appropriate for batch-to-batch operations.

3 Numerical optimization of batch processes

This section formulates the dynamic optimization problem and reformulates it as a static optimization problem that can be solved on a batch-to-batch manner.

3.1 Problem formulation: Dynamic optimization

We will consider the following constrained dynamic optimization problem:

$$\min_{\mathbf{u}[0,\, t_f]} \ \phi(\mathbf{x}(t_f)), \tag{8.8}$$

$$\text{s.t.} \quad \dot{\mathbf{x}}(t) = \mathbf{F}(\mathbf{x}(t), \mathbf{u}(t)), \quad \mathbf{x}(0) = \mathbf{x}_0, \tag{8.9}$$

$$\mathbf{S}(\mathbf{x}(t), \mathbf{u}(t)) \leq \mathbf{0}, \tag{8.10}$$

$$\mathbf{T}(\mathbf{x}(t_f)) \leq \mathbf{0}, \tag{8.11}$$

where ϕ is the terminal-time cost functional to be minimized, $\mathbf{x}(t)$ is the n-dimensional vector of state profiles with the known initial conditions $\mathbf{x}_0$, $\mathbf{u}(t)$ is the m-dimensional vector of input profiles, $\mathbf{S}$ is the n_S-dimensional vector of path constraints that hold during the interval $[0,\ t_f]$, and $\mathbf{T}$ is the n_T-dimensional vector of terminal constraints that hold at the final time t_f, with t_f being either free or fixed.[b]

3.2 Reformulation of a dynamic optimization problem as a static optimization problem

Numerically, the optimization problem (8.8)–(8.11) can be solved by control vector parameterization (CVP) [14], whereby the infinite-dimensional inputs $\mathbf{u}(t)$ are parameterized using a finite number of parameters $\boldsymbol{\pi}$. Typically, the inputs are divided into small intervals, and a polynomial approximation of the inputs is used within each interval. The input parameter vector $\boldsymbol{\pi}$ can also include switching times between intervals. It follows that the inputs can be expressed as $\mathbf{u}[0,\, t_f] = \mathcal{U}(\boldsymbol{\pi}, t)$ and the optimization performed with respect to $\boldsymbol{\pi}$, where $\boldsymbol{\pi}$ is the n_π-dimensional vector of *constant* decision variables. The difficulty with such a parameterization is that the path constraints $\mathbf{S}(\mathbf{x},\mathbf{u})$ have to be discretized as well, resulting in as many constraints as there are discretization points (typically a large number). However, if structural information regarding the shape of the optimal solution is available, the path constraints can be handled differently using an alternative parameterization as explained next.

The solution to a constrained terminal-time dynamic optimization problem is typically discontinuous and consists of various arcs. For a given input, the ith arc $\eta_i(t)$ can be of two types:

1. $\eta_{i,path}$: Arc $\eta_i(t)$ is determined by an active path constraint (constraint-seeking arc).
2. $\eta_{i,sens}$: Arc $\eta_i(t)$ is not governed by an active path constraint, it is sensitivity seeking.

---

b. If t_f is free, it becomes a decision variable. Note that it is also possible to have additional *constant* decision variables. In this case, the decision variables would include the vector of input profiles $\mathbf{u}[0,\ t_f]$ and the vector of constant quantities, $\boldsymbol{\rho}$. This general case will not be considered here for simplicity of exposition, see Ref. [8] for details.

Let us assume that the path constraints are met using either (i) assignment of input values in the case of active input bounds or (ii) online feedback control via the measurement or estimation of the constrained quantities and their regulation to track the active values. The other parts of the inputs are parameterized as in CVP as $\mathbf{u}[0, t_f] = \mathcal{U}(\boldsymbol{\pi}, t)$. Since the path constraints are handled directly, they do not involve the decision variables $\boldsymbol{\pi}$. Note that the knowledge of the constraint-seeking arcs is key for this parameterization. The final values of the states, which affect the objective function and the terminal constraints, can be expressed in terms of the decision variables $\boldsymbol{\pi}$ as follows:

$$\mathbf{x}(t_f) = \mathbf{x}_0 + \int_0^{t_f} \mathbf{F}(\mathbf{x}(t), \mathcal{U}(\boldsymbol{\pi}, t))dt := \mathcal{F}(\boldsymbol{\pi}), \tag{8.12}$$

as t_f is either known or an element of $\boldsymbol{\pi}$. Hence, the dynamic optimization problem (8.8)–(8.11) can be reformulated as

$$\min_{\pi} \quad \phi(\mathcal{F}(\boldsymbol{\pi})), \tag{8.13}$$

$$\text{s.t.} \quad \mathbf{T}(\mathcal{F}(\boldsymbol{\pi})) \leq \mathbf{0}, \tag{8.14}$$

since the system dynamics are implicit in $\mathcal{F}(\boldsymbol{\pi})$ and the path constraints are taken care of as mentioned earlier. It follows that the dynamic optimization problem can be reformulated as the following static NLP:

$$\min_{\pi} \quad \Phi(\boldsymbol{\pi}), \tag{8.15}$$

$$\text{s.t.} \quad \mathbf{G}(\boldsymbol{\pi}) \leq \mathbf{0}. \tag{8.16}$$

The scalar cost function to be minimized is written $\Phi(\boldsymbol{\pi}) := \phi(\mathcal{F}(\boldsymbol{\pi}))$, and the n_g-dimensional vector of constraints is written generically $\mathbf{G}(\boldsymbol{\pi}) := \mathbf{T}(\mathcal{F}(\boldsymbol{\pi}))$.

3.3 Batch-to-batch solution: Static optimization

Consider next the static optimization problem (8.15) and (8.16) and define the Lagrangian function

$$\mathcal{L}(\boldsymbol{\pi}) := \Phi(\boldsymbol{\pi}) + \boldsymbol{\mu}^T \nabla \mathbf{G}(\boldsymbol{\pi}), \tag{8.17}$$

where $\boldsymbol{\mu}$ denotes the n_g-dimensional vector of Lagrange multipliers associated with the constraints. With the assumption that the cost function Φ and the constraint functions $\mathbf{G}$ are differentiable, one can write the Karush-Kuhn-Tucker (KKT) necessary conditions of optimality given in Table 8.1. The upper part of this table shows the formal KKT conditions. However, from a practical point of view, if the set of active constraints is known, these conditions can be simplified as follows:

TABLE 8.1 KKT for a static optimization problem [15].

	Constraints	Sensitivities
Formally	$\mathbf{G} \leq \mathbf{0}$	$\frac{\partial \mathcal{L}}{\partial \boldsymbol{\pi}} = \mathbf{0}$
	or $\mu_j G_j = 0,\ \mu_j \geq 0,\ j = 1, \ldots, n_g$	
If active set known	$\mathbf{G}_a = \mathbf{0}$	$\left.\frac{d\Phi}{d\boldsymbol{\pi}}\right\vert_{\mathbf{G}_a=\mathbf{0}} = \mathbf{0}$

- Since the Lagrange multipliers of the nonactive constraints are identically equal to zero, one can discard the nonactive constraints and generate the conditions $\mathbf{G}_a = \mathbf{0}$ for the $n_{g,a}$ active constraints, where the subscript $(\cdot)_a$ denotes a quantity associated with the active constraints of problem (8.15) and (8.16).
- Considering only the active constraints, the sensitivity conditions can be written:

$$\frac{\partial \mathcal{L}}{\partial \boldsymbol{\pi}} = \frac{\partial \Phi}{\partial \boldsymbol{\pi}} + \boldsymbol{\mu}_a^T \frac{\partial \mathbf{G}_a}{\partial \boldsymbol{\pi}} = \mathbf{0}, \tag{8.18}$$

which indicates that the cost gradient $\frac{\partial \Phi}{\partial \boldsymbol{\pi}}$ must be aligned with a linear combination of the gradients of the active constraints, namely, $\boldsymbol{\mu}_a^T \frac{\partial \mathbf{G}_a}{\partial \boldsymbol{\pi}}$. Equivalently, one can say that the cost gradient must belong to the null space of the gradients of the active constraints. This null space can be computed as follows:

$$\frac{\partial \mathbf{G}_a}{\partial \boldsymbol{\pi}} \mathbf{N} = \mathbf{0}, \tag{8.19}$$

where $\mathbf{N} \in \mathbb{R}^{n_u \times n_r}$ is the null-space matrix defined by Eq. (8.19), with $n_r = n_u - n_{g,a}$ the dimension of the null space of $\frac{\partial \mathbf{G}_a}{\partial \boldsymbol{\pi}}$. It follows that

$$\frac{\partial \mathcal{L}}{\partial \boldsymbol{\pi}} \mathbf{N} = \frac{\partial \Phi}{\partial \boldsymbol{\pi}} \mathbf{N} = \left.\frac{d\Phi}{d\boldsymbol{\pi}}\right\vert_{\mathbf{G}_a=0} = \mathbf{0}, \tag{8.20}$$

which expresses the fact that these sensitivities represent the cost variations with respect to $\boldsymbol{\pi}$ when the active constraints are met. Note that this reduced set of KKT conditions does not involve any Lagrange multipliers $\boldsymbol{\mu}$.

3.4 Effect of plant-model mismatch

Numerical optimization relies on a model of the process. If an accurate model is available, one can reach plant optimality by solving the dynamic optimization problem (8.8)–(8.11) or, alternatively, the reformulated static optimization problem (8.15) and (8.16). However, in practice, since the model is typically only an approximation to the real process, it cannot by itself drive the *plant* to optimality. This requires the use of plant measurements, which lead to the concept of RTO. RTO encompasses several adaptation strategies [16], with the one based on feedback control being detailed in the following section.

4 Feedback-based optimization of uncertain batch processes

Plant optimality can be reached by enforcing the *plant* NCO that are associated with the dynamic optimization problem at hand or its reformulation as a static optimization problem. The feedback structure is determined offline on the basis of the best available model. Note that this model need not be very accurate; it can be a relatively simple representation of the process as the main requirement is not model accuracy but rather *model adequacy*.[c] We next detail the offline and the real-time activities that are required to implement feedback-based optimizing control.

4.1 Offline activity: Determine the feedback structure

The procedure involves two steps, namely, the characterization of the solution and the design of the feedback structure. These steps are described next.

4.1.1 Characterization of the solution

One uses an "adequate" model and numerically solves the dynamic optimization problem (8.8)–(8.11). As mentioned earlier, the computed solution will not be optimal for the plant due to plant-model mismatch. However, at this stage, one is mainly interested in *characterizing* the solution, that is, in identifying the types and sequence of arcs that constitute the solution. In particular, the path constraints that are active at the optimum, expressed formally as $\mathbf{S}_a(\mathbf{x}(t),\mathbf{u}(t)) = \mathbf{0}$, must be known as this information is used to reformulate the dynamic optimization problem as a static problem. Note that a path constraint is often active over some time interval $[t_1, t_2]$ and not necessarily for the entire batch duration $[0, t_f]$. In this case, however, t_1 and t_2 need not be known a priori as they can be adjusted via batch-to-batch control. Furthermore, the reliability of the types and sequence of arcs can be verified offline via robustness analysis, that is, by perturbing the model and making sure that, qualitatively, the same types and sequence of arcs result. With this information, the designer has a parameterized

c. The property that ensures that a model-based optimization problem is able to determine the optimal inputs for the plant is referred to as "model adequacy" [4, 17]. For a given model-based RTO scheme, a model is adequate if the RTO scheme is capable of predicting the correct set of active *plant constraints* and the correct directions of *plant gradients*, that is, the correct set of *plant* NCO. We will see later how model adequacy can be enforced with the RTO scheme labeled "modifier adaptation" [18].

model of the solution, called *solution model*, that is key for designing the feedback structure.

4.1.2 Design of the feedback structure

The solution model is a parameterized approximation to the optimal inputs, which include (i) the types and sequence of arcs, (ii) the input parameters as degrees of freedom (or manipulated variables [MVs]) that can be used for control, and (iii) the corresponding NCO (or controlled variables [CVs]). Note that there are as many NCO as there are degrees of freedom in the optimization problem, thus making the system of equations perfectly determined. It is important to understand that the solution model is simply a tool that helps solve the problem at hand in an efficient way, and there might be several perfectly valid solution models for a given problem. Hence, the designer will be able to propose alternative (from simplest to more complex) solution models to tackle the design of the multivariable control problem. This aspect, which represents the main strength of the approach labeled *NCO tracking* [8], is discussed in Ref. [11] and illustrated in the first case study therein.

In practice, the objective is to develop an easily implementable, yet efficient, control scheme. Enforcing plant optimality via feedback control requires having appropriate CVs and MVs. The CVs encompass (i) the active path constraints $\mathbf{S}_a(t) = \mathbf{0}$ that are handled online within batch, and (ii) the KKT conditions for the reformulated static optimization problem given in Table 8.1. The MVs consist of (i) some of the inputs, say the within-batch inputs $\mathbf{u}_{wb}(t)$, to track $\mathbf{S}_{p,a}(t) = \mathbf{0}$ in

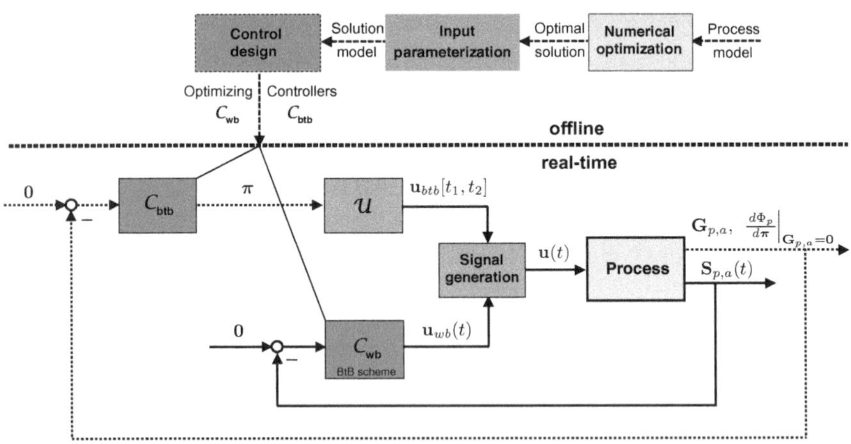

FIG. 8.2 NCO-tracking scheme. The offline activities include the generation of a solution model via numerical optimization and input parameterization, and the design of NCO-tracking controllers. The real-time activities involve within-batch (*solid lines*) and batch-to-batch (*dotted lines*) control to enforce the plant NCO, namely, $\mathbf{S}_{p,a}(t) = \mathbf{0}$ within batch, and $\mathbf{G}_{p,a} = \mathbf{0}$ and $\left.\frac{d\Phi_p}{d\pi}\right|_{\mathbf{G}_{p,a}=\mathbf{0}} = \mathbf{0}$ at batch end. The input signals $\mathbf{u}_{wb}(t)$ and $\mathbf{u}_{btb}[t_1, t_2]$ combine to generate the process inputs $\mathbf{u}(t)$.

certain time intervals,[d] and (ii) the input parameters $\boldsymbol{\pi}$, from which one can reconstruct the batch-to-batch inputs $\mathbf{u}_{btb}[t_1,t_2] = \mathcal{U}(\boldsymbol{\pi},t)$ as per Eq. (8.6). The corresponding feedback-based optimizing scheme is shown schematically in Fig. 8.2.

Remark 1 A key feature of NCO tracking is the possibility to choose appropriate MVs and CVs by introducing *approximations* to the solution. Hence, the choice of MVs and CVs is not made once for all, but it may require iterations. In fact, the appropriateness of these choices is often key to the success of NCO tracking in practice. These are true engineering decisions that are made with plant optimality in mind.

Remark 2 An interesting feature of this approach is the role played by the *process model* and the *solution model*. In a traditional optimization approach, the process model plays a key role, and much effort is made to obtain an "accurate" process model. Often, approximations are introduced without being able to assess their effects on the computed solution. In contrast, in NCO tracking, the process model needs to be adequate, but not necessarily accurate, that is, the requirement on the process model is much less. The main engineering decision is then made with respect to the expected solution, for which the effect of the approximations can de directly assessed.

Remark 3 NCO-tracking schemes attempt to meet the active constraints and push certain sensitivities to zero. Like with any control scheme, one cannot guarantee, in the presence of disturbances, that the set points (here the NCO) are met at all times. However, the presence of feedback helps *reduce the sensitivity* to these disturbances. One finds here all the features of multivariable feedback control, including the need for appropriate pairing of MVs and CVs.

Remark 4 Optimization of batch processes involves both path and terminal objectives as per Table 8.1. Hence, optimal operation of a single batch would typically require a model to predict the terminal objectives. When no model is available, one can rely on *batch-to-batch operation*, for which the reformulation of the optimization problem becomes handy.

4.2 Real-time activities: Implement feedback control

Optimizing feedback control is implemented according to Fig. 8.2 using both within-batch control (solid lines) and batch-to-batch control (dotted lines). These two complementary implementations are discussed in the following section.

4.2.1 Within-batch feedback

Within-batch feedback corresponds to standard multivariable control. Each active constraint $S_{p,a,j} = 0, j = 1,\ldots,n_{S_a}$, is kept active for a given time interval that is known at the beginning of the batch (either from prior knowledge or

d. The subscript $(\cdot)_p$ denotes a plant quantity.

computed via batch-to-batch control). When some of the control loops interact, decoupling control may be necessary [19].

4.2.2 Batch-to-batch feedback

The objective of batch-to-batch feedback is to enforce the KKT conditions in Table 8.1. There are different ways of doing it based on the information and the tools that are available. The resulting schemes differ with respect to the following issues, formulated here as questions: is the active set known, can the gradients be computed at the end of each batch, and can numerical optimization be performed between batches? These cases are described next and summarized in Table 8.2. We will only provide key references to the underlying schemes as these are well documented in the literature.

1. *Active set known, gradients available, use control (NCO tracking) [9, 20].* The KKT conditions consist of $\mathbf{G}_{p,a} = \mathbf{0}$ and $\left.\frac{d\Phi_p}{d\boldsymbol{\pi}}\right|_{\mathbf{G}_{p,a}=\mathbf{0}}$ that are computed at batch end. While $\mathbf{G}_{p,a}$ can be directly measured or estimated, the gradient $\left.\frac{d\Phi_p}{d\boldsymbol{\pi}}\right|_{\mathbf{G}_{p,a}=\mathbf{0}}$ requires system excitation through variation of $\boldsymbol{\pi}$ and measurement of the resulting cost Φ_p. Once the plant values of $\mathbf{G}_{p,a}$ and $\left.\frac{d\Phi_p}{d\boldsymbol{\pi}}\right|_{\mathbf{G}_{p,a}=\mathbf{0}}$

TABLE 8.2 Various batch-to-batch schemes.

	Known active set	BtB gradients required	Control or optimization	Plant optimality
NCO tracking	Yes	Yes	C	Yes
Constraint tracking	Yes	–	C	Yes
Self-optimizing control	Yes	–	C	Approximate
Modifier adaptation	–	Yes	O	Yes
Constraint adaptation	–	–	O	Yes

NCO tracking: Plant optimality assuming the active set is correct and the plant constraints and gradients are measured/estimated correctly. *Constraint tracking*: Plant optimality assuming the active set is correct, the optimal solution is fully determined by active constraints and the plant constraints are measured correctly. *Self-optimizing control*: Works well if the active set is correct and the plant outputs are measured correctly; yet the solution is only approximate because the controlled variables **c** are generated using the model and not the plant. *Modifier adaptation*: Plant optimality assuming the plant constraints and the gradients are measured/estimated correctly. *Constraint adaptation*: Plant optimality if the optimal solution is fully determined by active constraints and the plant constraints are measured correctly.

have been estimated, one can implement optimizing feedback by forcing these values to their set points of **0**.

2. *Active set known, no gradient, use control (CT) [21].* In constraint tracking (CT), the constraints expected to be active at the optimum are measured and controlled so as to be met after a few batches. Note that CT is a special case of NCO tracking, whereby only active constraints are tracked.
3. *Active set known, no gradient, use control (SOC) [22, 23].* Here, we are not concerned with active constraints, as it is assumed that these constraints are met, for example, via CT. Self-optimizing control forces the reduced gradients to zero via feedback control. Therein, the KKT conditions $\left.\frac{d\Phi_p}{d\boldsymbol{\pi}}\right|_{\mathbf{G}_{p,a}=\mathbf{0}}$, which cannot be computed at batch end, are replaced by the sensitivity variables $\mathbf{c}$, which are linear combinations of the inputs and outputs and are relatively insensitive to the uncertainty in the model parameters $\boldsymbol{\theta}$. These linear combinations are generated *offline* on the basis of the sensitivity matrix $\frac{\partial^2 \mathcal{L}}{\partial \boldsymbol{\pi} \partial \boldsymbol{\theta}}$. Feedback control is then used to keep these variables constant at the optimal values computed using the nominal model.
4. *Active set unknown, gradients available, use optimization (MA) [5, 18].* Modifier adaptation (MA) has been developed for this specific case. At the kth iteration, one needs to compute the zeroth-order modifiers $\boldsymbol{\epsilon}_k := \mathbf{G}_p - \mathbf{G}$, and the first-order modifiers $\boldsymbol{\lambda}_k^{\Phi} := \frac{d\Phi_p}{d\boldsymbol{\pi}} - \frac{d\Phi}{d\boldsymbol{\pi}}$ and $\boldsymbol{\lambda}_k^{G} := \frac{d\mathbf{G}_p}{d\boldsymbol{\pi}} - \frac{d\mathbf{G}}{d\boldsymbol{\pi}}$. While $\mathbf{G}_p$ can be directly measured or estimated, the gradients $\frac{d\Phi_p}{d\boldsymbol{\pi}}$ and $\frac{d\mathbf{G}_p}{d\boldsymbol{\pi}}$ require plant excitation through variation of $\boldsymbol{\pi}$ and measurement of the resulting cost Φ_p and $\mathbf{G}_p$.
5. *Active set unknown, no gradient, use optimization (CA) [24, 25].* When the optimal solution is fully determined by active constraints, there is no need to compute first-order modifiers. The resulting scheme, labeled constraint adaptation (CA), enforces the n_u most restrictive constraints. Note that, unlike in CT, the active set needs not be known in CA. Furthermore, plant optimality is reached via optimization, not feedback control.

5 Illustrative example: Batch distillation column

This case study illustrates the use of both constraint tracking and NCO tracking for the optimization of a batch distillation column.

5.1 Industrial batch distillation column

The separation of liquid mixtures into their various components using distillation is one of the major operations in the process industry. In particular, batch distillation is of considerable interest due to its flexibility in handling various mixtures and different product specifications in the same equipment, which is a big advantage in response to today's frequently changing production objectives.

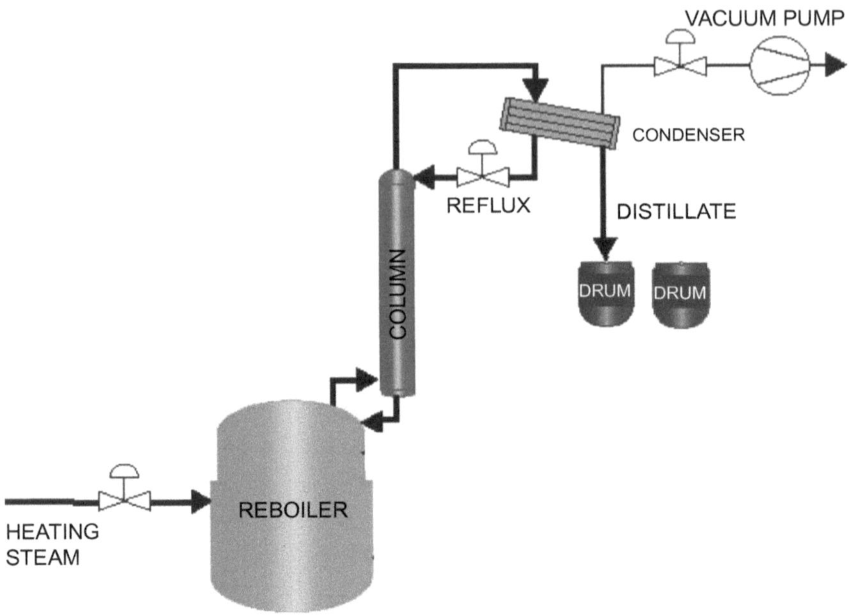

FIG. 8.3 Batch distillation column with the reboiler, the column, the condenser, and two accumulation pots.

A typical batch distillation process consists of a column, a reboiler with heating system, a condenser, a vacuum pump, and accumulation pots, as illustrated schematically in Fig. 8.3. The reboiler heats up the mixture to its boiling point and generates the gas phase inside the column, while the condenser produces the liquid phase in the top of the column. The column has several trays, or is composed of a packing structure, on which equilibrium is established between the gas and liquid phases in order to enrich the gas phase with the lighter component, while the heavier components condensate into the liquid phase [26]. The temperature and pressure gradients along the column result in a separation between lighter and heavier components in the top and bottom of the column, respectively. The liquid stream from the condenser is separated into (i) the reflux stream sent back to the column and (ii) the distillate stream collected as various fractions in specific accumulation pots.

The separation into fractions plays an important role in batch distillation of multicomponent mixtures. A fraction represents the amount of distillate collected in an accumulation pot before switching to the next fraction or stopping the distillation. The following two terms are used to specify the objective of a fraction:

- *Recovery*: Mass fraction of a species in the pot with respect to its initial charge.
- *Purity*: Mass fraction of a species with respect to the total mass in the pot.

In a so-called "impurity fraction," one tries to remove from the mixture as much impurity as possible, while losing as little product as possible. An impurity fraction is typically taken at the beginning of a distillation and between two product fractions. In a so-called "product fraction," the objective is to recover as much product as possible with specified purity requirements.

This study uses a two-component mixture that is representative of the separation found in the fragrance industry. The mixture contains the impurity A and the product B, with the boiling temperatures of 100°C and 120°C, respectively. Separation is conducted at constant low pressures between 11 mbar at the column bottom and 5 mbar at the column top. The objective is to minimize the batch time needed to recover in the reboiler a desired (large) amount of product B with given (high) purity. These product objectives can be reformulated as recovery objectives for the impurity fraction, namely, a desired (high) recovery of impurity with a desired (low) loss of product in the pot.

5.2 Process model

The simple dynamic shortcut model developed for this application [27] includes (i) overall material balance equations for the $S = 2$ species in the mixture, (ii) algebraic equations to calculate the physical properties [26, 28], (iii) equations based on the Antoine and Raoult relationships, and (iv) the Fenske-Underwood-Gilliland (FUG) model that allows estimating the distillate composition from the reboiler composition [29]. The MV (input) that is adjusted for optimization is the distillate flowrate, $w_D(t)$. Assuming negligible molar holdup in the column and in the condenser, the overall material balance can be written as

$$\frac{d\boldsymbol{n}_B(t)}{dt} = -\boldsymbol{x}_0(t)u_D(t), \qquad \boldsymbol{n}_B(t_0) = \boldsymbol{n}_{B,0}, \tag{8.21}$$

where $\boldsymbol{n}_B$ denotes the S-dimensional vector of numbers of moles in the reboiler, $\boldsymbol{n}_{B,0}$ is the corresponding numbers of moles of initial charge, $\boldsymbol{x}_0$ is the S-dimensional distillate molar fraction vector, u_D is the distillate molar flowrate, and t is the distillation time. The numbers of moles can be converted to masses upon using the S-dimensional vector of molecular weights $\boldsymbol{M}_w$:

$$\boldsymbol{m}_B(t) = diag(\boldsymbol{M}_w)\boldsymbol{n}_B(t), \tag{8.22}$$

$$\boldsymbol{m}_B(0) = diag(\boldsymbol{M}_w)\boldsymbol{n}_B(t_0), \tag{8.23}$$

$$\overline{M}_w(t) = \boldsymbol{M}_w^T \boldsymbol{x}_0(t), \tag{8.24}$$

$$w_D(t) = u_D(t)\overline{M}_w(t), \tag{8.25}$$

with $\overline{M}_w$ representing the average distillate molecular weight and w_D is the distillate mass flowrate. The mixture boiling temperature is computed using the Antoine and Raoult equations, a system of $N+2$ implicit equations that describe the equilibrium on the N theoretical trays and the reboiler ($k = 0$ for the column top and $k = N + 1$ for the reboiler):

$$\log \boldsymbol{p}_{s,k}(t) = \boldsymbol{a}_1 - \left(T_k(t)\boldsymbol{I}_S + diag(\boldsymbol{a}_3) \right)^{-1} \boldsymbol{a}_2, \tag{8.26}$$

$$\boldsymbol{p}_{p,k}(t) = diag\left(\boldsymbol{x}_k(t) \right) \boldsymbol{p}_{s,k}(t), \tag{8.27}$$

$$p_k(t) = \boldsymbol{1}_S^T \ \boldsymbol{p}_{p,k}(t), \tag{8.28}$$

where $\boldsymbol{p}_{s,k}$ and $\boldsymbol{p}_{p,k}$ represent the S-dimensional vectors of saturated and partial pressures, respectively, T_k is the temperature, p_k is the pressure, $\boldsymbol{x}_k$ is the S-dimensional molar fraction vector, $\boldsymbol{a}_{1,2,3}$ are the S-dimensional vectors of Antoine coefficients, and $\boldsymbol{1}_S$ is the S-dimensional vector with 1 as entries.

The distillate molar fraction $\boldsymbol{x}_0$ is computed using the FUG model:

$$\boldsymbol{x}_0(t) = diag\left(\overline{\boldsymbol{\alpha}}(t)^{N_{min}(t)} \right) \boldsymbol{x}_B(t) / \boldsymbol{1}_S^T diag\left(\overline{\boldsymbol{\alpha}}(t)^{N_{min}(t)} \right) \boldsymbol{x}_B(t), \tag{8.29}$$

where $\overline{\boldsymbol{\alpha}} = \boldsymbol{f}(T)$ represents the S-dimensional relative volatility vector, $N_{min} = f(w_D, \Delta p)$ is the minimal number of theoretical trays, and $\boldsymbol{x}_B = \boldsymbol{f}(\boldsymbol{n}_B)$ is the S-dimensional reboiler molar fraction vector.

The equations to compute the temperatures, pressures, and compositions represent a system of differential-algebraic equations (DAE) that are solved by using an appropriate DAE solver. The S-dimensional vector of pot masses of the impurity fraction is computed from material balances:

$$\boldsymbol{m}_p(t) = \boldsymbol{m}_B(t_0) - \boldsymbol{m}_B(t), \tag{8.30}$$

with the total mass being $m_t(t) = \boldsymbol{1}_S^T \boldsymbol{m}_p(t)$, The corresponding S-dimensional pot mass fraction and recovery vectors can be computed as

$$z(t) \ = \boldsymbol{m}_p(t) / m_t(t), \tag{8.31}$$

$$\boldsymbol{Y}(t) \ = diag^{-1}\left(\boldsymbol{m}_B(0) \right) \boldsymbol{m}_p(t). \tag{8.32}$$

5.3 Input parameterization of the impurity fraction

The optimization problem can be formulated for the impurity fraction as follows:

$$\begin{aligned} \min_{w_D(t),\, t_f} \quad & t_f, \\ \text{s.t.} \quad & \text{DAE model } (8.21) - (8.32) \\ & Y_A(t_f) \geq Y_{A,min}\ (99.5\%) \\ & Y_B(t_f) \leq Y_{B,max}\ (10\%) \\ & w_D(t) \geq w_{D,min}\ (0.5\text{kg/h}), \end{aligned} \tag{8.33}$$

where t_f is final batch time. The results of numerical optimization are shown in Fig. 8.4. The minimal batch time is $t_f^* = 5$ h, with the recovery requirements on

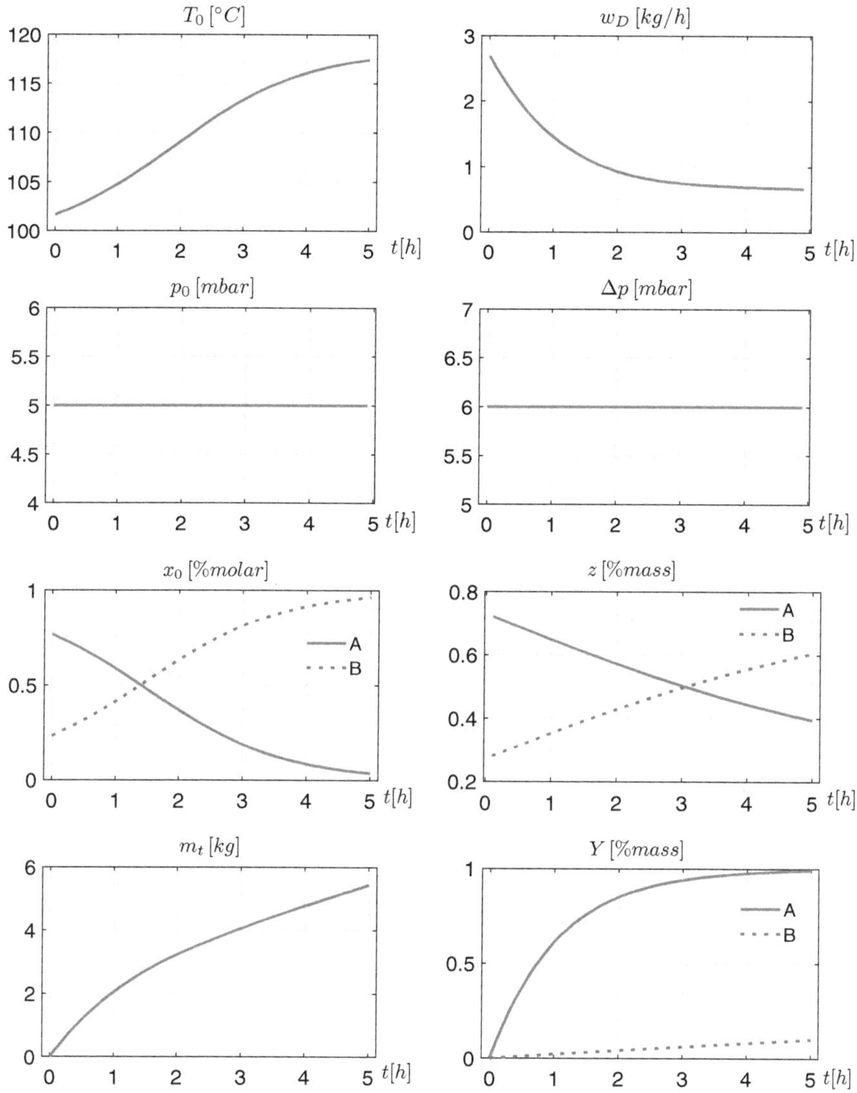

FIG. 8.4 Optimal profiles of the impurity fraction. Note that the optimal solution is characterized by $Y_A(t_f) = Y_{A,min}$ and $Y_B(t_f) = Y_{B,max}$.

$Y_A(t_f)$ and $Y_B(t_f)$ being met exactly. With two recovery constraints active at the optimum, it is proposed to parameterize the distillate flowrate using two parameters, namely, via the linear function:

$$w_D(t) = w_{D,0} + \delta t, \tag{8.34}$$

with the initial flowrate, $w_{D,0}$, and the slope, δ, as illustrated in Fig. 8.5.

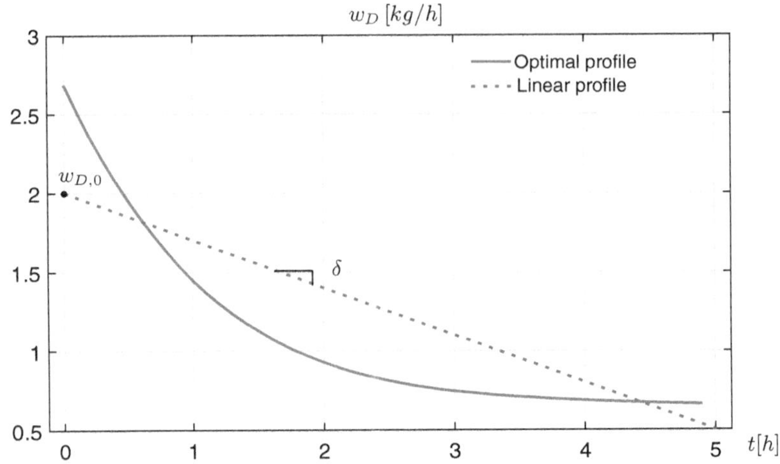

FIG. 8.5 w_D-profiles: Optimal profile and approximated linear profile with the two parameters $w_{D,0}$ and δ.

This way, one could meet the two active constraints simply via CT. However, the number of degrees of freedom can be reduced by one using a feature specific to batch distillation. Unlike problem formulation (8.33), where the final time t_f is a decision variable, t_f can be determined upon reaching a prespecified amount of distilled mass, namely, $m_{t,f} = f(\boldsymbol{m}_B(0), Y_{A,min}, Y_{B,max})$. It follows that, for a given $\boldsymbol{m}_B(0)$, this relationship links $m_{t,f}$ to the two recovery requirements. Hence, upon specifying $m_{t,f}$, the number of effective degrees of freedom in the optimization problem is reduced by one, that is, the two active constraints can be met using a single degree of freedom as detailed next. Numerical optimization of problem (8.33) with the decision variables $w_{D,0}$ and δ leads to a batch time of 5.15 h, that is, only 3% longer than true optimality, which can typically be neglected in industrial applications.

5.4 Control design and performance

Consider the gain matrix $\mathbf{K}_a$ that (locally) relates the two input parameters $\boldsymbol{\pi} := [w_{D,0}, \delta]^T$ to the two actively constrained outputs $\mathbf{G}_a := [Y_A(t_f), Y_B(t_f)]^T$. Since these two outputs are linearly dependent, this gain matrix has rank 1 and exhibits the following singular-value decomposition:

$$\mathbf{K}_a := \frac{\partial \mathbf{G}_a}{\partial \boldsymbol{\pi}} = [\mathbf{u}_1 \ \ \mathbf{u}_2] \begin{bmatrix} s_1 & 0 \\ 0 & 0 \end{bmatrix} [\mathbf{v}_1 \ \ \mathbf{v}_2]^T. \tag{8.35}$$

The free direction in the space of constrained outputs is along $\mathbf{u}_1$, which allows generating the controlled variable $c(t_f) = \mathbf{u}_1^T \begin{bmatrix} Y_A(t_f) \\ Y_B(t_f) \end{bmatrix}$ and the corresponding set point $c^{sp} = \mathbf{u}_1^T \begin{bmatrix} Y_A, min \\ Y_B, max \end{bmatrix}$.

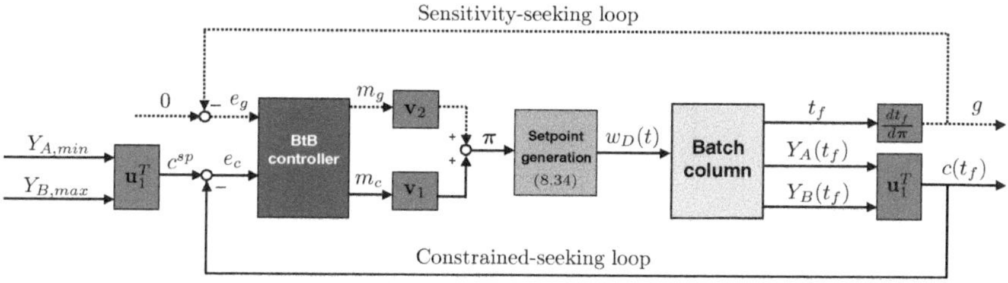

FIG. 8.6 BtB control with a constraint-seeking and a sensitivity-seeking loops and the inputs $\boldsymbol{\pi} := [w_{D,0}, \delta]^T$. The constraint-seeking loop uses the measurements $Y_A(t_f)$ and $Y_B(t_f)$ to construct the controlled variable $c(t_f)$ and generate the MV m_c. The sensitivity-seeking loop uses the gradient $g := \left.\frac{dt_f}{d\boldsymbol{\pi}}\right|_{\mathbf{G}_a=\mathbf{0}}$ to generate the MV m_g. These two MVs act in the directions $\mathbf{v}_1$ and $\mathbf{v}_2$, respectively, that are orthogonal to each other.

Constraint tracking and NCO tracking are implemented via BtB control as shown in Fig. 8.6 and discussed in the following sections. As process repeatability is assumed in this example, all batches start with the same initial conditions.

5.4.1 Constraint tracking

In the case of CT, only the constraint-seeking loop is active. The controlled variable $c(t_f)$ is computed from the batch-end measurements $Y_A(t_f)$ and $Y_B(t_f)$. The BtB controller processes the control error e_c to compute the manipulated variable m_c, which generates the input parameters $w_{D,0}$ and δ along the direction $\mathbf{v}_1$, that is, $\boldsymbol{\pi} = \mathbf{v}_1 m_c$. The setpoint generator uses that information to compute the distillate flowrate $w_D(t)$.

The performance of CT is shown in Fig. 8.7. The initial values of the input parameters are chosen so that the initial profile of $w_D(t)$ is significantly different from the optimal one. CT is able to significantly reduce the batch time from 8.5 to 5.5 h. In fact, since initially the recovery indices are better than required (higher recovery of impurity and lower loss of product), the BtB controller increases the distillate profile by increasing both $w_{D,0}$ and δ, thereby reducing the batch time. The active constraints are reached after five batches. Furthermore, the batch time of 5.5 h is rather close to the optimal value of 5 h, thus illustrating the efficiency of CT, that is, RTO without the sensitivity-seeking loop.

5.4.2 NCO tracking

NCO tracking includes both the constraint-seeking and the sensitivity-seeking loops, with their effects being additive. The constraint-seeking loop is designed as for CT. The sensitivity-seeking loop uses an estimate of the gradient $g := \left.\frac{dt_f}{d\boldsymbol{\pi}}\right|_{\mathbf{G}_a=\mathbf{0}}$ to compute the manipulated variable m_g, which generates a contribution to the input parameters $w_{D,0}$ and δ along the direction $\mathbf{v}_2$. Hence,

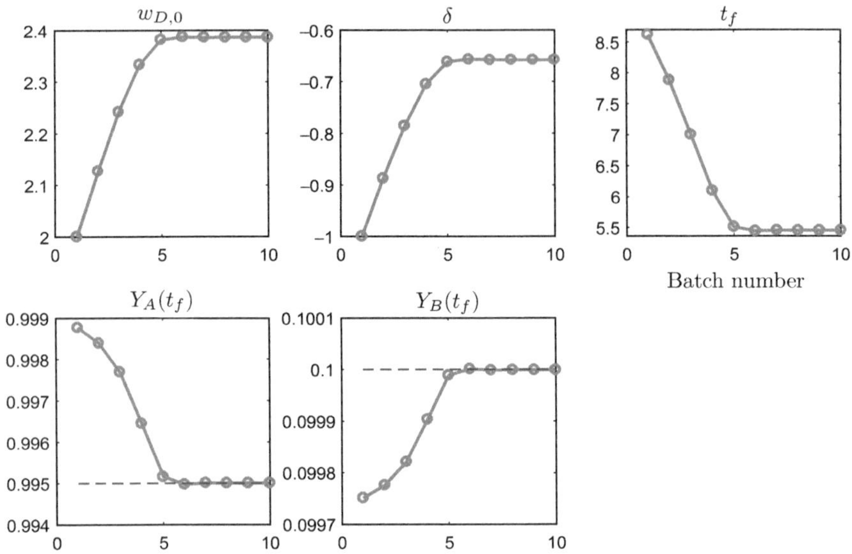

FIG. 8.7 BtB optimization using constraint tracking.

together, one has $\boldsymbol{\pi} = \mathbf{v}_1 m_c + \mathbf{v}_2 m_g$. The setpoint generator uses that information to compute the distillate flowrate $w_D(t)$. In addition to reaching the recovery requirements, this scheme will explicitly minimize batch time.

The optimization performance is shown in Fig. 8.8 for the same initial distillate profile as shown in Fig. 8.7. Convergence is reached after seven batches. Note that the sensitivity-seeking loop reduces the batch time by driving its gradient to zero (last subplot). The resulting batch time of 5.15 h corresponds to the result of numerical optimization with the linearly parameterized distillate profile. The difference of 0.15 h to the true optimum is due to the approximation introduced by the parameterization of $w_D(t)$ in Fig. 8.5.

5.4.3 *Practical aspects*

This distillation example illustrates the basic version of the NCO-tracking strategy developed by EPFL and Online Control. This strategy has been refined over the years to face the most common problems found in industrial practice. For example, additional difficulties such as the lack of process repeatability, the complexity of the mixture, and the need of cut conditions for each fraction have also been addressed. A strategy similar to the one described in this chapter has been tested on an industrial plant, showing up to 30% reduction in batch time for a constant-pressure application. Considering that further improvement can be expected by letting the pressure vary with time, one realizes how much potential lies in these measurement-based optimization schemes.

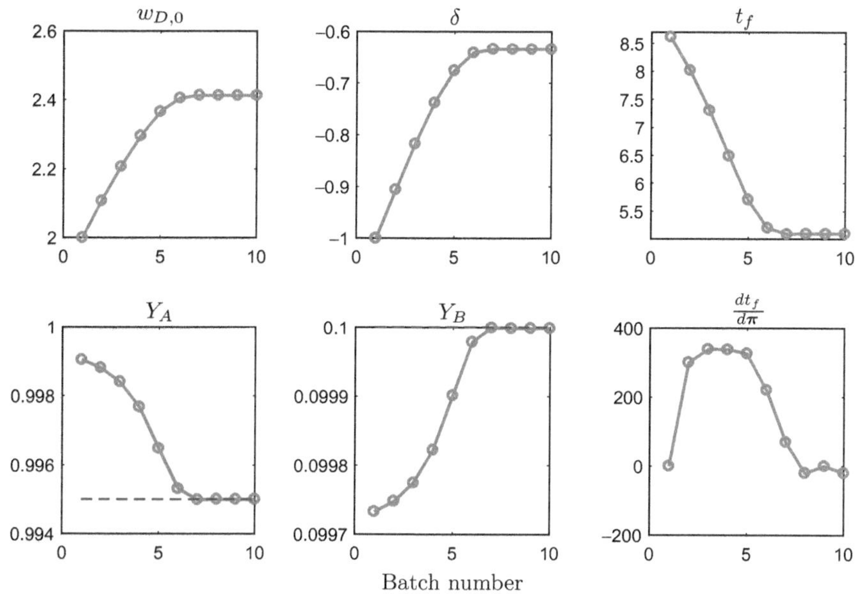

FIG. 8.8 BtB optimization using NCO tracking.

In this study, the NCO-tracking scheme with the sensitivity-seeking loop offers a slightly better performance in terms of batch time minimization, but it requires the evaluation of an experimental gradient at batch end. Hence, as the improvement over CT is not significant, it is not necessarily the scheme one would choose. It turns out that focusing exclusively on meeting active constraints has become a very appealing approach in industrial practice!

A more comprehensive distillation strategy that involves the pressure and pressure drop as additional MVs has also been developed by Online Control in the framework of the industrial software system O_Distill. This strategy is of particular interest to deal with difficult separations. More information is available in http://onlinecontrol.ch/?page_id=256.

6 Conclusions

This chapter has analyzed the optimal operation of batch and semibatch processes. It has been shown that incorporating measurements in the optimization framework helps improve the performances of chemical processes. The various RTO methods differ in the way measurements are used and the inputs adjusted to reject the effect of uncertainty. Measurements can be utilized to iteratively (i) update the states or the parameters of the model that is used for optimization or (ii) adjust the inputs via feedback control to enforce the NCO. It has been argued that the latter technique has the ability to reject the effect of uncertainty in the form of plant-model mismatch and process disturbances.

While several of the techniques presented here have been implemented successfully in industrial practice, many challenging problems still remain. Here are a few of them:

- Advances in optimization techniques in general, and in model predictive control in particular, have undoubtedly influenced industrial implementation. However, these techniques depend on the availability of process models of reasonably high fidelity and modest complexity, which is difficult to achieve in practice. Methods of nonlinear model reduction capable of representing complex process dynamics effectively with reduced-order, control-relevant models will provide a significant boost to the industrial application of advanced control and optimization techniques.
- While advances in state estimation have enabled the estimation of infrequently measured product characteristics, these estimates can never completely replace the actual measurements themselves. Advances in sensors, analyzers, and ancillary measurement technology will be required to make measurements of product properties more frequently available than is currently possible.
- As the structure of process control systems gets more complex, analyses of model dynamics, overall system stability, and achievable performance will be essential, especially for providing guidance in selecting the best alternative for each problem.

References

[1] D. Bonvin, Control and optimization of batch processes, in: J. Baillieul, T. Samad (Eds.), Encyclopedia of Systems and Control, Springer, 2014, pp. 1–6.

[2] G. François, D. Bonvin, Measurement-based real-time optimization of chemical processes, in: S. Pushpavanam (Ed.), Control and Optimisation of Process Systems, Advances in Chemical Engineering, 44, Elsevier, 2013, pp. 1–50.

[3] T.E. Marlin, A.N. Hrymak, Real-time operations optimization of continuous processes, in: AIChE Symposium Series—CPC-V, vol. 93, 1997, pp. 156–164.

[4] J.F. Forbes, T.E. Marlin, J.F. MacGregor, Model adequacy requirements for optimizing plant operations, Comput. Chem. Eng. 18 (6) (1994) 497–510.

[5] A.G. Marchetti, B. Chachuat, D. Bonvin, Modifier-adaptation methodology for real-time optimization, Ind. Eng. Chem. Res. 48 (13) (2009) 6022–6033.

[6] S. Skogestad, Self-optimizing control: the missing link between steady-state optimization and control, Comput. Chem. Eng. 24 (2000) 569–575.

[7] K.B. Ariyur, M. Krstic, Real-Time Optimization by Extremum-Seeking Control, John Wiley, New York, 2003.

[8] B. Srinivasan, D. Bonvin, Real-time optimization of batch processes via tracking the necessary conditions of optimality, Ind. Eng. Chem. Res. 46 (2) (2007) 492–504.

[9] G. François, B. Srinivasan, D. Bonvin, Use of measurements for enforcing the necessary conditions of optimality in the presence of constraints and uncertainty, J. Process Control 15 (6) (2005) 701–712.

[10] D. Bonvin, Optimal operation of batch reactors—A personal view, J. Process Control 8 (5–6) (1998) 355–368.

[11] D. Bonvin, G. François, Control and optimization of batch chemical processes, in: S. Rohani (Ed.), Process Control, Coulson and Richardson's Chemical Engineering, vol. 3B, Coulson and Richardson, 2017.

[12] U. Diwekar, Batch Processing: Modeling and Design, CRC Press, 2014.

[13] B. Srinivasan, D. Bonvin, Controllability and stability of repetitive batch processes, J. Process Control 17 (2007) 285–295.

[14] V.S. Vassiliadis, R.W.H. Sargent, C.C. Pantelides, Solution of a class of multistage dynamic optimization problems. 1. Problems without path constraints, Ind. Eng. Chem. Res. 33 (9) (1994) 2111–2122.

[15] M.S. Bazarra, H.D. Sherali, C.M. Shetty, Nonlinear Programming: Theory and Algorithms, second, John Wiley & Sons, New York, 1993.

[16] B. Chachuat, B. Srinivasan, D. Bonvin, Adaptation strategies for real-time optimization, Comput. Chem. Eng. 33 (10) (2009) 1557–1567.

[17] A.G. Marchetti, T. de Avila Ferreira, S. Costello, D. Bonvin, Modifier adaptation as a feedback control scheme, Ind. Eng. Chem. Res. 59 (6) (2020) 2261–2274.

[18] A.G. Marchetti, G. François, T. Faulwasser, D. Bonvin, Modifier adaptation for real-time optimization—Methods and applications, Processes 4 (4) (2016) 55.

[19] D.E. Seborg, T.F. Edgar, D.A. Mellichamp, F.J. Doyle, Process Dynamics and Control, John Wiley, New York, 2010.

[20] J.V. Kadam, W. Marquardt, B. Srinivasan, D. Bonvin, Optimal grade transition in industrial polymerization processes via NCO tracking, AIChE J. 53 (3) (2007) 627–639.

[21] D. Bonvin, B. Srinivasan, Optimal operation of batch processes via the tracking of active constraints, ISA Trans. 42 (1) (2003) 123–134.

[22] S. Skogestad, Plantwide control: the search for the self-optimizing control structure, J. Process Control 10 (5) (2000) 487–507.

[23] V. Alstad, S. Skogestad, Null space method for selecting optimal measurement combinations as controlled variables, Ind. Eng. Chem. Res. 46 (3) (2007) 846–853.

[24] B. Chachuat, A. Marchetti, D. Bonvin, Process optimization via constraints adaptation, J. Process Control 18 (2008) 244–257.

[25] T. de Avila Ferreira, Z. Wuillemin, T. Faulwasser, C. Salzmann, J. Van Herle, D. Bonvin, Enforcing optimal operation in solid-oxide fuel-cell systems, Energy 181 (2019) 281–293.

[26] R.H. Perry, D.W. Green, Perry's Chemical Engineers' Handbook, McGraw-Hill Professional, 2007.

[27] G. Rizzi, Modeling and Real-Time Optimization of Batch Distillation Processes (Master's thesis), University of Pisa, Italy, 2017.

[28] H. Kister, Distillation Design, McGraw-Hill Professional, 1992.

[29] S. Sundaram, L.B. Evans, Shortcut procedure for simulating batch distillation operations, Ind. Eng. Chem. Res. 32 (3) (1993) 511–518.

Chapter 9

On economic operation of switchable chlor-alkali electrolysis for demand-side management

Kosan Roh[a,b], Luisa C. Brée[a], Karen Perrey[c], Andreas Bulan[c], and Alexander Mitsos[a,d,e]

[a]*Process Systems Engineering (AVT.SVT), RWTH Aachen University, Aachen, Germany,* [b]*Department of Chemical Engineering and Applied Chemistry, Chungnam National University, Daejeon, Republic of Korea,* [c]*Covestro Deutschland AG, Leverkusen, Germany,* [d]*JARA-ENERGY, Aachen, Germany,* [e]*Institute of Energy and Climate Research: Energy Systems Engineering (IEK-10), Forschungszentrum Jülich GmbH, Jülich, Germany*

Nomenclature

Sets

$C=\{Cl_2, H_2, O_2, H_2O\}$	(index c) the set of components
$E=\{1,2\}$	(index e) the set of linearization intervals of the linearized electricity demand
$M=\{$H$_2$Mode, O$_2$Mode, cleanH$_2$-O$_2$, cleanO$_2$-H$_2\}$	(index m) the set of modes
$T=\{1,2,\ldots,t_n\}$	(index t) the set of all timesteps
$X=\{$et, au, sw$\}$	(index x) the set of equipment

Binary variables

$k_{m,e,t}$	indicates whether a linearization interval e of the electricity demand of a mode m is active at time t
$y_{m,t}$	determines whether mode m is active at time t
$z_{mm',t}$	indicates whether there is a transition from mode m' to mode m from time $(t-1)$ to t

Simulation and Optimization in Process Engineering. https://doi.org/10.1016/B978-0-323-85043-8.00007-6

Continuous variables

C_D [€/week]	weekly depreciation cost
$C_{OP,t}$ [€/h]	operating cost at time t
C_{TO} [€/week]	total production cost
$\dot{m}_{c,t}$[t/h]	production rate of component c at time t
$\dot{m}^{lin}_{Cl_2,m,e,t}$[t/h]	Cl_2 production rate of mode m for linearization power demand in interval e at time t
$P_{m,t}$ [kW]	electrical power demand of mode m at time t if this mode was active
α [−]	sizing factor

Parameters

$a_{m,e}$ [kWh/t]	gradient of linearized electricity demand of mode m in linearization interval e
$b_{m,e}$ [kW]	y-axis interception of linearized electricity demand of mode m in linearization interval e
$c_{m,e}$ [kW]	parameter adjusting the electricity demand of mode m in linearization interval e to the electrolyzer area
$\underline{\dot{m}}_{Cl_2}$, $\overline{\dot{m}}_{Cl_2}$ [t/h]	lower and upper bounds of the Cl_2 production rate if a reactive mode is active
$\Delta\dot{m}_{Cl_2}$[t/h]	maximal allowable production increase/decrease per timestep
p_c [€/t]	market prices for component c
$p_{el,t}$ [€/kWh]	spot price for electricity at time t
T [h]	time horizon
T_{OH} [h]	the number of annual operating hours
$\overline{\alpha}$[−]	an upper bound of the sizing factor α
β_x [MM€]	A capital investment for equipment x
γ_x [−]	1-year depreciation factor for equipment x
$\theta_{mm'}$ [−]	minimal mode duration of mode m' after switching from mode m to m' given in number of time steps
θ_{ramp} [−]	necessary time steps to ramp production from $\overline{\dot{m}}_{Cl_2}$ to $\underline{\dot{m}}_{Cl_2}$

1 Introduction

Renewable energy resources that are intermittently available have been penetrating the power grid. As a result, temporal fluctuations in electricity supply and prices get stronger and thus cause risk to destabilize the power grid instability. To overcome this challenge, flexible utilization of electric energy at power-intensive plants (i.e., varying the production level) has been recognized as an attractive solution for improving the economics of power consumers and/or reducing carbon footprint as well as stabilizing the power grid [1–4]. This operation strategy is commonly called demand-side management (DSM) or demand response. DSM is worthwhile for various power-intensive industrial plants such as air separation units [5–7], seawater reverse osmosis [8], aluminum production processes [9], glass production processes [10], and chlor-alkali (CA) processes [11,12]. These processes can economically benefit from DSM by participating in either electricity spot markets or grid balancing markets, or both [9].

Chlor-alkali electrolysis (CAE) is one of the power-intensive chemical plants. It produces the base chemicals chlorine (Cl_2) and sodium hydroxide (NaOH) via electrolysis of brine. There are three variants of CAE in worldwide commercial use, which differ in how ions are separated: diaphragm, amalgam, and membrane types [13]. CAE is particularly promising and desirable for DSM. The fast dynamics of CAE [13] make chlorine production suitable for varying the operational level. Also, CAE has a large potential for offering operational flexibility. For example, CAE in Germany consumed 4.3% of the German industrial electricity consumption in 2017 [14–16]. DSM applications to CAE have been investigated by a few studies, such as dynamic modeling and simulation of conventional CAE [17,18] and its downstream processes [19,20]; and optimization of the operation of conventional CAE [11,18,21] and switchable CAE [12,22,23].

In this chapter, our focus is on switchable CAE and its application to DSM. A CA process with a bifunctional electrode can be operated in different operational modes, namely hydrogen (H_2) mode and oxygen (O_2) mode, having different power demands while keeping chlorine and sodium hydroxide production at the same level. Switching the operation between the two modes according to current power prices can make DSM of the CA process beneficial, even with a limited capacity of chlorine storage [24] and already high capacity utilization of the plant [25]. In Section 2, we introduce the switching operation of CAE along with its advantages and limitations. In Section 3, we present a mathematical formulation for determining the optimal operation profile and the optimal overcapacity of switchable CAE. In Section 4, summarizing key findings from our previous work [12,22,23,26], we examine the optimal operation and sizing of switchable CAE for the purpose of DSM under various scenarios. In particular, we compare the economic performance with other flexibility options applicable to the chlorine industry.

2 Operational mode switching of chlor-alkali electrolysis

The operation strategies presented in this chapter focus on the membrane-based CAE as it dominates modern installed capacity among three variants [27]. As described in Fig. 9.1, in all strategies, the anode compartment (left side of both cells) is fed with saturated brine to produce chlorine and sodium ions (Na^+) when electric power is applied. Na^+ migrate through the membrane to the cathode compartment where water is fed. At the standard cathode (STC, the left), water is reduced to hydrogen and hydroxide ions (OH^-). The OH^- and migrated

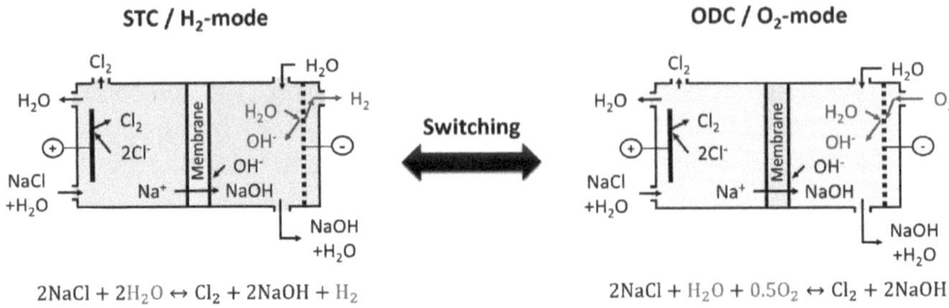

FIG. 9.1 Schematic diagram of CA electrolysis with STC (H_2-mode) and ODC (O_2-mode) and the respective chemical reactions. *(Modified from K. Roh, L.C. Brée, K. Perrey, A. Bulan, A. Mitsos, Flexible operation of switchable chlor-alkali electrolysis for demand side management, Appl. Energy 255 (2019) 113880. https://doi.org/10.1016/j.apenergy.2019.113880.)*

Na^+ form sodium hydroxide while the hydrogen is obtained as a gaseous byproduct.

Instead of STCs, implementing oxygen-depolarized cathodes (ODCs) allows direct contact of the liquid electrolytic phase with a gas phase at the cathode [28,29]. This alternative reduces the cell potential and thus the electricity demand due to a change in side reactions. As illustrated in Fig. 9.1 (the right), the anode compartment and the ion-exchange membrane for the setup with ODCs are identical to those with STCs. However, on the cathode side, gaseous oxygen and liquid sodium hydroxide are separated by the ODC. It allows oxygen diffusion into a reaction zone where the oxygen-reduction reaction can occur. In contrast to the conventional setup with STCs, hydrogen byproduct is not produced anymore.

Due to the difference of the chemical reactions, the total cell potential of a membrane CAE with STC and ODC measured at commercial plants is $U_{\text{Cell, STC}} \approx 3.0$ V and $U_{\text{Cell, ODC}} \approx 2.0$ V, respectively, with a current density of $6\,\text{kA/m}^2$ [28,29]. Hence, the power demand for the CAE with ODCs is lower than that with STCs by up to 30% theoretically and up to 25% in practice [30]. The aforementioned two cathodic operation principles can be combined in a bifunctional electrode [31]. The bifunctional cathode produces hydrogen when no oxygen is fed into the electrolyzer like an STC, here called H_2-mode in which hydrogen can be produced as a byproduct and sold, or works as an ODC when oxygen is supplied to the cell, here called O_2-mode. This bifunctionality gives us the possibility to switch the operation of CAE between the two modes. The superiority of the operation modes, i.e., which operation is economically more favorable, highly depends on prices of electricity and hydrogen, while prices of oxygen and water are less influential. Fig. 9.2 shows breakeven prices of electricity and hydrogen, at which the operation would be switched from one mode to the other. Utilization of the bifunctional electrode for switchable operation is preferred when an average price of electricity gets close to the breakeven price line. Furthermore, larger fluctuations in electricity prices

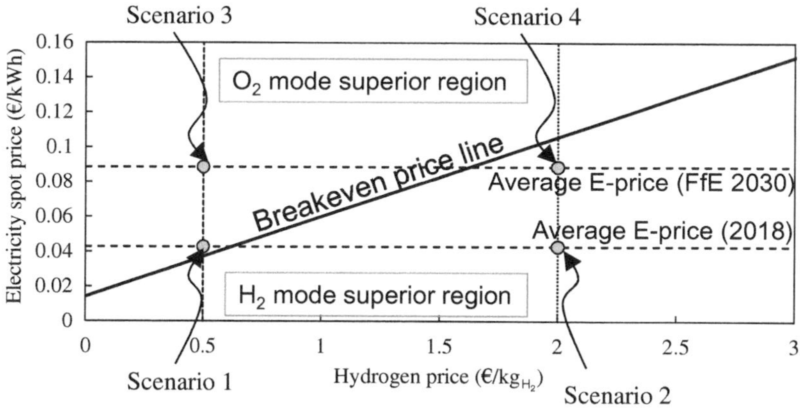

FIG. 9.2 Breakeven prices *(solid line)* at which both modes result in the same operating costs during operation at steady-state. 0.05 €/kg_{O_2} and 0.01 €/kg_{H_2O} are assumed for oxygen and water prices, respectively. The H_2-mode is superior to the O_2-mode on the right-hand side of the breakeven price line. *Four circles* indicate the electricity and hydrogen price combinations used in the four scenarios [32,33] analyzed in Section 4.2. *(From L.C. Brée, A. Bulan, R. Herding, J. Kuhlmann, A. Mitsos, K. Perrey, K. Roh, Techno-economic comparison of flexibility options in chlorine production, Ind. Eng. Chem. Res. 59 (26) (2020) 12186–12196. https://doi.org/10.1021/acs.iecr.0c01775. Credit: Copyright 2021 American Chemical Society.)*

around the breakeven price line bring us more opportunities to take benefit from switching of the operational mode.

Switching operation of CAE with the bifunctional electrode can avoid several challenges when DSM is pursued. Since most industrial CA processes are operated at over 95% capacity utilization [25], they can reduce the power load but cannot increase it substantially. Therefore, the aggregate demand for chlorine cannot be satisfied when aiming for DSM at the electricity spot market. CA processes would require additional investments for load shifts by oversizing the whole plant [25]. Capital investment for the plant oversizing depends on the preconditions of the plant site [34]. In particular, enlarging chlorine storage is very costly or even prohibited due to the strict regulations for ensuring safety [24]. If we cannot store chlorine for DSM, the downstream processes of CAE need to adjust its operation [19]. However, in the case of switchable CAE, the electricity consumption can be changed unlimitedly long even under limited chlorine storage capacity.

The challenge of the switching operation of CAE is that hydrogen and oxygen molecules have to be strictly separated to avoid any potential explosive reactions. To this end, a cleaning process (see Fig. 9.3) removing hydrogen or oxygen from the cathode chamber by purging inert gas such as nitrogen is essential when switching from one operational mode to the other. This processing step results in downtime of chlorine production. In order to guarantee a constant supply of chlorine (and sodium hydroxide) for downstream processes,

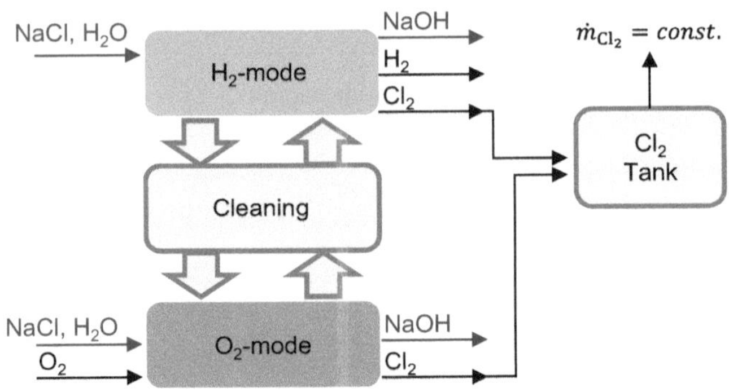

FIG. 9.3 Schematic diagram of switching operation of chlor-alkali electrolysis. *(Modified from L. C. Brée, K. Perrey, A. Bulan, A. Mitsos, Demand side management and operational mode switching in chlorine production, AIChE J. 65 (7) (2019) e16352. https://doi.org/10.1002/aic.16352.)*

intermediate chlorine storage is required. Also, we should substantially invest in retrofitting the electrolyzers as well as building additional infrastructure for supplying the cleaning gas and oxygen feedstock (if an STC-based CA plant exists) [22].

3 Mathematical formulation for optimal sizing and operation of switchable chlor-alkali electrolysis

We briefly present the mathematical models for optimal sizing and operation of switchable chlor-alkali electrolysis with respect to electricity price fluctuations. We adopt a lumped and quasisteady-state model with discrete-time variables. The model is comprised of four submodels including operational mode transition constraints, mass balances, ramping constraints, and cost functions.

3.1 Operational mode transition

This submodel is the core of the whole mathematical model. It allows an accurate and also efficient modeling of the operational mode switching using a discrete-time representation. For the implementation of a switchable operation, three constraints are imposed:

$$\sum_{m} y_{m,t} = 1 \quad \forall t \in T, \tag{9.1}$$

$$\sum_{m'} z_{m'm,t-1} - \sum_{m'} z_{mm',t-1} = y_{m,t} - y_{m,t-1} \quad \forall m \in M, t \in T, \tag{9.2}$$

$$y_{m',t} \geq \sum_{k=1}^{\theta_{mm'}} z_{mm',t-k} \quad \forall m, m' \in M, t \in T. \tag{9.3}$$

Binary variables y are introduced for each mode m showing its (in)active status. Eq. (9.1) assures that only one of all modes m is active at each time point t. Here, four possible modes $m \in \{H_2\text{Mode}, O_2\text{Mode}, \text{clean}H_2\text{-}O_2, \text{clean}O_2\text{-}H_2\}$ are considered. The first two modes correspond to chlorine production in the respective modes while the last two modes do the cell cleaning. The binary variable $z_{m'm,\,t}$ is 1 if and only if a transition from mode m to mode m' occurs from time $(t-1)$ to time t. These transition variables can be coupled with $y_{m,\,t}$ by Eq. (9.2) [35]. Setting some of $z_{mm',\,t}$ to zero, e.g., $z_{m^1m^2,\,t}$ suppresses the switching from mode m^1 to mode m^2. This assures that an operational mode m^1 does not directly follow the other operational mode m^2. To guarantee a minimum duration of each mode, Eq. (9.3) is imposed [36], where the CA plant has to stay in mode m^1 for at least $\theta_{mm'}$ time periods after switching from mode m to mode m'. This inequality constraint is required, e.g., for (i) guaranteeing complete gas removal during the cell cleaning and (ii) keeping the newly activated reactive mode lasting for a certain duration after switching, in order to prevent the whole plant from damage due to frequent switching [22].

3.2 Mass balance

The chlorine production rate $\dot{m}_{Cl_2,m,t}$ is bounded as

$$\alpha \underline{\dot{m}}_{Cl_2} y_{m,t} \leq \dot{m}_{Cl_2,m,t} \leq \alpha \overline{\dot{m}}_{Cl_2} y_{m,t} \quad \forall m \in M, t \in T, \tag{9.4}$$

where a continuous variable α is a sizing factor of the CAE with given bounds as $1 \leq \alpha \leq \overline{\alpha}$. We can optionally optimize α if optimal overcapacity of the switchable CAE should be determined, e.g., Refs. [23,26]. $\underline{\dot{m}}_{Cl_2}$ and $\overline{\dot{m}}_{Cl_2}$ indicate the minimum and maximum production rates in the two reactive modes when α is 1. The mass flow rate of other components including hydrogen, oxygen, brine, and sodium hydroxide during the two reactive modes is stoichiometrically calculated (see chemical reactions in Fig. 9.1). The mass flow rates of each component in each reactive mode m become zero if the respective mode is inactive (i.e., $y_{m,\,t} = 0$). The material content of the chlorine storage tank at t is calculated via simple mass balances. The content is limited by maximum (i.e., the storage capacity) and minimum bounds. We add an end-point constraint that enforces the chlorine content in the storage tank at the last timestep t_n to be the same as the content at t_0, which is assumed to be 50% of the storage capacity.

3.3 Power demand

As shown in Fig. 9.4, we apply a piecewise affine linear approximation to the underlying power demand function (derived from the power density data [37]) for the reactive modes by subdividing the chlorine production rate into multiple intervals e:

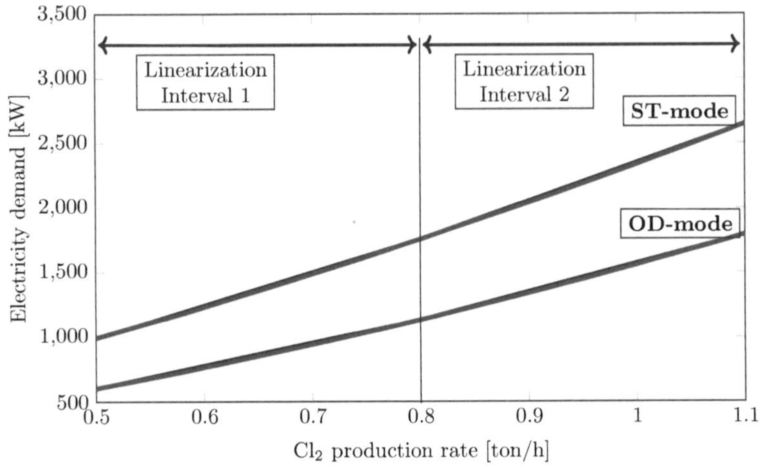

FIG. 9.4 Assumed electricity demand subject to the produced amount of Cl_2 for both modes *(red, gray in print version)* and the respective regression curves *(black)*. ST-mode and OD-mode correspond to H_2-mode and O_2-mode, respectively. *(From L.C. Brée, K. Perrey, A. Bulan, A. Mitsos, Demand side management and operational mode switching in chlorine production, AIChE J. 65 (7) (2019) e16352. https://doi.org/10.1002/aic.16352.)*

$$P_{m,t} = \sum_e \left(a_{m,e} \dot{m}^{\text{lin}}_{\text{Cl}_2,m,e,t} + b_{m,e} k_{m,e,t} + c_{m,e} \alpha k_{m,e,t} \right) \quad \forall m \in M, t \in T, \tag{9.5}$$

where $a_{m,\ e}$, $b_{m,\ e}$, and $c_{m,\ e}$ are parameters and $k_{m,\ e,\ t}$ indicates whether an interval e of the linearized electricity demand of a mode m is active at time t.

3.4 Ramping constraints

Ramping constraints are imposed on the chlorine production rate to limit its maximum change over a time step and a minimal active production rate:

$$\Delta \dot{m}_{\text{Cl}_2} = \frac{\alpha \left(\overline{\dot{m}}_{\text{Cl}_2} - \underline{\dot{m}}_{\text{Cl}_2} \right)}{\theta_{\text{ramp}}}, \tag{9.6}$$

$$\dot{m}_{\text{Cl}_2,t+1} \leq \dot{m}_{\text{Cl}_2,t} + \sum_m y_{m,t} \Delta \dot{m}_{\text{Cl}_2} + \left(1 - \sum_m y_{m,t} \right) \alpha \underline{\dot{m}}_{\text{Cl}_2} \quad \forall t \in T \setminus \{t_n\}, \tag{9.7}$$

$$\dot{m}_{\text{Cl}_2,t+1} \geq \dot{m}_{\text{Cl}_2,t} - \sum_m y_{m,t+1} \Delta \dot{m}_{\text{Cl}_2} - \left(1 - \sum_m y_{m,t+1} \right) \alpha \underline{\dot{m}}_{\text{Cl}_2} \quad \forall t \in T \setminus \{t_n\}, \tag{9.8}$$

$$\dot{m}_{\text{Cl}_2,t} \geq \sum_m \alpha \underline{\dot{m}}_{\text{Cl}_2} y_{m,t} \quad \forall t \in T, \tag{9.9}$$

where θ_{ramp} denotes the minimum number of time steps necessary to ramp the production between the maximum and minimum chlorine production rates.

3.5 Cost function

We calculate operating and depreciation costs of switchable CAE. The operating costs are calculated by

$$C_{\mathrm{OP},t} = \sum_{c} p_c \dot{m}_{c,t} + p_{\mathrm{el},t}(P_{\mathrm{H_2Mode},t} + P_{\mathrm{O_2Mode},t}) \quad \forall c \in \{\mathrm{H_2O}, \mathrm{H_2}, \mathrm{O_2}\} \quad \forall t \in T, \tag{9.10}$$

where p_c is a market price of component c and $p_{\mathrm{el},\,t}$ is an electricity spot price at time t. Here, the hydrogen is sold and thus represents negative costs. We exclude the cost terms of chlorine, brine, and sodium hydroxide from the cost calculation since their production rates are the same in both operational modes.

The depreciation costs are calculated by

$$C_{\mathrm{D}} = \alpha \frac{T}{T_{\mathrm{OH}}} \sum_{x} \beta_x \gamma_x \quad \forall x \in \{\mathrm{et}, \mathrm{au}, \mathrm{sw}\}, \tag{9.11}$$

where β_{et}, β_{au}, and β_{sw} are the capital investment coefficients for (i) installing the conventional CAE, (ii) enhancing the capacity of auxiliary units in up- and downstream, and (iii) installing pipelines and storage tanks for O_2 and cleaning gas supplies and retrofitting to the bifunctional electrode, respectively. The capital investment is assumed to be linearly proportional to the capacity of CAE. T and T_{OH} are the number of simulation horizon hours and annual operating hours, respectively. γ_x is the 1-year depreciation factor.

We then calculate the total production cost of switchable CAE over a given time horizon by

$$C_{\mathrm{TO}} = \sum_{t=0}^{t_n} C_{\mathrm{OP},t} + C_{\mathrm{D}}. \tag{9.12}$$

We minimize the total production cost while optimizing the plant operation and, optionally, overcapacity.

The presented models are elaborated in Brée et al. [22] and Roh et al. [23].

4 Case study

This section illustrates flexible operation of switchable chlor-alkali electrolysis in case studies. We first discuss the optimal operation for a fixed overcapacity. Then, we compare the economic performance of switchable CAE with other flexibility options in the chlorine industry. Finally, we allow for optimizing the plant overcapacity in addition to the operation in order to acquire further operational flexibility.

4.1 Optimal operational behavior of switchable chlor-alkali electrolysis

We consider a weekly electricity price profile for the year 2030 in Germany (with an average price of 0.0981 €/kWh) [22]. We assume that weekly electricity prices are perfectly forecastable. This simplification is intended to examine the ideal economic potential of the operational mode switching of CAE. As shown in Fig. 9.2, this price combination makes the switching operation of CAE economically favorable as the electricity spot price frequently crosses the breakeven price for the given hydrogen price (2 €/kg_{H_2}). The CA plant is assumed to be 2 oversized by 10%, i.e., the sizing factor α is not optimized but rather only the plant operation. All the other underlying assumptions such as the chlorine demand, chlorine storage capacity, minimum cleaning duration, and maximal ramping speed are the same as in Ref. [22]. We minimize the total operating cost ($\min \sum_{t=0}^{t_n} C_{OP,t}$). The optimal operation profile and active mode are determined by solving a mixed-integer linear program (MILP) in General Algebraic Modeling System (GAMS Version 28) with CPLEX 12.

As shown in Fig. 9.5, the operational mode switches between the two modes depending on the prices of electricity. Starting in the H_2-mode, when the electricity price increases, the chlorine production rate is decreased until it reaches the minimum operation level. Then, the plant is switched via the cleaning mode to the O_2-mode. When the electricity price drops, the plant operation is turned back to the maximum operation level to meet the chlorine demand downstream as well as refilling the chlorine storage tank. The total downtime of the CA plant is 50 min, which is required for the sum of all cleaning intervals.

As shown in Fig. 9.6, the flexible operation (combination of switching and variable operation) of switchable CAE (SW-OV-CA) can save both electricity and operating costs compared to the steady and variable operation with STCs (S-STC and OV-STC). The steady and variable operation with ODCs (S-ODC and OV-ODC) shows lower electricity consumption than both the conventional and switching cases due to the lower power demand. However, more savings in operating costs can be achieved by SW-OV-CA than S-ODC and OV-ODC because the bifunctional electrode can combine the benefits of both operational modes. Overall, the highest reduction in operating costs can be reached (5.86% compared to the steady operation with STCs) via flexible operation of SW-OV-CA.

4.2 Comparison of switchable chlor-alkali electrolysis to other flexibility options

We compare the economic performance of the switchable chlor-alkali electrolysis (SW-OV-CA) with other flexibility options applicable to the chlorine industry. Fig. 9.7 shows seven modifications to a conventional CA plant that operates at a steady-state (S-STC). The modifications include the oversizing

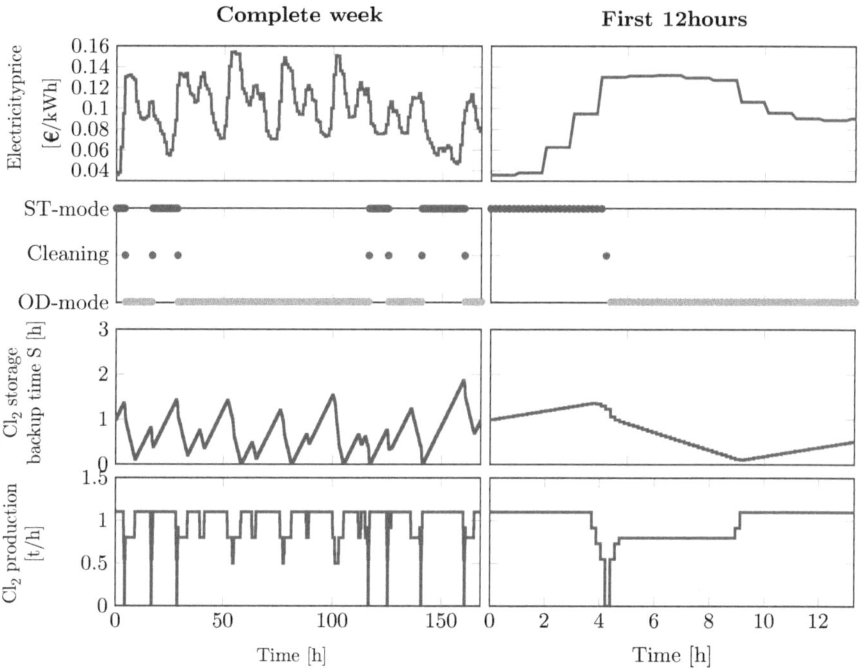

FIG. 9.5 Switchable and variable operation for the future electricity price profile. The first row presents the electricity price, the second-row shows which mode is active at which time step, the third row presents the content of the chlorine storage tank and the fourth row presents the chlorine production rate. The left column presents the results of the whole considered time horizon. The right column shows the optimal operation during the first 12 h of the time horizon. ST-mode and OD-mode correspond to H_2-mode and O_2-mode, respectively. *(Modified from L.C. Brée, K. Perrey, A. Bulan, A. Mitsos, Demand side management and operational mode switching in chlorine production, AIChE J. 65 (7) (2019) e16352. https://doi.org/10.1002/aic.16352.)*

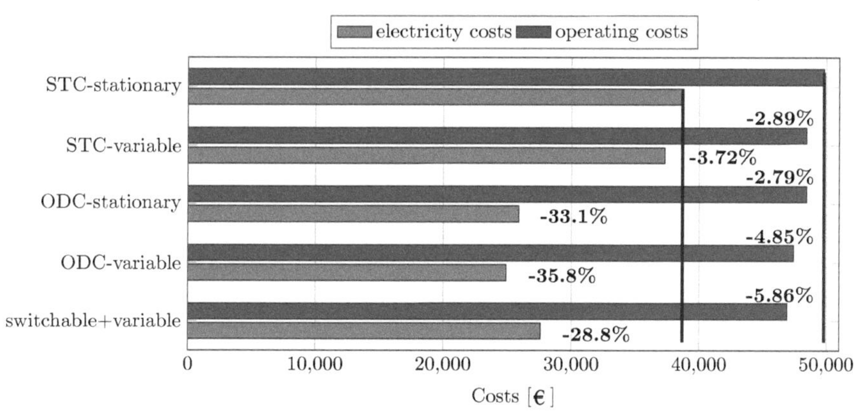

FIG. 9.6 Electricity cost savings and total operating cost savings compared to stationary operation of chlor-alkali electrolysis with STCs for the future electricity price profile. STC-stationary = S-STC; STC-variable = OV-ODC; ODC-stationary = S-ODC; ODC-variable = OV-ODC; switchable + variable = SW-OV-CA. *(Modified from L.C. Brée, K. Perrey, A. Bulan, A. Mitsos, Demand side management and operational mode switching in chlorine production, AIChE J. 65 (7) (2019) e16352. https://doi.org/10.1002/aic.16352.)*

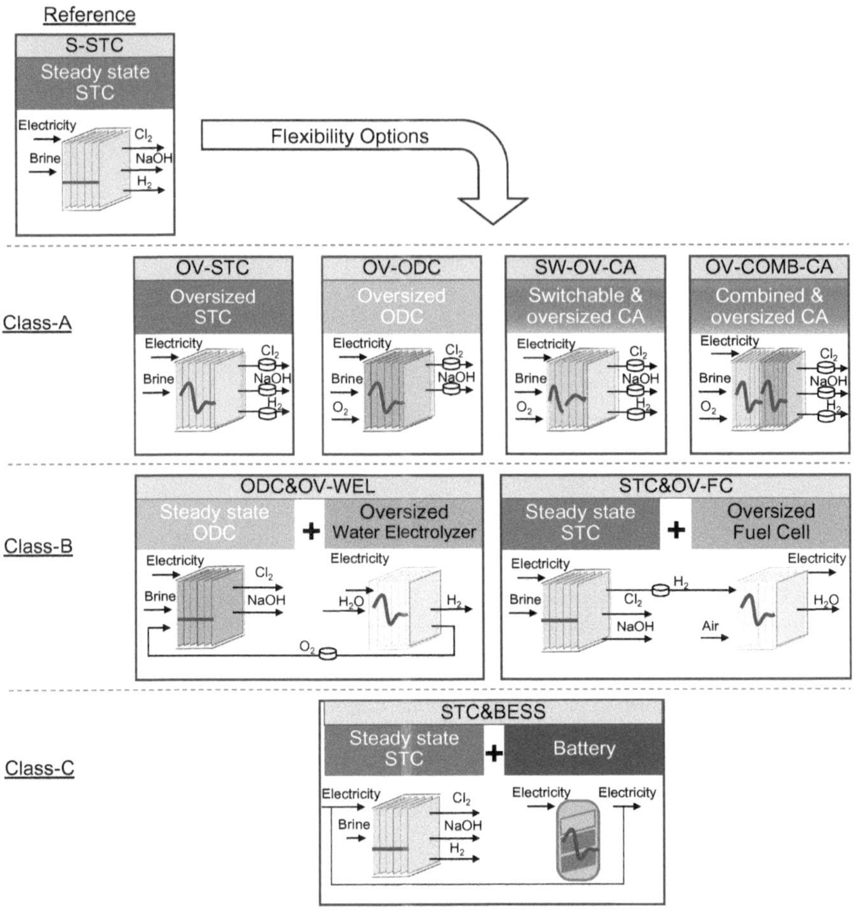

FIG. 9.7 Seven options of CA processes for providing flexibility. *(Reproduced with permission from L.C. Brée, A. Bulan, R. Herding, J. Kuhlmann, A. Mitsos, K. Perrey, K. Roh, Techno-economic comparison of flexibility options in chlorine production, Ind. Eng. Chem. Res. 59 (26) (2020) 12186–12196. https://doi.org/10.1021/acs.iecr.0c01775. Credit: Copyright 2021 American Chemical Society.)*

of CAE cells (OV-STC, OV-ODC, OV-COMB-CA), replacement of electrodes (SW-OV-CA), and integration with flexible auxiliary units (ODC&OV-WEL, STC&OV-FC, STC&BESS). Each option is elaborated in Section 2 of Ref. [12].

We consider four scenarios with different combinations of electricity price profiles and hydrogen prices (see Fig. 9.2). For the given overcapacities of the CAEs or the flexible auxiliary units (2.5, 5, and 10 MW), we optimize the weekly operation profile of the flexible processing unit in each modification. We then evaluate the payout time (POT), which is the length of time required to recover the increased capital investment by annual operating cost savings with respect to S-STC [23]. We assume that the operating cost savings in the given time horizon, i.e., 1 week, are the same over the year. As shown in Fig. 9.8, SW-OV-CA seems to be a good compromise. It ranks above the average in most scenarios in terms of POT due to the largest operating cost savings. Other options are beneficial in limited situations only (e.g., OV-STC and OV-ODC) or suffer from their long POTs due to the high investment costs

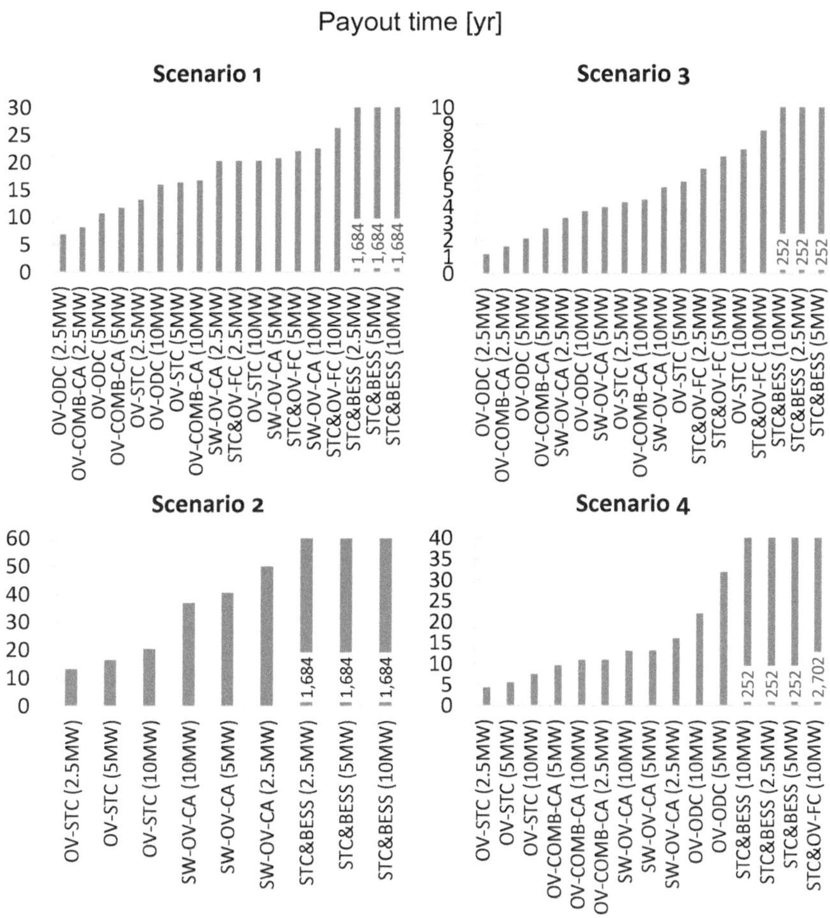

FIG. 9.8 Merit-order of the payout times considering the investment costs and operating cost savings w.r.t the reference case (S-STC) for various electricity price profiles and hydrogen prices. We omit the bars for those processes that cannot save operating costs with respect to the reference case. *(Reproduced with permission from L.C. Brée, A. Bulan, R. Herding, J. Kuhlmann, A. Mitsos, K. Perrey, K. Roh, Techno-economic comparison of flexibility options in chlorine production, Ind. Eng. Chem. Res. 59 (26) (2020) 12186–12196. https://doi.org/10.1021/acs.iecr.0c01775. Credit: Copyright 2021 American Chemical Society.)*

for installing flexible auxiliary units (e.g., ODC&OV-WEL, STC&OV-FC, and STC&BESS).

4.3 Simultaneous optimization of plant oversizing and operation

We now optimize the overcapacity of switchable CAE as well as its operation profile while minimizing the total production cost (min C_{TO}). Since the problem turns into a mixed-integer nonlinear program (MINLP), we apply a decomposition method [26] to find the global optimum in a tractable time. We again consider the electricity price profile for the year 2030 [23]. Plant oversizing affects capital investment for not only purchasing new electrolyzers and retrofitting the process for the switchable operation but also increasing the capacity of up- and downstream units. Thus, we vary β_{au} value from zero to 8.4 € MM to see how much the up- and downstream unit capacities are influential. The results are set in comparison to the reference, which is a CAE equipped with STCs for which overcapacity and operation profile is optimized as well, i.e., OV-STC.

As shown in Table 9.1, the larger the plant capacity, the higher the savings in total production costs. Due to the depreciation costs added, the relative savings in production costs are lower than those in operating costs. Moreover, the optimal plant capacity highly depends on the value of β_{au}. Substantial investment in enlarging up- and downstream units leads to small plant capacity and even negative production cost savings. Although the switching operation can reduce both the power consumption and operating costs significantly, it is not as attractive as the load variation because of further investment for the plant retrofit. Hence, the switchable CAE would be economically favorable only if the up- and downstream units have been designed with enough capacities and/or lower capital cost for the plant retrofit is needed.

5 Conclusion

We introduced the switching operation of chlor-alkali electrolysis for demand-side management. With the bifunctional electrode, this novel operation allows for adjusting not only the operation level but rather the operational mode to the temporal variations in electricity prices. Up to 30% of the plant power capacity can be provided as flexibility potential for an unlimitedly long duration despite the limited chlorine storage capacity. We also presented mathematical formulation for optimal plant sizing and operation for given electricity price profiles. In the case studies, we found that the switching operation can save a significant amount of electricity and operating costs. This leads to more economic benefits than other flexibility options. Optimizing the plant overcapacity can further enhance the cost savings. However, the high investment cost for retrofitting the chlor-alkali plant for employing switchability and enlarging the auxiliary units could prolong the payout time.

TABLE 9.1 Results of simultaneous optimization of the CA plant oversizing and operation.

Capital investment for auxiliary units (MM €)	Switchable? (number of switching)	Cost-saving (%)		Depreciation cost addition (%)	Total production cost-saving (%)	Optimal oversizing (α, %)	Electricity consumption change (%)	Payout time (yr)
		Electricity	Operating					
0	Y (7)	24.12	9.59	11.60	2.15	30.02	−14.60	7.0
	N	10.71	8.13	6.07	4.19	29.99	−3.52	2.5
4.2	Y (7)	20.42	6.89	10.88	−0.15	14.44	−12.93	11.0
	N	7.54	5.73	8.37	1.81	13.93	−2.41	5.8
8.4	Y (3)	19.72	3.95	10.87	−1.02	4.98	−15.21	14.8
	N	4.03	3.06	4.92	0.84	4.92	−1.56	6.5

Nonswitchable indicates a variable operation of CAE with STCs (OV-STC).
Modified from K. Roh, L.C. Brée, K. Perrey, A. Bulan, A. Mitsos, Flexible operation of switchable chlor-alkali electrolysis for demand side management, Appl. Energy 255 (2019) 113880. https://doi.org/10.1016/j.apenergy.2019.113880.

Further R&D is highly desired for making this novel technology applicable. For example, we should experimentally determine the optimal cleaning strategy, cell configuration, and cleaning duration that will ensure safe operation. Also, the environmental impacts of the switching operation on the energy system should be analyzed. Applying a rolling horizon approach may describe the operational behavior of the CA plant more accurately because the electricity spot prices in Germany for the next day are known only. Finally, the long-term performance of switchable chlor-alkali electrolysis over the plant lifetime should be analyzed because hydrogen price and electricity price, and fluctuation are anticipated to change considerably.

Acknowledgments

The authors gratefully acknowledge the financial support of the Kopernikus-project "SynErgie" by the Federal Ministry of Education and Research (BMBF) and the project supervision by the project management organization Projektträger Jülich (PtJ).

References

[1] A. Mitsos, N. Asprion, C.A. Floudas, M. Bortz, M. Baldea, D. Bonvin, A. Caspari, P. Schäfer, Challenges in process optimization for new feedstocks and energy sources, Comput. Chem. Eng. 113 (2018) 209–221, https://doi.org/10.1016/j.compchemeng.2018.03.013.

[2] M. Paulus, F. Borggrefe, The potential of demand-side management in energy-intensive industries for electricity markets in Germany, Appl. Energy 88 (2) (2011) 432–441, https://doi.org/10.1016/j.apenergy.2010.03.017.

[3] M.T. Kelley, R. Baldick, M. Baldea, Demand response operation of electricity-intensive chemical processes for reduced greenhouse gas emissions: application to an air separation unit, ACS Sustain. Chem. Eng. 7 (2) (2019) 1909–1922, https://doi.org/10.1021/acssuschemeng.8b03927.

[4] N. Baumgärtner, R. Delorme, M. Hennen, A. Bardow, Design of low-carbon utility systems: exploiting time-dependent grid emissions for climate-friendly demand-side management, Appl. Energy 247 (2019) 755–765, https://doi.org/10.1016/j.apenergy.2019.04.029.

[5] Q. Zhang, I.E. Grossmann, C.F. Heuberger, A. Sundaramoorthy, J.M. Pinto, Air separation with cryogenic energy storage: optimal scheduling considering electric energy and reserve markets, AIChE J. 61 (5) (2015) 1547–1558. arXiv:0201037v1 https://doi.org/10.1002/aic.14730.

[6] A. Caspari, C. Offermanns, P. Schäfer, A. Mhamdi, A. Mitsos, A flexible air separation process: 2. Optimal operation using economic model predictive control, AIChE J. 65 (11) (2019), https://doi.org/10.1002/aic.16721.

[7] R.C. Pattison, C.R. Touretzky, T. Johansson, I. Harjunkoski, M. Baldea, Optimal process operations in fast-changing electricity markets: framework for scheduling with low-order dynamic models and an air separation application, Ind. Eng. Chem. Res. 55 (16) (2016) 4562–4584.

[8] A. Ghobeity, A. Mitsos, Optimal time-dependent operation of seawater reverse osmosis, Desalination 263 (1–3) (2010) 76–88, https://doi.org/10.1016/j.desal.2010.06.041.

[9] P. Schäfer, H.G. Westerholt, A.M. Schweidtmann, S. Ilieva, A. Mitsos, Model-based bidding strategies on the primary balancing market for energy-intense processes, Comput. Chem. Eng. 120 (2019) 4–14, https://doi.org/10.1016/j.compchemeng.2018.09.026.

[10] K. Seo, T.F. Edgar, M. Baldea, Optimal demand response operation of electric boosting glass furnaces, Appl. Energy 269 (2020) 115077, https://doi.org/10.1016/j.apenergy.2020.115077.

[11] C. Babu, S. Ashok, Peak load management in electrolytic process industries, IEEE Trans. Power Syst. 23 (2) (2008) 399–405, https://doi.org/10.1109/TPWRS.2008.920732.

[12] L.C. Brée, A. Bulan, R. Herding, J. Kuhlmann, A. Mitsos, K. Perrey, K. Roh, Techno-economic comparison of flexibility options in chlorine production, Ind. Eng. Chem. Res. 59 (26) (2020) 12186–12196, https://doi.org/10.1021/acs.iecr.0c01775.

[13] T.F. O'Brien, T.V. Bommaraju, F. Hine, Handbook of Chlor-Alkali Technology, Developments in Hydrobiology, Springer, Dordrecht, 2007.

[14] F. Klaucke, T. Karsten, F. Holtrup, E. Esche, T. Morosuk, G. Tsatsaronis, J.-u. Repke, Demand response potenziale in der chemischen industrie, Chem. Ing. Tech. 89 (9) (2017) 1133–1141, https://doi.org/10.1002/cite.201600073.

[15] Euro Chlor, Chlor-Alkali Industry Review 2017/2018—Manufacturing & Applications, 2019. https://chlorineindustryreview.com/manufacturing-applications/. (Accessed 18 February 2020).

[16] K. Appunn, F. Bieler, J. Wettengel, Germany's Energy Consumption and Power Mix in Charts, 2018. https://www.cleanenergywire.org/factsheets/germanys-energy-consumption-and-power-mix-charts. (Accessed 1 July 2019).

[17] T. Budiarto, E. Esche, J. Repke, E. Leksono, Dynamic model of chloralkali membrane process, Procedia Eng. 170 (2017) 473–481, https://doi.org/10.1016/j.proeng.2017.03.076.

[18] J.I. Otashu, M. Baldea, Demand response-oriented dynamic modeling and operational optimization of membrane-based chlor-alkali plants, Comput. Chem. Eng. 121 (2019) 396–408, https://doi.org/10.1016/j.compchemeng.2018.08.030.

[19] C. Hoffmann, J. Weigert, E. Esche, J.-U. Repke, Towards demand-side management of the chlor-alkali electrolysis: dynamic, pressure-driven modeling and model validation of the 1,2-dichloroethane synthesis, Chem. Eng. Sci. 214 (2020) 115358, https://doi.org/10.1016/j.ces.2019.

[20] F. Klaucke, C. Hoffmann, M. Hofmann, G. Tsatsaronis, Impact of the chlorine value chain on the demand response potential of the chloralkali process, Appl. Energy 276 (2020) 115366, https://doi.org/10.1016/j.apenergy.2020.115366.

[21] X. Wang, H. Teichgraeber, A. Palazoglu, N.H. El-Farra, An economic receding horizon optimization approach for energy management in the chlor-alkali process with hybrid renewable energy generation, J. Process Control 24 (8) (2014) 1318–1327, https://doi.org/10.1016/j.jprocont.2014.04.017.

[22] L.C. Brée, K. Perrey, A. Bulan, A. Mitsos, Demand side management and operational mode switching in chlorine production, AIChE J. 65 (7) (2019) e16352, https://doi.org/10.1002/aic.16352.

[23] K. Roh, L.C. Brée, K. Perrey, A. Bulan, A. Mitsos, Flexible operation of switchable chlor-alkali electrolysis for demand side management, Appl. Energy 255 (2019) 113880, https://doi.org/10.1016/j.apenergy.2019.113880.

[24] F. Holtrup, Potenzial für Demand Side Management der Energieintensiven Industrie in Deutschland—Eine Kostenbetrachtung am Beispiel der Chlor-Alkali-Elektrolysen, Tech. rep., Weltenergierat Deutschland e. V., Berlin, 2015. https://www.weltenergierat.de/wp-content/uploads/2014/03/2016-01-DSM-Papier-v8.pdf. (Accessed 24 January 2021).

[25] F. Ausfelder, A. Seitz, S. von Roon, Flexibilitätsoptionen in der Grundstoffindustrie, DECHEMA, Gesellschaft für Chemische Technik und Biotechnologie eV, 2018.

[26] K. Roh, L.C. Brée, K. Perrey, A. Bulan, A. Mitsos, Optimal oversizing and operation of the switchable chlor-alkali electrolyzer for demand side management, in: A.A. Kiss, E. Zondervan, R. Lakerveld, L. Özkan (Eds.), The 29th European Symposium on Computer

Aided Process Engineering, Elsevier B.V, Eindhoven, 2019, pp. 1771–1776, https://doi.org/10.1016/B978-0-128-18634-3.50296-4.
[27] Euro Chlor, Chlorine Industry Review 2015–2016. Brussels, Belgium, 2016. http://www.eurochlor.org/download-centre/the-chlorine-industry-review.aspx. (Accessed 31 July 2017).
[28] J. Jung, S. Postels, A. Bardow, Cleaner chlorine production using oxygen depolarized cathodes? A life cycle assessment, J. Clean. Prod. 80 (2014) 46–56, https://doi.org/10.1016/j.jclepro.2014.05.086.
[29] I. Moussallem, J. Jörissen, U. Kunz, S. Pinnow, T. Turek, Chlor-alkali electrolysis with oxygen depolarized cathodes: history, present status and future prospects, J. Appl. Electrochem. 38 (9) (2008) 1177–1194, https://doi.org/10.1007/s10800-008-9556-9.
[30] thyssenkrupp, Solutions Provided by NaCl-ODC Electrolysis Technology, 2020. https://www.thyssenkrupp-uhde-chlorine-engineers.com/en/products/chlor-alkali-electrolysis/odc-technology. (Accessed 12 March 2021).
[31] A. Bulan, R. Weber, F. Bienen, Difunctional Electrode and Electrolysis Device for Chlor-Alkali Electrolysis. Patent WO 2017/174563 A1, 2017.
[32] Agora Energiewende, 2019. https://www.agora-energiewende.de/de/themen/-agothem-/Produkt/produkt/76/Agorameter/. (Accessed 15 November 2019).
[33] T. Kern, F. Böing, S. von Roon, Merit Order Netz-Ausbau 2030—Preiszeitreihe Basisszenario, Forschungsgesellschaft für Energiewirtschaft GmbH, München, 2018. https://www.ffe.de/themen-und-methoden/speicher-und-netze/521-merit-order-netz-ausbau-2030-mona-2030. (Accessed 24 January 2021).
[34] K. Arnold, T. Janßen, L. Echternacht, S. Höller, T. Voss, K. Perrey, Flexind: Flexibilisation of Industries Enables Sustainable Energy Systems, 2016, final report, Wuppertal, Germany https://wupperinst.org/fa/redaktion/downloads/projects/FlexInd_report.pdf. (Accessed 24 January 2021).
[35] S. Mitra, L. Sun, I.E. Grossmann, Optimal scheduling of industrial combined heat and power plants under time-sensitive electricity prices, Energy 54 (2013) 194–211, https://doi.org/10.1016/j.energy.2013.02.030.
[36] Q. Zhang, I.E. Grossmann, Enterprise-wide optimization for industrial demand side management: fundamentals, advances, and perspectives, Chem. Eng. Res. Des. 116 (2016) 114–131, https://doi.org/10.1016/j.cherd.2016.10.006.
[37] Personal Communication with Covestro Deutschland AG, 2018.

Chapter 10

Optimal experiment design for dynamic processes

Satyajeet Bhonsale, Philippe Nimmegeers, Simen Akkermans, Dries Telen, Ioanna Stamati, Filip Logist, and Jan F.M. Van Impe
BioTeC+, Department of Chemical Engineering, Technology Campus Gent, KU Leuven, Ghent, Belgium

1 Introduction

Advanced process control begins with a good and useful process model. According to Wieland et al. [1], a useful process model has three properties: (i) it captures the (typically nonlinear) process dynamics and describes the data with reasonable accuracy, (ii) allows testing of experimentally unrealizable (often due to plant safety) scenarios, and (iii) provides insight into the system under consideration. Developing a model that fulfills these criteria is a tedious and iterative process that follows a cycle illustrated in Fig. 10.1 [2]. The first and foremost input required to begin the modeling process is the modeling goal. The necessity to develop a model might arise from the need to gain process insight, optimize, or control. Along with the modeling goal, any a priori knowledge and experimentally available data constitute the three information sources in a model building cycle [3]. Based on these sources, the modeler makes the choice of variables (inputs/outputs), and the model type (linear/nonlinear, static/dynamic, etc.). This is included in the frame definition. Furthermore, based on all the information available, the modeler must determine the model complexity. This involves choosing the mathematical structure of the model. The mathematical structure can be based on physical laws (called white-box models), empirical correlations (black-box models), or a combination of both (leading to grey-box models). White-box models rely solely on the physical understanding of the process (a priori knowledge, i.e., the left side of the diagram) and utilize experimental data only to validate the model. Black-box models are constructed without relying on any process understanding and use experimental data (i.e., the right side of the diagram) to determine the relationship between the input and output variables. A grey-box model incorporates

Simulation and Optimization in Process Engineering. https://doi.org/10.1016/B978-0-323-85043-8.00010-6

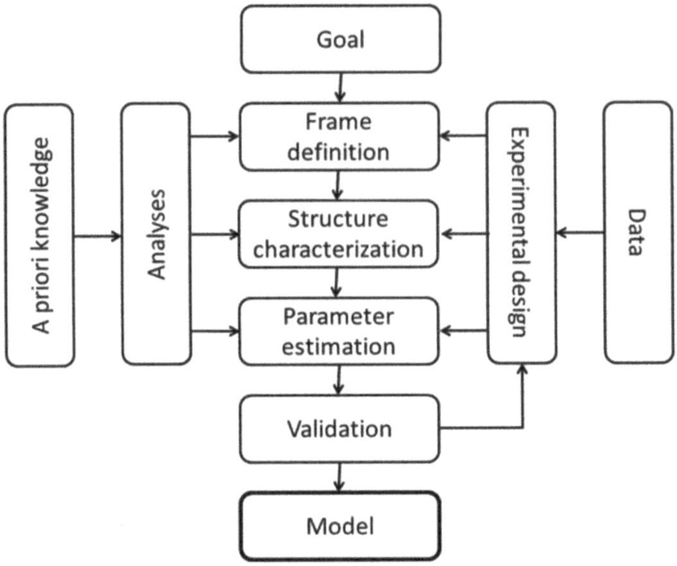

FIG. 10.1 The modeling cycle [2].

as much a priori knowledge as possible while using empirical data-driven (or physically motivated) relations in absence of such knowledge [4].

In most cases, the process industry relies on black-box or grey-box models for simulation, optimization, and control. Black-box models are regularly deployed on a localized scale to tune PID-controllers (e.g., transfer function models) or for Dynamic Matrix Control, etc. For advanced process control (e.g., nonlinear model predictive control, real-time optimization, etc.) more detailed grey-box models are used. Simulation packages for operator training typically use (almost) white-box models. The choice of the model structure depends on the process and the modeling goal [5]. Model structures can be classified as dynamic or static, and linear or nonlinear. Static models do not describe the temporal evolution of the process. As such, these models are not common in process control. Dynamic models are typically expressed as ordinary differential equations (ODEs). When spatio-temporal evolution of the process needs to be modeled, partial differential equations (PDEs) are used. Solving PDEs normally involves numerical approximations that convert the PDEs into a (typically very large) set of ODEs. Depending on the interdependency of the states and the controls, a model structure can be either linear or nonlinear. In principle, the model structure needs to be inferred from both the available data and prior knowledge (see Fig. 10.1). However, in practice, this is a very hard problem. Therefore, the model structure characterization problem is quite often relaxed into a model structure discrimination problem: upon defining an appropriate norm, the best possible model structure is selected from a predefined set of competing models. Note that there is no guarantee that the most suitable process model is included in this set.

Grey-box and black-box models require some parameters to be estimated from experimental data. A useful model must ensure that its parameters are structurally and practically identifiable. Structural identifiability, as the name suggests, is related to the model structure. A parameter is structurally nonidentifiable if even a large change in its value does not significantly influence the model behavior [6]. Typically, structural nonidentifiability arises due to excessive parameterization of the model or inclusion of dynamics which cannot be discerned experimentally [7]. Thus, it is imperative to ensure that the model structure chosen not only characterizes the process but also leads to an identifiable parameter set. Practical identifiability is concerned with the accuracy of the parameters. A parameter is practically nonidentifiable if small changes in its value do not impact the model behavior. Structural identifiability is a prerequisite for practical identifiability. The accuracy and hence the reliability of these parameters (and subsequently the model) depends on the quantity and quality of experimental data available. Practical nonidentifiability can be resolved by appropriate choice of measurement locations, sampling times, better instrumentation, and thorough collection of informative experimental data [8]. However, experimentation is often cost- and labor-intensive. It is not only the quantity of data collected but also the quality, i.e., the information content of the data, which renders a parameter practically identifiable.

To avoid unnecessary costs and waste of resources and time, the domain of Design of Experiments (DoE) aims to provide a crucial statistical link between the experiment and model [9]. Conventional DoE ignores the available model structure and tries to establish a black-box link between the inputs (called factors) and outputs (called responses). Classic examples of DoE are the factorial designs that require the experimenter to set the inputs to different levels and observe the process response. By varying multiple inputs at different levels, the interaction between the inputs can also be investigated. To reduce the number of experiments, other experiment designs such as the central composite design, Box-Behnken, Latin Hypercube, etc. are used. A comparison of various DoE approaches for models describing the effect of environmental conditions on microbial growth is reported in Mertens et al. [10]. They conclude that central composite design leads to high uncertainty in their parameter estimates, while the Box-Behnken design gives the most accurate estimates. The DoE approach is optimized for a second-order response surface, and completely ignores any a priori knowledge of model structure. Thus, such designs are not always suitable for models which describe process dynamics.

To ensure that the model describes the process accurately, it is quintessential that the model structure is considered while designing the experiments. In this chapter, we follow Franceschini and Macchietto [9] and characterize model-based experiment design (MBDoE) as the approach in which explicit model equations and current parameter estimates are utilized to design the next experiment. When (dynamic) models are utilized in an optimization framework to determine a (dynamic) input profile for the next experiment, we term it optimal experiment design (OED). OED can be tailored toward discriminating between multiple model structures (OED/SD) or toward obtaining the most accurate parameter estimates (OED/PE). The first section of the chapter describes the formulation of OED problems for model discrimination. Typically, these designs rely on maximizing a measure of distance between two competing models. The next section discusses OED formulations for parameter estimation. As these designs rely on optimizing a scalar criterion based on the variance-covariance matrix, various approaches to compute the variance-covariance are discussed. Finally, some advanced topics like robust OED (i.e., incorporating parametric uncertainty in the experiment design) and multi-criteria OED are discussed.

In this chapter, a generalized process model (M) described using a system of ODEs is given by

$$\frac{dx(t)}{dt} = f(x(t), u(t), w(t), \theta, t)$$

$$y(t) = h(x(t), v(t))$$

where $x \in \mathbb{R}^{n_x}$ is the time-varying state vector (e.g., temperature, concentration in well-mixed reactors, or the spatial discretization of a partial differential representation of a tubular reactor), $u \in \mathbb{R}^{n_u}$ is the vector of all controls or inputs that can be manipulated in an experiment, $\theta \in \mathbb{R}^{n_p}$ is the set of parameters that need to be estimated from variables that can be measured. These variables are contained in the vector $y \in \mathbb{R}^{n_y}$. The complete definition of the process model also requires all the initial conditions $x_0 = x(t=0)$ to be known. The process is associated with process noise $w(t)$, which describes the unmodelled variabilities of the process due to, e.g., the process's inherent stochasticity. Furthermore, no measurement sensor is perfect. The measurement itself is associated with measurement noise $v(t)$. The measurement noise can be characterized by its variance-covariance matrix C_y. The samples from the process are collected in a vector y_m. In any given experiment, n_s samples are collected at discrete sampling points.

2 Optimal experiment design for model structure discrimination

In many cases, multiple mathematical structures can be used to model a process or a part of the process. For example, in bioprocesses, the specific microbial growth rate can be modeled via different kinetic expressions such as the Monod kinetics, Haldane kinetics, etc. When multiple such structures are available, a model adequacy test needs to be performed to determine which model structure is the best suited. Model adequacy can be determined via statistical tests like χ^2-test [11] or by using criteria like the (corrected) Akaike information criterion [12]. In some cases, model adequacy tests cannot reliably determine which structure is most suitable with the available experimental data. Then, experiments tailored for model discrimination need to be designed. This is called Optimal Experiment Design for Model Structure Discrimination (OED/SD). Let us assume that a process can be modeled by two candidate models $y = M_1(u, \theta_1, t)$ and $y = M_2(u, \theta_2, t)$. The n_t process measurements are collected in the vector y_m. In general, the OED/SD is formulated as an optimization problem that maximizes a functional of the distance($D_{1,\ 2}$) between the two models.

$$u^* = \arg\max_u \sum_{k=0}^{n_t} D_{1,2}(u, t_k)$$

In many cases, the choice of the sampling instance can also be included as an optimization variable.

The simplest objective function for OED/SD is the Eulerian distance between the two model outputs for only one experiment [13]. This is given by

$$D_{1,2}(u, t_k) = d_{1,2}(u, \theta_1, \theta_2, t_k)^\top \cdot d_{1,2}(u, \theta_1, \theta_2, t_k)$$

where,

$$d_{1,2}(u, \theta_1, \theta_2, t_k) = (M_1(u, \theta_1, \mathrm{t_k}) - M_2(u, \theta_2, \mathrm{t_k}))$$

The criterion proposed by Box and Hill [14], considers uncertainty—both on measurements and the model output. This is useful when large output uncertainties can mask the differences between the two models. The criterion is based on designing new experiments that maximize the entropy, an information-theory-based measure of uncertainty.

The T-criterion is based on the a priori assumption that one of the models (M_1) is true with known parameter values (θ_1^*) [15,16]. For dynamic experiments, the criterion is given by

$$D_{1,2} = \min_{\theta_2} \sum_{i=1}^{n_e} w_i \sum_{k=0}^{n_t} \left\| M_1\left(u_i, \theta_1^*, t_k\right) - M_2(u_i, \theta_2, t_k) \right\|^2$$

This criterion then has to be maximized, i.e., the OED/SD problem is a max-min problem that can be computationally challenging.

The criterion introduced by Munack [17] is based on maximizing the Mahalanobis distance between subsequent parameter estimates. In the OED/SD procedure, new simulation experiments are performed using both models. Then sequentially assuming one of the models to be true, the parameters of the other model are estimated. The Mahalanobis distance, given by $\delta^2 = (\theta - \theta^*)^\top \cdot S^{-1} \cdot (\theta - \theta^*)$ is computed for all cases, and the smallest δ^2 is then maximized to design new experiments. Here $S = \frac{\partial M_i}{\partial \theta_i}$ is the model output sensitivity with respect to the parameters.

Finally, the Schwaab criterion [18,19] maximizes a weighted distance between the two model outputs. It is given by

$$D_{1,2}(u,t_k)=d_{1,2}(u,\theta_1,\theta_2,t_k)^{\top}\cdot V_{1,2}^{-1}(u,t_k)\cdot d_{1,2}(u,\theta_1,\theta_2,t_k)$$

The posterior covariance matrix of differences between the model predictions $V_{1,2}$ is computed from the individual covariance matrix (V_i) of the model prediction and the covariance matrix (V) of experimental deviations. These individual covariance matrices are computed from the sensitivity (S).

$$V_{1,2}=2V+V_1+V_2$$

$$V_i=S_i\cdot V_{\theta_i}\cdot S_i^{\top}$$

$$V_{\theta_i}=\left[S_i^{\top}\cdot V^{-1}\cdot S_i+V_{\theta,i}(u_{n_e})\right]^{-1}$$

The posterior covariance matrix of model parameters $V_{\theta,i}$ consists of two parts, the covariance of the newly designed experiment and the current covariance matrix till experiment n_e.

2.1 OED/SD in practice

In predictive microbiology, microbial growth is typically described by the Baranyi-Roberts model [20].

$$\frac{dn(t)}{dt}=\frac{Q(t)}{Q(t)+1}\cdot\mu_{\max}(T(t))\cdot[1-\exp(n(t)-n_{\max})]$$

$$\frac{dQ(t)}{dt}=\mu_{\max}(T(t))\cdot Q(t)$$

with $n(t)$ the natural logarithm of the cell density at time t. n_0, $n_{\max}$ are the initial and maximum values of cell density, and $\mu_{\max}$ is the temperature-dependent maximum specific growth rate of the microorganism. $Q(t)$ is a measure of the physiological state of the cell. The temperature dependence of the maximum specific growth rate is described by a variety of secondary models. The cardinal temperature model with inflection (CTMI) [21] is a widely accepted temperature model. For some microorganisms, an adjusted version of the CTMI (aCTMI) model might be more accurate [22,23]. The difference between the models is depicted in Fig. 10.2.

Stamati et al. [24], design experiments to discriminate between these two models for *Escherichia coli* K12 MG1655. The design is based on the method proposed by Schwaab et al. [19]. However, even before designing optimal experiments for discrimination, few preliminary experiments are required to obtain a first estimate of model parameters. The preliminary experiments need to be informative for both CTMI and aCTMI models. Thus, the preliminary

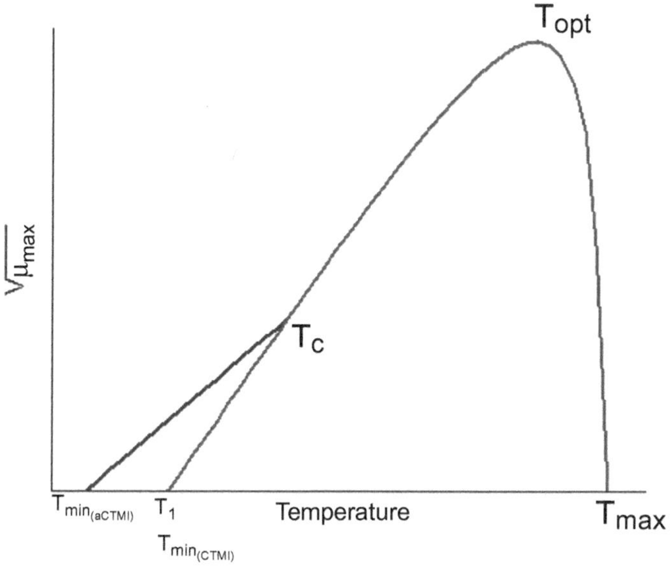

FIG. 10.2 Square root of maximum specific growth rate as a function of temperature modeled by CTMI *(red, gray in print version)* and aCTMI *(blue, dark gray in print version). (From I. Stamati, et al., Optimal experimental design for discriminating between microbial growth models as function of suboptimal temperature: from in silico to in vivo, Food Res. Int. 89 (2016) 689–700.)*

experiment is designed using a multiobjective framework for experiment design (explained in a later section).

With two different models estimated from the preliminary experiment, the approach proposed by Schwaab et al. [19] is followed to design new experiments for discrimination. With only one experiment, sufficient discrimination could not be achieved. Thus, the re-estimated parameters were used again to design a new optimal experiment. In total four discrimination experiments are designed iteratively. The in vivo application of some of these experiments is repeated to overcome practical issues (e.g., contamination after some time, no sampling at night, etc.). The experimental temperature profiles and the resulting cell densities are depicted in Fig. 10.3. At every stage, the model adequacy is tested using values of the Akaike criterion (AIC), the weighted least square errors (WLSE), and χ^2-test errors. These values after model fitting with all the eight experiments are reported in Table 10.1.

It can be concluded that for the microorganism used, the aCTMI model is more accurate than the CTMI model. With eight experiments, sufficient discrimination is achieved. Now with the model structure chosen, optimal experiments can be designed to maximize the accuracy of the parameter estimate. This is called optimal experiment design for parameter estimation.

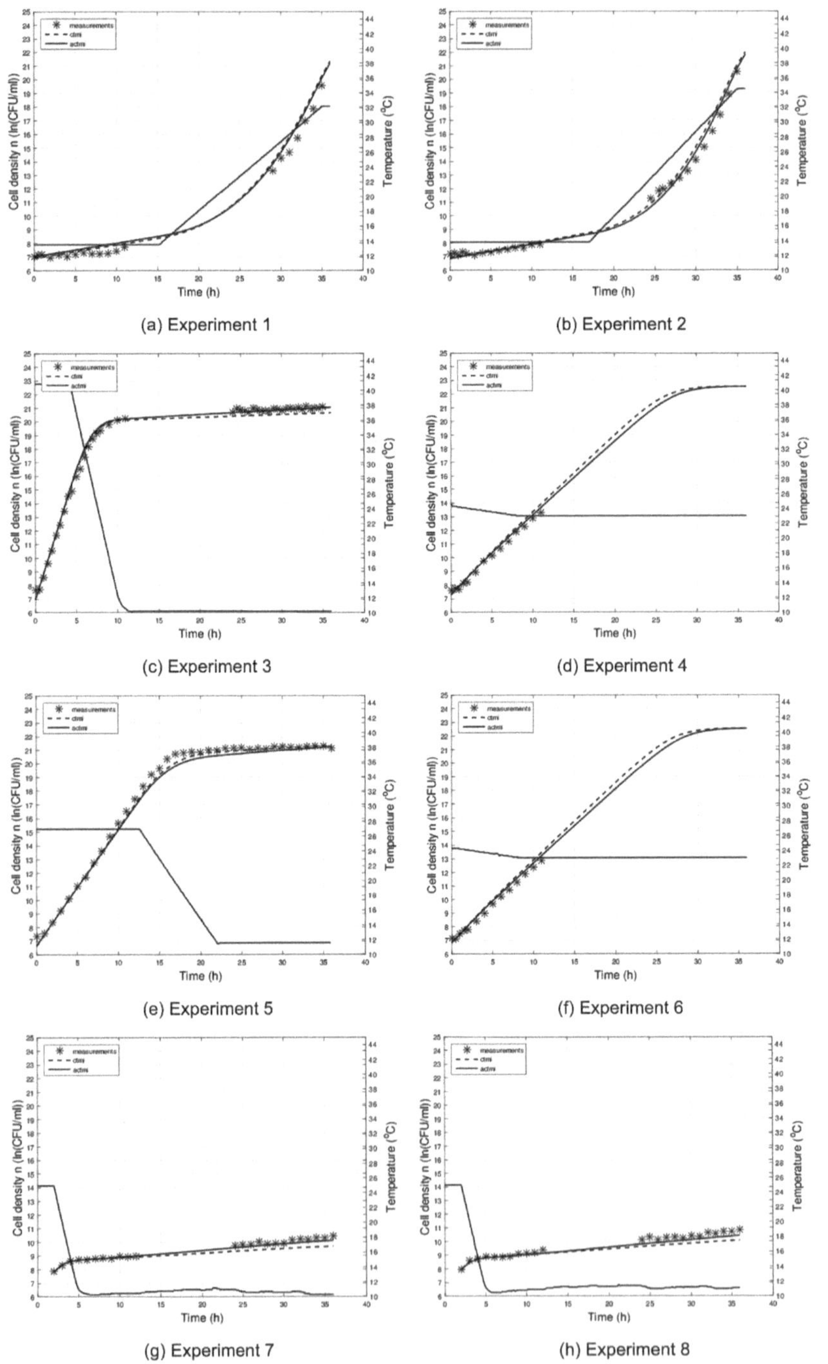

FIG. 10.3 Preliminary experiments (A, B) and the discrimination experiments (C–H) and measured data points. *(From I. Stamati, et al., Optimal experimental design for discriminating between microbial growth models as function of suboptimal temperature: from in silico to in vivo, Food Res. Int. 89 (2016) 689–700.)*

TABLE 10.1 AIC and WLSE values after model fit with eight experiments [24].

Model	AIC	WLSE ($\sigma = 0.1808$)	WLSE ($\sigma = 0.5874$)	χ^2
CTMI	−102	1105	105	125
aCTMI	−134	755	72	123

3 Optimal experiment design for parameter estimation

Once the appropriate model structure is chosen, the parameters in the model should be estimated as accurately as possible. The parameters are estimated using an optimization routine which minimizes an objective $J(\theta)$. A common framework for parameter estimation is based on minimizing the sum of squared errors [6]. Here, a quadratic objective function is used for parameter estimation. The quadratic cost describes the variance-covariance weighted distance between the experimental data (y_m) and the simulated results. This is expressed as

$$J(\theta) = (y_m - y(\theta))^\top \cdot C_y^{-1} \cdot (y_m - y(\theta))$$

The estimate of the parameter is then given by

$$\hat{\theta} = \arg\min_\theta J(\theta)$$

Given the process and measurement noise, the estimated parameters are associated with a specific uncertainty. This parametric uncertainty is quantified by the parameter variance-covariance matrix C_θ. As mentioned earlier, a useful model needs to be practically identifiable. Practically identifiable parameters are associated with narrow individual confidence intervals, i.e., with small variance on the parameter. Thus, increasing the accuracy of the parameter estimate is synonymous with reducing the value of the parameter variance-covariance matrix. The aim of optimal experiment design for parameter estimation (OED/PE) is to determine a (dynamic) input profile ($u(t)$) which optimizes a scalar criterion based on the parameter variance-covariance matrix. Often, sampling rates and experimental initial conditions are also included as parameters to be optimized in the OED. As discussed, formulating an OED/PE problem requires the computation of: (i) the variance-covariance matrix, and (ii) scalar criteria based on the variance-covariance matrix. In the coming sections, various approaches for these computations are discussed. The formulation of OED/PE as an optimal control problem using the Fisher Information Matrix is presented along with a few commonly used scalar criteria.

3.1 Computing parameter variance-covariance matrix

In this section, various approaches to determine the parameter variance-covariance matrix will be presented. The classical Fisher information approach is discussed first, followed by other derivative-free approaches.

3.1.1 Fisher information matrix approach

Through measurement variance-covariance matrix and parameter sensitivity matrix, the Fisher Information Matrix (FIM, $I(\theta)$) summarizes the information about measurement accuracy and the influence of a parameter on the output. It is defined as

$$I(\theta, t) = \sum_{t=0}^{n_t} S_t^{\top} \cdot C_y^{-1} \cdot S_t$$

where S_t is the output sensitivity with respect to the parameter at the sampling point t. This sensitivity matrix is defined as

$$S_t = \begin{bmatrix} \frac{\partial y_1}{\partial \theta_1} & \cdots & \frac{\partial y_1}{\partial \theta_{n\theta}} \\ \vdots & \ddots & \vdots \\ \frac{\partial y_{n_y}}{\partial \theta_1} & \cdots & \frac{\partial y_{n_y}}{\partial \theta_{n\theta}} \end{bmatrix}$$

As the output vector y depends on the dynamic state, the FIM can be computed alongside the model simulation by adding the following matrix differential equations

$$\frac{d}{dt}\frac{\partial x(t)}{\partial \theta} = \frac{\partial f(\cdot)}{\partial x} \cdot \frac{\partial x(t)}{\partial \theta} + \frac{\partial f(\cdot)}{\partial \theta}$$

$$\frac{d}{dt} I(\theta, t) = \left(\frac{\partial x(t)}{\partial \theta}\right)^{\top} \cdot \left(\frac{\partial h(x)}{\partial x}\right)^{\top} \cdot C_y^{-1} \left(\frac{\partial x(t)}{\partial \theta}\right) \cdot \left(\frac{\partial h(x)}{\partial x}\right)$$

with initial conditions

$$\frac{\partial x}{\partial \theta}(0) = \frac{\partial x_0}{\partial \theta}$$

$$I(\theta, 0) = 0$$

If the measurement noise is independent and follows a normal distribution, the least square estimator is equivalent to the maximum likelihood estimator (MLE). The MLE is a consistent estimator with its covariance bounded below through the Cramer-Rao bound

$$C_\theta \geq I^{-1}(\theta t_f)$$

Another useful property of the MLE is that asymptotically (i.e., $n_t \rightarrow \infty$) the Cramér-Rao bound becomes equality, and the inverse of the FIM can be used as an approximation of the parameter variance-covariance matrix.

3.1.2 Direct mapping through parameter estimation

This approach introduced by Schenkendorf et al. [25] is based on the sigma point approach [26]. The sigma points method relies on sampling to estimate the expectation and the variance of a nonlinear transformation ($Y = F(\gamma)$) of a random variable (γ). The random variable is represented by sampling points called the sigma points. The sigma points are computed based on the variance (V_γ) of the random variable

$$\gamma_\sigma = \gamma \pm \sigma, \text{ where, } \sigma \leftarrow \sqrt{(n+\kappa)V_\gamma}$$

For n uncertain parameters, the set of sigma points contains $2n+1$ elements. This set is then transformed by the nonlinear map to obtain a set of responses,

$$Y_0 = F(\gamma_0), \text{and, } Y_\sigma = F(\gamma_\sigma)$$

The expectation and the variance-covariance matrix are computed using this response set.

$$\overline{Y} = \frac{1}{n+\kappa}\left[\kappa Y_0 + \frac{1}{2}\sum^{2n} Y_\sigma\right]$$

$$V_Y = \frac{1}{n+\kappa}\left[\kappa\left(Y_0 - \overline{Y}\right)\left(Y_0 - \overline{Y}\right)^\top + \frac{1}{2}\sum^{2n}\left(Y_\sigma - \overline{Y}\right)\left(Y_\sigma - \overline{Y}\right)^\top\right]$$

In the application to OED, the noise on the measurement vector is considered to be the random variable [25], i.e.,

$$\gamma = \left[\nu_{1(0)}, \cdots, \nu_{n_y(0)}, \cdot, \nu_{1(n_t)}, \cdots, \nu_{n_y(n_t)}\right]$$

The entire parameter estimation procedure is the nonlinear map which makes the estimated parameters to be considered as the responses, i.e.,

$$Y = \hat{\theta}$$

As the measurement noise is assumed to follow zero-mean Gaussian distribution with known variance (C_y), the sigma points are realized for the ($n_y \times n_t$) dimensional noise vector γ leading to ($2(n_y \times n_t)+1$) measurement realizations

$$\gamma_0 = h(x, \theta)$$

$$\gamma_\sigma = h(x, \theta) \pm \sigma$$

For each of these measurement realizations, parameter estimation is carried out to obtain $(2(n_y \times n_t)+1)$ parameter estimates which are used to determine the variance-covariance matrix.

3.1.3 Other computational approaches to approximate variance-covariance matrix

Heine et al. [27] present an approach based on a maximum a priori update of the parameter vector via a Bayesian formulation. The parameter variance-covariance for a dynamic experiment is computed as

$$C_\theta(t) = C_{\theta 0} - C_{\theta,x}(t) \cdot C_x^{-1} \cdot C_{\theta,x}(t)$$

$C_{\theta 0}$ is the parameter variance-covariance matrix known a priori. The two other matrices, the cross-covariance between parameters and states ($C_{\theta,\ x}$) and variance-covariance of the states (C_x), need to be determined. These matrices can be determined via Monte Carlo simulations using random parameter samples, Kalman equations using a linear approximation, or via the unscented transformation (i.e., the sigma point approach) discussed earlier. In Heine et al. [26], a Stirling polynomial approximation of the model response is used with $2n_\theta+1$ sigma points.

Telen et al. [28] describe a formulation where the variance-covariance matrix is computed via the solution of a Riccati equation. This formulation is based on a linear approximation of the original nonlinear least squares-based parameter estimation problem. The linearized problem leads to a linear-quadratic optimal control problem which can be solved via Pontryagin's Maximum Principle. The main advantage of this formulation is that the process noise can be explicitly accounted for. Furthermore, the Riccati equation formulation can also be used to compute the cross-covariance matrices required in the approach proposed by Heine et al. [27].

3.2 OED/PE as an optimal control problem

Once the parameter variance-covariance matrix is computed, the OED problem can be cast as an optimization problem with a scalar objective (Φ) based on it. The optimization problem can be described as.

$$\min \Phi\left(C_\theta\left(t_f\right)\right)$$

subject to,

$$\frac{dx(t)}{dt} = f(x(t), u(t), w(t), \theta, t)$$

$$y(t) = h(x(t), v(t))$$

$$0 \geq c_p(x(t), u(t), w(t), \theta, t)$$

where $c_p(\cdot)$ represents any constraints on the process dynamics. Based on the approach used to compute the variance-covariance matrix, additional constraints need to be added. In the maximum a priori update approach, the dynamics induced at the additional $2n_\theta + 1$ sigma points are added along with the equations for determination of the variance-covariance, and cross-covariance matrix. The direct mapping approach leads to a bilevel optimization problem in which $2n_y n_t + 1$ parameter estimation problems need to be solved at every iteration making it computationally very challenging.

If the FIM approach is used the above optimal control problem is further constrained by the sensitivity matrix dynamics. The resulting optimization is given by

$$\min \Phi\left(I\left(t_f\right)\right)$$

subject to,

$$\frac{dx(t)}{dt} = f(x(t), u(t), w(t), \theta, t)$$

$$y(t) = h(x(t), v(t))$$

$$0 \geq c_p(x(t), u(t), w(t), \theta, t)$$

$$\frac{d}{dt}\frac{\partial x(t)}{\partial \theta} = \frac{\partial f(t)}{\partial x} \cdot \frac{\partial x(t)}{\partial \theta} + \frac{\partial f(t)}{\partial \theta}$$

$$\frac{d}{dt} I(\theta, t) = \left(\frac{\partial x(t)}{\partial \theta}\right)^{\top} \cdot \left(\frac{\partial h(x)}{\partial x}\right)^{\top} \cdot C_y^{-1} \cdot \left(\frac{\partial x(t)}{\partial \theta}\right) \cdot \left(\frac{\partial h(x)}{\partial x}\right)$$

with the appropriate initial conditions.

3.2.1 *Optimization criteria for OED/PE*

A variety of scalar criteria based on either the FIM or the variance-covariance are reported in the literature. Some of the most common criteria are described below, and their geometrical interpretation is illustrated in Fig. 10.4.

1. A-criterion: The A-criterion minimizes the volume of the hyper box which bounds the variance ellipsoid. The A-criterion is equivalent to minimizing the arithmetic mean of all the parameters' errors. Mathematically, it is given by

$$\min\left(\text{trace}\left(I\left(t_f\right)^{-1}\right)\right) \text{ or } \min\left(\text{trace}\left(C_\theta\left(t_f\right)\right)\right)$$

2. D-criterion: The D-criterion aims to minimize the volume of the variance ellipsoid. The volume of the ellipsoid is proportional to the determinant

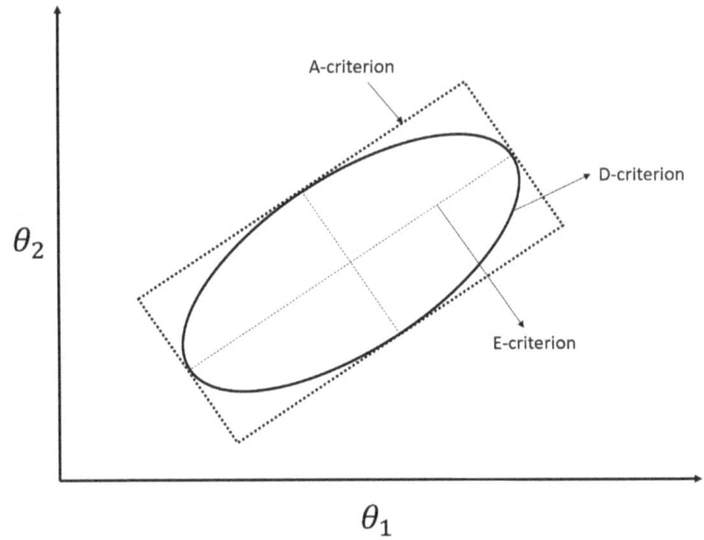

FIG. 10.4 Illustration of OED criteria with the joint confidence region of two-parameter estimates.

of the variance-covariance matrix. The D-criterion corresponds to minimizing the geometric mean of all the parameters' errors. Mathematically, it is given by

$$\min\left(-\det\left(I\left(t_f\right)\right)\right) \text{ or } \min\left(\det\left(C_\theta\left(t_f\right)\right)\right)$$

3. E-criterion: The E-criterion aims to minimize the largest axis of the variance ellipsoid. This corresponds to minimizing the largest parameter error or correlation As the axes of the variance ellipsoid are given by the eigenvalues of the variance-covariance matrix, the E-criterion is defined mathematically as

$$\min\left(\lambda_{\min}\left(I\left(t_f\right)\right)\right) \text{ or } \min\left(\lambda_{\max}\left(C_\theta\left(t_f\right)\right)\right)$$

4. Modified E-criterion: This criterion aims to achieve a (hyper)spherical joint confidence region, implying the minimization of correlation between parameters. An advantage of this criterion is that the theoretical optimum is known to be 1, i.e., a perfectly (hyper)spherical joint confidence region. Mathematically, this is equivalent to minimizing the condition number (ratio of the largest and smallest eigenvalues)

$$\min\left(\frac{\lambda_{\max}\left(I\left(t_f\right)\right)}{\lambda_{\min}\left(I\left(t_f\right)\right)}\right) \text{ or } \min\left(\frac{\lambda_{\min}\left(C_\theta\left(t_f\right)\right)}{\lambda_{\max}\left(C_\theta\left(t_f\right)\right)}\right)$$

3.3 OED/PE in practice

Van Derlinden et al. [23] demonstrate the applicability of OED/PE to maximize the accuracy of model parameters of a specific microbial growth rate model. For *E. coli* K12 MG 1655. As mentioned in a previous example, the temperature dependency of the maximum specific growth rate can be modeled using a CTMI model. For temperature T between the minimum ($T_{\min}$) and maximum ($T_{\max}$), this is described by

$$\mu_{\max} = \mu_{opt} \frac{(T - T_{\min})^2 (T - T_{\max})}{(T_{opt} - T_{\min})((T - T_{opt})(T_{opt} - T_{\min}) - (T_{opt} - T_{\max})(T_{opt} + T_{\min} - 2T))}$$

The CTMI model has four parameters that need to be estimated, $\theta = [\mu_{opt}, T_{opt}, T_{\min}, T_{\max}]$.

To design an experiment, Van Derlinden et al. [23] consider parameter pairs. Thus, for four parameters six parameter pairs are defined and the temperature profile is optimized to estimate each of the parameter pairs sequentially. For any given pair, the other parameters are assumed to be perfectly known. The D-criterion is used to design the experiments. To ensure that the designed experiment is practically feasible, several constraints are imposed on the optimization problem. First, the temperature is bounded between a low and a high value. These values are determined through physical intuition. For example, the lowest temperature is constrained to be 15°C to ensure that the distinction between the microbial lag phase and exponential growth phase is clear. As abrupt temperature change is known to induce a transient disturbance in the microbial evolution, a second constraint binds the consecutive temperature jumps to be within 5°C.

The optimal temperature profiles for the experiments are depicted by the dashed line in Fig. 10.5. These profiles are applied to a computer-controlled benchtop bioreactor and the resulting cell densities are determined via plate counting the aseptically acquired samples. In the figure, the dots represent the measured cell density. Note that in this case, the sampling times are not optimized. They are chosen after the optimal profile is determined such that all the informative zones are covered.

The parameters estimated from these experiments show good statistical quality and are characterized by less than 5% uncertainty error. Furthermore, the developed model is also shown to be transferable to another experimental dataset. The model predictions, along with their 95% confidence and prediction intervals are depicted in Fig. 10.6.

4 Advanced developments in optimal experiment design

4.1 Robust optimal experiment design for parameter estimation

As the FIM and the variance-covariance matrix depend on the parameters, a priori parameter knowledge (i.e., the current available parameter estimate) is

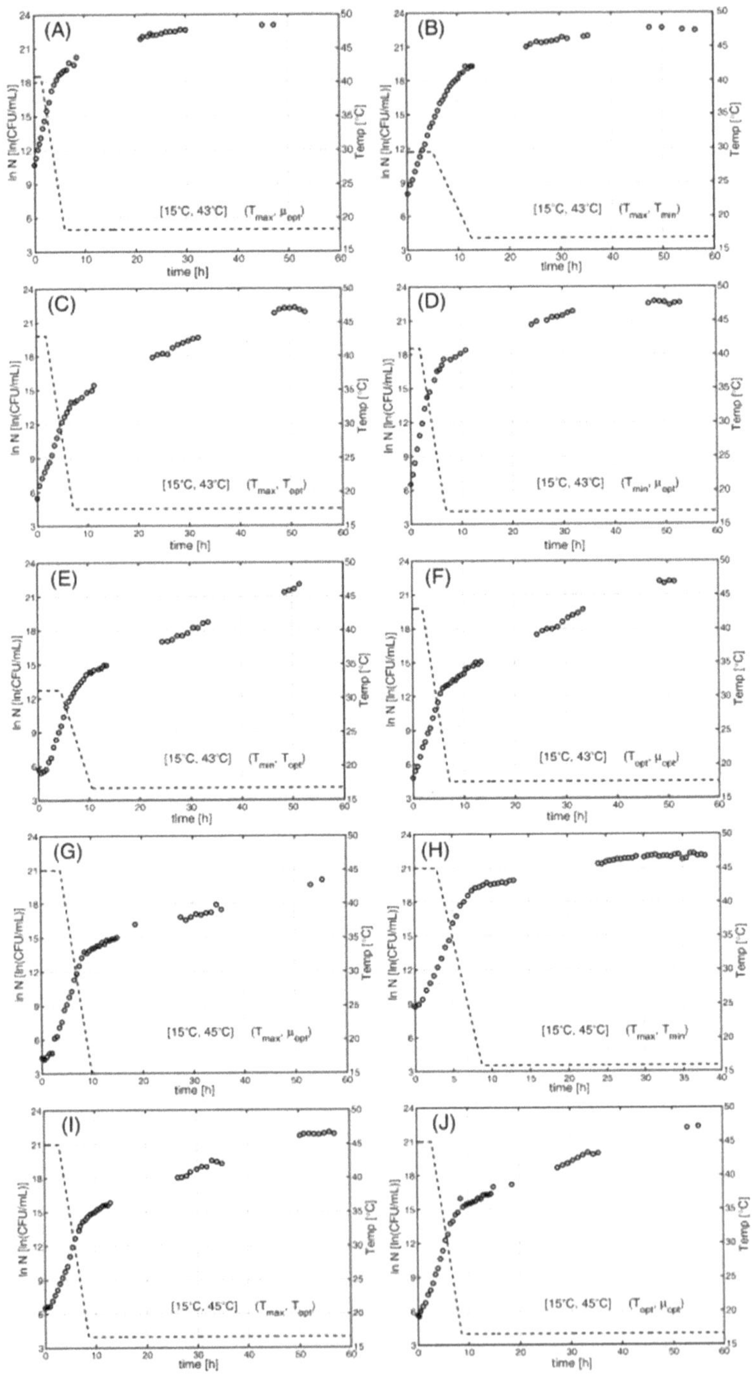

FIG. 10.5 Optimal temperature profiles designed for each parameter pair and resulting cell density data. *(From E. Van Derlinden, K. Bernaerts, J.F. Van Impe, Accurate estimation of cardinal growth temperatures of* Escherichia coli *from optimal dynamic experiments, Int. J. Food Microbiol. 128 (2008) 89–100.)*

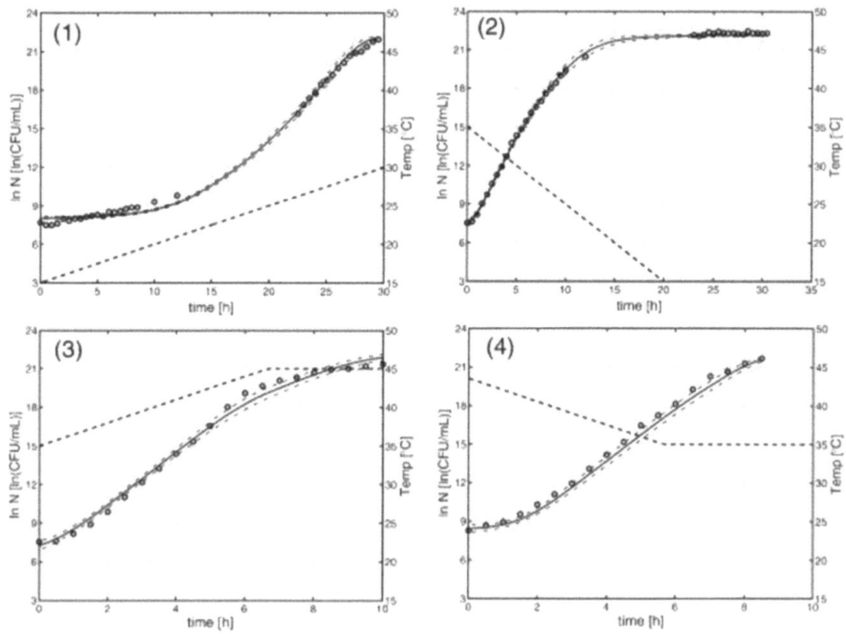

FIG. 10.6 Validation of CTMI model parameters using independent datasets. *(From E. Van Derlinden, K. Bernaerts, J.F. Van Impe, Accurate estimation of cardinal growth temperatures of* Escherichia coli *from optimal dynamic experiments, Int. J. Food Microbiol. 128 (2008) 89–100.)*

unavoidable. This parameter estimate (or an initial guess) also comes with underlying uncertainties, and as such any experiment designed based on these estimates may result in significantly lower information content. Thus, experiments must be designed to ensure that enough information is gathered regardless of the true parameter value. This is termed robustness with respect to information content. A practical approach to this problem is to alternate between parameter estimation and OED (with the current best estimate). This process is tedious, time-consuming, and expensive. Furthermore, experiments designed using uncertain parameters can lead to violations of any constraints imposed on the system. Thus, the parametric uncertainty should be considered to design experiments that are robust with respect to the constraints.

One approach to design robust experiments is to cast the OED problem as a maximin problem. If the current parameter estimate (θ^*) is normally distributed with variance C_{θ^*}, a confidence ellipsoid with confidence γ can be constructed as

$$\|\theta - \theta^*\|_{C_{\theta^*}}^2 \leq \beta$$

The robust OED problem can be reformulated by replacing the objective and constraints in the nominal OED problem as

$$\min_{u(\cdot)} \max_{\|\theta-\theta^*\|^2_{C_{\theta^*}} \leq \beta} \Phi(I(t_f))$$

subject to,

$$0 \geq \max_{\|\theta-\theta^*\|^2_{C_{\theta^*}} \leq \beta} c_p(x(t), u(t), w(t), \theta, t)$$

Essentially, this formulation is the worst-case approach. The inner optimization can be solved explicitly using a linear approximation [29]. A first-order Taylor series expansion transforms the inner problem into a convex quadratically constrained linear problem whose solution is given by Nimmegeers et al. [30].

$$\Phi(I(t_f)) + \sqrt{\beta}\left\|\frac{d\Phi(I(t_f))}{d\theta}\right\|_{C_\theta}$$

Similarly, the constraints can be formulated as

$$0 \geq c_p(\cdot) + \sqrt{\beta}\left\|\frac{dc_p(\cdot)}{d\theta}\right\|_{C_\theta}$$

Although straightforward, this approach requires the computation of the sensitivity of the FIM with respect to the parameters. These sensitivities can be obtained by finite differences or automatic differentiation.

A second approach is to utilize stochastic information objectives and reformulate the constraints as probabilistic chance constraints. This approach is termed stochastic OED. The known probability density of the uncertainties is used to compute the expected information content and form chance constraints. The nominal OED problem is reformulated in this form as

$$\min_{u(\cdot)} \mathbb{E}[\Phi(I(t_f))] + \alpha_\Phi \sqrt{\mathbb{V}\mathrm{ar}(\Phi(I(t_f)))}$$

subject to,

$$\epsilon \geq \Pr[0 \leq c_p(x(t), u(t), w(t), \theta, t)]$$

These chance constraints are computationally challenging [31] and are approximated as deterministic constraints using the Cantelli-Chebyshev inequality

$$0 \geq \mathbb{E}[c_p(\cdot)] + \alpha_c \sqrt{\mathbb{V}\mathrm{ar}(c_p(\cdot))}$$

This approach requires the expectation and the variance of the objective and constraints to be computed. This can be obtained via the polynomial chaos expansion (PCE). The polynomial chaos expansion aims to approximate the model response by an infinite sum of orthogonal polynomials. The use of orthogonal polynomials simplifies the moment calculation, and the expectation and variance can be computed directly from the coefficients of the expansion. Further details of the PCE-based approach for stochastic optimization can be found in Nimmegeers et al. [30,32].

The stochastic OED approach is preferred when robustness with respect to information content is important. The minimax approach leads to overly conservative experiments with low information content. When robustness with respect to constraint violation is desired the choice of the approach depends on the computational complexity of the problem. Monte Carlo studies show that both robust and stochastic OED problems significantly reduce constraint violations, although the stochastic OED performs slightly better. However, the PCE-based formulation of stochastic OED is much more computationally expensive.

Consider the Lotka-Volterra fishing problem from Sager et al. [33], Telen et al. [34], and Nimmegeers et al. [30]. This model is used to track a predetermined steady state for the predator and prey population in which the decision to fish is considered to be a binary decision variable. For implementation purposes, a relaxed version of the problem has been solved and the strategy of Sager et al. [33], has been used to connect the optimal control values to binary values. The model equations are:

$$\frac{dx_1}{dt} = x_1 - p_1 x_1 x_2 - 0.4 x_1 u$$

$$\frac{dx_2}{dt} = -x_2 + p_2 x_1 x_2 - 0.2 x_2 u$$

where x_1 is the biomass of the prey and x_2 the biomass of the predator which is both considered to be measurable. The symbol u is the fishing control. The initial conditions are set to: $x_1(0) = 0.5$ and $x_2(0) = 0.7$, the final time is fixed at $t_f = 12$ and, the assumed nominal parameter values are $p_1 = 1$ and $p_2 = 1$. The parameter variance-covariance matrix is assumed to be

$$C_\theta = \begin{bmatrix} 0.01 & 0 \\ 0 & 0.01 \end{bmatrix}$$

In this example of robust and stochastic OED, the aim is to develop an optimal fishing strategy $u(t)$ and sampling strategy $w(t)$ to estimate the parameters p_1 and p_2 in the prey and predator mass balances related to the interaction between predator and prey. The number of measurements is limited to six measurements. The goal in this case study is to maximize the information content as expressed by the minimum eigenvalue of the Fisher Information Matrix (E-criterion).

Three scenarios have been studied in this case study to investigate the influence of accounting for the variance on the information content during the experiment design: nominal OED, PCE-based stochastic OED (with expected value

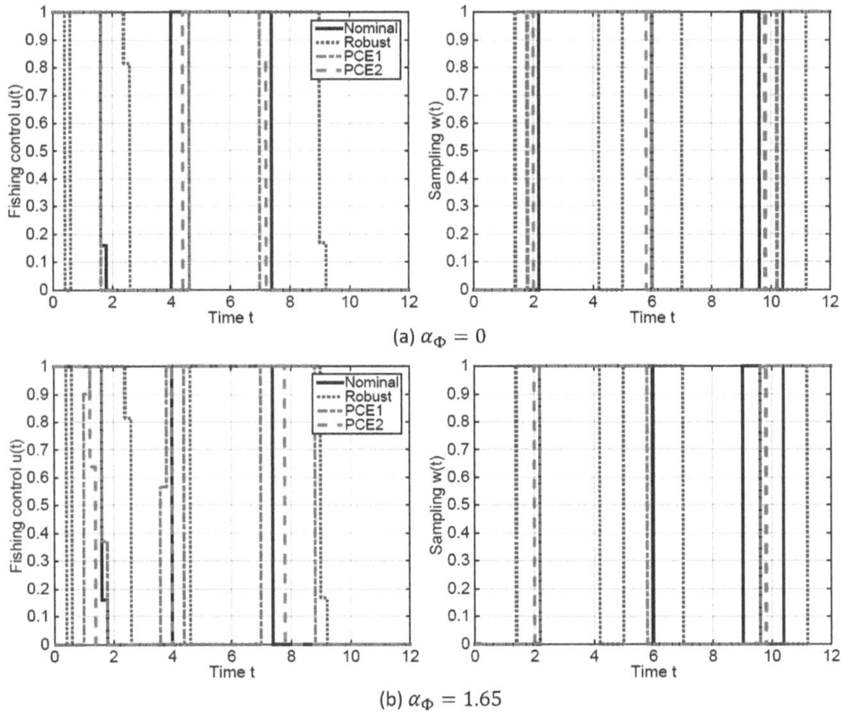

FIG. 10.7 Fishing control $u(t)$ and sampling action $w(t)$ profiles for the E-optimal (Nominal), approximate robust (Robust), and PCE-based stochastic (PCE1, PCE2) experiment designs. *(From P. Nimmegeers, et al., Optimal experiment design under parametric uncertainty: a comparison of a sensitivities based approach versus a polynomial chaos based stochastic approach, Chem. Eng. Sci. 221 (2020) 115651.)*

OED, i.e., $\alpha_\Phi = 0$ and $\alpha_\Phi = 1.65$) and an approximate robust design in which a 95% confidence region is considered (i.e., $\gamma = 6$). Results for a normal parametric uncertainty distribution are presented for the parameters p_1 and p_2.

To illustrate the difference between the implemented strategies, the control profiles for $\alpha_\Phi = 0$ and $\alpha_\Phi = 1.65$, i.e., the fishing control $u(t)$ and the sampling action $w(t)$, are depicted in Fig. 10.7A and B, respectively.

When the parameters of the system would be exact, there is a slight loss in information content when using the stochastic approach compared with the nominal case of approximately 5% (PCE2) and 10% (PCE1) for $\alpha_\Phi = 0$ and a loss in the information content of approximately 32% (PCE1), 3% (PCE2) for $\alpha_\Phi = 1.65$. The loss in information content when comparing the approximate robust approach with the nominal case is dramatic (approximately 80%). This is caused by the large back-off caused by the norm $\left\| \frac{d\Phi(I(t_f))}{d\theta} \right\|_{C_\theta}$.

In general, it is concluded that approximate robust designs are too conservative and result in very low information content. The nominal design performs better than the stochastic approach in the neighborhood of the nominal parameter values. In other regions, the information content for the nominal design drops sharply while this is not the case for the stochastic approach, indicating that the stochastic approach allows obtaining informative experiments for a wider range of parameter values.

The computation time of the different approaches is closely related to the number of states. The PCE2 requires a higher computation time than the other approaches due to the higher number of states that are involved in the system, i.e., six times the number of states in the nominal case. The PCE1 approach requires three times the number of nominal states. The approximate robust approach needs a higher computation time than the nominal case due to the evaluation of the norm for which an additional effort in terms of automatic differentiation is needed.

4.2 Multicriterion optimal experiment design

Each of the information criteria described before have their own strengths and weaknesses. As such, the choice of the criterion remains a question. Furthermore, different criteria are often conflicting. In practice, both experiments that lead to decorrelated as well as accurate parameters are desired. One approach designs an experiment with respect to one criterion while constraining the other criterion to a predetermined value [8]. In another approach design, the experiment with respect to a weighted sum of the two criteria. To incorporate parameter decorrelation as well as global parameter accuracy, the weighted sum of D-criterion and the modified E-criterion can be used [35].

$$\Phi = (1-a)\cdot \log\left(\frac{\lambda_{\max}\left(I\left(t_f\right)\right)}{\lambda_{\min}\left(I\left(t_f\right)\right)}\right) - a\cdot \log\left(\det\left(I\left(t_f\right)\right)\right)$$

The decimal logs ensure proper magnitudes for the two criteria for better scaling. If a two-parameter system is considered, the FIM is a 2×2 matrix and the above criterion reduces to

$$\Phi = (1-a)\cdot \log\left(\frac{\lambda_{\max}(\cdot)}{\lambda_{\min}(\cdot)}\right) - a\cdot \log\left(\lambda_{\max}(\cdot)\lambda_{\min}(\cdot)\right) = \log\left(\frac{\lambda_{\max}(\cdot)^{1-2a}}{\lambda_{\min}(\cdot)}\right)$$

When $a=0$, it represents the modified E-criterion, for $a=0.5$ it represents the E-criterion, and for $a=1$ it represents the D-criterion.

Multiobjective optimization (MOO) can be used to obtain insight into the information trade-off between multiple different criteria. The result of

MOO is not a single solution, but a set of equally optimal solutions called the Pareto set. A variety of solution strategies exist to solve a MOO problem. Some common strategies are the Weighted Sum, Normal Boundary Intersection [36], Enhanced Normalized Normal Constraints [37]. A Pareto front allows the decision-maker to visualize the trade-off between the multiple conflicting criteria. Telen et al. [34] propose a MOO-based strategy to allow the experimenter to choose a criterion. This strategy is depicted in Fig. 10.8.

Recall the Lotka-Volterra fishing problem example from the previous section. Contrary to the previous section, the sampling strategy is not considered as an experimental input to be optimized and different OED objective functions (A, D, E, ME, M) are studied and compared with the tracking objective (tracking both states at 1) and an anticorrelation criterion [38]. In Fig. 10.9 a radar chart is shown, indicating the different optimal experiments for the Lotka-Volterra case study.

Different groups can be distinguished based on Fig. 10.9:

- A- and D-optimal designs: similar in all, except the correlation value.
- E-, ME-, and M-optimal designs: in the respective criteria these are ranked in the top three. For A- and D-optimal designs, they are scored similarly. Note that the overall information content is lower than for the A- and D-criteria, as these experiments focus on the most uncertain parameter.
- Correlation-based designs: Scores the worst on the other experiments, except for the ME-optimal design. Tracking objective: Not an optimal experiment design, serves as a benchmark.

As an example, the proposed procedure of Telen et al. [34] is illustrated for a bi-objective case in which the A- and E-criterion are selected as objective functions and for which steps 6 and 7 in the procedure (Fig. 10.8) are elaborated. The E-criterion can help in overcoming the poor performance of the A-criterion with respect to correlation. The resulting Pareto front is depicted in Fig. 10.10. From the state profiles in Fig. 10.11, it is clear that both experiments result in poor tracking (left and middle Fig. 10.11).

The inclusion of the E-criterion reduces the uncertainty of the most uncertain parameter, at the cost of slightly increasing the uncertainty of the most certain parameter. The joint confidence intervals are shown in Fig. 10.12. Almost circular, but large, joint confidence is obtained for the E-optimal design. When the emphasis is on globally accurate parameter estimates, the joint confidence region is completely encompassed by the one for the E-optimal design. The joint confidence region of the A-optimal design is more tilted, indicating a higher correlation. This shows the strength of combining several OED criteria by multiobjective optimization, i.e.: a better conditioning and less correlation in this case study.

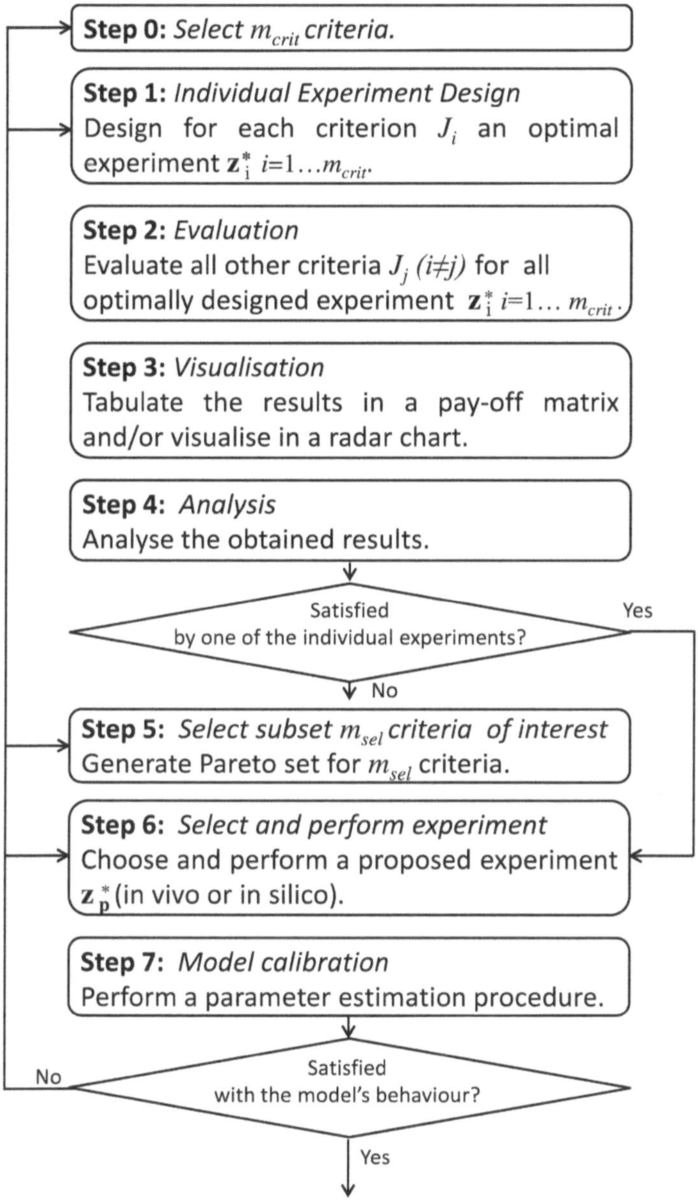

FIG. 10.8 Procedure for multiobjective OED. *(From from D. Telen, et al., Optimal experiment design for dynamic bioprocesses: a multi-objective approach, Chem. Eng. Sci. 78 (2012) 82–97.)*

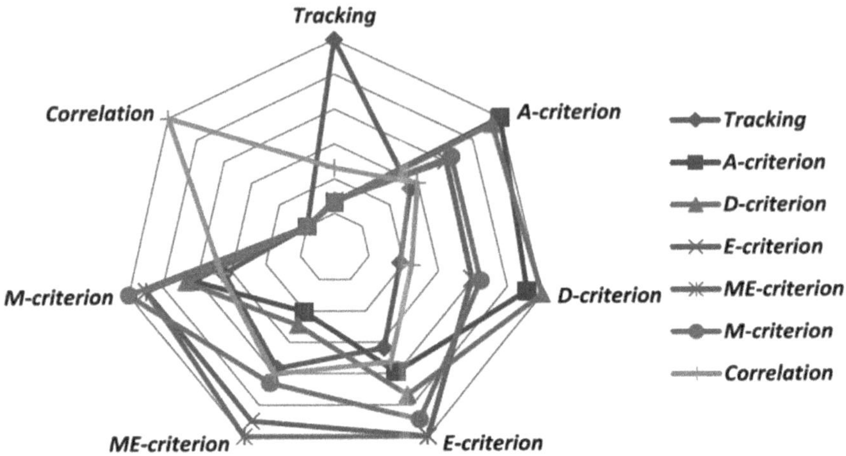

FIG. 10.9 Radar chart with the different optimally designed experiments. *(From D. Telen, et al., Optimal experiment design for dynamic bioprocesses: a multi-objective approach, Chem. Eng. Sci. 78 (2012) 82–97.)*

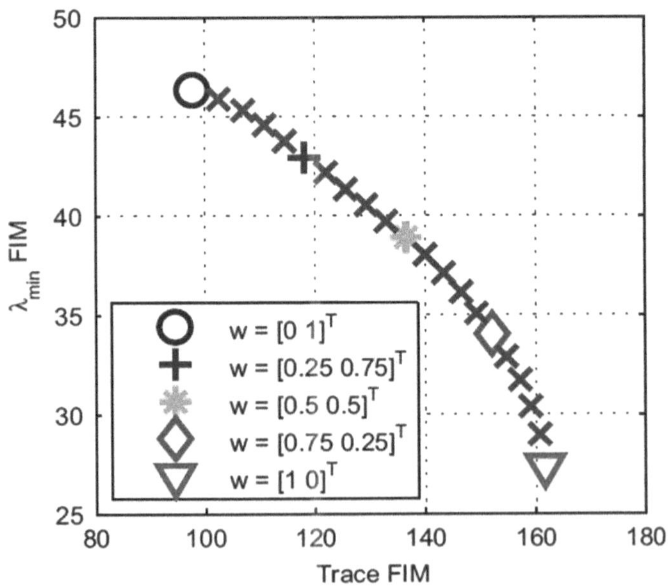

FIG. 10.10 Pareto front for the trade-off between A- and E-criterion. *(From D. Telen, et al., Optimal experiment design for dynamic bioprocesses: a multi-objective approach, Chem. Eng. Sci. 78 (2012) 82–97.)*

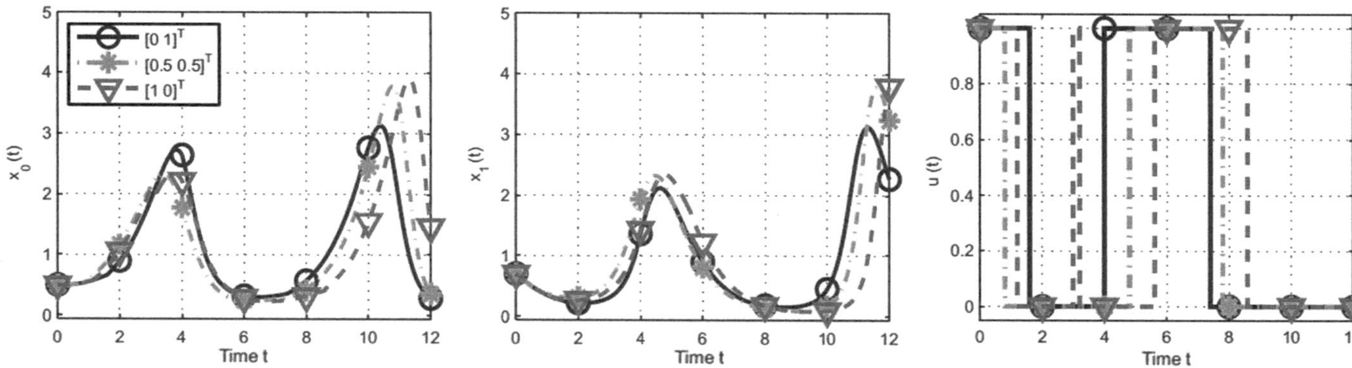

FIG. 10.11 State (left and middle) and control (right) profiles for the Lotka-Volterra case study. *(Modified from D. Telen, et al., Optimal experiment design for dynamic bioprocesses: a multi-objective approach, Chem. Eng. Sci. 78 (2012) 82–97.)*

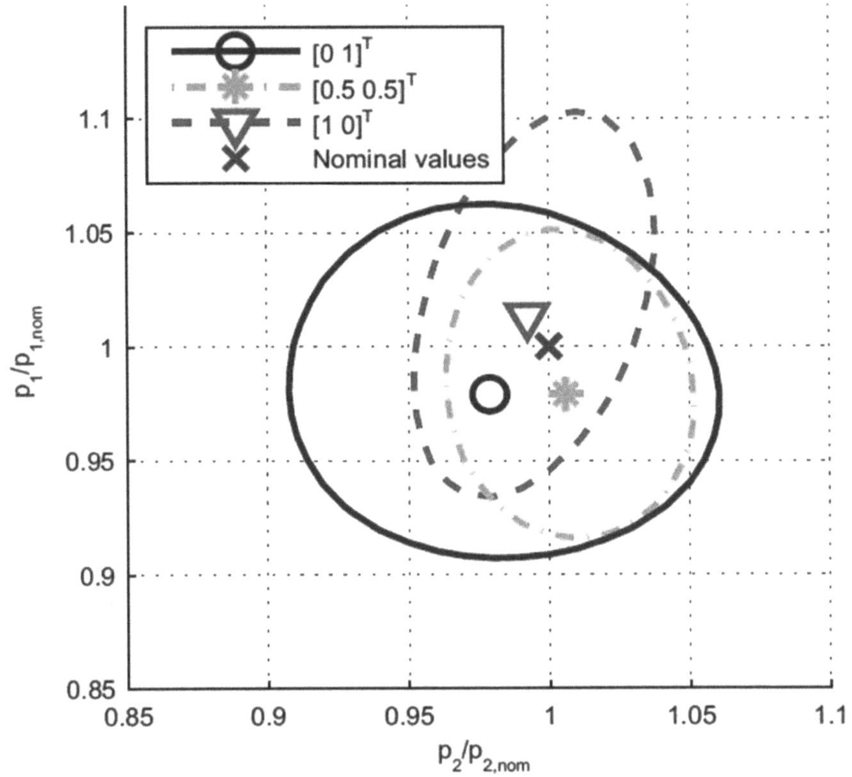

FIG. 10.12 95%-joint confidence regions for different weights of the A- and E-criteria criterion. *(From D. Telen, et al., Optimal experiment design for dynamic bioprocesses: a multi-objective approach, Chem. Eng. Sci. 78 (2012) 82–97.)*

5 Conclusions

This chapter discussed the formulation of optimal experiment design for model structure discrimination (OED/SD) and parameter estimation (OED/PE) as optimization problems. In OED/SD the distance between the two model outputs is maximized, while in OED/PE a scalar criterion of the Fisher Information Matrix or variance-covariance matrix is optimized. The various approaches to compute these matrices are discussed. Finally, the OED/PE problem is formulated as an optimal control problem and can be solved using a variety of tools. Casiopea (https://casiopeia.readthedocs.io/en/latest/index.html) is a dedicated tool for OED based on CasADi [39]. The OED formulated as an optimal control problem can be solved using tools such as ACADO [40], Pyomo [41], Muscod based TACO-toolkit [42], etc. Pomodoro is a CasADi based toolkit used to solve (multiobjective) optimal control problems. Pomodoro [43] has been used to solve OED problems both with single objective and conflicting multiple objectives [30,34,44]. A major advantage of Pomodoro is its Pareto-Browser that allows the decision-maker to visualize the Pareto front and interact with the front. The visualization and interaction can aid the user to make more insightful decisions on the choice of the design criteria. Pomodoro is available freely for academic use through the following link: https://cit.kuleuven.be/biotec/software/pomodoro.

Optimal experiment design is a powerful tool that can be used to reduce the experimental effort required to build process models. As demonstrated earlier in the chapter, OED can be deployed to discriminate between multiple possible model structures or to increase the accuracy of parameter estimates. In either of the cases, experiment design, experimentation, and parameter estimation form an iterative process. This has been highlighted by the case study on specific microbial growth rate modeling described earlier. Apart from the microbiological applications, OED has demonstrated its utility across the (bio) chemical process sector. For example, in De-Luca et al. [45] a single optimal experiment is computed to obtain statistically relevant parameter estimates of a freeze-drying model which involves a complex interaction between heat and mass transfer. Similar applications have been reported on catalytic reactor models [46], physiological models (e.g., glucose homeostasis models for Type-1 diabetes [47]), and bioprocess models [48,49].

References

[1] F.-G. Wieland, et al., On structural and practical identifiability, Curr. Opin. Syst. Biol. 25 (2021) 60–69.

[2] J. Spriet, G. Vansteenkiste, Computer-Aided Modelling and Simulation, Academic Press, London, New York, 1982.

[3] E. Balsa-Canto, A.A. Alonso, J.R. Banga, An iterative identification procedure for dynamic modeling of biochemical networks, BMC Syst. Biol. 4 (2010).

[4] L. Ljung, System Identification: Theory for the User, Prentice Hall, 1998.

[5] D. Bonvin, et al., Linking models and experiments, Ind. Eng. Chem. Res. 55 (2016) 6891–6903.

[6] E. Walter, L. Pronzato, Identification of Parametric Models From Experimental Data, Springer-Verlag, London, 1997.

[7] P. Nimmegeers, et al., Identifiability of large-scale non-linear dynamic network models applied to the ADM1-case study, Math. Biosci. 288 (2017) 21–34.

[8] S.S. Bhonsale, B. Stokbroekx, J. Van Impe, Assessment of the parameter identifiability of population balance models for air jet mills, Comput. Chem. Eng. 143 (2020) 107056.

[9] G. Franceschini, S. Macchietto, Model-based design of experiments for parameter precision: state of the art, Chem. Eng. Sci. 63 (2008) 4846–4872.

[10] L. Mertens, E. Van Derlinden, J.F. Van Impe, Comparing experimental design schemes in predictive food microbiology: optimal parameter estimation of secondary models, J. Food Eng. 112 (2012) 119–133.

[11] B. Donckels, et al., Performance assessment of the anticipatory approach to optimal experimental design for model discrimination, Chemom. Intell. Lab. Syst. 110 (2012) 20–31.

[12] D. Anderson, K.P. Burnham, Model Selection and Multi-Model Inference: A Practical Information-Theoretic Approach, Springer Nature, 2003.

[13] W.G. Hunter, A.M. Reiner, Designs for discriminating between two rival models, Technometrics 7 (1965) 307–323.

[14] G.E.P. Box, W.J. Hill, Discrimination among mechanistic models, Technometrics 9 (1967) 57–71.

[15] A.C. Atkinson, V.V. Fedorov, The design of experiments for discriminating between two rival models, Biometrika 62 (1975) 57–70.

[16] D. Ucinski, B. Bogacka, T-optimum designs for discrimination between two multiresponse dynamic models, J. R. Stat. Soc. Ser. B (Stat. Methodol.) 67 (2005) 3–18.

[17] A. Munack, Some improvements in the identification of bioprocesses, IFAC Proc. Vol. 25 (1992) 89–94.

[18] B. Donckels, et al., An anticipatory approach to optimal experimental design for model discrimination, Chemom. Intell. Lab. Syst. 95 (2009) 53–63.

[19] M. Schwaab, J.L. Monteiro, J.C. Pinto, Sequential experimental design for model discrimination, Chem. Eng. Sci. 63 (2008) 2408–2419.

[20] J. Baranyi, T.A. Roberts, A dynamic approach to predicting bacterial growth in food, Int. J. Food Microbiol. 23 (1994) 277–294.

[21] L. Rosso, J.R. Lobry, J.P. Flandrois, An unexpected correlation between cardinal temperatures of microbial growth highlighted by a new model, J. Theor. Biol. 162 (1993) 447–463.

[22] S. Bajard, et al., The particular behaviour of *Listeria monocytogenes* under sub-optimal conditions, Int. J. Food Microbiol. 29 (1996) 201–211.

[23] E. Van Derlinden, K. Bernaerts, J.F. Van Impe, Accurate estimation of cardinal growth temperatures of *Escherichia coli* from optimal dynamic experiments, Int. J. Food Microbiol. 128 (2008) 89–100.

[24] I. Stamati, et al., Optimal experimental design for discriminating between microbial growth models as function of suboptimal temperature: from in silico to in vivo, Food Res. Int. 89 (2016) 689–700.

[25] R. Schenkendorf, M. Mangold, A. Kremling, Optimal experimental design with the sigma point method, IET Syst. Biol. 3 (2009) 10–23.

[26] S. Julier, J.K. Uhlmann, A General Method for Approximating Nonlinear Transformations of Probability Distributions, Technical Report, Robotics Research Group, University of Oxford, 1996.

[27] T. Heine, M. Kawohl, R. King, Derivative-free optimal experimental design, Chem. Eng. Sci. 63 (2008) 4873–4880.

[28] D. Telen, et al., Optimal experiment design under process noise using Riccati differential equations, J. Process Control 23 (2013) 613–629.

[29] S. Körkel, et al., Numerical methods for optimal control problems in design of robust optimal experiments for nonlinear dynamic processes, Optim. Methods Softw. 19 (2004) 327–338.

[30] P. Nimmegeers, et al., Optimal experiment design under parametric uncertainty: a comparison of a sensitivities based approach versus a polynomial chaos based stochastic approach, Chem. Eng. Sci. 221 (2020) 115651.

[31] A. Mesbah, S. Streif, A probabilistic approach to robust optimal experiment design with chance constraints, IFAC-PapersOnLine 48 (2015) 100–105.

[32] P. Nimmegeers, et al., Dynamic optimization of biological networks under parametric uncertainty, BMC Syst. Biol. 10 (2016) 86.

[33] S. Sager, et al., Numerical methods for optimal control with binary control functions applied to a Lotka-Volterra type fishing problem, in: Lecture Notes in Economics and Mathematical Systems, Springer-Verlag, 2009, pp. 269–289.
[34] D. Telen, et al., Optimal experiment design for dynamic bioprocesses: a multi-objective approach, Chem. Eng. Sci. 78 (2012) 82–97.
[35] K.J. Versyck, J.F. Van Impe, Trade-offs in design of fed batch experiments for optimal estimation of biokinetic parameters, in: Proceedings of the 1998 IEEE International Conference on Control Applications (Cat. No.98CH36104), vol. 1, 1998, pp. 51–55.
[36] I. Das, J.E. Dennis, Normal-boundary intersection: a new method for generating the Pareto surface in nonlinear multicriteria optimization problems, SIAM J. Optim. 8 (1998) 631–657.
[37] A. Messac, A. Ismail-Yahaya, C.A. Mattson, The normalized normal constraint method for generating the Pareto frontier, Struct. Multidiscip. Optim. 25 (2003) 86–98.
[38] G. Franceschini, S. Macchietto, Novel anticorrelation criteria for model-based experiment design: theory and formulations, AIChE J. 54 (2008) 1009–1024.
[39] J.A.E. Andersson, et al., CasADi—a software framework for nonlinear optimization and optimal control, Math. Program. Comput. 11 (2019) 1–36.
[40] B. Houska, H.J. Ferreau, M. Diehl, ACADO toolkit—an open source framework for automatic control and dynamic optimization, Optimal Control Appl. Methods 32 (2011) 298–312.
[41] M.L. Bynum, et al., Pyomo–Optimization Modeling in Python, third ed., Springer Science & Business Media, 2021.
[42] C. Kirches, S. Leyffer, TACO: a toolkit for AMPL control optimization, Math. Program. Comput. 5 (2013) 227–265.
[43] S. Bhonsale, et al., Pomodoro: a novel toolkit for dynamic (multiobjective) optimization, and model based control and estimation, IFAC-PapersOnLine 51 (2018) 719–724.
[44] D. Telen, et al., Approximate robust optimization of nonlinear systems under parametric uncertainty and process noise, J. Process Control 33 (2015) 140–154.
[45] R. De-Luca, et al., Accelerating the development and transfer of freeze-drying operations for the manufacturing of biopharmaceuticals by model-based design of experiments, Ind. Eng. Chem. Res. 59 (2020) 20071–20085.
[46] V. Prasad, D.G. Vlachos, Multiscale model and informatics-based optimal design of experiments: application to the catalytic decomposition of ammonia on ruthenium, Ind. Eng. Chem. Res. 47 (2008) 6555–6567.
[47] F. Galvanin, et al., Optimal design of clinical tests for the identification of physiological models of type 1 diabetes mellitus, Ind. Eng. Chem. Res. 48 (2009) 1989–2002.
[48] V. Abt, et al., Model-based tools for optimal experiments in bioprocess engineering, Curr. Opin. Chem. Eng. 22 (2018) 244–252.
[49] J.H. Kim, J.M. Lee, Efficient online model-based design of experiments via parameter subset selection for batch dynamical systems, Comput. Chem. Eng. 121 (2019) 646–653.

Chapter 11

Characterization of reactions and growth in automated continuous flow and bioreactor platforms—From linear DoE to model-based approaches

Tilman Barz[a,b], Julian Kager[c], Christoph Herwig[d], Peter Neubauer[b], Mariano Nicolas Cruz Bournazou[b,e], and Federico Galvanin[f]

[a]Center for Energy, AIT Austrian Institute of Technology GmbH, Vienna, Austria, [b]KIWI-biolab, Department of Biotechnology, Technische Universität Berlin, Berlin, Germany, [c]Competence Center CHASE GmbH, Linz, Austria, [d]Institute of Chemical, Environmental and Bioscience Engineering, Technische Universität Wien, Vienna, Austria, [e]DataHow AG, Dübendorf, Switzerland, [f]Department of Chemical Engineering, University College London (UCL), London, United Kingdom

1 Introduction

Strategies to maximize the information outcome of laboratory experiments are relevant in any field where experimentation is time-consuming and/or costly. In the pharmaceutical industry statistical design of experiments (DoE) is an established technique used to develop biological or chemical reaction processes [1–5]. DoE has shown to noticeably reduce the experimental effort compared to conventionally or heuristically designed experiments. However, DoE uses relatively simple static and linear regression equations (the model) to predict the effects of changes in the independent variables (the factors) on dependent variables (the responses). Due to these limitations in the choice of a model, it is not well suited to handle dynamic experiments, where both factors and responses can be complex time profiles [4]. This becomes evident as the technical possibilities in modern laboratories (e.g., automation, robotics, miniaturization, sensors and analytics, high throughput) allow for highly complex and parallelized experiments where the processes are nonlinear and dynamic. Model-based Optimal Experimental Design (mbOED) methods were explicitly developed to incorporate mechanistic nonlinear and dynamic process models.

Simulation and Optimization in Process Engineering. https://doi.org/10.1016/B978-0-323-85043-8.00014-3

They are therefore ideally suited for the characterization of reactions and growth in synthetic chemistry and biotechnology.

mbOED has been an active area of research, see, e.g., the reviews in Refs. [4,6–11]. While it is mostly used for the precise estimation of model parameters, it can also be used to design experiments for model discrimination, i.e., to distinguish between alternative model structures. Another use case is to improve prediction accuracy, i.e., the reduction of uncertainties in model prediction [6,11–14]. In recent years important advances have been made on mbOED for nonlinear dynamic systems, and for large-scale applications, see Ref. [11], covering a wide variety of fields and disciplines: control, mechanical, electrical and chemical engineering, thermodynamics, medicine, geophysics, biology, pharmacology, chemistry, and mechanics. Despite this noticeable progress

> only about 4% of the 77 practical applications of DoE published in technical journals between 2001 and 2005 and reviewed in Ref. [15] were mbOED (optimal) [16].

Moreover, considering applications of mbOED to cellular and biological processes and systems, reported between 2000 and 2013, it can be noted that

> fewer than 15% of the 85 reviewed articles contained experimental data. The majority of the developed methods and examples were evaluated by simulations (in silico) [16].

This contribution reviews the actual degree of synergy between lab automation and optimal experimental design techniques for the characterization of reactions and growth in synthetic chemistry and biotechnology. We take a practical point of view and focus on recent demonstrations of automated optimal experiments in miniaturized continuous flow and fed-batch reactor platforms (Section 2). Sections 2.1 and 2.2 give examples and discuss basic characteristics of these platforms which can efficiently generate experimental data for screening, process characterization, optimization of reaction and growth, and scale-up. Section 2.3 reviews recent applications of classical DoE, model-free, and mbOED for the systematic planning of experiments. The focus is on the dimension and complexity of the experimental space, the models, and collected data. Section 3 discusses special aspects and challenges regarding the sequential planning of dynamic experiments, the identification of biochemical reaction models, the computation of optimal designs, and implementation in software packages. Section 4 presents an industry view on the application of model-based approaches in automated platforms and an exemplary case study. A discussion and conclusions are given in Section 5.

2 Miniaturized platforms and applications

The automation of chemical and biological laboratories has rapidly gained popularity in research and industry [17–19]. Many laboratory tasks can be performed by automated robots (moving arms, liquid handlers, storage and shaking, labeling, etc.). Online and inline analytics for reaction monitoring together with miniaturization and parallelization enables high throughput experimentation.

2.1 Continuous-flow microreactor platforms in synthetic chemistry

Continuous-flow reactors have been employed in a wide variety of situations from process monitoring [20] to screening of operating conditions [21]. These systems are constituted by four essential elements:

- A reagents delivery system;
- A flow reactor;
- An inline measurement system;
- A digital device to control data acquisition, reactor conditions, and reagents delivery.

The reaction monitoring can be carried out by offline gas chromatography (GC) or high-pressure liquid chromatography (HPLC) analysis and relies on rapid and effective quenching. Data acquisition times can be reduced by spectroscopic detection techniques for online or inline reaction monitoring, e.g., using fluorescence, ultraviolet–visible (UV/vis), near- and mid-infrared (IR), benchtop nuclear magnetic resonance (NMR), and Raman spectroscopies [22–24].

Figs. 11.1 and 11.2 show a sketch and a picture of a continuous-flow platform. The exemplary system includes a 2m-long tube PEEK reactor with a 250μm internal diameter (Agilent Technologies, Palo Alto, USA) placed in a stirred oil bath heated by a rope heater. The small diameter ensures the reactor behaves as a plug flow reactor in the range of liquid flowrates used (5–100μL min^{-1}) allowing high rates of heat transfer so that the reaction mixture reaches reaction temperature (at reactor inlet) or room temperature (at reactor outlet) very quickly. Two syringe pumps filled with different reagent concentrations are used to control the residence time and feed concentration (neMESYS low-pressure module, Cetoni GmBH, Korbussen, Germany). The reagents are mixed at a four-way junction before reaching the oil bath ensuring they are well mixed before entering the reactor. A sensor is connected inline with the pressure vessel to provide an accurate reading of the back-pressure and a second sensor is placed inline to measure the upstream pressure and hence the pressure drop. The reactor outlet is connected to a six-way switching valve of an automatic sampler-dilutor connected to an HPLC.

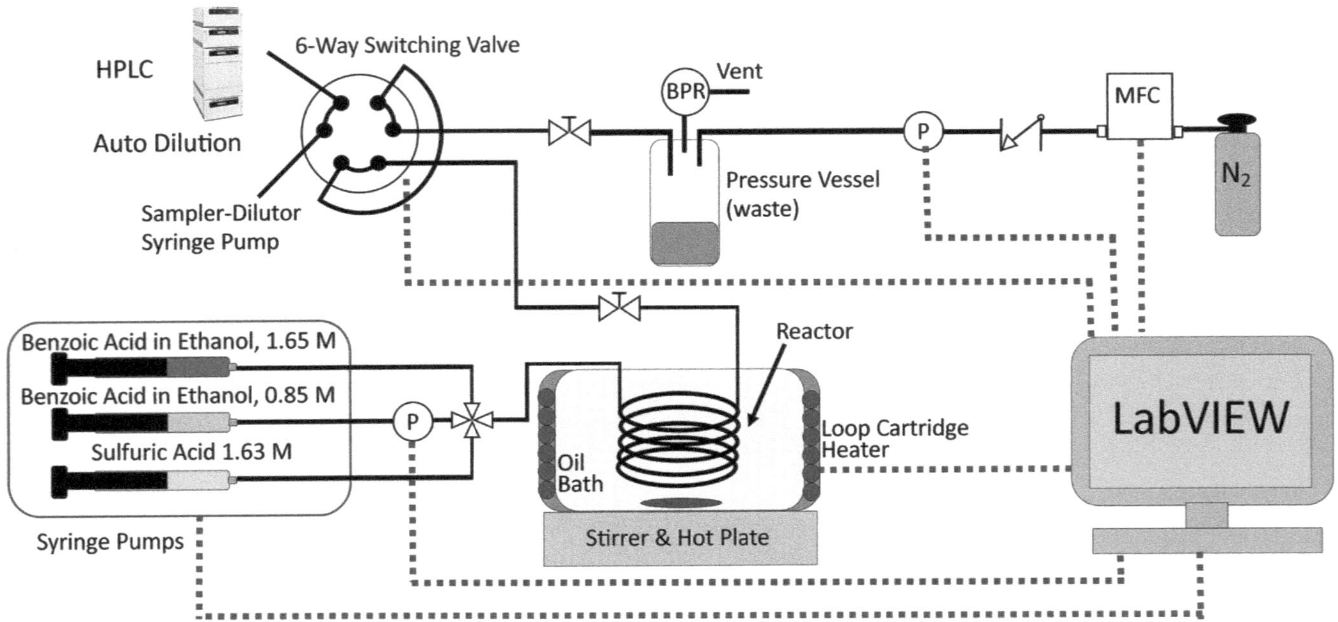

FIG. 11.1 Continuous-flow reactor platform developed in Ref. [25] for the esterification of benzoic acid with ethanol using sulfuric acid as a homogenous catalyst. *Red dashed lines* (*dark gray* in print version) indicate LabVIEW (National Instruments Corp., Austin, Texas, USA) controls the equipment, *blue dashed lines* (*gray* in print version) indicate LabVIEW reads the measurement from the equipment. *BPR*: back-pressure regulator; *MFC*: mass flow controller; *P*: pressure sensor. *(From C. Waldron, A. Pankajakshan, M. Quaglio, E. Cao, F. Galvanin, A. Gavriilidis, An autonomous microreactor platform for the rapid identification of kinetic models, React. Chem. Eng. 4 (2019) 1623–1636.)*

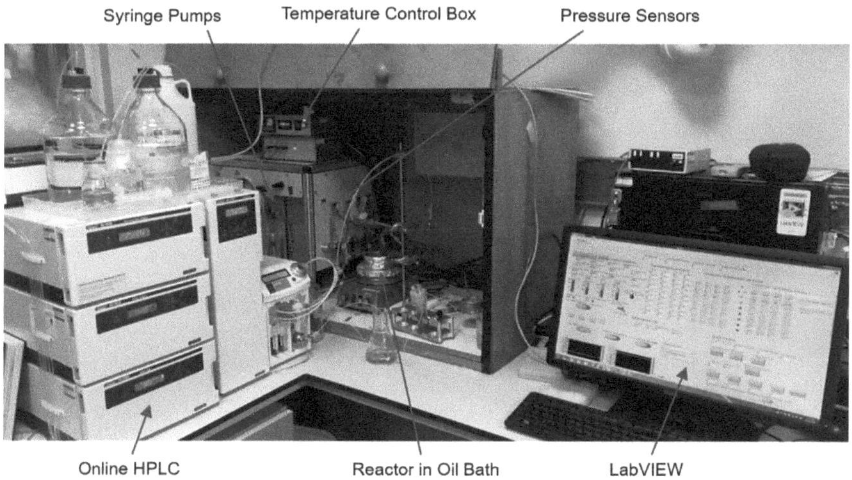

FIG. 11.2 Continuous-flow reactor system at the Department of Chemical Engineering (UCL, United Kingdom).

2.1.1 Operation

Continuous-flow reactors are usually operated under constant conditions (at steady-state) mainly because of the ease of control and operation and the consistency of data obtained at stationary conditions [26]. If the measurements are separated at least by three hydrodynamic residence times of the plug flow reactor it is reasonable to assume that they represent individual batches [27], see Fig. 11.3.

For sufficiently narrow residence time distributions this assumption also holds operating flow reactors under transient flow conditions. Hereby, the waiting times between measured data points can be eliminated and the amounts of starting materials can be reduced [23,28]. In Ref. [27] it is reported that the generation of 30 data points requires 5 h by steady-state experiments, while experiments under dynamic conditions generate 336 data points in just 1.5 h. The extension of this approach to experiments with constant temperature and flow ramp rates has been proposed recently, see Section 3.1.

2.2 Bioreactor platforms with automatic liquid handling

Mini-bioreactor platforms are mainly used for the optimization of cultivation and production media and the cultivation conditions. These systems are combined with technologies for automated liquid handling (sampling, analysis, nutrient addition, and pH titration) and work with peristaltic pumps, flow cells, or pipetting robots [30]. There are several commercial solutions available with reactor volumes ranging from a few μL up to approx. 250 mL. Table 11.1 gives some references highlighting the capabilities of these platforms. At present, the

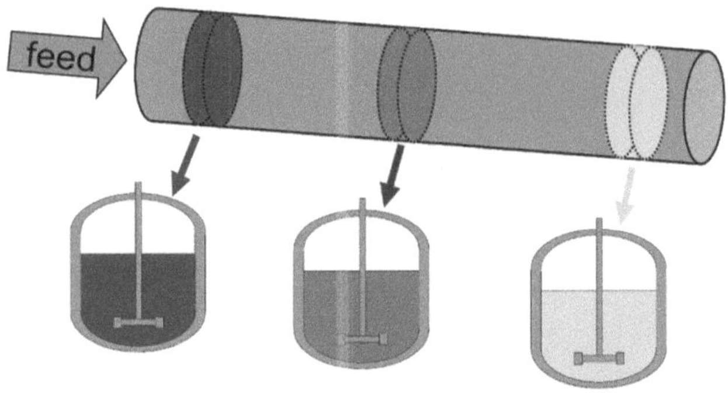

FIG. 11.3 For a plug flow each fluid element passing the reactor represents a batch reactor with a specific residence time. *(Modified from C. Waldron, A. Pankajakshan, M. Quaglio, E. Cao, F. Galvanin, A. Gavriilidis, Model-based design of transient flow experiments for the identification of kinetic parameters, React. Chem. Eng. 5 (2020) 112–123.)*

ambr250 (Sartorius AG, Göttingen, Germany) is the most used high throughput mini-bioreactor system in the industry [31]. Similar systems also commercially available are: DASbox (Eppendorf AG, Hamburg, Germany), BioXplorer (h.e. l. Group, London, United Kingdom), Multifors 2 (Infors AG, Bottmingen, Switzerland), bioREACTOR (2mag AG, Munich, Germany), or solutions building on shaken well plates such as the bioLector (m2p-labs GmbH, Baesweiler, Germany). The most important capabilities of these devices are:

- feeding and other control strategies similar to industrial processes (except in bioREACTOR),
- stirrers (except for the bioLEctor) and high oxygen transfer coefficients,
- online measurements (e.g., pH, dissolved oxygen) and automated sampling,
- most also include off-gas measurements.

At the μL scale, experiments are carried out in shaken well-plates with up to 114 parallel batch cultivations and duration times between 1 and 14 days. Because of the small working volumes, the cultivation conditions are usually rather simple (batches with different initial concentrations and variable end times, etc.) and samples for offline measurements of nutrients, biomass, and product are usually taken only once (see Table 11.1). Sensor spots at each well allow for quasi-continuous scattered light and fluorescence measurements of dissolved oxygen and pH, for some systems even of biomass and product formation. These systems can produce large amounts of data (20,000 points) in less than 1 day.

At the 10–15 mL scale, the platforms use up to 48 parallel reactors with gas-inducing stirrers, sensor spots for optical online measurements, and the possibility for more frequent automated offline sampling (up to 12 samples per reactor reported). These systems can be operated in fed-batch mode (mostly pulse feeding) and allow for basic control of temperature, agitation, oxygen, and pH. Fig. 11.4 shows a bioREACTOR 48 platform (2mag AG, Munich, Germany)

TABLE 11.1 Automated parallel bioreactor platforms with liquid handling for scale-down development and upstream process characterization.

Reference	Platform	Application	Measurements	Experiments
Huber et al. [32]	BioLector platform with 96 shaked microtiter plates (*190µL working volumes*), pipetting robot, and liquid-handling workstation	*E. coli* expressing the fluorescence protein EcFbFP	*Biomass and protein formation detected every 5min* by scattered light and fluorescent measurements in each well; *over 20,000 data points*	96 parallel experiments with varying inducer concentrations and induction times (with up to 6 copies); 12–18h cultivations
Hemmerich et al. [33]	BioLector platform with 48 shaked well Flowerplates (*800µL working volumes*); 2 liquid handling workstations	Genome-reduced *Corynebacterium glutamicum* strains engineered for Sec-mediated heterologous cutinase secretion	Quasi-continuous monitoring of biomass concentration via backscatter measurement; pH and DO via fluorescence sensor spots integrated at the bottom of each well; *up to six low-volume samples of 20µL taken per cultivation* for substrate and product concentration analysis	*2×48 experiments in 24h cultivations*
Amanullah et al. [34]	SimCell platform with 5 incubators each capable of holding up to 42 micro-bioreactor arrays (MBAs), each with 6 gas-permeable MB's (*600–700µL working volumes*) in rotary agitators	*E. coli* expressing branched-chain amino acids into recombinant pre-proinsulin	MBAs partially with DO sensors; measurement of biomass, pH, and DO performed once per day by manual transfer of the MBAs to a sensing station; offline samples: cell growth, metabolites, and protein concentration	*114 Parallel MBs in 19 MBAs*; six MBs manually harvested each day; total length 14days (one sample per reactor)

Continued

TABLE 11.1 Automated parallel bioreactor platforms with liquid handling for scale-down development and upstream process characterization—cont'd

Reference	Platform	Application	Measurements	Experiments
Faust et al. [35]	Bioreactor 48 with 48 parallel stirred reactors (*10 mL working volumes*) integrated in a liquid-handling system; gas-inducing stirrers; bolus feeding; *continuous feeding by micro-pumps* for dispensions of up to three liquids	*E. coli* high cell density cultivation for expression of two heterologous proteins	Fluorescence measurements for online DO and pH; atline measurements of dry cell weights (DCW) and OD600, and product (up to 12 samples per cultivation)	Up to 12 parallel experiments of duration up to 40 h; *smallest dosage volumes of one pump lift was about 1.3 μL*; average volume addition by liquid handler is 17 times higher than by micro-pump
Janzen et al. [36], similar setup e.g., in Haby et al. [37]	Bioreactor 48, 4 blocks each with 8 bioreactors (*8–15 mL working volumes*) with automated liquid-handling system; gas-inducing stirrers; intermittent additions of inducer, titration agent, feed medium, and for pH control	Screening of *E. coli* cells with two different expression systems and three different process conditions	Fluorescence measurements for online DO and pH; automated atline (around 7) biomass OD550 measurements; automated sample transfer for offline product titer analysis	Around 40 h cultivations; temperature (ramp) control; DO control cascade using stirrer speed and oxygen supply; capacity of *up to 96 cultivations per week in an industrial laboratory*
Janakiraman et al. [38]	Ambr15 platform with 24–48 mini-bioreactors (*10–15 mL working volumes*) on 2–4 culture stations; pitched blade impeller sparge tube for gassing; automated liquid handling; individual control of pH, DO, and feed for each reactor; temperature and agitation control per culture station	Recombinant CHO cell; IgG-expressing mammalian cell line	Optical sensors for pH and DO monitoring; daily sampling for collection of 13 process and product quality parameters	14-Day fed-batch culture; *36 experimental conditions performed in a single run*; temperature shift on day 6; pH and Do control, and a total feed percentage of 30%.

Tai et al. [39]	Ambr250 platform with 24 stirred bioreactors (*150mL working volumes*) with automated liquid handling system; up to four continuously pumped liquid feeds	*E. coli* fermentation process for recombinant protein production	Online data for DO, pH (and temperature); 96 samples (4 from each reactor) over an 18h time span	*24 Parallel runs* with different conditions derived from *10-factor screening design* for optimizing titer and truncated product
Kager et al. [40]	DASbox system with 4 parallel reactors extendable to 16 reactors (*250mL working volumes*); dynamic feedrate control possible; biomass softsensor integrated into DASware control software	Recombinant *E. coli* fed-batch process with multiple substrate feed	*Mixed substrate, air, and oxygen inflow; gas composition of the effluent gas*; feed pump rate; CO_2 and O_2 concentrations by off-gas analyzer	4 Parallel experiments with different lactose uptake; 32h operation; *Estimation of biomass formation rate* from carbon and oxygen balance

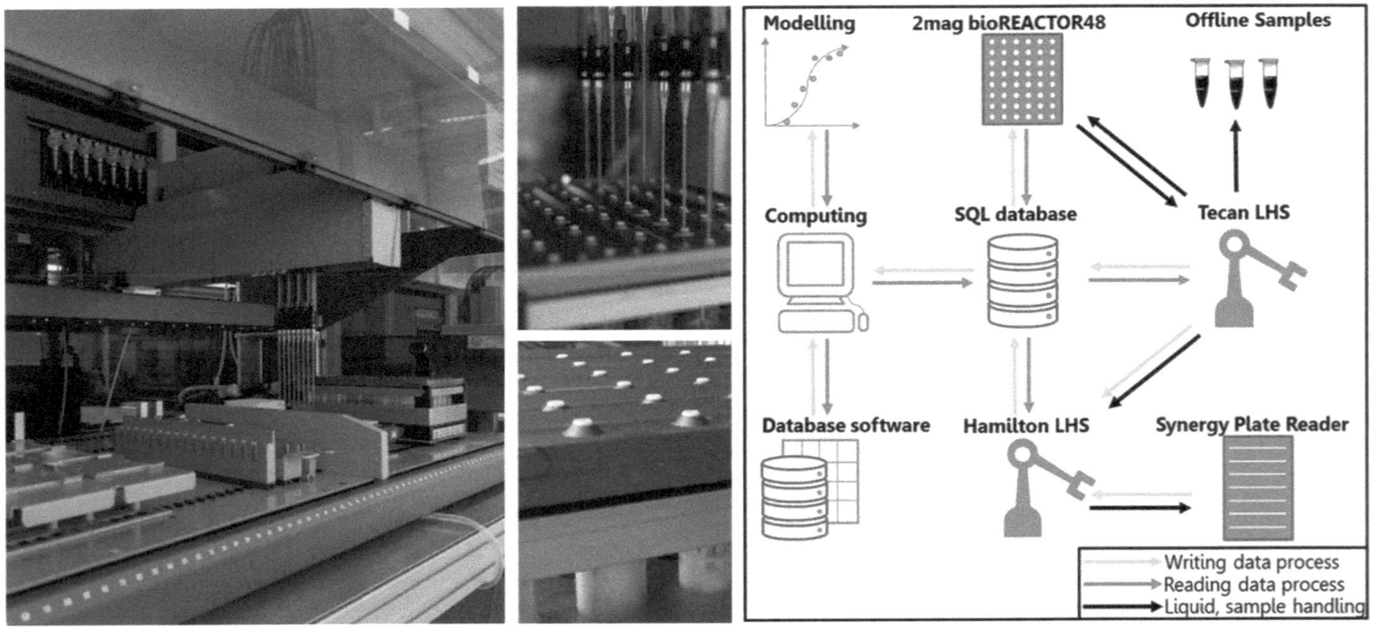

FIG. 11.4 BioREACTOR 48 platforms (2mag AG, Munich, Germany) with reactor working volumes between 8 and 15 mL, automated liquid handling, and an integrated plate reader for atline analytics at the Department of Biotechnology (TU Berlin, Germany).

operated by a Tecan liquid handler (Tecan Group Ltd., Männedorf, Switzerland) and a Microlab Star liquid handling station (Hamilton Company, Bonaduz, Switzerland) with an integrated plate reader for atline analytics of biomass, glucose, and acetic acid during cultivation.

Finally, with working volumes above 100 mL, a wider range of cultivation conditions with dynamic feed rates, high frequent sampling, and CO_2 and O_2 off-gas monitoring are possible. These platforms are extendable to 24 stirred reactors and can perform cultivations at conditions that are very close to industrial processes [41]. The possibility to connect a wider range of peripheral devices such as precise peristaltic pumps, balances, and additional sensors such as permittivity and in-situ spectroscopy are advantages over miniaturized systems at the 10–15 mL scale. Fig. 11.5 shows a system with two benchtop reactors with automatic sampling (highest frequency every 10 min), sampling treatment, and transfer to HPLC, enzymatic or a dry matter analyzer [42].

2.3 Applications of DoE, self-optimization, and mbOED—A bibliographical review

2.3.1 DoE for continuous flow reactions in synthetic chemistry

In synthetic chemistry application of DoE has greatly increased and is mainly used in the optimization of batch processes in lab contexts [1,3]. DoE applications typically involve the multivariate and broad (1) *screening of reaction parameters*, the (2) *evaluation of parameter effects*, and the (3) *optimization of reaction conditions*, e.g., through a surface response model [2]. In continuous flow, DoE is used less frequently and predominantly combined with offline analysis to design a sequence of steady-state points for optimizing reactor performance [1,43,44].

Ref. [1] gives a recent review on DoE applications published in Organic Process Research & Development Journal between 2003 and 2013. For continuous flow reactors both, full and fractional factorial designs have been employed mostly for the *optimization of reaction conditions*, such as conversion, yield, or reaction time. A few applications considered robustness tests for changes of optimal reaction conditions. While for screening up to 9 factors were considered, the typical flow applications consider

- rather few factors, between 2 and 4, mostly residence time, temperature, stoichiometry/ substrate concentrations, and pressure);
- the number of experiments ranges between 7 and 58.

Reaction times are typically fast with residence times often significantly below 10 min facilitating the rapid generation of kinetic data.

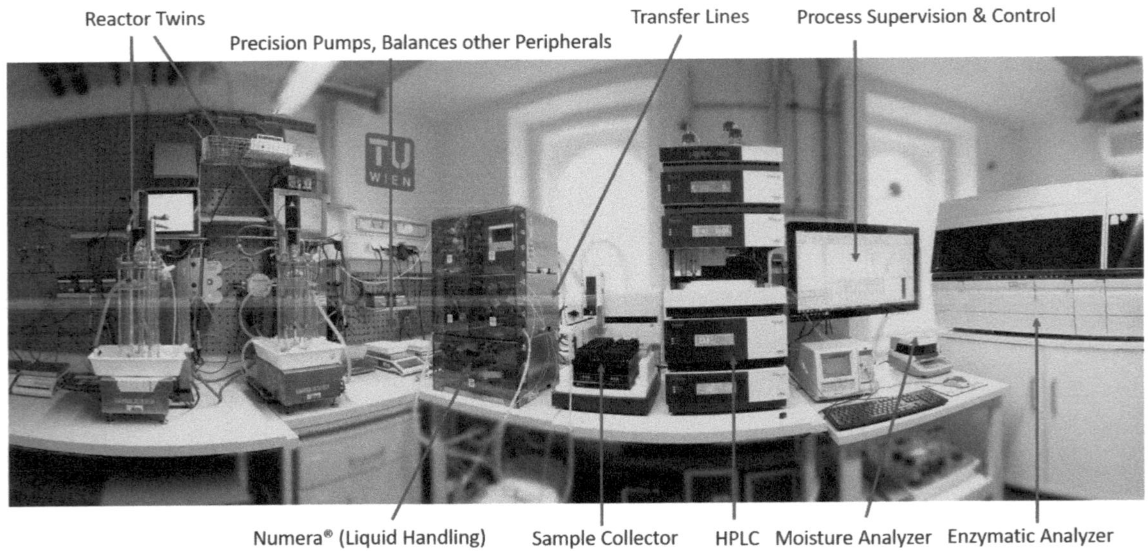

FIG. 11.5 Two parallel benchtop reactors (Infors AG, Bottmingen, Switzerland) of 2.7 L working volume connected to Numera liquid handling system (Securecell AG, Urdorf, Switzerland) for automated sampling, sample treatment (dilution & filtration) and transfer to several analytical devices [42] at the Institute of Chemical, Environmental and Bioscience Engineering (TU Wien, Austria).

2.3.2 Sequential self-optimization of continuous-flow reactions

Automated continuous flow microreactors incorporating inline (or online) analyzers for reaction monitoring have been used for online multivariable optimization [27]. So-called self-optimizing algorithms are straightforward to implement at automated platforms employing sequential DoE techniques [22,28,45–50]. In these autonomous reaction platforms after every experiment is terminated and new data is collected, specific algorithms are used to construct black-box representations of the system (without knowledge of the reaction kinetics or the mechanism) for designing the following experiment with the aim to optimize the reaction performance (i.e., conversion or yield), see, e.g., [51]. More recently also machine learning (ML) techniques exploiting Gaussian process (GP) regression and Bayesian optimization have been explored to optimize multistep processes with respect to multiple objectives, with a great potential to make substantial savings in time and resources [51,52].

Table 11.2 gives an overview of recent (later than 2010) experimental applications of self-optimization in continuous flow systems. It can be seen that:

- between 3 and 4 factors have been considered,
- relatively large numbers (up to 160) of static experimental conditions could be evaluated.

Using fully automated platforms the waiting times between measurements depend on the time for re-establishing steady-state conditions (determined by the thermal characteristics of the reactor), and the availability of the measurements analytical results. The latter can be significantly reduced using inline measurements, e.g., FT-IR spectroscopy, as, e.g., discussed in Refs. [27,49].

2.3.3 DoE for the development of biotechnological processes

A review on applications of DoE methodology in the development of biotechnological processes is given in Ref. [5] with a focus on the period after 1990 and covering the mainstream of bioprocess applications. Most applications concern the development and *optimization of media and mixture composition* for cultivation and production. The typical experiment for optimization of media composition is done in two steps: by first screening a variety of media ingredients by a two-level fractional factorial design approach, and second optimizing their levels by response surface methodology. The first (screening) step identifies a subset of a few important factors, the second step determines optimal factor values [5]. The reported applications consider between 3 and 15 factors mostly with the aim to maximize the cellular productivity, i.e., final product yield or final concentration.

Applications of DoE for the *optimization of process state variables and process operation* consider quantitative factors (controlled or uncontrolled) such as temperature, pH, and dissolved oxygen tension, as well as qualitative factor variables, e.g., different impeller turbines. Quantitative factors are often controlled

TABLE 11.2 Applications of self-optimization in automated continuous flow reactors for optimal synthesis of organic compounds.

Reference	Application	Aim	Controls	Measurements	Experiments	algorithm
McMullen et al. [47]	Knoevenagel condensation reaction; and oxidation of benzyl alcohol to benzaldehyde to benzoic acid	Maximize yield and throughput	2 and 4 Factors: temperature, residence time, 1 reactor inlet concentrations	Online HPLC	Between 13 and 36 experiments, and 46 sequential experiments	Nelder–Mead, Steepest Descent, and SNOBFIT
Parrott et al. [49]	Etherification reactions in supercritical carbon dioxide	Maximize yield	3 Factors: temperature, pressure, flow rate of the CO 2 pump	Online GLC	160 Samples for one optimization run, 35 h to complete	SMSIM Simplex Algorithm
Holmes et al. [46]	Synthesis of acrylamide/ synthesis of EGFR kinase inhibitor	Maximize yield	4 Factors: flow rate, temperature, 2 feed concentration	Atline HPLC	42 Separate experiments required, 26 h	Derivative-free SNOBFIT
Holmes et al. [22]	Reaction of methyl nicotinate with aqueous methylamine	Maximize yield	3 Factors: flow rate, temperature, feed concentration	Online mass spectrometry	21 Experiments, 12 h	Derivative-free SNOBFIT
Cherkasov et al. [48]	MBY semihydrogenation reaction	Maximize yield	3 Factors: substrate flow rates, catalyst poison, and IPA solvent	Online GC	Seems that more than 50 points were used.	derivative-free SNOBFIT
Fath et al. [53]	Reaction of benzaldehyde with benzylamine to form *n*-benzylidenebenzylamine	Maximize product concentration, quantity, minimize costs	3 Factors: stoichiometric ratio, residence time, temperature	Inline FT-IR spectroscopy	Between 20 and 27 experiments, relative fast reactions with residence times <6 min	Modified simplex and model-free DoE

at a predetermined set-point or follow the course of a preset (e.g., feeding) trajectory. Moreover, also optimal sequence and timing of operational actions, such as when to inoculate, harvest, or induce/transfect a culture can be considered. The identification of these states and operational parameters is often combined with the optimization of the media and mixture composition [5]. The typical applications are [5]:

- Optimization of yield, growth rate, and biomass concentration.
- Mostly 2–4 factors are considered at different levels, however, also 5 and maximum of 7 factors are reported.
- Fractional designs are used to reduce the number of experiments, where typical numbers are around 30 with a maximum of 38 experiments.

DoE for parallel cultivation platforms

In contrast to the sequential design strategies carried out in continuous flow reactors, bioprocess cultivations are often carried out in parallel. The development of DoE protocols for parallel miniaturized and automated platforms, such as 96- or 384-well plates, is discussed in Ref. [54]. According to the authors, the use of DoE in these so-called "constrained" platforms is scarce. A major hurdle is that a DoE approach generates a complex series of settings that lead to cumbersome ensuing lab work and an increased risk of introducing errors when, e.g., pipetting is done manually to individual wells. Automated systems with single tip pipette techniques are too slow, and traditional experimental designs yield non-compatible formats for multi-pipette systems. The authors propose the concept of Rectangular Experimental Designs for Multi-Unit Platforms which adjusts traditional DoE to the constraints of multi-pipettes and fixed plate formats, e.g., to the use of pipetting column and rows in parallel [54].

While experiments for the optimization of *media and mixture composition* are typically conducted using microplates and shaked flasks (e.g., 250-mL Erlenmeyer flasks placed on rotary shakers), experiments for the *optimization of process states and operation* are mostly conducted at the liter scale. In Ref. [5], an illustrative example is given for a multi-fermenter system consisting of 12 1-L vessels equipped with pH, dissolved oxygen, and temperature control. The fermenters were operated as a fed-batch based on a central composite design with 32 cultivations considering three factors, each at three levels: pH, temperature, cultivation time, and specific growth rate controlled by feeding. Response surface modeling was used to predict the expression rate of recombinant antibodies produced by *E. coli*.

2.3.4 Model-based OED for continuous flow reactions

Table 11.3 reviews recent mbOED applications for the development of chemical reaction kinetic models using automated continuous flow platforms. The review is limited to papers in scientific literature covering the years 2010–20.

TABLE 11.3 Applications with experimental data of mbOED applied for continuous flow microreactor platforms.

Reference	Application	Method	Model	Controls/factors	Experiments/ measurements	Algorithm
McMullen And Jensen [55](a); Reizman and Jensen [56] (b)	(a) Diels-Alder reaction of isoprene and maleic anhydride in DMF; (b) nucleophilic aromatic substitution of morpholine onto 2,4-dichloropyrimidine	D-optimal MBDoE for model discrimination and parameter precision, sequential approach	ODE models, up to 5 species, up to 8 parameters (Arrhenius reaction rate parameters for up to four different kinetic rate constants)	Static conditions, up to 4 factors (temperature, initial concentrations of 2 species, residence time)	Online HPLC, up to 4 species measured; up to 24 experiments with 12 MBDoe optimal experiments	MATLAB ODE and optimization solvers; approximate solution of MBDoE by selection from predefined experiments (screening guesses)
Echtermeyer et al. [57]	Pd-catalyzed C—H activation reaction; C(sp3)–H activation reaction to synthesize aziridines	D-optimal MBDoE for parameter precision, design-by-(parameter)-grouping to overcome correlation problems	10 ODE's model, 8 parameters (reaction rate constants and activation energies)	Static conditions, 5 factors (temperature, flow rate, 3 initial reactant and catalyst concentrations), and sampling times (and number of samples)	Online GC for 2 species; 8 experiments to estimate up to 3 parameters in a group; in total 71 samples	gPROMS dynamic simulation and MBDoE and parameter estimation suite

Waldron et al. [25,58]	Esterification between benzoic acid and ethanol with (a) heterogeneous Amberlyst-15 catalyst; (b) homogeneous catalyst (sulfuric acid)	(a) MBDoE for model discrimination and parameter precision; (b) sequential MBDoE for parameter precision	(a) 4 different ODE models with up to 6 parameters (adsorption and reaction rate constants); (b) 2 ODE's model, 2 Arrhenius parameters	Static conditions; (a-b) 3 factors (temperature, flow rate, and feed concentration)	Online HPLC for 2 species; experiment duration less than 7 min; up to 12 experiments with 4 MBDoE optimal experiments	Python's sequential least squares programming (SLSQP) algorithm and ODE solver; approximate solution of MBDoE
Aroh and Jensen [59]	Paal-Knorr cyclocondensation reaction for the synthesis of pyrroles; Paal-Knorr (one-step bimolecular) reaction	Heuristic experiment designs; individual and simultaneous fitting of up to 4 experiments	3 ODE's model; 5 rival kinetic model candidates, each with two Arrhenius parameters and different orders of reaction	Dynamic conditions, temperature ramps (up- and downwards), and residence time ramps (constant rate of changes)	Inline IR spectrometer for one reactant; 20 min experiment generates 150 data points	Simulation and parameter estimation using Matlab
Taylor et al. [60]	Reaction of phenol with acetyl chloride to form phenyl acetate; and paracetamol and consecutively diacetamate	Heuristic experiment designs for model discrimination using corrected Akaike's Information Criterion	ODE models; considering 3 and 63 rival reaction models for different one or two-step reactions; estimate reaction rate constants and activation energies	Dynamic conditions; flow ramp experiments (downwards) at different (constant) temperatures and for two different initial feed concentrations	Online HPLC for three reactants; 90–100 min experiments with sampling intervals between 4 and 6 min	MATLAB ODE solver and interior-point optimization algorithm

Continued

TABLE 11.3 Applications with experimental data of mbOED applied for continuous flow microreactor platforms—cont'd

Reference	Application	Method	Model	Controls/factors	Experiments/ measurements	Algorithm
Waldron et al. [29]	Benzoic acid and ethanol esterification with sulfuric acid as the catalyst	Sequential PE and MBDoE for parameter precision	Same model and 2 parameters as in Waldron et al. [2019b]	Dynamic conditions, simultaneous ramping of flowrate and temperature (downwards, constant rate of change), initial reactant concentration	Online HPLC for 2 reactants, sampling every 7 min	Python's Nelder–Mead simplex and SLSQP algorithm; globalization strategy by Latin Hypercube sampling
Schaber et al. [61]	Reaction of phenylisocyanate with *t*-butanol	MBDoE for model discrimination and parameter precision	1D dynamic (transport diffusion) PDE model; 5 rival kinetic model candidates, each with two Arrhenius parameters	Dynamic conditions; residence times of 1–10 min (flow rate steps and ramps) and temperature ramps between 50 and 130°C, individual flow rates of 3 reactants	Online FT-IR spectroscopy for 2 reactants; 3.5 h dynamic experiment with (continuous) measurements sampled at 377-time points	Jacobian simulation software linked with SNOPT gradient-based optimization solver

The reported activities aim at the *identification of models and model parameters for optimization of reaction conditions and scale-up predictions*. In these works, algorithms for parameter estimation and mbOED are employed to select the best model among a set of given model structures (i.e., model discrimination) [55,58,60] and/or to improve the statistical quality of the parameter estimates for a given model structure [25,55,57,58,61,62].

All models consider low-dimensional reaction networks, typically with Arrhenius-type reaction terms. The most complex model in Ref. [56] considers five species and eight parameters describing four different rate constants. In Ref. [60] the authors discriminate between 63 rival reaction models for relatively simple one or two-step reactions.

The first three applications in Table 11.3 use static conditions where each (steady-state) experiment is defined by 3–5 factors. Keeping temperature and inlet concentrations constant, one steady-state point is established usually after three residence times. The remaining four applications design dynamic conditions for flow and temperature (some using heuristic instead of optimal designs).

The steady-state experiments (between 12 and 71 static conditions) are designed following a sequential approach, similar to the self-optimization experiments in Section 2.3.2. The experiments under dynamic conditions can generate a high number of data points (each equivalent with one steady-state experiment), e.g., 150 points in 20 min [59]. Therefore, already one or two experiments can be sufficient for precise estimation of kinetic parameters, and data points can be planned simultaneously [29]. Moreover, using ramps with a constant slope for the parametrization of dynamic conditions results in a relatively low-dimensional experiment design space which makes the numerical solution of the mbOED problem trackable. For example, considering initial values and constant slopes for temperature and flow together with initial reactant concentrations, one or two simultaneously planned experiments are defined by 5–8 parameters [29]. Ref. [61] evaluates the performance for dynamic conditions by manipulation of individual reactant feed concentrations, see Section 3.1 for a critical discussion of the results.

2.3.5 Model-based OED for bioreactor cultures

Ref. [16] provides an interesting review on the application of mbOED to cellular and biological processes and systems, including signaling pathways, bioreactor cultivations and pharmacokinetics, pharmacodynamics, and clinical applications. The authors prioritize recent applications reported between 2000 and 2013. Nearly all process models are represented by ordinary differential equations (ODEs) with fewer than 5% of the models being differential algebraic equations (DAEs), stochastic, integro-differential equations, or partial differential equations (PDEs) [16]. 19 out of 85 papers are counted in the biomass and bioreactor category where the maximum number of model equations was relatively low not exceeding 10 ODEs. The average number of unknown model parameters seems to be fairly consistent across all application areas and is typically <10. More complex models with up to 103 ODE's are reported for cell signaling pathways. The maximum number of parameters (>300) is found for a mammalian cell culture model. However, these complex models are used primarily to illustrate and motivate advances to mbOED techniques [16].

An interesting experimental application for the identification of catalyzed reactions in a cell signaling model is presented in Ref. [63]. The authors apply an iterative robust mbOED for three consecutive experiments optimizing simple time-dependent (piecewise constant) concentration profiles. The model consists of 4 ODE's and 6 system parameters (plus additional control and calibration parameters) and is calibrated using data from live-cell fluorescence microscopic images analysis in single-cell assays.

Table 11.4 reviews recent (not earlier than 2010) mbOED applications for the identification of growth models using automated bioreactor platforms. The reported activities envision the development of predictive models for the optimization of cultivation conditions in so-called scale-down experiments, where the main purpose is to simulate representative conditions of real large-scale bioprocesses, in order to, e.g., investigate effects of oxygen limitations, excess glucose concentrations, or pH fluctuations [41,64–66]. All applications use shaked flasks or stirred tank parallel cultivation platforms in the mL to L scale. Parallel cultivations are typically carried out in duplicates or triplicates, to account for culture variability, or to decrease the uncertainty in the data. E.g., [67] uses 32 parallel 1 mL reactors in a 48 microtiter plate with online oxygen measurements and atline concentration measurements. The experimental designs are defined by static conditions (usually temperature, feed concentration, and flow rate), dynamic conditions (timing and concentrations of feed pulses), and initial conditions (e.g., initial batch concentrations, volumes), and experimental duration and sampling time points.

Static and initial conditions are usually defined by a low number (<10) of corresponding design variables. Dynamic conditions are realized by high-frequency pulse-based feeding, where each feed pulse can be defined by individual concentrations and/or volumes. Applying these dynamic feeding strategies individually for 4 parallel reactors can result in comparatively large design space with 248 variables [68]. Similarly, considering the optimization of sampling time points for atline measurements can increase the design space considerably, see Ref. [69].

In Table 11.4 mostly ODE models are used with <10 equations and between 10 and 30 parameters, except for the mammalian cell culture model with 28 differential and algebraic equations in Ref. [69].

TABLE 11.4 Applications with experimental data of mbOED applied for automated shaked flasks and mini-bioreactor platforms.

Reference	Application	Platform	Method	Model	Controls, experiment design	Measurements, experiments	Software, algorithm
Schmideder et al. [70]	(Batch and) continuous cultivation of an *E. coli* strain	48 Stirred-tank bioreactors modified for continuous operation, 10 mL scale	Heuristic design, parameter estimation, Monte Carlo Bootstrap analysis for parameter uncertainty quantification	2 Algebraic (steady-state) equations, Monod growth model with 4 parameters	Static conditions, 10 dilution rates	Atline 10 samples (20 data points for biomass and glucose concentrations at steady state); 10 parallel continuous cultivations; duration at least five residence times	MATLAB's lsqnonlin for parameter estimation
Habicher et al. [67]	Continuous fed-batch cultivation of a protease producing *Bacillus licheniformis* culture	48 Microtiter plates, 1 mL scale, with polymer glucose release	Heuristic design, parameter estimation, reproducibility checks for up to 32 parallel cultivations, cross-validation experiments	3 ODE's, Monod-type model with 12 parameters, 2 parameters estimated	Initial conditions, 1 factor (3 levels initial biomass concentration) for parameter estimation; 2 factors (3 levels for filling volume and osmotic pressure); polymer-based controlled-release fed-batch	Online oxygen transfer rate; atline sampling for biomass (OD 600), pH, protease activity, and glucose concentration; 32 parallel cultivations with triplicates	MATLAB's fmincon algorithm with global search function

Continued

TABLE 11.4 Applications with experimental data of mbOED applied for automated shaked flasks and mini-bioreactor platforms—cont'd

Reference	Application	Platform	Method	Model	Controls, experiment design	Measurements, experiments	Software, algorithm
Hernandez Rodriguez et al. [71]	CHO cell culture seed train; batch production on different scales of a therapeutic recombinant protein	Parallel shaked flasks, different scales with 40, 70, 300, and 1500 mL volumes	Heuristic design; Bayesian parameter estimation using Markov chain Monte Carlo method, definition of prior parameter distributions; credible intervals for prediction	7 Monod-type ODE's; 17 parameters (including initial concentration), reduced by 6 based on identiability analysis and expert information	Same static conditions for all seed train cultivations; controlled temperature, CO_2, humidity, seeding VCD; 1 factor (4 volume levels)	Atline sampling, 4–5 measurements per cultivation for viable cell density, viability, concentrations of glucose, glutamine, lactate, ammonia; > 200 data points for 4 parallel cultivations (each duplicated); individual analysis, cultivation time spans of 264 h	Matlab implementation of Bayessian framework

Möller et al. [72]	Batch and fed-batch cultivation of antibody-producing CHO	Parallel shaked flasks, 80mL scale	Heuristic experiment design for parameter precision, model-assisted DoE on a reduced experimental design search space	9 ODE's, Monod-type model with 16 parameters; no identifiability analysis	Static and initial conditions, 3 factors (feed start, feed glutamine concentration, feed rate), 12 factors in total for 4 parallel experiments	Atline sampling (every 12h) with 9 measurements (in total 99 data points); 4 parallel experiments, around 168h cultivation time	MATLAB's Nelder Mead Simplex algorithm; DoE Software (Design-Expert 9, Statcon, USA)
Kontoravdi et al. [69]	Fed-batch cultivation of antibody-producing mammalian cell cultures	Shaked flask, 200mL scale	D-optimal MBDoE for parameter precision; parameter identifiability analysis using global sensitivities; cross-validation	28 Differential and algebraic equations containing 30 parameters; 7 identifiable parameters	Conditions: initial glucose, (constant) feed glucose and glutamine concentrations (3), timing and volumes of 7 feed pulses (14), measurement times (22), experiment duration (1); (static, controlled) temperature and CO2 air concentration; in total ~40 variables	Atline sampling with 5 measurements (in total 110 data points) for nutrient, glutamine, ammonia, viable and total cell, and extracellular MAb concentrations; 168h cultivation time	gPROMS experiment design and parameter estimation suited using RQPD sequential quadratic programming code

Continued

TABLE 11.4 Applications with experimental data of mbOED applied for automated shaked flasks and mini-bioreactor platforms—cont'd

Reference	Application	Platform	Method	Model	Controls, experiment design	Measurements, experiments	Software, algorithm
Cruz Bournazou et al. [68]; Barz et al. [73]	*E. coli* fed-batch process with intermittent feeding	48 Stirred-tank bioreactors, automated liquid handling, 10 mL scale	A-optimal MBDoE, online sequential approach with embedded local parameter set selection	6 ODE's, Monod-type model with 17 parameters, 1 additional (oxygen transfer) parameter for each parallel reactor; in total 23 identifiable parameters	Dynamic conditions: 3 volume portions per feed pulse (enzyme, glucose, acetate), in total 248 decisions for 4 parallel designs each with 18 pulses; static conditions: controlled pH and temperature	Online dissolved oxygen; atline sampling for biomass, acetate, and glucose (in total 432 data points for 8 reactors and 18 samples per cultivation plus continuous oxygen signals); initial design and 3 re-designs	MATLAB, CVODES solver (SundialsTB); TOMLAB's clsSolve and SNOPT implementation for parameter estimation and experimental design

2.3.6 *Model based OED dimension and complexity*

Based on Section 2.3.4 and 2.3.5, it can be said that for continuous-flow systems relatively simple low-dimensional reaction networks with maximum five species are considered. Arrhenius-type reaction equations are used. The complexity of the Monod-type growth models used for batch and fed-batch cultivations is similar. With exception of [61,69] all reported applications use ODE models with less than 10 equations, less than 10 parameters, some growth models use <30 parameters. These numbers fairly coincide with the ones reported for mbOED applications in the literature.[a]

The numbers of factors (or experiment design variables) and experiments are quite similar to numbers and factors considered in DoE applications in Section 2.3.1–2.3.3 and in other engineering fields[b]: The typical number of experiments is well below 100 (highest numbers are reported for self-optimizing platforms), the number of factors is between 2 and 4 (maximum 7).

2.4 Summary

Continuous-flow reactor platforms are usually used after the first screening of important reaction parameters, thus for the optimization of a small number of design variables for the optimization of reaction conditions and scale-up predictions. Flow reactor systems are mostly operated under static experimental conditions or for simple dynamic experimental conditions, e.g., constant flow rate and temperature ramps. They generate a sequence of steady-state flow data points. DoE methods and especially sequential self-optimization and sequential mbOED methods have been successfully applied (see Section 2.3). The relatively low dimension of the models and experiment design space makes the numerical solution of mbOED problems tractable.

In contrast, experiments in parallel miniaturized bioreactor platforms are mostly planed using static DoE's for high throughput in strain screening and media optimization. Important process operation parameters such as time or state-dependent variations on temperature, pH, induction, etc. are rarely considered. Sequential self-optimization algorithms have not been adopted mostly because the cultivation and sample analysis times are significantly larger compared with reactions in synthetic chemistry (hours to days compared with second to minutes), and because these systems almost exclusively

a. Ref. [16] reviews mbOED applications between 2000 and 2013 in the biomass and bioreactor category. Accordingly, the maximum number of model equations was relatively low not exceeding 10 ODE's, the average number of unknown model parameters is typically <10.

b. Ref. [15] reviews 77 practical applications of DoE reported in technical journals between 2001 and 2005 mainly in materials science, in mechanical and chemical engineering and in the industrial sector. It turns out that: In 77% of the cases analyzed the aim of the DoE was reached with a number of experiments less than or equal to 30. The average number of factors is 5.06 with a standard deviation of 2.45. No paper was found in which more than 12 factors have been taken into account.

use batch or fed-batch disposable (single-use) reactors which are currently not automized for the operation with consecutive experiments. Experimental demonstrations exploiting the full potential of mbOED are scarce.

3 Special aspects and challenges

3.1 Static vs dynamic experimental conditions

While the methods for computing dynamic mbOED have been thoroughly studied in theory, they are rarely adopted for practical applications. The challenges related to the design and analysis of dynamic experiments are discussed in the following.

3.1.1 *Continuous-flow reactors*

In continuous flow, the measurements are usually collected at different steady-state experimental conditions, see also Section 2.1.1. However, assuming ideal plug flow behavior (see Section 2.1.1), information might be gathered also from measurements at multiple time intervals during one experiment under dynamic conditions, see Table 11.3 for examples.

In Ref. [61] the authors present data for heuristic and mbOED experiments defined by piecewise constant and piecewise linear functions for the temperature and the feed flow rates of three different reactants, see Fig. 11.6 (left). In contrast to all other contributions in Table 11.3, in Ref. [61] the plug flow assumption is dropped. The authors develop a dynamical model of the PFR represented by a partial differential equation (PDE) system to consider temperature and concentration profiles in time and the axial spatial dimension.

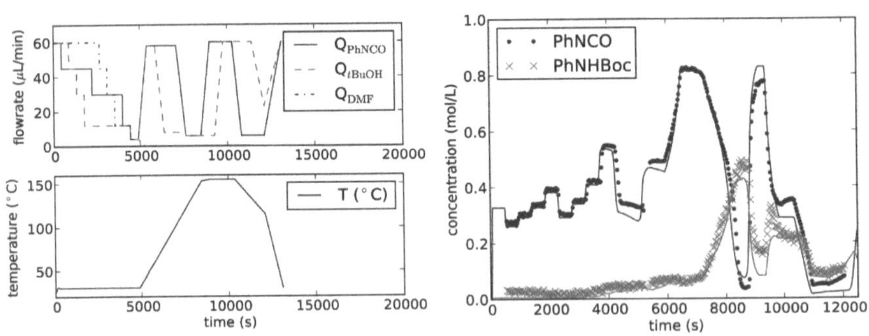

FIG. 11.6 Dynamic experiment design for the identification of a kinetic model using a continuous-flow reactor platform. Left: Experimental conditions are defined by ramps and step changes for the reactor temperature and the individual flow rates of three reactants (and the corresponding residence time). Right: Points show experimental data from IR for two reactants and results from the best-fitting dynamic model. *(From S.D. Schaber, S.C. Born, K.F. Jensen, P.I. Barton, Design, execution, and analysis of time-varying experiments for model discrimination and parameter estimation in microreactors, Org. Process Res. Dev. 18 (2014) 1461–1467.)*

The results indicate imperfect fits, see Fig. 11.6 (right). While these problems can be attributed to neglected reactions they can also result from structural errors in the reactor model regarding the dynamics of heating, and control errors such as oscillations in the flow rates of the reactants delivered by multiple syringe pumps. Therefore, in Ref. [61] the authors recommend additional initial concentration experiments to verify the accuracy of the (ramp) controls, and to verify proper synchronization between experimental conditions and IR data measurement. In addition, the adequacy of the modeling of the fast dynamics of the flow system needs to be carefully checked, especially the modeling of all volume elements, including dead volumes, such as reactor and quench portion, or tubing between quench and IR flow cell [59,61].

3.1.2 Batch- and fed-batch bioreactors

In the linear DoE, every single reactor run delivers a single response obtained from one sample point usually at the end of the batch. Although the previously discussed parallelization and miniaturization of bioreactors enable HT screening and the execution of larger experimental designs there is still the desire to get the most out of every single experiment [74]. The so-called intensified DoE uses hybrid modeling techniques for the analysis of experiments with intra-experimental shifts of process parameters and multiple sample points [75]. Besides the advantage of including several different predefined conditions within one experiment, possible time and irreversible memory effects can potentially hinder the evaluation of such experiments.

For process characterization also dynamically changing conditions can be considered including shifts, pulses, oscillations, etc. [76]. E.g., in Ref. [77] the authors deduced the influence of temperature on growth by a slow temperature change at constant glucose concentration, see Fig. 11.7. An important prerequisite to extract information during these dynamic changes is their inclusion in the response function. As this is not the case in linear regression models such as Multiple Linear Regression (MLR) or Partial Least Squares Regression, elemental balancing, kinetic or hybrid models [78] are needed for evaluation.

3.1.3 Measurement frequency, measurement errors and optimal sampling

Conducting experiments under dynamic conditions can significantly increase the information content in the data. Online and automated atline analytics, such as in-situ spectroscopy as well as liquid handlers connected to enzymatic analyzers or HPLC's are powerful methods for monitoring of (bio-) catalyzed reactions and cultivations [79]. Such intensified setups generate a mix of dense and sparse data from different sources which makes the quantification of systematic and random measurement errors and the weighting of measurements in a regression problem difficult. An interesting approach is the combined modeling of the

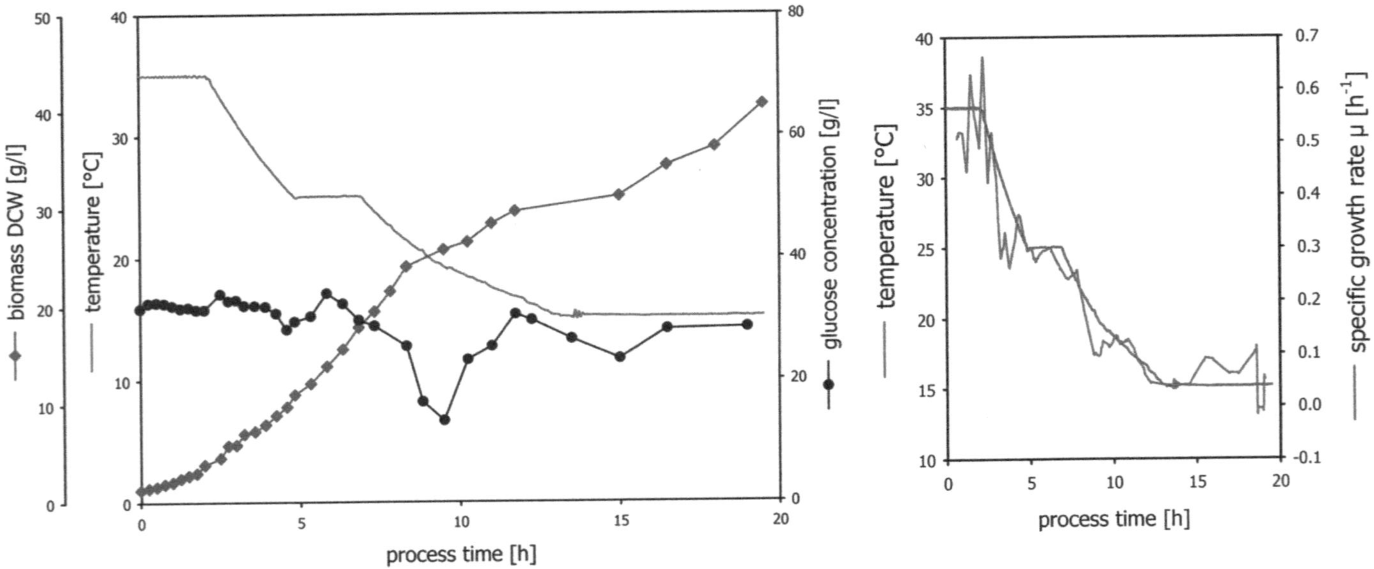

FIG. 11.7 Dynamic investigation of temperature effect on specific growth taken from [77]. Besides highly frequent offline sampling of biomass and glucose (left) the growth rate was estimated by elemental balancing considering online measurements of consumed oxygen and produced CO_2 (right). *(From P. Sagmeister, T. Langemann, P. Wechselberger, A. Meitz, C. Herwig, A dynamic method for the investigation of induced state metabolic capacities as a function of temperature, Microb. Cell Fact. 12 (2013) 1–11.)*

reaction and the data generating process, see Ref. [80] for applications with spectroscopic measurement data.

Because of the small reactor volumes in miniaturized bioreactor platforms, the number of samples is often limited. Therefore optimal sampling time selection or event-triggered sampling are of central importance to capture the process dynamics [81]. Selecting time points where the sensitivity of the measured states with respect to the parameters is high can significantly reduce the uncertainties in parameters obtained from the solution of a regression problem [82].

3.2 Sequential planning and updating in mbOED

The models in DoE are linear in parameter models. These linear models are very powerful for statistical modeling and for the description of static systems that have a fairly convex input–output relationship in the corresponding design space. Unfortunately, this is rarely the case for chemical or biological reaction networks. Here the models are typically represented by systems of nonlinear differential equations. Accordingly, the models and corresponding regression problems are nonlinear in parameters.

For linear (in parameter) models optimal experiment designs can be computed without any knowledge of the real parameter values. In contrast, for nonlinear (in parameter) models, the experimenter faces

> *"…an awkward situation. Designing an efficient experiment requires knowledge of the parameter, but the purpose of the experiment is to generate data to yield parameter estimates!"* [83].

Wrong assumptions or outdated parameter estimates may therefore severely affect the quality of an experimental design. This is the case if parameters differ from the initial (or prior) estimate for which an optimal experiment was designed and realized. To increase the robustness against parametric uncertainties, the mbOED is usually implemented following a sequential approach, see Fig. 11.8.

Sequential mbOED techniques, even termed adaptive experiment design techniques in system identification [85], have been specifically proposed to exploit the progressive increase of the information resulting from a dynamic experiment while the experiment is running [86–88]. Every time a new informative sampling point is acquired from a measurement system, the new information is used to update the model parameters by solution of an online parameter estimation problem. Based on the new parameters the future experimental conditions are then redesigned by solution of an online mbOED problem (Fig. 11.9). The sequential solution of both problems progressively improves the reliability of the model and the robustness of the experiment design.

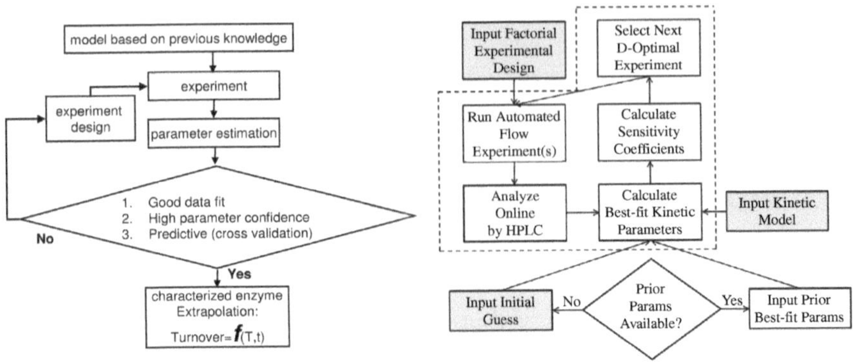

FIG. 11.8 Flowcharts for sequential experiment planning in the context of mbOED for parameter precision. Left: Application in biochemistry for enzyme characterization in an automated parallel platform using 96-well microtiter plates of 200 μL working volume [84]. Right: Application in organic chemistry for automated identification of reaction kinetic parameters in a silicon microreactor with a channel cross-sectional dimension of 500 μm × 400 μm [56]. Both procedures make use of prior information on the kinetic model and its parameters. *(Left: From K. Rachinskiy, H. Schultze, M. Boy, J. Büchs, 'Enzyme Test Bench': a biochemical application of the multi-rate modeling, in: International Workshop on Multi-Rate Processes and Hysteresis, IOP Publishing, 2008, p. 12021; Right: Modified from B.J. Reizman, K.F. Jensen, An automated continuous-flow platform for the estimation of multistep reaction kinetics, Org. Process Res. Dev. 16 (2012) 1770–1782.)*

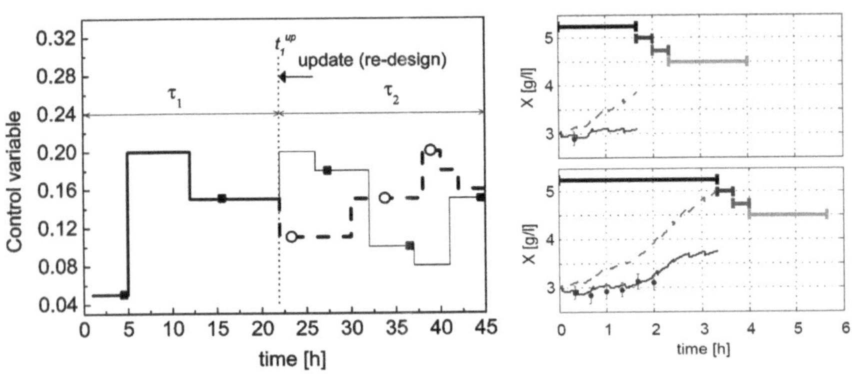

FIG. 11.9 Applications of online model-based re-design. Left: Change from a primitive control trajectory (*solid line*) to an updated control trajectory (*dashed line*) [86]. Right: Evolution of automated robotic cultivations at the example of cell turbidity in one mini-bioreactor in a 6 h experiment. *Red dashed* (*gray* in print version): simulation with current parameter estimates; blue lines: simulation after parameter re-estimation; *blue dots* (*gray* in print version): measured data. The bars represent the horizon of: available online and atline data (*black*); the time required to compute the parameter estimation (*blue*, *gray* in print version); the time for ED optimization (*red*, *gray* in print version); and horizon length of the experimental re-design (*green*, *light gray* in print version) [68]. *(Left: From F. Galvanin, M. Barolo, F. Bezzo, Online model-based redesign of experiments for parameter estimation in dynamic systems, Ind. Eng. Chem. Res. 48 (2009) 4415–4427; Right: From M.N. Cruz Bournazou, T. Barz, D. Nickel, D.C. López Cárdenas, F. Glauche, A. Knepper, P. Neubauer, Online optimal experimental re-design in robotic parallel fed-batch cultivation facilities for validation of macro-kinetic growth models using E. coli as an example, Biotechnol. Bioeng. 114 (2017) 610–619.)*

3.2.1 Robustness issues

The sequential mbOED may still be heavily affected by parameter uncertainties, especially at the beginning of an experiment where measurement information is scarce, and when no prior information about parameters is available (i.e., priors from preliminary screening experiments) [88,89]. To tackle this issue a robust mbOED problem can be formulated. Hereby, instead of considering a given nominal parameter value for optimization, the mbOED is computed for the "worst" values of parameters from confidence regions, see, e.g., [89–91] for further details.

The computation of robust mbOED can also be highly relevant for problems with (state and/or output) constraints, e.g., to ensure consistent drug quality and safe operations. Accounting for the effects of uncertainty results in the formulation of so-called chance constraints, which are satisfied with the desired probability level, see Ref. [92,93] for applications in the context of mbOED.

3.2.2 Termination criteria

Different termination criteria are conceivable. From the practical viewpoint the iterations are stopped when the maximal allowed experimental budget is reached (e.g., time, reactants), or when it is not to be expected that further experimentation leads to significant new information (e.g., the informative design space is explored and mbOED algorithm proposes to repeat input combinations). Other criteria monitor the convergence (measured as the relative change between iterations) of parameter estimates, corresponding 95% confidence intervals, or of the experiment design criterion [55,58].

3.3 Parameter identifiability

The reliability of estimated parameters is frequently affected by problems of structural and practical identifiability [94]. Causes of poor identifiability in nonlinear and linear system identification are [95]: (*i*) over-parameterized process models; (*ii*) model mismatch (i.e., candidate model structures are not adequate to represent the system); (*iii*) limited quantity/quality of experimental data; (*iv*) correlations in parameters and/or experimental data; (*v*) parameters with neglectable contribution on observed variables (i.e., low parameter sensitivity); (*vi*) inappropriate numerical initialization of model parameters. Structural identifiability is a property of the model structure. It assumes perfect identification conditions and is mainly affected by (points i, ii, iv). Practical identifiability tests if the information content in the measurements is enough to find accurate and precise parameter estimates. That means even if a model structure satisfies the requirements for a-priori identifiability [94], it may still be impossible or extremely challenging to precisely estimate its parameters by fitting available noisy experimental data [96].

The rank of the parameter estimation problem is a typical measure for structural identifiability. The condition number indicates the influence of mismeasurement on parameter estimates. For ill-conditioned problems (with a high condition number) the optimization process (numerical solution) amplifies the measurement errors. The solution of ill-conditioned model identification problems may pose significant challenges even to state-of-the-art algorithms [95,97]. Possible issues are inaccurate/unstable solutions, convergence issues, and/or numerical instability during online model identification tasks. These issues may lead to numerical failures in the course of an autonomous experimental campaign, to a model with poor extrapolation properties, and/or high uncertainties in parameter estimates. Improving the robustness of automated model identification platforms toward identifiability is key to further promote their employment in the discovery and study of chemical and biological reaction kinetics.

Potential remedies to non-identifiability are the following [98,99]:

- *Regularization-based (RG) methods*: Regularization involves the introduction of a bias in the parameter estimates with the aim of constraining their variance and consequently reducing the condition number associated with the parameter estimation problem. Potential regularization techniques to be used are: (i) Tikhonov regularization; (ii) truncated singular value decomposition; and iii) parameter subset selection and design-by-grouping method. The evaluation of these techniques in the context of sequential mbOED can be found e.g., in Ref. [57,88,95].
- Reparametrization-based (RP) methods [100]: The aim of reparametrization is to transform the original parameter space into a robust parameter space where mbOED and parameter estimation problems can be solved more effectively on well-conditioned objective functions.

The use of RP approaches for autonomous sequential model identification has been recently proposed as a valid computational tool for transforming the identification and experiment design problems into well-conditioned problems, see Fig. 11.10 for an illustration. The result is an increased stability of the parameter estimation in an RP-based experimental campaign compared to a conventional non-RP campaign. The aforementioned approaches effectively improve the robustness of operations in autonomous microreactor platforms for model identification [58].

3.4 Bayesian statistics

Following the sequential mbOED design approach (see Section 3.1) parameters and corresponding parameter uncertainties are updated (re-estimated) as the results of experiments come in. This update of information makes use of the Bayes theorem which combines prior knowledge (e.g., from last experiments) and evidence (in form of real-world data), leading to posterior (updated)

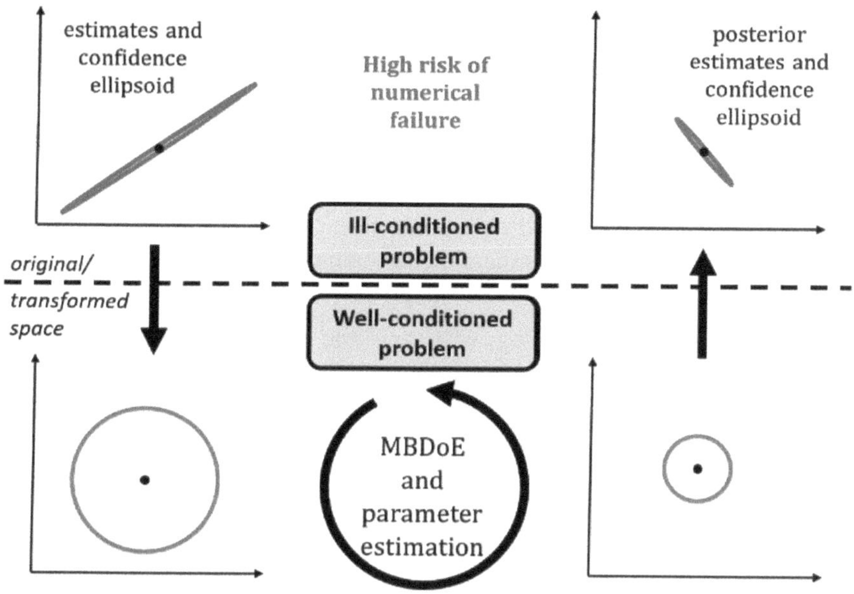

FIG. 11.10 Reparametrization approach for handling non-identifiability: The original ill-conditioned mbOED problem is transformed into a well-conditioned problem. The numerical solution is computed in the transformed "robust" parameter space. Results are analyzed after back-transformation. The effect on the parametric uncertainty is illustrated by confidence regions for two-parameter estimates. The values of these two parameters cannot be uniquely estimated (identifiability problem) in the original parameter space where the parameter confidence ellipsoids are characterized by extremely high eccentricity, see Ref. [62] for details.

knowledge [7]. In the sequential mbOED the amount of information that is gained by the experiment equals the reduction in uncertainty from the prior to the posterior (parameter) distribution. The maximization of this information gain is the purpose of each sequential update [101]. As the prior information is not affected by the re-design, the information gain is in the (new) designed experiments, i.e., in the new data [29,88]. The consideration of prior information in the PE yields the so-called maximum a posteriori (MAP) estimation [56]. Assuming Gaussian priors in MAP estimation corresponds to Tikhonov regularization technique (deviations from the expected value of the prior are penalized by a quadratic cost in the objective function of the PE problem). Thus, sequential design techniques and MAP are also a measure to handle ill-conditioned problems.

Bayesian estimation and experimental design, see Ref. [7], have gained popularity over the last decades, however, applications to actual experiments still lag behind the theory. This is most probably because prior information on the uncertainty of the model parameters is usually difficult to obtain, especially as the design optimization occurs before the data are collected [101]. That is why the prior information is often neglected [4], and optimal designs are based on the likelihood following the Frequentist approach [83].

An interesting experimental application of sequential Bayesian mbOED for parameter identification incorporating global sensitivity analysis is presented in Ref. [102]. Cultivations with yeast are conducted in a benchtop fermenter optimizing dynamic glucose feeding profiles and choosing the best sampling schedule to excite the relevant metabolic pathways. Another experimental application is presented in Ref. [71] where the authors apply Bayesian PE for model identification (see Table 11.4) and discuss the quantification of prior knowledge including uncertainties using expert and literature information, see Fig. 11.11.

3.5 Mathematical modeling, software and algorithms

The term DoE usually refers to a design that minimizes the parameter uncertainty of a linear (in parameters) regression model mostly assuming a linear or quadratic term structure [103]. The bibliographic review in Section 2.3 reveals that most standard full and fractional factorial and mixture designs are used. They are defined by the number of factors, the range of experimental conditions, and/or discrete possible values or "levels" of these factors. When resources are limited or there exist constraints on certain combinations of factor settings, e.g., as reported for parallel automated platforms by Ref. [54], so-called computer-aided designs (such as D-optimal designs) can be applied. Generally, it can be said that as long as the design space is not excessively large and constraint patterns are regular, linear DoE's are computed routinely by practitioners without expert knowledge on numerical optimization within the moderate computational time on standard PCs. Statistical DoE methods are increasingly used in industrial R&D processes and are familiar to most lab leaders [1]. Implementations exist in several proprietary software packages with Graphical User Interface, such as Design Expert, MODDE, DoE Fusion Pro, STAVEX, Minitab, and JMP, but also in numerous open-source libraries for the programming languages R (see, e.g., https://cran.r-project.org/web/views/ExperimentalDesign.html), and in the commercial programming environments SAS and MATLAB.

Section 2.3.2 also reports the application of so-called self-optimization methods for the automated sequential optimization of continuous flow processes. These methods belong to the class of so-called derivative-free optimization algorithms, which do not require knowledge of first or second derivates of the response surface, see Ref. [104] and refs. therein. In Ref. [47] the authors present an interesting comparison of the performance of three alternative algorithms: the Nelder–Mead Simplex algorithms which is a direct local search algorithm and based on the evaluation of objective function values in the reaction parameter space; the Steepest Decent algorithm in combination with a response surface modeling technique to approximate gradients and compute search directions; and the SNOBFIT algorithm [105] which constructs local linear and quadratic surrogate models and also incorporates a global search strategy by sampling unexplored regions of the parameter space. It is noted that all

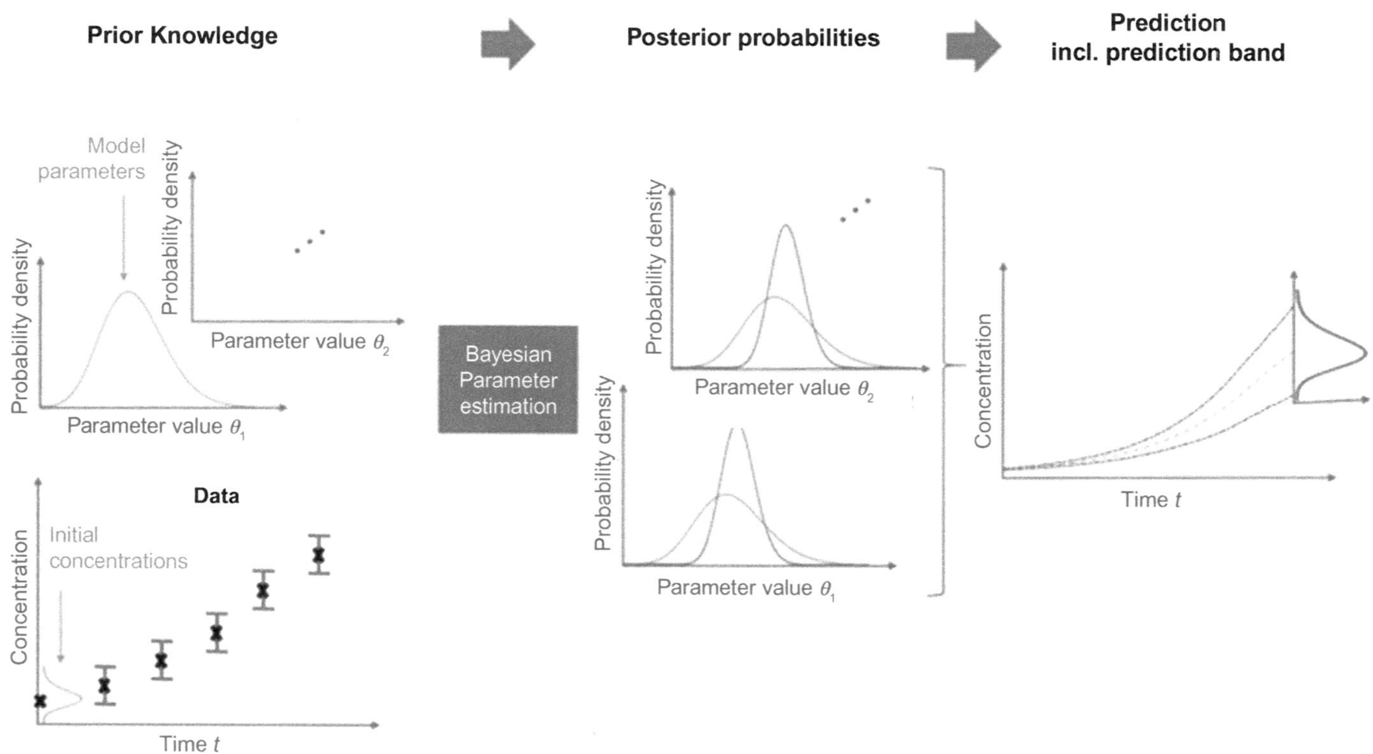

FIG. 11.11 Bayesian framework for model fitting and parameter estimation dealing with uncertainty in measurements and model parameters. The framework is applied to a nonlinear kinetic cell culture model using an MCMC method, see Ref. [71] for details. *(From T. Hernández Rodríguez, C. Posch, J. Schmutzhard, J. Stettner, C. Weihs, R. Pörtner, B. Frahm, Predicting industrial-scale cell culture seed trains—a Bayesian framework for model fitting and parameter estimation, dealing with uncertainty in measurements and model parameters, applied to a nonlinear kinetic cell culture model, using an MCMC method, Biotechnol. Bioeng. 116 (2019) 2944–2959.)*

applications in Table 11.2 use customized optimization algorithms written in MATLAB and are almost exclusively based on the SNOBFIT algorithm, see Fig. 11.12 for an illustration.

In the same way as for linear DoE's, also model-based DoE's are based on a statistical analysis of computed solutions to a regression problem, i.e., the identification of model parameters from noisy experimental data. The optimization criteria used in the mbOED are usually complex nonlinear functions of the unknown parameters and the experiment design variables [90,106]. For dynamic experiments, the design variables can generally be represented by arbitrary (continuous-) time-varying functions, such as piecewise constant, piecewise linear, or sums of basis functions (e.g., flow rates, temperatures), discrete-time variables (e.g., pulse-based feeding, sampling times), and parameters and conditions (e.g., initial concentration).

Compared with a numerical simulation, the numerical solution of nonlinear dynamic mbOED problems is significantly more complex and computationally expensive. This critically depends on the number of experiments, design variables and parameters. Using a high-performance desktop computer, and the state-of-the-art approach for numerical solution, see Ref. [107], the solution for a problem with 270 design variables and 6 parameters can easily take several hours [88]. It is obvious that considering cultivations in parallel platforms these numbers are easily exceeded, see, e.g., [68,69]. To cope with this, corresponding large-scale mbOED problems can be solved only approximately, or their dimension (number of design variables) can be reduced considering limited receding horizons and adopting the online sequential re-design method [68]. Moreover, for small-scale mbOED problems, it is also possible to completely avoid the use of optimization algorithms by a selection of the best experimental design (with the maximum objective function value) from a set of previously defined screening guesses (e.g., generated by a discretization of the experimental space) [55,56].

Most applications in Section 2.3.2 use customized codes in MATLAB and Python for solving the differential equation models (simulation problem) which are then linked either to nonlinear least-squares fitting algorithms or general (gradient-based or derivative-free) nonlinear programming algorithms (parameter estimation problem). Fewer applications use dedicated commercial simulation software which includes procedures for estimating (kinetic) parameters, such as DynoChem (Scale-up Systems Ltd., Ireland) in Ref. [27], Jacobian (RES Group, Inc., Needham, MA, US) in Ref. [61]. It is noted that in Section 2.3.2 gPROMS is the only commercial software that directly supports mbOED allowing for the consideration of all relevant experimental conditions, see Ref. [57,69].

In the review on mbOED within the cellular process related literature [16], the authors identify seven (free and commercial) software programs that directly perform mbOED, and six programs that support the development of mbOED algorithms by built-in tools for simulation, parameter estimation, sensitivity analysis, and general optimization algorithms.

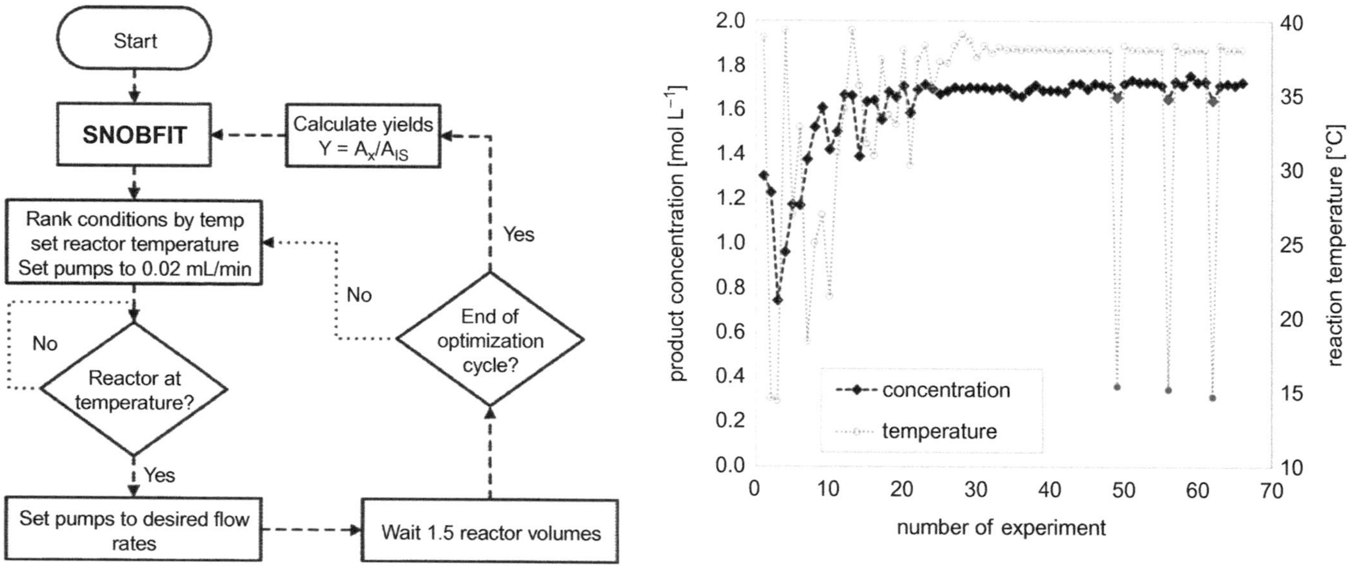

FIG. 11.12 Left: Self-optimization algorithm applied for the maximation of the reaction yield in an autonomous continuous-flow reactor system. The SNOBFIT algorithm automatically adjusts the flow rate (residence time and feed concentrations) and reaction temperature. Yields are calculated at steady-state from online HPLC. Right: Real-time response of Simplex algorithm with analysis of the robustness against malfunction of equipment (breakdown of temperature control marked by *red points* (*gray* in print version)). *(Left: From N. Holmes, G.R. Akien, A.J. Blacker, R.L. Woodward, R.E. Meadows, R.A. Bourne, Self-optimisation of the final stage in the synthesis of EGFR kinase inhibitor AZD9291 using an automated flow reactor, React. Chem. Eng. 1 (2016) 366–371; Right: From V. Fath, N. Kockmann, J. Otto, T. Röder, Self-optimising processes and real-time-optimisation of organic syntheses in a microreactor system using Nelder–Mead and design of experiments, React. Chem. Eng. 5 (2020) 1281–1299.)*

In the academic context several tools and algorithms have been developed: (i) for kinetic parameter identification and uncertainty quantification, e.g., with focus on the estimation of kinetic parameters from datasets including spectra (see Ref. [80] for a recent review of available tools); (ii) for the quantification of rate constants in cell signaling dynamic models (see Ref. [108] or [109] for a general overview on software packages suitable for bioprocess modeling).

From the programs that directly perform mbOED only gPROMS, DAKOTA and R seem to be under continuous development, see Ref. [110] for a list of relevant packages in R for DoE for nonlinear models (https://cran.r-project.org/web/packages/ICAOD/index.html). Other interesting packages that incorporate mbOED not cited in Ref. [16] are VPLAN [90], EasyFit [106], GPdoemd [111], and Chemasim [112].

4 Industry view

In the pharmaceutical industry, especially for parallel bioreactor platforms for scale-down model development and upstream process characterization, the necessity of mbOED tools is becoming evident in view of the fact that expensive laboratories are:

(i) on hold despite full automation due to a lacking experimental plan;
(ii) generating expensive data with low information content, and
(iii) having a large failure rate in the experiments.

A clear example of the increasing awareness in the industry is the initiative from Merck as part of the "always curious" program [113]. Within this program, the software is being developed to enable the optimal design of experiments and operation of an ambr250 parallel reactor platform with perfusion modules. User-friendliness and operation by non-experts in mbOED are claimed to be the main challenges.

Researchers at BASF report the application of mbOED methods to support screening and characterization of the long-term stability of enzymes for biocatalysis [84,114,115]. Experimental analysis is carried out for 30 candidate enzymes identified by a primary screening of protein libraries. Accelerated lifetime experimental tests were carried out on a so-called "Enzyme Test Bench" in 96-well microtiter plates (200 μL working volume) applying short-term experiments (several hours duration) under partly extreme conditions. The platform (similar to the platforms in Section 2.2) allows for fully automated parallel experiments with online reaction monitoring by fluorescent measurements, e.g., hydrolysis reactions by pH-shift and oxidation reactions by DOT, mixing and aeration as well as temperature control. In the first step, the 30 candidate enzymes were tested in parallel (3 replicates) using heuristic experimental designs. Second, three promising candidates were selected and individually characterized using mbOED method and kinetic models for thermal deactivation to optimize *process state variables and process operation* (compare with

Section 2.3). Three sequential optimal experiments were planned each with three or four consecutive temperature ramps and different initial reactant concentrations, compare with Section 3.1. The total effort is estimated at 115 h, of which more than 50% (60 h) are assigned to the modeling effort. The authors state that doing the same experiments in a standard (non-optimal) setting at a liter scale, the effort would be prohibitively high with an estimated effort of 11,100 h. This is mostly because of less parallelization in the first screening step; largely increased efforts for manual experiment preparation and conduction; and non-optimal experimental conditions requiring larger experiment run times. This case study nicely underlines that mbOED techniques not only reduce experimental efforts, they also support a systematic and consistent characterization [116].

4.1 *mbOED* software, flexibility, usability, and required expert knowledge

Ready-made mbOED software for the routine application in industrial screening does not exist, see also Section 3.5. This is because mbOED software needs to offer high flexibility in the choice of algorithm settings to be adapted to a specific use case. While commercial software packages such as gPROMS are used by many practitioners for their appealing user interface, easily implemented solution strategies, and customer service, these software packages represent a compromise between user-friendliness, performance, and/or flexibility [80,116].

The design and development of mbOED software packages (such as VPLAN and PARFIT) need a continuous and close collaboration between industry and academia, see e.g., [116]. Still, these packages have many parts of "expert knowledge" to be used efficiently and they require a high level of numerical understanding and special training. Some software solutions aiming to find a trade-off between the added value of model-based tools and user-friendliness and flexibility are appearing in the market (e.g., process data management and analytics software Werum inCyght from Körber Group, Hamburg, Germany; or DataHowlab from DATAHOW AG, Zürich, Switzerland). They combine mechanistic models that are general to a family of processes and data-driven models to fit specific applications.

Typical commercial automation systems are cheap and excellent for simple or complex narrowly defined tasks (controlling an HPLC system with multiple detectors). While the re-configuration and re-connection of the hardware of (miniaturized) automated systems to perform a different process typically is a matter of hours or days, the adaptation of automation software for control and monitoring requires significantly more time [48]. Therefore, a critical success factor of integrated software solutions for (model-based) process data management and analytics is their flexibility and (in view of the autonomous operation) their high reliability and stability.

5 Discussion and conclusions

Methods for designing optimal experiments replace intuitive parts in the experimental characterization with numerical procedures. The results are not only better solutions, but the systematic, standardized, and automated processing of results also increases consistency, and finally throughput. To date, purely data-driven methods (such as classical DoE methods) have failed to cope with the scarcity of data and the complexity of biochemical reaction systems, demonstrating that mechanistic modeling, engineering expertise, and advanced mbOED techniques are needed.

As already noted in 2013 in Ref. [16]: While the methods for computing mbOED have been thoroughly studied in theory, practical applications that demonstrate the added value are scarce. Regarding the integration of mbOED method in automated miniaturized platforms toward an autonomous operation it can be said that applications for continuous-flow platforms considering small numbers (3–5) of static or simple dynamic experimental conditions, and small reaction kinetic models (up to five species) mark the state-of-the-art.

Continuous-flow platforms are designed to generate a sequence of (quasi) steady-state flow data points and are therefore perfectly suited for the fully automated sequential planning, identification, and optimization of reactions. For batch and fed-batch reactor platforms, a mix of static and dynamic conditions including the timing of operational actions needs to be considered, which easily leads to a 40-dimensional design space for one reactor, parallelization further increases the dimension and complexity of related mbOED problems. The characterization of biochemical reactions under dynamic conditions can bring a significant information gain, however, at a high price: It means significantly increased efforts for the accurate realization and control of dynamic conditions, for the verification of proper synchronization between all actors, sensors, and the database, and for the accurate modeling of the dynamic response of the platform, considering, e.g., all volume elements, delays, sensor responses, etc.

And there exist more challenges related to the efficient (approximate) solution of related problems, the convergence of adaptive elements in sequential frameworks, and the explicit consideration of parametric uncertainties. Before they can be widely adopted in industry, the numerical programs need to be integrated into usable and flexible software tools for automation, data processing, and management. The application range and flexibility of the software might be widened by combining mechanistic and data-driven modeling techniques.

While these rather technical challenges might be solved with sufficient time and resources, there are certain tasks in the workflow for mbOED which are difficult or even impossible to formalize, to translate into an algorithm and to code, and therefore hard to fully automate. These tasks are linked to expert knowledge, such as the definition of limitations and constraints of experimental conditions, and iterations between model selection, model reduction, and model fitting, or the evaluation of results under multiple criteria. Thus, companies willing to implement mbOED tools need to carry out significant internal changes including (i) acquisition of the required know-how and experts, (ii) major changes in the data handling methods, accompanied by all its related risks, (iii) testing, validating, and implementing the novel tools, and finally (iv) assuring sufficient and skilled personnel to maintain and adapt the tools to future challenges.

References

[1] S.A. Weissman, N.G. Anderson, Design of experiments (DoE) and process optimization. A review of recent publications, Org. Process Res. Dev. 19 (2015) 1605–1633.

[2] P.M. Murray, F. Bellany, L. Benhamou, D.-K. Bučar, A.B. Tabor, T.D. Sheppard, The application of design of experiments (DoE) reaction optimisation and solvent selection in the development of new synthetic chemistry, Org. Biomol. Chem. 14 (2016) 2373–2384.

[3] A. Gioiello, V. Mancino, P. Filipponi, S. Mostarda, B. Cerra, Concepts and optimization strategies of experimental design in continuous-flow processing, J. Flow Chem. 6 (2016) 167–180.

[4] G. Franceschini, S. Macchietto, Model-based design of experiments for parameter precision: state of the art, Chem. Eng. Sci. 63 (2008) 4846–4872, https://doi.org/10.1016/j.ces.2007.11.034.

[5] C.-.F. Mandenius, A. Brundin, Bioprocess optimization using design-of-experiments methodology, Biotechnol. Prog. 24 (2008) 1191–1203.

[6] E. Walter, L. Pronzato, Qualitative and quantitative experiment design for phenomenological models—a survey, Automatica 26 (1990) 195–213, https://doi.org/10.1016/0005-1098(90)90116-Y.

[7] K. Chaloner, I. Verdinelli, Bayesian experimental design: a review, Stat. Sci. 10 (3) (1995) 273–304.

[8] A.F. Emery, A.V. Nenarokomov, Optimal experiment design, Meas. Sci. Technol. 9 (1998) 864.

[9] I. Ford, D.M. Titterington, C.P. Kitsos, Recent advances in nonlinear experimental design, Technometrics 31 (1989) 49, https://doi.org/10.1080/00401706.1989.10488475.

[10] D.M. Steinberg, W.G. Hunter, Experimental design: review and comment, Technometrics 26 (1984) 71–97, https://doi.org/10.1080/00401706.1984.10487928.

[11] A. Attia, A. Alexanderian, A.K. Saibaba, Goal-oriented optimal design of experiments for large-scale Bayesian linear inverse problems, Inverse Prob. 34 (2018) 95009.

[12] F. Jost, M. Kudruss, S. Körkel, S.F. Walter, A computational method for key-performance-indicator-based parameter identification of industrial manipulators, J. Math. Ind. 7 (2017) 1–25, https://doi.org/10.1186/s13362-017-0039-7.

[13] A. O'Hagan, J.F.C. Kingman, Curve fitting and optimal design for prediction, J. R. Stat. Soc. B. Methodol. 40 (1978) 1–42.

[14] B. Houska, D. Telen, F. Logist, M. Diehl, J.F.M. van Impe, An economic objective for the optimal experiment design of nonlinear dynamic processes, Automatica 51 (2015) 98–103.

[15] L. Ilzarbe, M.J. Álvarez, E. Viles, M. Tanco, Practical applications of design of experiments in the field of engineering: a bibliographical review, Qual. Reliab. Eng. Int. 24 (2008) 417–428.

[16] A. Chakrabarty, G.T. Buzzard, A.E. Rundell, Model-based design of experiments for cellular processes, Wiley Interdiscip. Rev. Syst. Biol. Med. 5 (2013) 181–203.

[17] T.L. Bailey, A. Nieto, P.H. McDonald, A nonradioactive high-throughput screening-compatible cell-based assay to identify inhibitors of the monocarboxylate transporter protein 1, Assay Drug Dev. Technol. 17 (2019) 275–284.
[18] M. Dörr, M.P.C. Fibinger, D. Last, S. Schmidt, J. Santos-Aberturas, D. Böttcher, A. Hummel, C. Vickers, M. Voss, U.T. Bornscheuer, Fully automatized high-throughput enzyme library screening using a robotic platform, Biotechnol. Bioeng. 113 (2016) 1421–1432.
[19] K. Culbreath, H. Piwonka, J. Korver, M. Noorbakhsh, Benefits derived from full laboratory automation in microbiology: a tale of four laboratories, J. Clin. Microbiol. 59 (2021), e01969-20.
[20] T.C. Malig, J.D.B. Koenig, H. Situ, N.K. Chehal, P.G. Hultin, J.E. Hein, Real-time HPLC-MS reaction progress monitoring using an automated analytical platform, React. Chem. Eng. 2 (2017) 309–314.
[21] B. Walsh, J.R. Hyde, P. Licence, M. Poliakoff, The automation of continuous reactions in supercritical CO 2: the acid-catalysed etherification of short chain alcohols, Green Chem. 7 (2005) 456–463.
[22] N. Holmes, G.R. Akien, R.J.D. Savage, C. Stanetty, I.R. Baxendale, A.J. Blacker, B.A. Taylor, R.L. Woodward, R.E. Meadows, R.A. Bourne, Online quantitative mass spectrometry for the rapid adaptive optimisation of automated flow reactors, React. Chem. Eng. 1 (2016) 96–100.
[23] S. Schwolow, F. Braun, M. Rädle, N. Kockmann, T. Röder, Fast and efficient acquisition of kinetic data in microreactors using in-line Raman analysis, Org. Process Res. Dev. 19 (2015) 1286–1292.
[24] A.-C. Bédard, A. Adamo, K.C. Aroh, M.G. Russell, A.A. Bedermann, J. Torosian, B. Yue, K. F. Jensen, T.F. Jamison, Reconfigurable system for automated optimization of diverse chemical reactions, Science 361 (2018) 1220–1225.
[25] C. Waldron, A. Pankajakshan, M. Quaglio, E. Cao, F. Galvanin, A. Gavriilidis, An autonomous microreactor platform for the rapid identification of kinetic models, React. Chem. Eng. 4 (2019) 1623–1636.
[26] C.A. Hone, N. Holmes, G.R. Akien, R.A. Bourne, F.L. Muller, Rapid multistep kinetic model generation from transient flow data, React. Chem. Eng. 2 (2017) 103–108.
[27] V. Fath, P. Lau, C. Greve, N. Kockmann, T. Röder, Efficient kinetic data acquisition and model prediction: continuous flow microreactors, inline Fourier transform infrared spectroscopy, and self-modeling curve resolution, Org. Process Res. Dev. 24 (2020) 1955–1968.
[28] J.S. Moore, K.F. Jensen, "Batch" kinetics in flow: online IR analysis and continuous control, Angew. Chem. 126 (2014) 480–483.
[29] C. Waldron, A. Pankajakshan, M. Quaglio, E. Cao, F. Galvanin, A. Gavriilidis, Model-based design of transient flow experiments for the identification of kinetic parameters, React. Chem. Eng. 5 (2020) 112–123.
[30] F. Kong, L. Yuan, Y.F. Zheng, W. Chen, Automatic liquid handling for life science: a critical review of the current state of the art, J. Lab. Autom. 17 (2012) 169–185.
[31] A. DePalma, Scaledown keeps processes on the up and up, Genet. Eng. Biotechnol. News 37 (2017) 1–18.
[32] R. Huber, D. Ritter, T. Hering, A.-K. Hillmer, F. Kensy, C. Müller, L. Wang, J. Büchs, Robolector–a novel platform for automated high-throughput cultivations in microtiter plates with high information content, Microb. Cell Fact. 8 (2009) 42.
[33] J. Hemmerich, N. Tenhaef, C. Steffens, J. Kappelmann, M. Weiske, S.J. Reich, W. Wiechert, M. Oldiges, S. Noack, Less sacrifice, more insight: repeated low-volume sampling of microbioreactor cultivations enables accelerated deep phenotyping of microbial strain libraries, Biotechnol. J. 14 (2019) 1800428.

[34] A. Amanullah, J.M. Otero, M. Mikola, A. Hsu, J. Zhang, J. Aunins, H.B. Schreyer, J.A. Hope, A.P. Russo, Novel micro-bioreactor high throughput technology for cell culture process development: reproducibility and scalability assessment of fed-batch CHO cultures, Biotechnol. Bioeng. 106 (2010) 57–67.
[35] G. Faust, N.H. Janzen, C. Bendig, L. Römer, K. Kaufmann, D. Weuster-Botz, Feeding strategies enhance high cell density cultivation and protein expression in milliliter scale bioreactors, Biotechnol. J. 9 (2014) 1293–1303.
[36] N.H. Janzen, G. Striedner, J. Jarmer, M. Voigtmann, S. Abad, D. Reinisch, Implementation of a fully automated microbial cultivation platform for strain and process screening, Biotechnol. J. 14 (2019) 1800625.
[37] B. Haby, S. Hans, E. Anane, A. Sawatzki, N. Krausch, P. Neubauer, M.N. Cruz Bournazou, Integrated robotic mini bioreactor platform for automated, parallel microbial cultivation with online data handling and process control, SLAS Technol. 24 (2019) 569–582.
[38] V. Janakiraman, C. Kwiatkowski, R. Kshirsagar, T. Ryll, Y.-.M. Huang, Application of high-throughput mini-bioreactor system for systematic scale-down modeling, process characterization, and control strategy development, Biotechnol. Prog. 31 (2015) 1623–1632.
[39] M. Tai, A. Ly, I. Leung, G. Nayar, Efficient high-throughput biological process characterization: definitive screening design with the Ambr250 bioreactor system, Biotechnol. Prog. 31 (2015) 1388–1395.
[40] J.J. Kager, J. Fricke, U. Becken, C. Herwig, A generic biomass soft sensor and its application in bioprocess development, Eppendorf Apple Note No. 357 (2017) 206–213.
[41] V. Sandner, L.P. Pybus, G. McCreath, J. Glassey, Scale-down model development in ambr systems: an industrial perspective, Biotechnol. J. 14 (2019) 1700766.
[42] A. Hofer, S. Hauer, P. Kroll, J. Fricke, C. Herwig, In-depth characterization of the raw material corn steep liquor and its bioavailability in bioprocesses of Penicillium chrysogenum, Process Biochem. 70 (2018) 20–28.
[43] O.W. Gooding, Process optimization using combinatorial design principles: parallel synthesis and design of experiment methods, Curr. Opin. Chem. Biol. 8 (2004) 297–304.
[44] B. McKay, M. Hoogenraad, E.W. Damen, A.A. Smith, Advances in multivariate analysis in pharmaceutical process development, Curr. Opin. Drug Discov. Devel. 6 (2003) 966–977.
[45] D.C. Fabry, E. Sugiono, M. Rueping, Self-optimizing reactor systems: algorithms, on-line analytics, setups, and strategies for accelerating continuous flow process optimization, Isr. J. Chem. 54 (2014) 341–350.
[46] N. Holmes, G.R. Akien, A.J. Blacker, R.L. Woodward, R.E. Meadows, R.A. Bourne, Self-optimisation of the final stage in the synthesis of EGFR kinase inhibitor AZD9291 using an automated flow reactor, React. Chem. Eng. 1 (2016) 366–371.
[47] J.P. McMullen, K.F. Jensen, An automated microfluidic system for online optimization in chemical synthesis, Org. Process Res. Dev. 5 (2010) 1169–1176.
[48] N. Cherkasov, Y. Bai, A.J. Expósito, E.V. Rebrov, OpenFlowChem–a platform for quick, robust and flexible automation and self-optimisation of flow chemistry, React. Chem. Eng. 3 (2018) 769–780.
[49] A.J. Parrott, R.A. Bourne, G.R. Akien, D.J. Irvine, M. Poliakoff, Self-optimizing continuous reactions in supercritical carbon dioxide, Angew. Chem. Int. Ed. 50 (2011) 3788–3792.
[50] D. Cortés-Borda, E. Wimmer, B. Gouilleux, E. Barré, N. Oger, L. Goulamaly, L. Peault, B. Charrier, C. Truchet, P. Giraudeau, An autonomous self-optimizing flow reactor for the synthesis of natural product carpanone, J. Org. Chem. 83 (2018) 14286–14299.
[51] A.D. Clayton, J.A. Manson, C.J. Taylor, T.W. Chamberlain, B.A. Taylor, G. Clemens, R.A. Bourne, Algorithms for the self-optimisation of chemical reactions, React. Chem. Eng. 4 (2019) 1545–1554.

[52] A.M. Schweidtmann, A.D. Clayton, N. Holmes, E. Bradford, R.A. Bourne, A.A. Lapkin, Machine learning meets continuous flow chemistry: automated optimization toward the Pareto front of multiple objectives, Chem. Eng. J. 352 (2018) 277–282.
[53] V. Fath, N. Kockmann, J. Otto, T. Röder, Self-optimising processes and real-time-optimisation of organic syntheses in a microreactor system using Nelder–Mead and design of experiments, React. Chem. Eng. 5 (2020) 1281–1299.
[54] M. Olsson, E. Johansson, M. Berntsson, L. Eriksson, J. Gottfries, S. Wold, Rational DOE protocols for 96-well plates, Chemom. Intel. Lab. Syst. 83 (2006) 66–74.
[55] J.P. McMullen, K.F. Jensen, Rapid determination of reaction kinetics with an automated microfluidic system, Org. Process Res. Dev. 15 (2011) 398–407.
[56] B.J. Reizman, K.F. Jensen, An automated continuous-flow platform for the estimation of multistep reaction kinetics, Org. Process Res. Dev. 16 (2012) 1770–1782.
[57] A. Echtermeyer, Y. Amar, J. Zakrzewski, A. Lapkin, Self-optimisation and model-based design of experiments for developing a C–H activation flow process, Beilstein J. Org. Chem. 13 (2017) 150–163.
[58] C. Waldron, A. Pankajakshan, M. Quaglio, E. Cao, F. Galvanin, A. Gavriilidis, Closed-loop model-based design of experiments for kinetic model discrimination and parameter estimation: benzoic acid esterification on a heterogeneous catalyst, Ind. Eng. Chem. Res. 58 (2019) 22165–22177.
[59] K.C. Aroh, K.F. Jensen, Efficient kinetic experiments in continuous flow microreactors, react, Chem. Eng. 3 (2018) 94–101.
[60] C.J. Taylor, M. Booth, J.A. Manson, M.J. Willis, G. Clemens, B.A. Taylor, T.W. Chamberlain, R.A. Bourne, Rapid, automated determination of reaction models and kinetic parameters, Chem. Eng. J. 413 (2020) 127017.
[61] S.D. Schaber, S.C. Born, K.F. Jensen, P.I. Barton, Design, execution, and analysis of time-varying experiments for model discrimination and parameter estimation in microreactors, Org. Process Res. Dev. 18 (2014) 1461–1467.
[62] M. Quaglio, C. Waldron, A. Pankajakshan, E. Cao, A. Gavriilidis, E.S. Fraga, F. Galvanin, An online reparametrisation approach for robust parameter estimation in automated model identification platforms, Comput. Chem. Eng. 124 (2019) 270–284.
[63] S. Bandara, J.P. Schlöder, R. Eils, H.G. Bock, T. Meyer, Optimal experimental design for parameter estimation of a cell signaling model, PLoS Comput. Biol. 5 (2009), e1000558.
[64] P. Neubauer, E. Anane, S. Junne, M.N. Cruz Bournazou, Potential of integrating model-based design of experiments approaches and process analytical technologies for bioprocess scale-down, in: Advances in Biochemical Engineering/Biotechnology, Springer Berlin Heidelberg, Berlin, Heidelberg, 2020, pp. 1–28.
[65] H. Noorman, An industrial perspective on bioreactor scale-down: what we can learn from combined large-scale bioprocess and model fluid studies, Biotechnol. J. 6 (2011) 934–943.
[66] F.R. Schmidt, Optimization and scale up of industrial fermentation processes, Appl. Microbiol. Biotechnol. 68 (2005) 425–435.
[67] T. Habicher, E.K.A. Rauls, F. Egidi, T. Keil, T. Klein, A. Daub, J. Büchs, Establishing a fed-batch process for protease expression with *Bacillus licheniformis* in polymer-based controlled-release microtiter plates, Biotechnol. J. 15 (2020) 1900088.
[68] M.N. Cruz Bournazou, T. Barz, D. Nickel, D.C. López Cárdenas, F. Glauche, A. Knepper, P. Neubauer, Online optimal experimental re-design in robotic parallel fed-batch cultivation facilities for validation of macro-kinetic growth models using *E. coli* as an example, Biotechnol. Bioeng. 114 (2017) 610–619, https://doi.org/10.1002/bit.26192.
[69] C. Kontoravdi, E.N. Pistikopoulos, A. Mantalaris, Systematic development of predictive mathematical models for animal cell cultures, Comput. Chem. Eng. 34 (2010) 1192–1198.

[70] A. Schmideder, T.S. Severin, J.H. Cremer, D. Weuster-Botz, A novel milliliter-scale chemostat system for parallel cultivation of microorganisms in stirred-tank bioreactors, J. Biotechnol. 210 (2015) 19–24.
[71] T. Hernández Rodríguez, C. Posch, J. Schmutzhard, J. Stettner, C. Weihs, R. Pörtner, B. Frahm, Predicting industrial-scale cell culture seed trains—a Bayesian framework for model fitting and parameter estimation, dealing with uncertainty in measurements and model parameters, applied to a nonlinear kinetic cell culture model, using an MCMC method, Biotechnol. Bioeng. 116 (2019) 2944–2959.
[72] J. Möller, K.B. Kuchemüller, T. Steinmetz, K.S. Koopmann, R. Pörtner, Model-assisted design of experiments as a concept for knowledge-based bioprocess development, Bioprocess Biosyst. Eng. 42 (2019) 867–882.
[73] T. Barz, A. Sommer, T. Wilms, P. Neubauer, M.N.C. Bournazou, Adaptive optimal operation of a parallel robotic liquid handling station, IFAC-PapersOnLine 51 (2018) 765–770.
[74] R. Muñoz-Tamayo, P. Martinon, G. Bougaran, F. Mairet, O. Bernard, Getting the most out of it: optimal experiments for parameter estimation of microalgae growth models, J. Process Control 24 (2014) 991–1001.
[75] B. Bayer, G. Striedner, M. Duerkop, Hybrid modeling and intensified DoE: an approach to accelerate upstream process characterization, Biotechnol. J. 15 (2020) 2000121.
[76] O. Spadiut, S. Rittmann, C. Dietzsch, C. Herwig, Dynamic process conditions in bioprocess development, Eng. Life Sci. 13 (2013) 88–101.
[77] P. Sagmeister, T. Langemann, P. Wechselberger, A. Meitz, C. Herwig, A dynamic method for the investigation of induced state metabolic capacities as a function of temperature, Microb. Cell Fact. 12 (2013) 1–11.
[78] D. Solle, B. Hitzmann, C. Herwig, M. Pereira Remelhe, S. Ulonska, L. Wuerth, A. Prata, T. Steckenreiter, Between the poles of data-driven and mechanistic modeling for process operation, Chem. Ing. Tech. 89 (2017) 542–561.
[79] S. Kara, J.J. Mueller, A. Liese, Online analysis methods for monitoring of bioprocesses, Chim. Oggi 29 (2011) 38–41.
[80] C. Schenk, M. Short, J.S. Rodriguez, D. Thierry, L.T. Biegler, S. García-Muñoz, W. Chen, Introducing KIPET: a novel open-source software package for kinetic parameter estimation from experimental datasets including spectra, Comput. Chem. Eng. 134 (2020), 106716.
[81] D. De Pauw, P. Vanrolleghem, Nesting genetic algorithms to solve a robust optimal experimental design problem, in: 3rd Biennial Meeting of the International Environmental Modelling and Software Society, 29, International Environmental Modelling and Software Society (iEMSs), 2006.
[82] Z. Kutalik, K.-H. Cho, O. Wolkenhauer, Optimal sampling time selection for parameter estimation in dynamic pathway modeling, Biosystems 75 (2004) 43–55.
[83] P. Chaudhuri, P.A. Mykland, Nonlinear experiments: optimal design and inference based on likelihood, J. Am. Stat. Assoc. 88 (1993) 538–546.
[84] K. Rachinskiy, H. Schultze, M. Boy, J. Büchs, 'Enzyme Test Bench': a biochemical application of the multi-rate modeling, in: International Workshop on Multi-Rate Processes and Hysteresis, IOP Publishing, 2008, p. 12021.
[85] H. Hjalmarsson, From experiment design to closed-loop control, Automatica 41 (2005) 393–438.
[86] F. Galvanin, M. Barolo, F. Bezzo, Online model-based redesign of experiments for parameter estimation in dynamic systems, Ind. Eng. Chem. Res. 48 (2009) 4415–4427, https://doi.org/10.1021/ie8018356.
[87] J.D. Stigter, D. Vries, K.J. Keesman, On adaptive optimal input design: a bioreactor case study, AICHE J. 52 (2006) 3290–3296.

[88] T. Barz, D.C. López C., M.N.C. Bournazou, S. Körkel, S.F. Walter, Real-time adaptive input design for the determination of competitive adsorption isotherms in liquid chromatography, Comput. Chem. Eng. 94 (2016) 104–116.
[89] R. De-Luca, F. Galvanin, F. Bezzo, A methodology for direct exploitation of available information in the online model-based redesign of experiments, Comput. Chem. Eng. 91 (2016) 195–205.
[90] S. Körkel, E. Kostina, H.G. Bock, J.P. Schlöder, Numerical methods for optimal control problems in design of robust optimal experiments for nonlinear dynamic processes, Optim. Methods and Softw. 19 (2004) 327–338.
[91] C.R. Rojas, J.S. Welsh, G.C. Goodwin, A. Feuer, Robust optimal experiment design for system identification, Automatica 6 (2007) 993–1008.
[92] F. Galvanin, M. Barolo, F. Bezzo, S. Macchietto, A backoff strategy for model-based experiment design under parametric uncertainty, AICHE J. 56 (2010) 2088–2102.
[93] A. Mesbah, S. Streif, A probabilistic approach to robust optimal experiment design with chance constraints, IFAC-PapersOnLine 48 (2015) 100–105.
[94] S. Vajda, H. Rabitz, E. Walter, Y. Lecourtier, Qualitative and quantitative identifiability analysis of nonlinear chemical kinetic models, Chem. Eng. Commun. 83 (1989) 191–219.
[95] D.C. López C., T. Barz, S. Körkel, G. Wozny, Nonlinear ill-posed problem analysis in model-based parameter estimation and experimental design, Comput. Chem. Eng. 77 (2015) 24–42, https://doi.org/10.1016/j.compchemeng.2015.03.002.
[96] P. Nimmegeers, J. Lauwers, D. Telen, F. Logist, J. van Impe, Identifiability of large-scale non-linear dynamic network models applied to the ADM1-case study, Math. Biosci. 288 (2017) 21–34, https://doi.org/10.1016/j.mbs.2017.02.008.
[97] A. White, M. Tolman, H.D. Thames, H.R. Withers, K.A. Mason, M.K. Transtrum, The limitations of model-based experimental design and parameter estimation in sloppy systems, PLoS Comput. Biol. 12 (2016), https://doi.org/10.1371/journal.pcbi.1005227, e1005227.
[98] T.A. Johansen, On Tikhonov regularization, bias and variance in nonlinear system identification, Automatica 33 (1997) 441–446, https://doi.org/10.1016/S0005-1098(96)00168-9.
[99] P.C. Hansen, Rank-Deficient and Discrete Ill-Posed Problems: Numerical Aspects of Linear Inversion, Society for Industrial and Applied Mathematics, 1998.
[100] A.K. Agarwal, M.L. Brisk, Sequential experimental design for precise parameter estimation. 1. Use of reparameterization, Ind. Eng. Chem. Process. Des. Dev. 24 (1985) 203–207.
[101] Y. Bard, Nonlinear Parameter Estimation, Academic Press, New York, 1974.
[102] M.F. Luna, E.C. Martínez, Optimal design of dynamic experiments in the development of cybernetic models for bioreactors, Chem. Eng. Res. Des. 136 (2018) 334–346.
[103] D.C. Montgomery, Design and Analysis of Experiments, John Wiley & Sons, 2017.
[104] L.M. Rios, N.V. Sahinidis, Derivative-free optimization: a review of algorithms and comparison of software implementations, J. Glob. Optim. 56 (2013) 1247–1293.
[105] W. Huyer, A. Neumaier, SNOBFIT—stable noisy optimization by branch and fit, ACM Trans. Math. Softw. 35 (2008) 1–25.
[106] K. Schittkowski, Experimental design tools for ordinary and algebraic differential equations, Ind. Eng. Chem. Res. 46 (2007) 9137–9147.
[107] S.F. Walter, A. Schmidt, S. Körkel, Adjoint-based optimization of experimental designs with many control variables, J. Process Control 24 (2014) 1504–1515, https://doi.org/10.1016/j.jprocont.2014.06.019.
[108] E.D. Mitra, W.S. Hlavacek, Parameter estimation and uncertainty quantification for systems biology models, Curr. Opin. Syst. Biol. 18 (2019) 9–18.

[109] J. Hemmerich, N. Tenhaef, W. Wiechert, S. Noack, pyFOOMB: Python framework for object oriented modeling of bioprocesses, Eng. Life Sci. 21 (2020) 242–257.
[110] U. Grömping, CRAN Task View, Design of Experiments (DoE) & Analysis of Experimental Data, 2020. https://cran.r-project.org/web/views/ExperimentalDesign.html.
[111] S. Olofsson, L. Hebing, S. Niedenführ, M.P. Deisenroth, R. Misener, GPdoemd: a Python package for design of experiments for model discrimination, Comput. Chem. Eng. 125 (2019) 54–70.
[112] N. Asprion, R. Böttcher, J. Mairhofer, M. Yliruka, J. Höller, J. Schwientek, C. Vanaret, M. Bortz, Implementation and application of model-based design of experiments in a flowsheet simulator, J. Chem. Eng. Data 65 (2019) 1135–1145.
[113] M. Sokolov, Project: Machine Learning and Predictive Modeling for the Control of Continuous Processes Using Digital Twins, DataHow A.G. Zürich, Switzerland (2021).
[114] K. Rachinskiy, H. Schultze, M. Boy, U. Bornscheuer, J. Büchs, "Enzyme Test Bench," a high-throughput enzyme characterization technique including the long-term stability, Biotechnol. Bioeng. 103 (2009) 305–322.
[115] K. Rachinskiy, M. Kunze, C. Graf, H. Schultze, M. Boy, J. Büchs, Extension and application of the "enzyme test bench" for oxygen consuming enzyme reactions, Biotechnol. Bioeng. 111 (2014) 244–253.
[116] K. Rachinskiy, Entwicklung einer modellbasierten Hochdurchsatztechnik zur Enzymcharakterisierung mit Schwerpunkt auf Langzeitstabilität, 2008.

Chapter 12

Product development in a multicriteria context

Philipp Süss[a], Gregor Foltin[a], Melanie Heidgen[a], David Hajnal[b], Jorge Diaz[b], Hergen Schultze[b], Jochen Gattermayer[b], and Stefan Lehner[b]
[a]*Fraunhofer Institute for Industrial Mathematics (ITWM), Kaiserslautern, Germany,* [b]*BASF SE, Ludwigshafen, Germany*

1 Introduction

In this work, we understand product development as the process of finding a synthesis of ingredients that result in a marketable and profitable commodity. More specifically, we focus on a laboratory setting which was given the task to develop a new formula for a product like paint or glue with certain characteristics. Approaches in the industry to this challenge range from a scientifically unstructured trial-and-error experimentation to a carefully constructed Design of Experiments [1,2] to find interesting mixtures of ingredients.

From an initial set of experimental data, it is further common practice to fit some model functions and make predictions about values not yet sampled to be then verified by further experiments. In this work we formalize this procedure and carry the approach one step further: we introduce Qritos, a completely data-driven decision-support system for multicriteria product design. The system assists in determining critical production process variables we call **inputs**. The resulting product quality is measured by quantifiable **properties**.

Inputs to create a product can be ingredients or production process configurations. Ingredients are specified by their volume or weight and process configurations defining relevant settings of a processing step. They are given in units like time, pressure, or temperature (e.g., time and temperature settings for an oven to dry a certain part). Together, these inputs make up the **variables** of the product design.

By changing the variables of the product design, the product will exhibit varying properties. These properties are measured in units such as weight, stability, absorption rates, or drying times under certain conditions. We assume these **outcomes** to be quantifiable in a laboratory.

Simulation and Optimization in Process Engineering. https://doi.org/10.1016/B978-0-323-85043-8.00001-5

Qritos is envisioned to be used by product designers in the early phases of development. More specifically, we assume that there are no numerical models to predict all relevant product properties from a certain input configuration or the product designer wants to take a purely data-driven approach. Further, there are certain **specifications** the final product should meet. Specifications are given in terms of the measurable product properties and are of the following forms:

- Measured values for property A should be as small as possible (e.g., weight, cost)
- Measured values for property B should be as large as possible (e.g., stability, heat resistance)
- Measured values for property C should ideally have value X (e.g., time to dry, absorbance rate)

It is very likely that there exist some conflicts within the product specifications: an ideal product where all measured properties attain their best physically possible values is impossible to produce. Usually, there is at least a **trade-off** between properties measuring the marketable qualities and production cost to consider. It is also often the case that some specifications can be fulfilled better only at the cost of others.

The product designer is presented with two challenges:

1. Find out how the inputs are related to each product property, and
2. Explore and, ultimately, balance the trade-offs among conflicting product specifications.

There is also the cost of experimenting with various input configurations and measuring the resulting properties to consider: the number of experiments run in a laboratory setting should be small.

In this work, we show how Qritos can be used to address these challenges. In the section Model Fitting, we address the process of generating experimental data fitting for the Qritos workflow and also explore other commonly used systems and tools currently used in practice. In the section on Multicriteria Optimization and Decision-Making, we introduce an optimization model for Pareto-optimal product design and how a user interacts with the system to balance trade-offs among conflicting product specifications. We put the entire Qritos workflow together for a real-world example to design a chemical formulation for exterior paint. Finally, we close this work with some discussion and outlook for the future development of this decision-support system.

2 Model fitting

Qritos is a data-driven application. Technically, known models of how the inputs relate to the outputs could be used as functions for the later optimization stage. However, one of the advantages of the software is that such models are not required. All functions used in the later stages are models fitted to data.

For this, we require experimental data in the form of $[X\ \ Y]$, where each row $\left(x^{(j)}\ \ y^{(j)}\right)$ is a set of input values $x^{(j)} := [x_{jk}]$ and the measured property values $y^{(j)} := [y_{ji}]$ as a result of the inputs. We describe the data generation process further below. The data is imported into the software where the inputs and outputs can be defined (Fig. 12.1).

For every property y_i, we fit a function F_i of the inputs x to this data. Qritos currently provides the options to fit simple linear multiple regression models for each property or fitting a multivariate partial least squares (PLS1 and PLS2) model. Once the choice is made in Qritos, the data is sent to the modeling software R [3] where the fitting is done. The results from the fit are sent back to Qritos so the user can view the results of the regression (Fig. 12.2).

2.1 Generating the data: Design of experiments

BASF as a chemical company is developing, producing, and selling a large variety of materials. The application properties of these materials, which can be tuned by changing the chemical recipes or processes, must satisfy specific customer needs. The occurrence of a large amount of input parameters leads to challenges in planning the necessary experiments needed for customer-specific product development. At BASF, the established DoE (Design of Experiments) [1,2] technique is widely used. An experimental design consists of a list of prescribed experiments at specific input parameter combinations which are needed to calibrate a postulated mathematical model function. After successful calibration and cross-validation, the model function is then used for predictions and optimization calculations.

Setting up optimal experimental designs requires carefully balancing two aspects: statistical optimality and technical feasibility. Statistical optimality criteria often include either implicitly or explicitly input variance maximization. For instance, the well-known D-Optimal [4] design for the exploration of a linear model will most likely tend to occupy the corner points of the multidimensional polyhedron defined by ranges of the input parameters, if no additional constraints are applied. Mathematically, this contributes to minimizing the uncertainty in measuring the effects of the input parameters on the application properties of the resulting material. Technically, however, the input variance maximization may lead to proposed experiments in which multiple input parameters exhibit simultaneously their extreme values, and carrying out those experiments may become technically impossible, e.g., due to safety issues.

One possible way to tackle the challenge described above is to use existing prior knowledge to formulate mathematical constraints which a priori rule out all experiments from an experimental design that appear to be technically infeasible. Since experienced technicians, engineers, scientists typically unify both practical and theoretical knowledge, one strategy followed at BASF materials

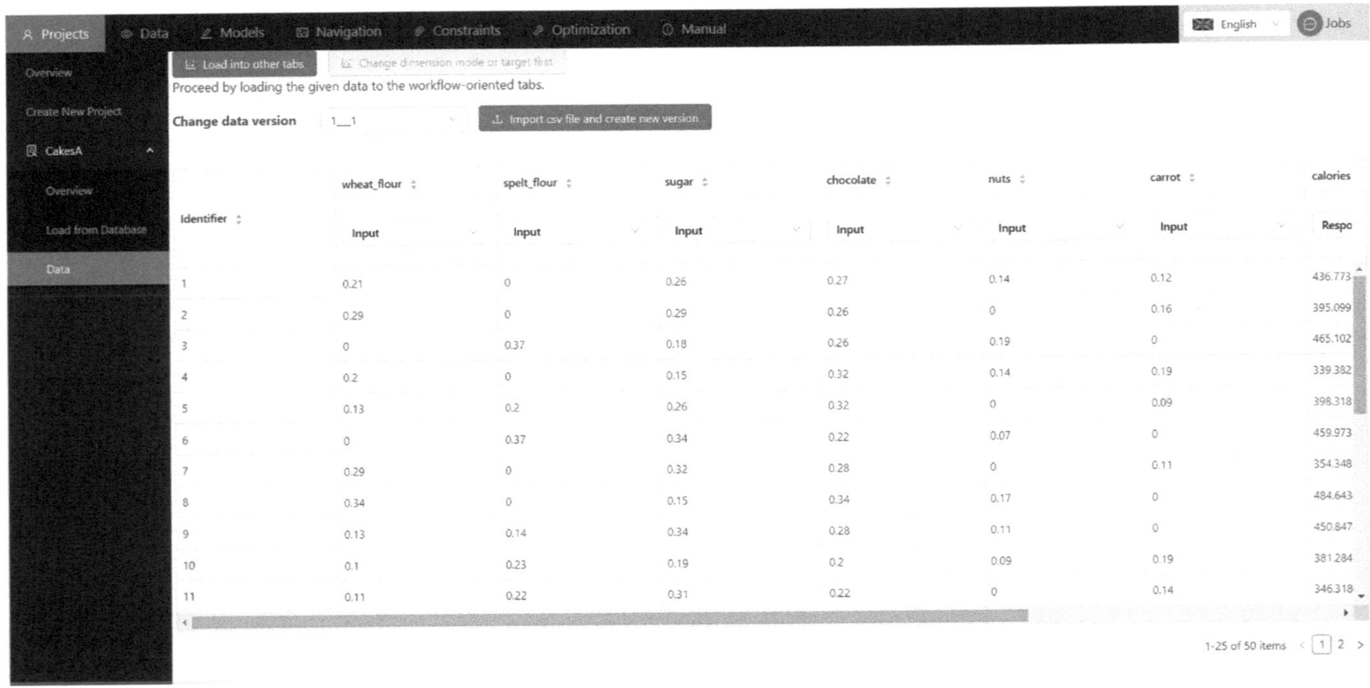

FIG. 12.1 Data input screen. Columns can be classified as inputs or responses (properties) to be minimized, maximized, or to be set to achieve a specific value.

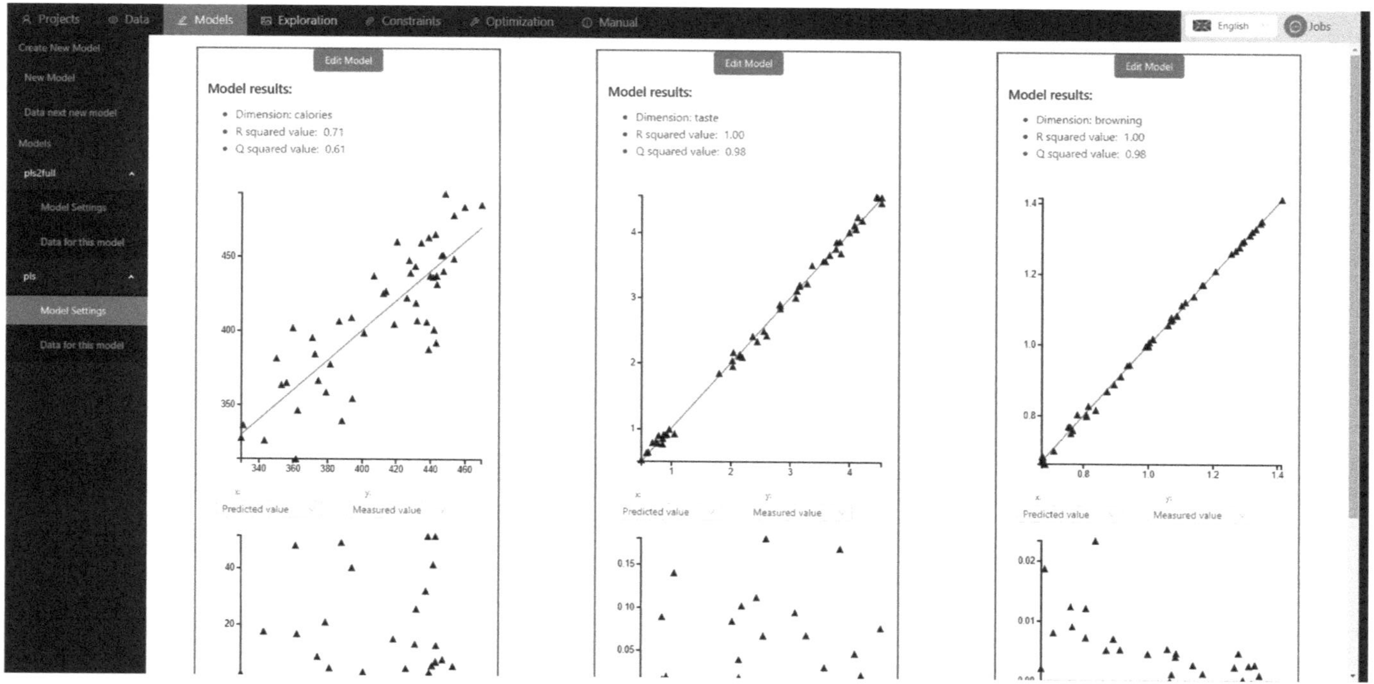

FIG. 12.2 Results of a PLS2 model for an example data set. More information on the coefficients for each model can be opened in a different view.

research is to formalize as much as possible of this existing knowledge in form of mathematical constraints to be used in the generation of D-Optimal designs. Technically, this can be carried out by using commercial DoE software like, e.g., MODDE Pro [5], but it requires interdisciplinary teamwork to collect and formalize as much information as possible. The resulting designs can be regarded as a careful balance between statistical optimality and a priori assumed technical feasibility.

As an example, a physically necessary constraint for the synthesis of polymers for adhesives shall be mentioned. Polymers for adhesives are often synthesized in an emulsion polymerization reaction by using a mixture of several monomers. One key physical parameter of the synthesized polymer is its so-called glass transition temperature, in the following denoted by Tg, which is the temperature at which the polymer undergoes a transition from a liquid state into an amorphous solid state upon cooling. It is generally known that a given polymer can only be used as an adhesive if Tg is located sufficiently below the desired application temperature of the adhesive. If, instead of a mixture of monomers, only one specific monomer is polymerized, then it forms a so-called homopolymer that has a specific Tg value. Typically, the values for Tg are known for all homopolymers corresponding to the monomers used in typical mixtures in polymerization reactions. Furthermore, the Tg value for the resulting polymer can be at least approximately calculated by applying a mixing rule known as the so-called Fox equation. For these cases, the Fox equation can be used to define constraint inequalities to rule out all experiments with infeasible Tg values in the generation of D-Optimal designs.

3 Multicriteria optimization and decision-making

In this section, we derive a **M**ulti**C**riteria **O**ptimization (MCO) problem for the product design problem. The goal is to use the models created in the last section to help the product designer balance all given specifications. The method is inspired by earlier works in radiotherapy treatment planning [6] and later work in chemical process design [7,8]. We employ multicriteria optimization to generate solutions that are *efficient*; we are interested in finding product variations where no single specification can be improved without deteriorating at least one other. For example, a product design where the weight (to be minimized) can be lowered without impairing any other quality measure is not efficient in the multicriteria sense and should be of no interest to the product designer.

Our method consists of two parts. First, the MCO problem is formulated using the models from the last section and the set of all efficient solutions is approximated numerically. This happens automatically and without user interaction. Second, the designer is presented with a visual interface to explore this set of solutions interactively and in an intuitive way: even for a large number of specifications (say, 10) the designer can "navigate" among interesting solutions using sliders on the screen. Each step is described in the following sections.

4 Approximating the set of efficient product designs

A generic MCO problem with n objectives is of the form

$$\min_{x \in \mathcal{X}} \begin{bmatrix} F_1(x) \\ \vdots \\ F_n(x) \end{bmatrix}$$

where $\mathcal{X}$ denotes the set of feasible inputs and can be defined by constraint functions $g_j(x)\ R\ 0$ (with $R \in \{\leq, \geq, =\}$) for some index set $j \in J$.

The variables of our MCO problem are given by the inputs and the multiple objectives result from combining the product specifications and the models created in the last section. Specifically, a model function f_i for property i as a function of all inputs x is

-. formulated as a minimization objective if the specification calls for a minimal value of f_i: $F_i(x) := f_i(x)$
-. formulated as a maximization objective if the specification calls for a maximal value of f_i: $F_i(x) := -f_i(x)$

If the specification calls for the achievement of a particular value t_i of f_i, we introduce two slack variables s_i^+ and s_i^-, add $F_i(x, s_i^+, s_i^-) := s_i^+ + s_i^-$ as a minimization objective and introduce the constraints $f_i(x) + s_i^+ - s_i^- - t_i = 0$ and s_i^+, $s_i^- \geq 0$ to $\mathcal{X}$. The term $s_i^+ + s_i^-$ measures the absolute deviation from the specified target value. While other formulations to minimize the deviation from a target value are possible, this formulation has the advantage that by only adding linear terms to the formulation, we don't make the optimization more difficult to solve numerically.

Finally, arbitrary constraints on inputs and outcomes are easily included to formulate $\mathcal{X}$ (Fig. 12.3). Common constraints on the inputs in product mixing problems are

- lower and upper bounds on the values for each input (if only to make sure that the optimization stays within the bounds of the data to avoid extrapolation of the fitted model functions), and
- concentration constraints of the form $\sum_k x_k = 1$ for some or all inputs x_k.

Calculating or approximating the set of all efficient solutions of our multicriteria product design problem is a numerically challenging problem. Many approaches exist in literature and the interested reader is referred to [9] and the references therein. For Qritos we use *sandwiching* methods for convex [10] and nonconvex [11] problems to attain rather sparse but good approximations of the solution set. Sandwiching methods generate upper and lower bounds for the set of efficient solutions and terminate only when the gap between these bounds is small. These methods are also used in chemical process design [7]. Aside from being fast, these methods are also *exact* methods, setting them apart from more heuristic approaches discussed in the literature [12].

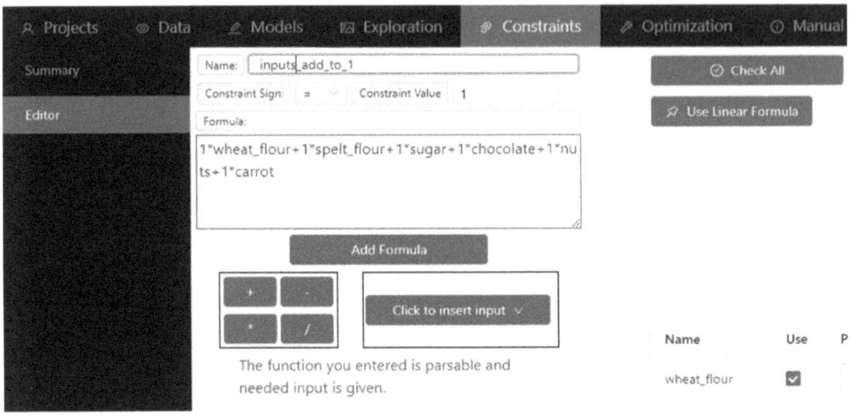

FIG. 12.3 Specifying constraints in Qritos.

5 Navigating the set of efficient product designs

Sandwiching methods attempt to achieve a sparse approximation of the Pareto set—only calculate as many points of the Pareto set to represent it well by the lower envelope of their convex hull. However, this set might still contain 100 or more points, depending largely on the number of objectives and the degree of correlation between them. The product designer should learn about the chances and limitations of what the data models currently predict to be possible. Qritos offers **Pareto navigation** [13,14] as a highly interactive visual tool to achieve this (Fig. 12.4).

In Pareto navigation, the decision-maker focuses on one possible solution and attempts to successively improve it. We start with a (near) efficient product design and display all quality measures and input configurations as predicted by the models from the previous chapter. All these values are shown on sliders, with the slider handles depicting the current values for inputs and outcomes. Everything is now interactive, and any value can be increased or decreased, and the effect of this modification can be observed by watching the other slider handles move. The user can now make improvements on the specifications and study how the inputs change, and the other quality measures improve or degrade. The designer can always stop and correct another slider or even go back to a previous design.

The decision-maker can also control the value ranges of any slider: restrictions can be placed on a maximum or minimum value for any objective or input slider. The tool then immediately shows which parts of other objectives and input configurations cannot be reached anymore. This enhances the understanding of the limitations of possible product configurations. While navigating using the slider handles is a direct manipulation of one particular design, restrictions answer more global questions like "What is still possible in terms of all inputs and objectives if I limit the weight of the product to a certain value? And how do these ranges change if I increase this restriction slightly?" (Fig. 12.5).

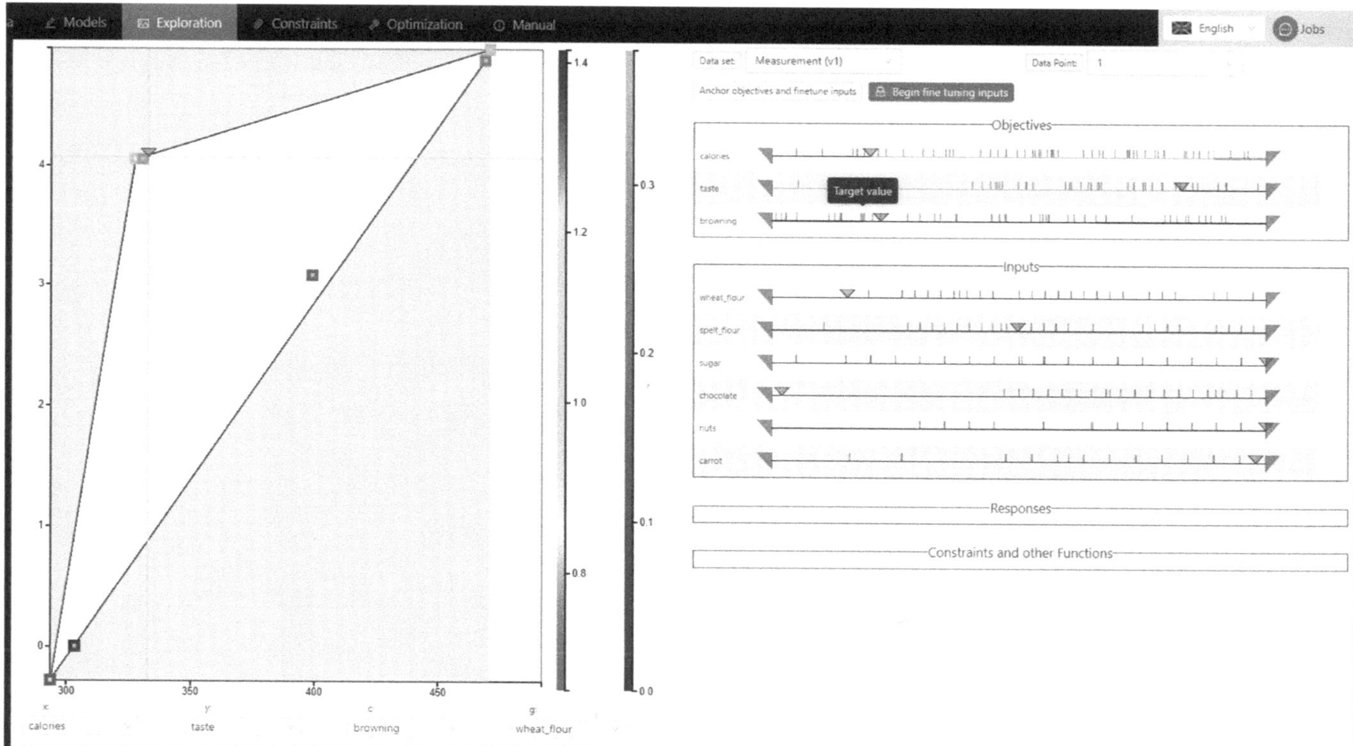

FIG. 12.4 The Pareto navigation in Qritos. The user can pull any input or objective slider to the right (downward-pointing triangle) and see the result in the customizable 2D plot to the left. At the moment, the objective plotted on the x-axis is to be minimized, and the objective on the y-axis is to be maximized. The third objective with a target value is shown as color code. The square points in the plot are the Pareto points that make up the approximation of this 3D MCO problem. Note that the three dimensions are projected onto the plane and the points in the center only appear to be inside the front in this projection—they are, in fact, efficient.

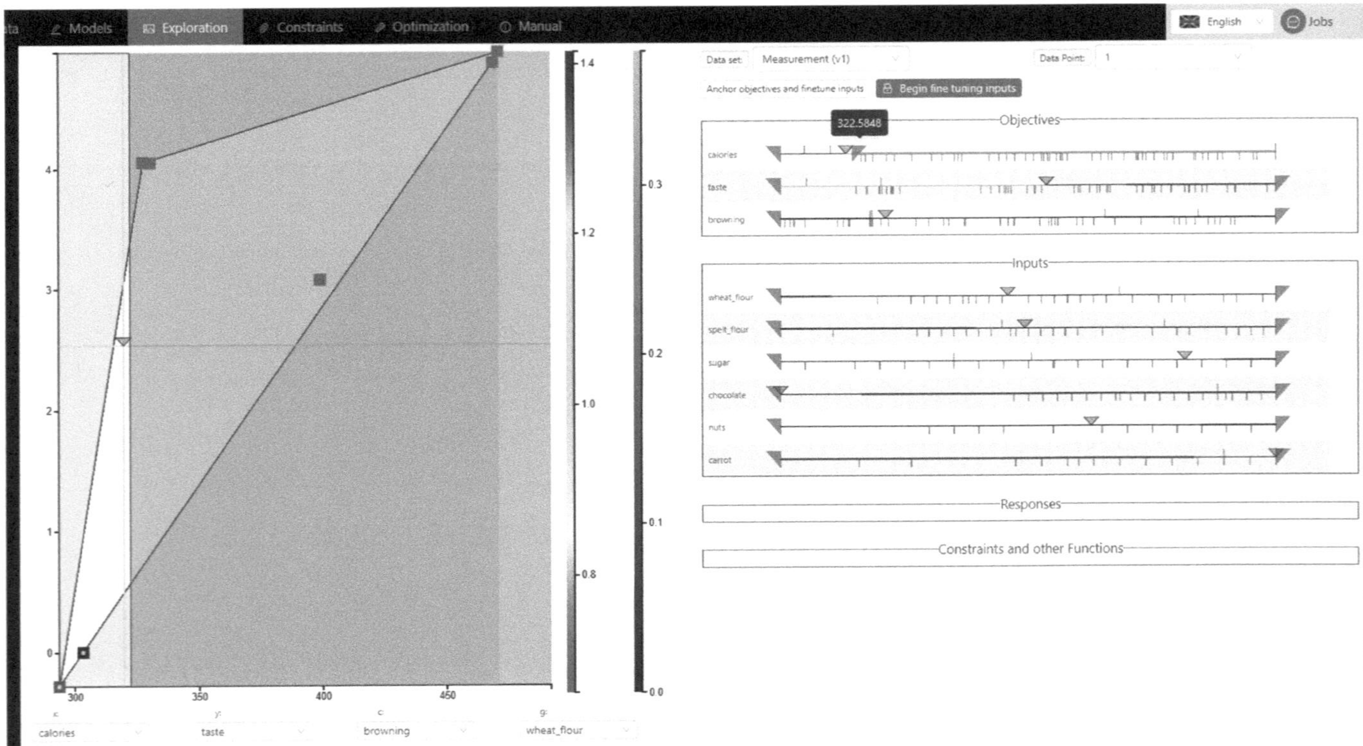

FIG. 12.5 The first objective is now restricted severely. It is immediately clear how this influences the best attainable values of the other objectives.

6 The role of Qritos in the design process

Navigating and restricting the Pareto set—or the approximation thereof—in this manner conveys a sense of this high-dimensional body and, ultimately, an understanding of the trade-offs involved in the product design. In real time, it can translate preferences conveyed by the user through slider adjustments into consequences and trade-offs involving the outcome and inputs.

This decision-making technique defers the choice of preferences as to what a good balance between the specifications should be to a point in time when the designer can actually see the consequences of these preferences. There is no guessing at a good balance without seeing any results from numerical simulation and optimization because all necessary calculations have been carried out during the Pareto set approximation phase. Questions like "How much does achieving the smallest possible weight cost me in terms of durability?" can be answered in real time using Pareto navigation. Because these questions can be answered fast, Pareto navigation encourages interactive decision-making sessions with all stakeholders involved in the design and production process (most important of which may be the paying customer).

Pareto navigation makes the limits of what can be achieved in terms of quality specifications transparent and decisions can be communicated clearly without having to resort to educated guesses. This is a scientific approach to decision-making because Pareto navigation shows measurably the consequences of a product design, and it can be repeated and verified by any stakeholder using the same data and models. The result should be a thorough understanding of how the decision on the final product properties came to be; especially the client who will finally purchase the product can appreciate why certain specifications were fulfilled the way they were.

Pareto navigation is the final stage in the decision-making tool Qritos. This step may result in a final product being produced. The decision-maker should pay attention to the resulting input configurations and possibly some known uncertainties. If the model uncertainty for one or several promising products is too high, the product designer could order a few more lab experiments to reduce the uncertainty in exactly those regions of interest. As such, Qritos can be used at very early stages of the design process: run a few experiments and measurements in the lab, use Qritos to find interesting compromises, validate the model predictions in the lab and continue.

7 Application: Designing an exterior paint recipe

We teamed up with the business unit of BASF Architectural Coatings to use Qritos to optimize an exterior paint formulation. The inputs are the content of each of the ingredients of the chemical recipe and the properties measured are the high-shear-rate viscosity, the mid-shear-rate viscosity, the capillary water uptake, and water vapor permeability.

7.1 Chemical recipe

The paint formulation under study consists of titanium dioxide, neutralizing agents, co-dispersants, polymer dispersions, fillers, defoamers, coalescent, biocides, water, rheology modifiers, and dispersants. These are the inputs to our formulation. Many of these components have fixed values that lead to a mix with the appropriate chemical and physical characteristics needed for a paint formulation. In this study, only the last two component types were varied in such a way that seven rheology modifiers and five dispersants were studied.

7.2 DoE

Rheology modifiers come in two classes denoted RMa and RMb, which can take ranges 4–7 and 5–15 parts per thousand, respectively. We used three different rheology modifiers of the class RMa (RMa1–RMa3) and four rheology modifiers of the class RMb (RMb1–RMb4). Similarly, five different dispersants denoted D1–D5 were used, within the range 2–10 parts per thousand. Given the simple characteristics of the experimental setup, constraints were dispensable.

Using MODDE 12, each of the three classes of inputs (RMa, RMb, D) were defined as qualitative types, where the number of individual components of each class determines a different level: three levels for RMa, four levels for RMb, and five levels for D. For each of the three qualitative classes, a quantitative factor with their corresponding range were created. This allows the design to assign values for the classes within the ranges of interest. Using a linear model and a D-Optimal design, the design matrix results in 23 experimental runs.

7.3 Laboratory measurements

The number of runs of the experimental design (23) was much lower than initially expected (around 30) leaving the initial resource availability of runs with a surplus of supplies and materials. These were fully utilized by conducting extra runs to incorporate two new rheology modifiers (RMc1, RMc2) for a total of 31 experimental runs. Given the different values used for the 14 components, the amount of water was adjusted to maintain all the values as parts per thousand, leading to a total of 15 inputs.

For each sample, the four key properties were measured. The assessment of the high-shear-rate viscosity (HSRv) was performed using a cone/plate type viscometer with an output in the range 0–5. Similarly, measurements of the mid-shear-rate viscosity (MSRv) were done with a Krebs Stormer viscometer for accurate direct reading in the range 40–141 Krebs units. The measurement of water vapor permeability (WVP) was performed via gravimetric methods using standard commercial instruments according to standard DIN 52 615. Similarly, capillary water uptake (CWU) was determined according to standard DIN 62 617.

7.4 Modeling

The formulation components that remained constant were aggregated into a single field named Fixed Parts. This field with a fixed value of 957 per thousand was excluded from the modeling. The four properties were modeled using the partial-least-squared (PLS) algorithm with 12 PLS components and only using linear effects of the 15 inputs.

The resulting model exhibits a poor cross-validated predictive power for capillary water uptake ($Q^2=0.28$) and unsatisfactory for mid-shear-rate viscosity ($Q^2=0.48$), whereas the results for the other two responses are much better ($Q^2 \geq 0.70$) (Fig. 12.6).

7.5 Optimization goals (product specifications)

An optimal formulation must satisfy strict quality conditions determined by the value of these four properties. A target value for the high-shear-rate viscosity guarantees the best user experience when applying the paint to a surface. Similarly, targeting a particular value for the mid-shear-rate viscosity will result in a product that can properly remain on the surface where it is being applied without dripping (too low viscosity) while at the same time being a liquid mix rather than a paste (too high viscosity).

For these reasons and following the highest market standards, we set the optimization goals to a target for the high-shear-rate viscosity of 2.5; whereas for the mid-shear-rate viscosity, we set a target of 110.

A high-quality paint must have high resistance to water diffusing into the application to keep the surface underneath free of moisture; therefore, the capillary water uptake (CWU) must be minimized.

Finally, an outstanding paint should also allow for any moisture to be able to escape through the application; hence, water vapor permeability (WVP) is maximized.

In this way, the optimization goals are set to design an optimal paint exhibiting physical characteristics that guarantee ideas user experience and optimize the physical properties of the coating keeping them dry and reducing the risk (or tendency) for contamination with algae and funguses.

Before running the optimization step, a constraint is included to guarantee that the results will represent parts per thousand for their direct use in the laboratory. Since the fixed parts constitute 95.7% of the formulation, the constraint is set so that the linear combination of all the factors is equal to 43 (in the units of part per thousand in the data set).

7.6 Decision-making process

The result of the optimization in Qritos is presented in Fig. 12.7. Some regions in the objective space appear excluded and the target values of the two

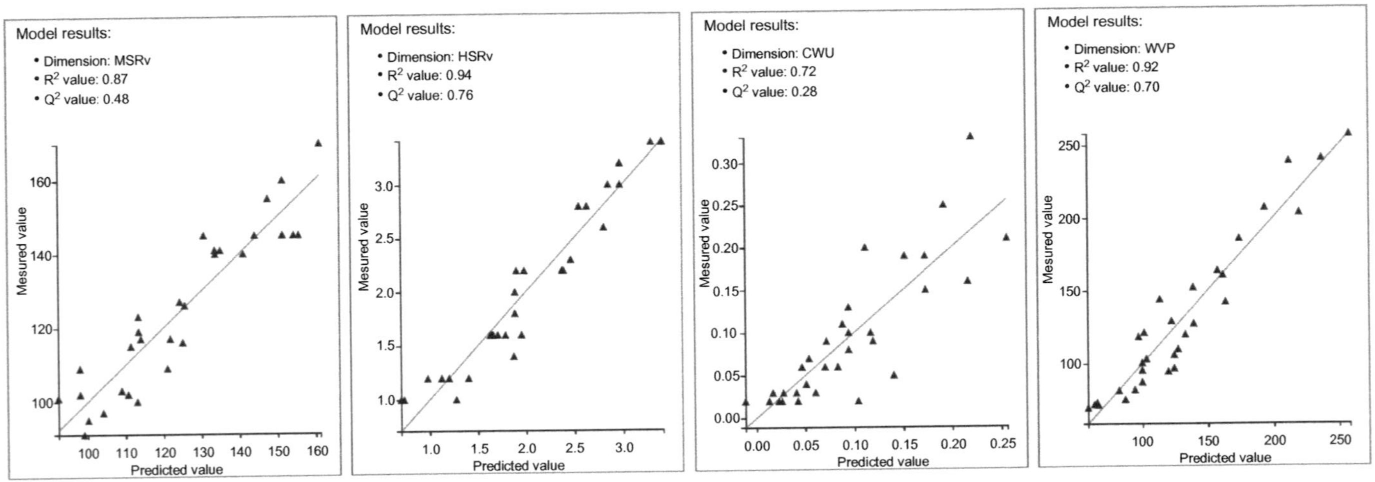

FIG. 12.6 Summary plots from statistical modeling.

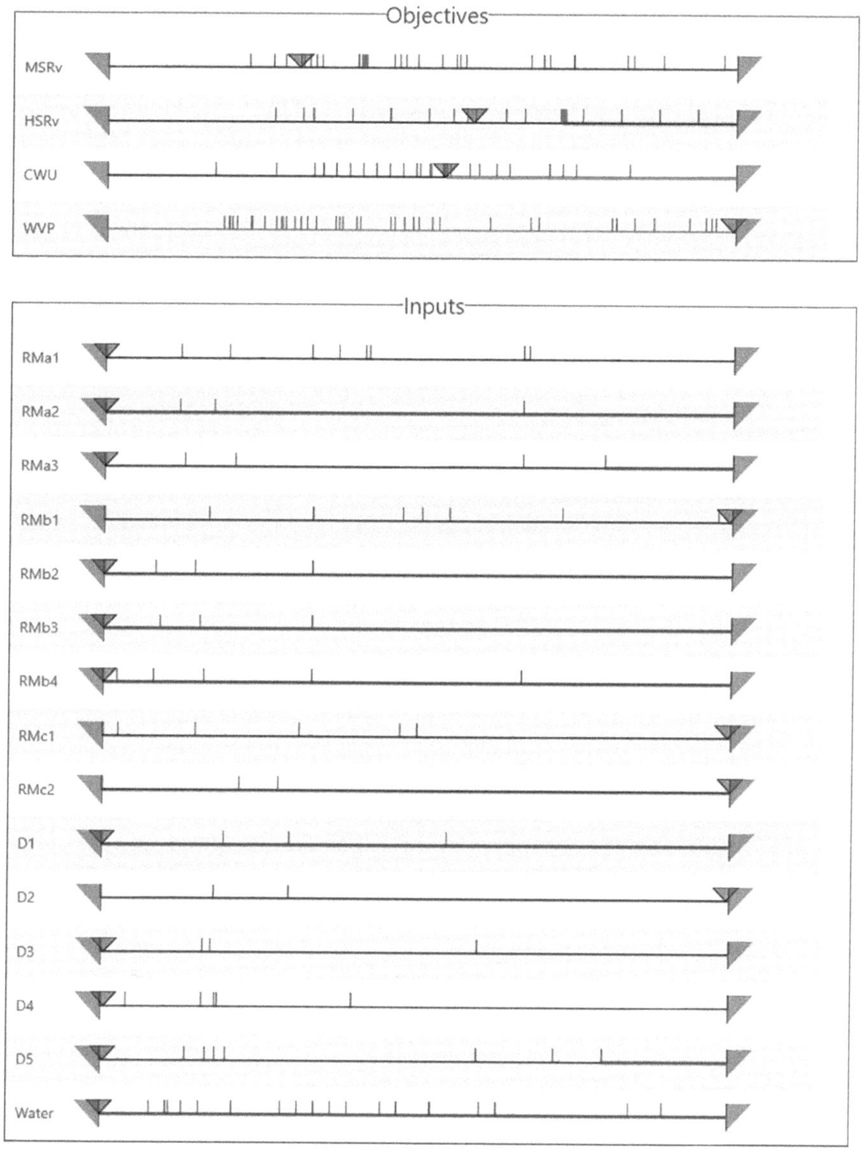

FIG. 12.7 Slider elements used for Pareto navigation. The *green triangles (gray in print version) pointing downwards* are used to manipulate the current solution and the *pink triangles (gray in print version) currently sitting on the edges of the sliders* can be used to restrict unwanted ranges on each slider. The *thicker vertical ticks* on the first two objective sliders indicate the best value for these outcomes. The *thinner ticks* indicate coordinates of the sample points (data).

viscosities are within the feasible region. At this point, the exploration of the objective space was done by an expert on paint formulations. Our colleague spent some time exploring extreme values for all the objectives while keeping track of the input sliders to know the characteristics of the formulation needed to reach the explored result. Some reasonably good results were found that required the admixture of several dispersants. This requirement can be a challenging request for paint manufacturers as they would require extra ingredients in their formulations. The expert then tuned the sliders to identify the compromise on capillary water uptake (CWU) and water vapor permeability (WVP) when trying to reach the target for the two viscosities. It took just a glance for our expert to verify that the concession required to reach the viscosity targets was not worth the loss in water vapor permeability. At last, our colleague decided that given the compromises identified, the ideal scenario would be the one with the highest water vapor permeability. Moving this slider to its upper limit led to the discovery of a very simple formulation requiring a single dispersant and the admixture of two classes of rheology modifiers. This configuration was selected to be tried in the laboratory, and the corresponding experimental procedure is now underway.

It should be emphasized how the sliders facilitated the expert the possibility of exploring a remarkable number of chemical recipes simply by adjusting the quantities in the objective space. In other words, an expert has the possibility of designing the properties of the desired product and Qritos returns the required concentration for each of the ingredients. This procedure for navigating the Pareto front in developing products not only guarantees optimal results and makes the decision of the amount of ingredients to try in the lab trivial, but it also saves time, effort, and resources. In the words of our expert after utilizing the tool "Improving the performance of a paint formulation is very time-consuming. The use of DoE combined with Qritos has enabled us to optimize our formulations but also our lab resources."

8 Outlook

For many practical applications, data-based modeling and model-based prediction are often carried out by models including a dimension reduction algorithm. Widely used examples are principal component regression or partial least squares models. In these models, so-called latent variables are formed and the target variables, which later serve as optimization objectives, are modeled as functions of these latent variables rather than as direct functions of the original input variables, such as chemical composition or process parameters. If the dimension of the space defined by the latent variables is smaller than the dimension of the space defined by the original input parameter for the considered system, then the model exhibits the feature of dimension reduction, and we call the model degenerated. This degeneracy has important technical consequences. If a real-world system is described by a degenerated model with sufficiently high accuracy, then it is possible to systematically change the original input parameters, such as chemical recipe or process, in such a way, that none of the target variables of interest are significantly changed. The space defined by the set of all possible accessible points in input space with this property is referred to as the *invariant subspace*. In the chemical industry, there is often the desire to find alternative chemical recipes or processes to an existing one without changing the application properties of the resulting material. To realize this, different techniques are investigated. One possibility is to apply, e.g., the D-Optimality principle on the invariant subspace in combination with constraints on the original input space to find a set of recipes with maximal chemical diversity, constant application properties, and technical feasibility. Another approach is to use the model functions themselves and fix their evaluations as constraints, effectively freezing the outputs to a target level. These constraints could be used in Pareto navigation and then only allow the inputs to change accordingly.

References

[1] G. Box, W. Hunter, J. Hunter, Statistics for Experimenters, Wiley, New York, 1978.

[2] R. Fisher, Statistical Methods, Experimental Design, and Scientific Inference, Oxford University Press, Oxford, 1990.

[3] R Core Team, R: A Language and Environment for Statistical Computing, R Foundation for Statistical Computing, Vienna, 2017. Von https://www.R-project.org/abgerufen.

[4] F. Triefenbach, Design of Experiments: The D-Optimal Approach and its Implementation as a Computer Algorithm, Umeå University, Sweden, 2008 (Bachelor thesis).

[5] Modde, MODDE 12 Design of Experiments Solution. Umetrics™ Suite of Data Analytic Solutions, 2017.

[6] K. Küfer, M. Monz, A. Scherrer, P. Süss, A. Fernando, A. Azizi Sultan, Multicriteria optimization in intensity modulated radiotherapy planning, in: P. Pardalos, E. Romeijn (Eds.), Handbook of Optimization in Medicine, Springer, Boston, MA, 2009.

[7] M. Bortz, J. Burger, N. Asprion, S. Blagov, R. Böttcher, U. Nowak, et al., Multi-criteria optimization in chemical process design and decision support by navigation on Pareto sets, Comput. Chem. Eng. 60 (10) (2014) 354–363.

[8] J. Burger, N. Asprion, S. Blagov, R. Böttcher, U. Nowak, M. Bortz, et al., Multi-objective optimization and decision support in process engineering—implementation and application, Chem. Ing. Tech. 86 (7) (2014) 1065–1072.

[9] M. Ehrgott, L. Shao, A. Schöbel, An approximation algorithm for convex multi-objective programming problems, J. Glob. Optim. 50 (2011) 397–416.

[10] J. Serna Hernandez, Multi-Objective Optimization in Mixed Integer Problems With Application to the Beam Selection Optimization Problem in IMRT (Dissertation), Technische Universität Kaiserslautern, 2011.

[11] K. Teichert, A Hyperboxing Pareto Approximation Method Applied to Radiofrequency Ablation Treatment Planning (Dissertation), Technische Universität Kaiserslautern, 2014.

[12] C. Coello, S. Brambila, J. Gamboa, M. Tapia, R. Gomez, Evolutionary multiobjective optimization: open research areas and some challenges lying ahead, Complex Intell. Syst. 6 (2020) 221–236.

[13] M. Hartikainen, K. Miettinen, K. Klamroth, Interactive Nonconvex Pareto Navigator for multi-objective optimization, Eur. J. Oper. Res. 275 (2019) 238–251.
[14] M. Monz, Pareto Navigation—Algorithmic Foundation of Interactive Multi-Criteria IMRT Planning, University of Kaiserslautern, Germany, 2006 (Dissertation).

Chapter 13

Dispatching for batch chemical processes using Monte-Carlo simulations—A practical approach to scheduling in operations

Heiner Ackermann[a], Michael Helmling[a], Stefan Hoeser[b], Neele Leithäuser[a], Miguel A. Romero-Valle[c], Carlos Tellaeche[c], and Christian Weiß[a]

[a]*Fraunhofer ITWM, Kaiserslautern, Germany,* [b]*BASF Italia S.p.A., Pontecchio Marconi, Italy,* [c]*BASF SE, Ludwigshafen, Germany*

1 Introduction

Operational scheduling is an important and challenging area of operations research. For many production plants, chemical and otherwise, good scheduling and dispatching strategies are crucial to maximize profits. Unfortunately, practical production environments are often so complex, that provably optimal solutions are nearly impossible to produce. The uncertainties inherent in most practical production settings pose an additional layer of difficulty. For operational scheduling, further challenges arise from having to handle sudden, unexpected disruptions, and from the need to compute good solutions in relatively short time lapses. This is why operational scheduling, in particular, is often hard to deal with algorithmically.

In this chapter, we study the problem of operational scheduling and dispatching for a specific batch chemical plant, owned by the BASF. This plant consists of a single preceding reactor (called the master reactor), which feeds several independent production lines. Details of the process layout and other important plant parameters are described in Section 1.1. Difficulties in the scheduling problem arise mainly from significant stochasticity in the reactor processing times and the absence of buffers, which could prevent stochastic

Simulation and Optimization in Process Engineering. https://doi.org/10.1016/B978-0-323-85043-8.00004-0

delays from accumulating by storing intermediate products between two reactors. To account for this, the plant is automated in such a way that important scheduling decisions can be made just-in-time before they are needed. This reduces the impact of stochastics on the solution quality. Of course, on the other hand, such just-in-time decision-making leaves only a little time to compute good scheduling proposals when they are needed.

Our proposed solution approach is shown in Fig. 13.1. First, we read out the current state of the plant from the plant automation and IT systems. Then, we use a Monte-Carlo simulation (with the minimum necessary amount of domain knowledge) to generate predictions of relevant equipment occupancies in the near future. These predictions are fed into an operational scheduling algorithm, which provides real-time decision support to plant operators. Finally, the prediction results and scheduling suggestions are played back to the operator via a dedicated graphical user interface, as well as written back into the plant automation system.

The purpose of this study is twofold. Firstly, we want to present a practical scheduling problem and its solution from mathematical modeling down to implementation at the actual plant. The main result of this case study is a working, real-time operative dispatching support application, which has been implemented at the BASF plant in Pontecchio and has been integrated with the on-site plant information management system (PIMS). In this pursuit, we have also developed tailor-made, operational scheduling algorithms for the specific plant structure described above. To the best of our knowledge, in the literature on scheduling algorithms, this setting has so far not been studied deeply either in theory or practice (see Section 1.2).

The second reason for this study is to outline the general methodology used to solve our practical problem. We believe that our approach of using Monte-Carlo simulations to feed predictions into an operational scheduling algorithm can work in many other settings as well, far beyond the special setup studied in this chapter. It should also be noted, that a Monte-Carlo simulation together with an automatic scheduling strategy, once implemented, can be used for many other purposes than just operative dispatching. We make some suggestions for such additional uses later in Sections 4 and 5.

This chapter is structured as follows. In the remainder of this section, we describe the problem setting in detail (Section 1.1) and give a brief overview of the related literature (Section 1.2). Then, in Section 2, we present the details of our solution approach, both the setup of the Monte-Carlo simulation as well as the implemented scheduling algorithm. Subsequently, we give insights into the transition from theory to a usable application for real-time operative dispatching running at a BASF site in Section 3. In Section 4, we briefly outline how the implemented methods can be used for other purposes than just operative scheduling/dispatching. Finally, conclusions to our work are presented in Section 5.

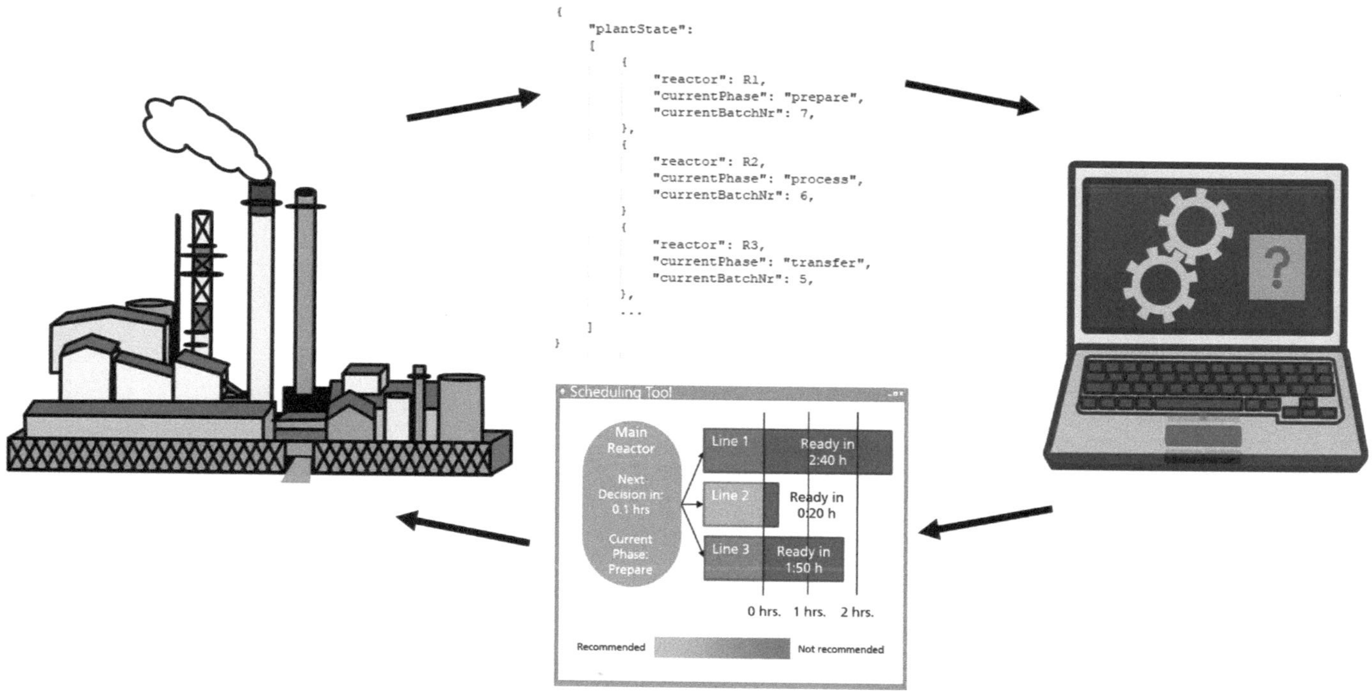

FIG. 13.1 Visualization of our proposed solution approach.

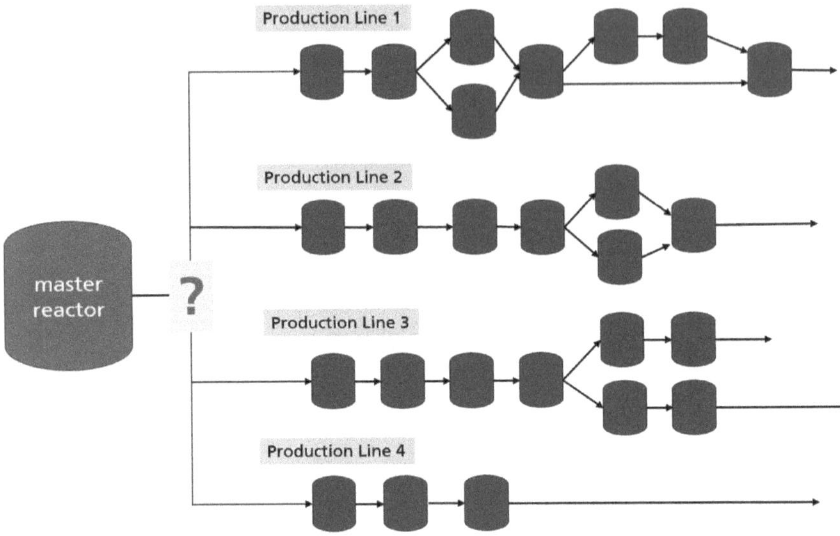

FIG. 13.2 Example for a network as described in Section 1.1. It is a simplified version of a real production asset operated by BASF.

1.1 Problem setting

The network of batch reactors we study in this chapter is shown in Fig. 13.2.

It consists of a single master reactor in the front serving a set $\mathcal{L}$ of several parallel production lines. The master reactor provides specific amounts and blends of the same source material to the production lines but can only serve one production line at a time. A production line $\ell \in \mathcal{L}$ can be viewed as a sequence of (partly parallel) reactors. Each line is designed and optimized for the production of an individual product.

Each reactor, both in the lines and the master reactor, continuously cycles through a number of phases. Each phase has a particular purpose and a related processing time. The phases can roughly be described as follows:

1. Maintenance (skipped, unless necessary);
2. Preparation;
3. Receive material;
4. Process material;
5. Send material forward;
6. Clean up.

Since there are no buffers, a reactor may be forced to pause in the third phase if no material is ready to be received (no predecessor is ready to send its current material forward) or in the fifth phase if no successor is currently ready to receive the material. If a reactor is forced to pause its cycle in such a manner, it is called an idle time.

After initially filling the target line from the master reactor, the product is pushed down the line automatically and no further scheduling decision is needed. Thus, the main scheduling decision is which production line to serve next from the master reactor. The scheduling objective is to achieve a certain minimum production target (batches produced during a given scheduling horizon) in each production line, as well as produce additional batches if possible.

The minimum production targets are part of the input of the scheduling problem and may change from one scheduling window to the next. They are derived from sales agreements and similar contractual obligations. Naturally, as contracts would otherwise incur penalties, the desired targets for a given time window are always lower than the maximum achievable throughput of a line during that time. Thus, the problem is in this sense always feasible. Note that in order to maximize throughput, no idle times should occur in the bottleneck reactor of any production line.

As mentioned above, difficulties arise from the large stochasticity of each individual reactor's processing times. This is caused by a large number of factors, e.g., the uncertainty in the various internal and external parameters that affect the chemical reaction rates, such as outer temperature, pressure, the quality of raw material, and many more, but also uncertainties in the supply chain, affecting the availability of raw materials, availability of waste disposal capacities, and so on. Formally, in our plant model, we specify a stochastic processing time for each reactor and each phase. See Section 2.1.1 for details on how such a stochastic processing time can be derived from data.

For the plant we study, the scheduling window set by the supply chain team is usually 1 week. However, due to the described stochastic effects, it is not useful to compute a dispatching schedule for a full week in advance. Experiments show that even looking a day ahead, the resulting schedule would be too volatile to be useful. Instead, thanks to plant automation, an operator can choose the next batch type just-in-time before actually filling new raw material into the master reactor and adjust the decision according to the momentary plant state, if necessary.

Still, choosing the next batch in the master reactor manually tends to require a lot of experience in practice. The stochastics in the phase processing times, as well as the size and complexity of the production lines, make it hard to estimate, without support, when a line will be ready to receive a new batch. If, after the processing time in the master reactor, the targeted line is not ready to receive, the batch has to wait in the master reactor, blocking it from continuing its cycle and potentially causing idle times in other lines.

1.2 Literature

We now briefly outline how the contents of this chapter tie in with the related literature. As we are considering a practical optimization problem, there are both theoretical and applied works that are relevant to our study. Therefore, this section is split up into two parts. In Section 1.2.1 we consider related work from the scheduling literature, while in Section 1.2.2 we discuss research on the scheduling of batch processes, with a focus on application. For the purpose of brevity, we restrict ourselves to reviewing those two branches of the literature and do not discuss other, less closely related areas such as general literature on batch processes, for example.

1.2.1 Machine scheduling

In this subsection, we will focus on the works most relevant to understand how our problem can be inserted in the broader context of scheduling literature. For a general introduction to scheduling see, e.g., Ref. [1].

In standard machine scheduling terminology, our problem can be described as a very specialized version of a job shop problem, where different jobs share a common first machine and then continue down in independent flow shop-like lines [2]. The downstream line to which a job enters is dependent on the product type to which it belongs. There may be several jobs associated with a single product type. The number of jobs that need to be scheduled for each product type may, for example, correspond to the production targets for that product.

Some authors have instead described the setting as a specialized version of flexible flow shop, where the first stage has only one machine and the later stages have specialized machines, one for each product type, see Refs. [3–7].

Either way, note that for the production lines, processing times are job-independent. Often, job or flow shops with this property are referred to as proportionate (see, e.g., Ref. [8]). For the master reactor, the processing times are also independent of the job but may depend on the chosen product type.

For a theoretical study, we can simplify our setting by contracting the independent production lines into single machines, where such a machine has as processing time the maximum cycle time of any reactor in the respective production line. In that case, the scheduling model consists of only two stages, with the master reactor at the first stage and one machine per production line at the second stage [2]. However, even in this simplified, two-stage model, to the best of our knowledge, the stochastic version of the problem has not been studied in the scheduling community.

It is noteworthy that the deterministic problem (where processing times are fixed) is easy to solve if there are unlimited buffers, but already appears to be rather more complex to analyze as soon as intermediate storage is no longer available. Indeed, many of the works cited above consider the variant with intermediate storage, but with job-dependent processing times (see, e.g., Refs. [4,6,7]). Again, to the best of our knowledge, no general complexity results are available for the version without intermediate storage, other than that it is strongly NP-hard to solve if the number of production lines is not fixed and can be arbitrary [2]. Of course, this is not a realistic assumption in practice.

Several different MIP formulations, as well as heuristics, have been analyzed for the deterministic version of our problem, and specialized polynomial-time algorithms have been developed for special cases (such as the case where production targets are the same for every product), see Ref. [2].

1.2.2 Scheduling chemical batch processes

Scheduling batch processes has a three decades-long history of study, with some works being even older. It is important to note, that scheduling batch processes can be understood in a strategic and an operative sense. In the strategic view, the question is to decide in which way one wants to schedule a plant (dependent, for example, on market demands, etc.), such that the plant can be engineered to accommodate for such a strategy, see, e.g., Ref. [9]. In this review, we focus on operative scheduling, where production for a given plant needs to be planned in an optimized way.

There are many papers and surveys on topics such as problem classification, mathematical modeling, and heuristics (see, e.g., Refs. [10–12] for surveys). All three papers propose classifications of problem settings based on how closely the produced products are related, availability of intermediate storage, as well as the optimization goal, among other factors. In those terms, our problem could be described as multipurpose (products deviate enough from each other that they need to traverse resources in different orders or need different subsets of resources altogether, see also Ref. [9]), without intermediate storage, with the goal to reach production targets and otherwise maximize throughput. Note though, that the plant we study is a very special kind of multipurpose. Indeed, in our setting different products nearly have completely independent production lines, apart from the initial master reactor.

Owing to the applied nature of much of the research, dealing with uncertainty has been one issue studied extensively over the years (for reviews see Refs. [13–15]). Traditionally, using rescheduling (or, sometimes, reactive scheduling), i.e., the computation of a new or partially new production schedule in the case of deviations, has drawn significant interest from researchers (see Refs. [12,14,16–18]). Robust solution approaches have also been considered [14]. Informally, a schedule is considered robust if it is resilient against deviations in the original scheduling data (such as processing times, etc.). Robust scheduling and rescheduling can also be combined, in order to be able to start with a schedule that is relatively protected against deviations and then do a rescheduling step only in the case of changes that are too extreme for the robust approach to effectively guard against [14,19]. However, due to the high amount of uncertainty present in our problem, these solution approaches are not useful for us: a robust schedule would either be too far from optimal because it has to be guarded against too many large deviations, or it would have to be rescheduled so often, that computing the original schedule is basically wasted.

More recently, there are two strings of literature that are highly relevant for the issues we consider in this chapter. One string of research studies the use of online scheduling to deal with the uncertainties inherent in the operation of chemical batch processes, see, e.g., Refs. [16,20,21]. The authors propose an online scheduling framework with a small rolling horizon H, in which the whole schedule is recomputed every $h \ll H$ time units. Note that our algorithms proposed in Section 2.2 are of this type, although our H is very small, as we only compute the next batch. The authors consider various aspects of this approach in several papers (e.g., Refs. [16,20,21]) and argue strongly for its use in practice.

The other string of literature interesting to our object of study considers the possibilities to enhance scheduling methodologies through the results of tailor-made simulations (e.g., Refs. [22,23]).

For strategic decision-making, the idea of using simulations to enhance scheduling decisions for batch processes has, in fact, emerged far earlier, see Ref. [24]. Here, the authors demonstrate how to find a good production schedule for a multiproduct plant by running a series of simulations for each dispatching strategy and picking the one with the best outcome. The authors do not consider uncertainties in Ref. [24]. They run multiple simulations, not in a Monte-Carlo fashion, but to test multiple different problem inputs.

Interestingly, Kuriyan et al. [24] already consider the idea to use simulations as part of their candidate algorithms (rather than only for evaluation of the algorithms), in order to derive better estimates for completion times in complicated process networks. This idea is actually one of the key features of the approach we propose in this paper. Unfortunately, at the time the paper was written, computers were not fast enough to run such simulations in a reasonable time frame to be used in operational algorithms (Kuriyan et al. [24] predict a runtime of at least 8500–17,000 seconds if they were to attempt this).

In the more recent papers (in particular Ref. [23]) the idea of enhancing scheduling algorithms via simulations is brought, partially, to the shop floor. In particular, techniques to use simulation results as part of a rescheduling strategy are considered. The author proposes to use a simulation in order to set up check-points in the future, where the plant state is compared to a predicted plant state and rescheduling is initiated in case deviations exceed a given threshold.

In addition, for a case study, Joglekar [23] suggests using a predicted finish time from a simulation in order to decide the start time of the subsequent batch, similar to what we propose in this chapter. However, there are three important differences to our approach. First, the case study from Ref. [23] considers only a single product batch process, so the only decision made by the rescheduling algorithm is when a new batch should be started, not to make a selection of a specific production line to fill. Second, Joglekar [23] does not use Monte-Carlo simulations for the predictions on the operational level and instead uses fixed values for processing times and other simulation parameters (even though variability exists in the case study process). Third, the simulation used in our approach is much less complex than the one used by Joglekar [23]. Only the path of a batch through the reactor network is needed, together with the processing times data of the reactor phases in order to learn the stochastic processing times. This means that our approach can work without special knowledge of the production recipe or any of the involved reactions, which may not always be readily available.

2 Proposed solution

Our goal is to establish a decision support system for operational scheduling at the plant described in Section 1.1. In this section, we present the two main algorithmic components, namely the Monte-Carlo simulation to predict relevant equipment ready times and the subsequent scheduling method. We first describe how to set up the Monte-Carlo simulation in Section 2.1. Then, in Section 2.2, we develop an online scheduling algorithm to compute the next best batch to be produced in the master reactor, based on predictions from the Monte-Carlo simulation.

2.1 Production line simulation

In this section, we describe the agent-based discrete-event simulation (DES, see Ref. [25]) we have developed for the simulation of individual production lines. That simulation is the key feature for obtaining line ready times $R_\ell(t)$ (the first time after t when line ℓ is ready to receive a new batch) required for the scheduling algorithms in Section 2.2.

The DES paradigm fits very well as it allows to efficiently simulate systems in which relevant *events* occur only at discrete, nonuniform points in time. In our case, each reactor is an agent that controls its phase cycle. Material transfer between reactors requires both participating reactors to be ready in the appropriate phase.

From a high-level perspective, the simulation allows to predict a production line's state at some future time t_f, given an initial state at time t_0. Alternatively, it can predict the (minimum) time $t^* > t_0$ at which some condition will be fulfilled, e.g., "the initial reactor is ready for in transfer."

Of course, due to the random volatility mentioned in the introduction, such a prediction is of a stochastic nature. Hence, we repeat the simulation multiple times in a Monte-Carlo fashion with different random seeds, which allows to estimate the distribution of the parameters of interest, such as $R_\ell(t)$.

In the following sections, we first describe how to model and obtain random phase durations needed to accurately simulate the plant. Then, we provide a brief overview of the structure of our simulation framework.

2.1.1 Random phase durations

Based on an evaluation of the historic reactor phase durations recorded in the PIMS, we have estimated probability distribution for each individual phase of the system (cf. Section 1.1). While the duration of some phases is (almost)

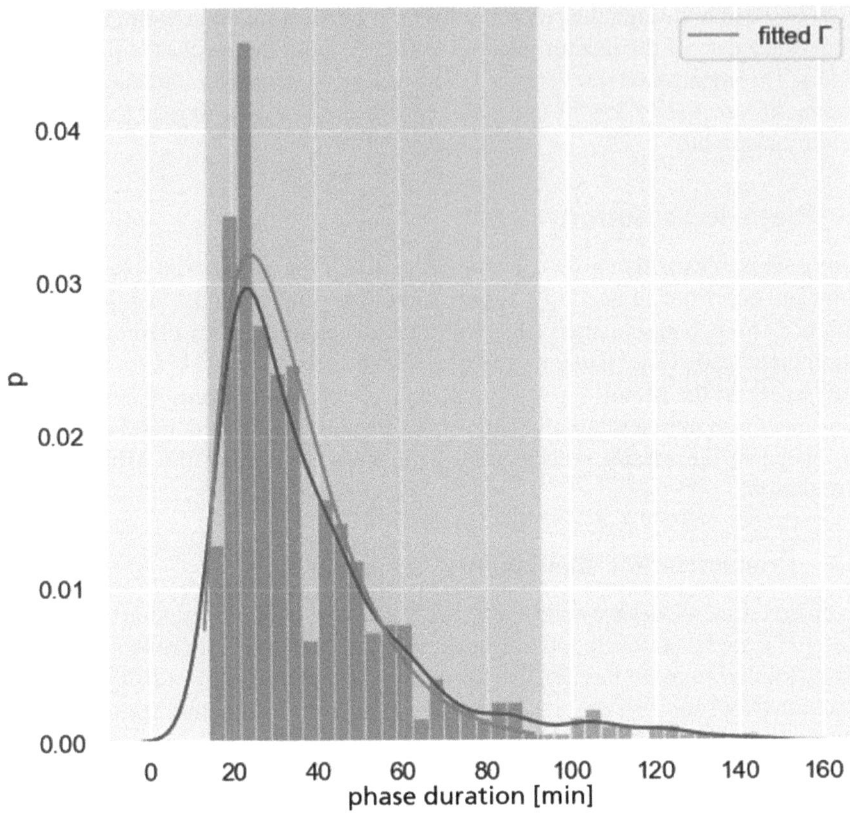

FIG. 13.3 Example reactor phase distribution. The *blue bars (gray in print version)* correspond to empirical durations, the *blue line (dark gray in print version)* is a standard kernel-based density estimation, and the *red line (gray in print version)* is the density of the fitted Gamma distribution.

constant, many others, especially the *process material* phases, exhibit substantial deviation. By manual testing, we found that the Gamma distribution with probability density $p_{\alpha,\beta}(x) = \frac{\beta^{\alpha} x^{\alpha-1} e^{-\beta x}}{\Gamma(\alpha)}$ allows for a good fit to the empirical distribution of processing times (cf. Fig. 13.3). This experimental observation is also in line with findings by, e.g., Juan et al. [26] or van den Akker and Hoogeveen [27], who also observes that Gamma distributions are useful for modeling stochastic processing times.

During the simulation, whenever a reactor enters a new phase, the system draws a random sample from the distribution we generated for this phase from the historic data. This random sample represents the processing duration for this run of the phase. We thereby assume that the phase durations are independent of each other; it is an open question whether this assumption is justified. Note that upon initialization, when a reactor has already been in phase p for time t, one needs to sample from the conditional distribution $P(x \mid x \geq t)$ to ensure that the sample is consistent with the observed state.

2.1.2 The simulation framework in a nutshell

Our simulation framework consists of two main components: a line state and a number of agents, which act upon that state to change it as the simulation proceeds.

The line state first and foremost consists of one reactor state for each reactor in the line. It can either be artificially generated, or it can be retrieved (historic or real-time) from the plant PIMS when the simulation starts. Each reactor state within the line state saves its respective reactor's current phase, together with the duration for which that phase has been active. It also saves the batch number and size of the batch, if a batch is currently processed inside the reactor. Finally, the reactor state maintains a current processing time sample for the active phase, so that the simulation can know when to switch the reactor into the next phase, and processing history, so that the simulation can be evaluated at the end of its runtime.

For each reactor state, the simulation has one reactor agent which acts solely upon that state. The main functions of a reactor agent are to move its reactor state into the next phase once its current phase has finished processing, and interact with predecessor and successor reactor agents in order to synchronize transfer times. Note that for more complicated transfer settings, involving more than one predecessor and successor reactors, some additional control agents are needed. For brevity, we omit the details here.

2.2 Scheduling

Now we turn to constructing a local, online decision algorithm, which at any time $t \in [0, 1 \text{ week}]$, computes the best next line to fill. For simplicity, in this section, we always assume that at time t the master reactor is ready to start a new batch. Otherwise, the approach described below still works, only the formulas get a little more complicated, as additional terms are needed to account for the master reactor not being ready.

In what follows, we describe several online dispatching algorithms (or, in scheduling terms, online construction heuristics) in increasing complexity. As possible inputs, an algorithm gets projections of the line ready times $R_\ell(t)$, $\ell \in \mathcal{L}$, obtained from the Monte-Carlo simulation. Furthermore, the algorithm should take as inputs production targets N_ℓ (the number of batches to be produced in line ℓ in a given week) and the number of batches in each line already produced in that same week at time $t \in [0, 1 \text{ week}]$, denoted by $n_\ell(t)$. The goal is to fulfill all production targets and, if possible, produce additional batches. This ensures that all contractual obligations are met and profits are increased beyond that, if possible.

First, consider the production targets N_ℓ and assume for a moment that our production was continuous. Then, for any given time t during the week and any line ℓ, we can compute the number of batches (including fractional batches) that should have been produced in order to exactly fulfill the target at the end of the week, assuming a constant production speed. We call this number the *continuous production requirement* $\mathrm{CP}_\ell(t)$. For $t \in [0, 1\text{ week}]$ we can compute $\mathrm{CP}_\ell(t)$ as

$$\mathrm{CP}_\ell(t) = \frac{N_\ell}{1\text{ week}} \times t. \tag{13.1}$$

We are also interested in the so-called production rating $\mathrm{Pr}_\ell(t)$ for line ℓ at time t, given by

$$\mathrm{Pr}_\ell(t) = \frac{n_\ell(t) - \mathrm{CP}_\ell(t)}{N_\ell}. \tag{13.2}$$

Note that if the production rating at time t is negative, then the production is running behind the continuous expectation. On the other hand, if the production rating is positive, then the production is ahead of its continuous expectation. Clearly, the lower the production rating, the more urgent it becomes to choose the respective line.

With these considerations, we can construct a first online algorithm as follows.

Algorithm 13.1 Given production targets N_ℓ and already produced batches $n_\ell(t)$ for each line $\ell \in \mathcal{L}$, we compute the next line to fill at time $t \in [0, 1\text{ week}]$ when the master reactor is ready to start a new batch via

1. for each line ℓ, compute the production rating $\mathrm{Pr}_\ell(t)$ using formulae (13.1) and (13.2);
2. choose the next batch to fill the line ℓ with the smallest production rating $\mathrm{Pr}_\ell(t)$.

Notice that for Algorithm 13.1, the Monte-Carlo simulation is not necessary. Only production targets, batches already produced, and the remaining time are needed as input. Obviously, we expect some decisions of this algorithm to be bad, as the algorithm completely ignores the current plant state.

We run simulations to study the quality of Algorithm 13.1. Our test plant consists of a master reactor M and four production lines $\ell \in \mathcal{L} = \{1, 2, 3, 4\}$, with the processing times based on the real plant operated by our industrial partners. We initialize the plant with 250 different start states. For each state, we run 10 simulations over 1 week each. Whenever a new batch needs to be initiated before the master reactor enters its third phase (receiving material), Algorithm 13.1 is used to decide which line should be filled next. The weekly targets used as input for the algorithm are shown in Table 13.1. To illustrate that the targets are reasonable, Table 13.1 also provides the weekly expected capacities of each line. These are obtained by considering the amount of batches a line could process in a week if the master reactor worked only for that line.

The results of the simulations are shown in Table 13.2. In addition, a box plot for the number of batches achieved in each line is given in

TABLE 13.1 Weekly targets and expected weekly capacities.

	Line 1	Line 2	Line 3	Line 4	Total
Target	21	38	14	6	79
Weekly expected capacity	24.07	43.47	16.74	9.35	93.63

TABLE 13.2 Simulation results for Algorithm 13.1.

	Line 1	Line 2	Line 3	Line 4	Total
Average number of batches	21.08	37.88	14.06	6.14	79.15
Batches (% of expected capacity)	87.57	87.13	83.96	65.61	84.48
Targets fulfilled on average	Yes	No	Yes	Yes	No
Targets fulfilled (% of runs)	95.24	80.80	99.44	100.00	80.80
Targets fulfilled (% of start states)	71.20	31.20	94.80	100.00	31.20

Fig. 13.4. Note that due to a concentration of throughput values around the median, the actual box is condensed to a line and only outliers are visualized separately. Fig. 13.6, at the end of this section, also shows a box plot for all three algorithms introduced in this section in comparison.

As we can see, while the average number of batches for Line 2 does not meet the weekly target, still Algorithm 13.1 manages to fulfill all weekly targets in 80.8% of the runs. The problematic line seems to be Line 2: indeed, considering Table 13.2 in detail, we can see that whenever Line 2 meets its target, all other lines do so, too (otherwise the percentage of runs where all targets are met would be lower than that for Line 2). This is easily explained: since Line 2 has the largest number of expected batches (nearly twice as many as the next line) and thus the smallest cycle time, waiting times due to scheduling errors are most impactful for Line 2. As mentioned before, we expect Algorithm 13.1 to make some bad decisions, as it does not take into account the plant state, thus it will sometimes ignore idle lines in favor of trying to fill a line which is still running. Such mistakes are very costly especially for lines with small cycle times, like Line 2.

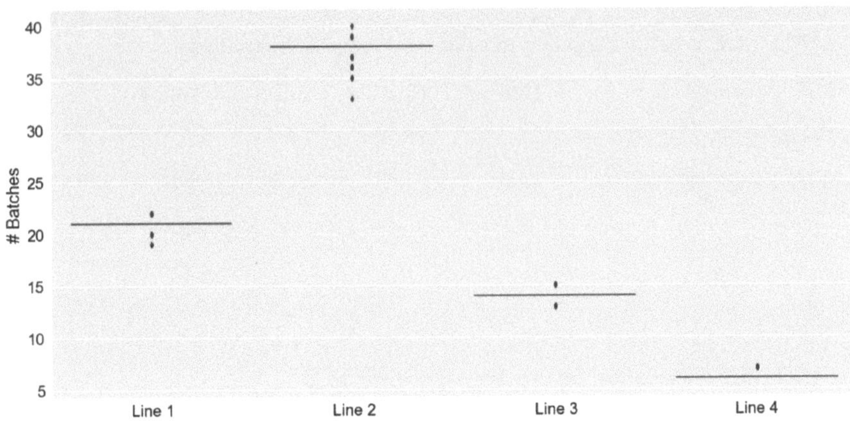

FIG. 13.4 Box plot showing the number of batches in each line achieved by Algorithm 13.1.

This can also be seen in Fig. 13.4. While in most of the runs we exactly meet the target for Line 2, there are some outliers as far down as only 33 batches, losing over 23% capacity in Line 2. Note also, that overall despite fulfilling all targets in 80.8% of the runs, we lose, on average, over 15% capacity in the whole plant and nearly 35% in Line 4. This, again, is due to unnecessary idle times induced by the algorithm ignoring the plant state.

Finally, recall that the contractual targets are chosen by the sales team in such a way that they align nicely with the engineering of the plant and that, in turn, the engineering of the plant is continually optimized, as far as possible, to align with market demand. If targets are not chosen to align roughly with expected line capacities, or in a more complicated setting, the performance of Algorithm 13.1 would drop significantly. Indeed, if we, for example, reduce targets for Line 2 and Line 3 by half, causing targets to no longer be (roughly) proportionate to the expected line capacities, we lose a lot more batches (even though the new targets are still met). With such disproportionate targets, experiments show that we lose over 30% of the total plant capacity, with Lines 2 and 3 producing less than half their capacity.

For comparison, we next consider an algorithm that takes only line ready times $R_\ell(t)$ at time t into account, ignoring production targets and currently finished batches. Recall that in addition to the line ready times, we also know the (expected) processing times of the master reactor for phase k when ℓ is the last chosen line, denoted by $p^M_{k,\ell}$. In what follows, recall the numbering of the reactor phases as discussed in Section 1.1. In order to ease notation, we use $p^M_{b,\ell} = p^M_{3,\ell} + p^M_{4,\ell}$ as the processing time of the master reactor while it has a batch (including the transfer-in phase and excluding the transfer-out phase) and $p^M_{c,\ell} = p^M_{5,\ell} + p^M_{6,\ell} + p^M_{1,\ell} + p^M_{2,\ell}$ as the processing time of the master reactor from the transfer-out phase until it is ready to start a new batch.

Using the projected line ready times and the processing times of the master reactor, we can compute, for each pair of lines line $\ell_1,\ell_2 \in \mathcal{L}, \ell_1 \neq \ell_2$, the idle time $I_{\ell_1,\ell_2}(t)$ produced on ℓ_2 lines, if ℓ_1 is chosen, via

$$I_{\ell_1,\ell_2}(t) = \max\left\{0, \max\left\{R_{\ell_1}(t), t+p^M_{b,\ell_1}\right\} + p^M_{c,\ell_1} + p^M_{b,\ell_2} - R_{\ell_2}(t)\right\}. \quad (13.3)$$

Here, the term $\max\left\{R_{\ell_1}(t), t+p^M_{b,\ell_1}\right\}$ describes the time until line ℓ_1 can receive the batch if chosen, while the term $p^M_{c,\ell_1} + p^M_{b,\ell_2}$ describes the time after that before line ℓ_2 could be filled. Note that if the master reactor is not ready to start a new batch at time t, then the first term changes to $\max\{R_{\ell_1}(t), R_M(t) + p^M_{b,\ell_1}\}$, where $R_M(t)$ is the time when the master reactor is projected to be ready to start the next batch.

Finally, for a line $\ell \in \mathcal{L}$, we denote by $I_\ell(t)$ the total sum of idle times produced on other lines, if ℓ is chosen, i.e.,

$$I_\ell = \sum_{\substack{\hat{\ell} \in \mathcal{L} \\ \hat{\ell} \neq \ell}} I_{\ell,\hat{\ell}}(t). \quad (13.4)$$

Then, we can define a new algorithm to choose a next batch at t in the following way.

Algorithm 13.2 Given predicted line ready times $R_\ell(t)$ for each line $\ell \in \mathcal{L}$, we compute the next line to fill at time $t \in [0, 1 \text{ week}]$ when the master reactor is ready to start a new batch via

1. for each line ℓ, compute the sum of idle times $I_\ell(t)$ produced on other lines if ℓ is chosen, using formulae (13.3) and (13.4);
2. choose the next batch to fill the line ℓ which causes the smallest sum of idle times on other lines $I_\ell(t)$.

Note that the computation time of Algorithm 13.2 depends on the number of Monte-Carlo runs used to predict line ready times. Therefore, there will be a tradeoff between exactness of the predicted ready times and computation speed of the algorithm.

Again, we run simulations to study the quality of the algorithm. We use the same plant and parameters as before. For the computation of line ready times $R_\ell(t)$ we use 100 Monte-Carlo runs at each decision. This number is small enough that in a practical implementation a result can be obtained in less than a minute. In fact, when the algorithm runs in practice, even 500 runs per decision to compute line ready times were acceptable with regard to computation time. Table 13.3 shows the simulation results for Algorithm 13.2.

TABLE 13.3 Simulation results for Algorithm 13.2.

	Line 1	Line 2	Line 3	Line 4	Total
Average number of batches	20.40	40.25	16.04	9.17	85.86
Batches (% of expected capacity)	84.77	92.58	95.82	98.05	91.65
Targets fulfilled on average	No	Yes	Yes	Yes	No
Targets fulfilled (% of runs)	46.04	99.96	100.00	100.00	46.00
Targets fulfilled (% of start states)	0.80	99.60	100.00	100.00	0.80

As can be seen, the loss of production compared to the line capacities is significantly smaller and we manage to use over 91% of the total plant capacity. However, we also reliably miss the target for Line 1. We miss the target in over 50% of all runs and only manage 2 out of 250 start states (0.8%) in which we meet the target of Line 1 for all runs of that start state. Line 1 is problematic due to two reasons. One, while targets are chosen roughly proportionate to the expected line capacities, Line 1 has a tighter target than Lines 3 and 4, thus in order to reach the target for Line 1, an algorithm has to "overvalue" choosing Line 1 slightly over those lines. Obviously, as Algorithm 13.2 ignores production targets, it does not do so, leading to the results at hand. Two, Line 1 is slightly more volatile in its processing times than the other lines relatively early in the line, therefore it is harder to predict when it may be empty exactly.

In order to incorporate the strengths of both previous algorithms and mitigate their weaknesses, we now construct a third algorithm as a weighted combination of Algorithms 13.1 and 13.2. First, we compute for each line $\ell \in \mathcal{L}$ the production rating $\mathrm{Pr}_\ell(t)$ and the sum of idle times on other lines $I_\ell(t)$ using Eqs. (13.1)–(13.4), respectively. Then, in order to properly combine the two values into a single line rating, we first project them onto the $[0,1]$-interval. Let $I_{\max}(t) = \max_{\ell \in \mathcal{L}} I_\ell(t)$ and $I_{\min}(t) = \min_{\ell \in \mathcal{L}} I_\ell(t)$. Then we compute the relative idle time rating of line ℓ by

$$I_\ell^r(t) = \frac{I_{\max}(t) - I_\ell(t)}{I_{\max}(t) - I_{\min}(t)}. \tag{13.5}$$

Note that $I_{\min}$ is projected onto 1 and $I_{\max}$ is projected onto 0, i.e., if line ℓ would be chosen by the second algorithm, then $I_\ell^r(t) = 1$.

Analogously, compute the relative production rating $\mathrm{Pr}_\ell(t)$ by projecting the production ratings onto the $[0,1]$-interval, i.e.

$$\mathrm{Pr}_\ell^r(t) = \frac{\mathrm{Pr}_{\max}(t) - \mathrm{Pr}_\ell(t)}{\mathrm{Pr}_{\max}(t) - \mathrm{Pr}_{\min}(t)}. \tag{13.6}$$

Here, $\mathrm{Pr}_{\max}(t) = \max_{\ell \in \mathcal{L}} \mathrm{Pr}_\ell(t)$ and $\mathrm{Pr}_{\min}(t) = \min_{\ell \in \mathcal{L}} \mathrm{Pr}_\ell(t)$. Note that again the minimum production rating $\mathrm{Pr}_{\min}(t)$ is projected onto 1, while the maximum production rating $\mathrm{Pr}_{\max}(t)$ is projected onto 0. Thus, if line ℓ would be chosen by the first algorithm, then $\mathrm{Pr}_\ell^r(t) = 1$. Note that (13.5) and (13.6) are similar to the definition of desirability from multicriteria optimization [28].

Now, in order to obtain a single, combined line rating $L_\ell(t)$ for each line ℓ, we simply combine the relative production rating and the relative idle time rating in a weighted sum, i.e.,

$$L_\ell(t) = w_{\mathrm{Pr}} \times \mathrm{Pr}_\ell^r(t) + w_{\mathrm{I}} \times I_\ell^r(t), \tag{13.7}$$

with a production weight w_{Pr} and an idle time weight w_{I}, such that $w_{\mathrm{Pr}} + w_{\mathrm{I}} = 1$.

Then, our third algorithm can be described as follows.

Algorithm 13.3 Given predicted line ready times $R_\ell(t)$, production targets N_ℓ and already produced batches $n_\ell(t)$ for each line $\ell \in \mathcal{L}$, we compute the next line to fill at time $t \in [0, 1 \text{ week}]$ when the master reactor is ready to start a new batch via

1. compute line ratings $L_\ell(t)$ for each line ℓ using formulae (13.5)–(13.7);
2. choose the next batch to fill the line ℓ with the largest line rating $L_\ell(t)$.

Of course, in addition to the production targets and the computed idle times, the result of Algorithm 13.3 depends on the choice of weights w_{Pr} and w_{I}. Note that setting $w_{\mathrm{Pr}} = 1$ and $w_{\mathrm{I}} = 0$ yields the same result as Algorithm 13.1 while setting $w_{\mathrm{Pr}} = 0$ and $w_{\mathrm{I}} = 1$ yields the same result as Algorithm 13.2.

As before, we run simulations on 250 start states, with 10 simulation runs per start state. We test values for $w_{\mathrm{Pr}} = \{0.05, 0.1, 0.15, \ldots, 0.5\}$ and $w_{\mathrm{I}} = 1 - w_{\mathrm{Pr}}$. Fig. 13.5 shows how the algorithms perform compared to each other. The y-axis shows the product of the average total number of batches produced and the part of runs where all targets were fulfilled. We call this product the algorithm rating of a given algorithm. This way of rating the different weight combinations is in line with the weighted product method to combine multiple criteria of different scales into one [28].

As we can see, if measured by the product of both targets, Algorithm 13.3 with $w_{\mathrm{Pr}} = 0.1$ and $w_{\mathrm{I}} = 0.9$ performs best. However, note that even the "worst-performing" algorithm in Fig. 13.5 still has a rating of around 82. Thus even that algorithm outperforms Algorithms 13.1 and 13.2 (63.95 and 39.5 respectively).

Now we consider the best performing weight combination $w_{\mathrm{Pr}} = 0.1$ and $w_{\mathrm{I}} = 0.9$ in comparison with Algorithms 13.1 and 13.2. Note that not only

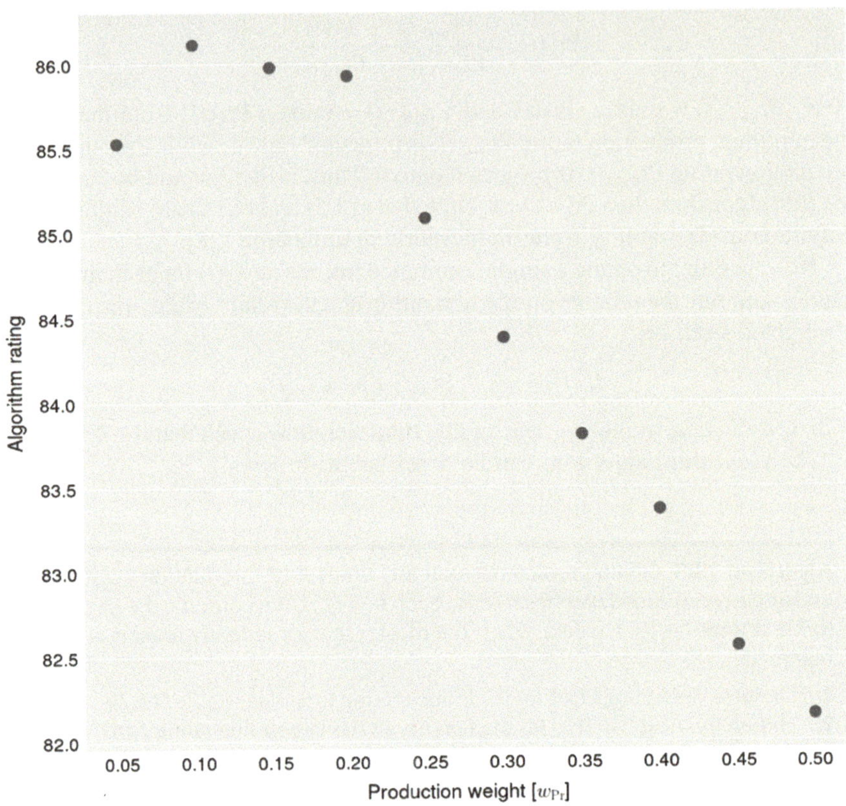

FIG. 13.5 Performance of different weighted algorithms, measured by the product of the average total number of batches produced and the part of runs where all targets were achieved.

are targets met with much greater reliability than for Algorithms 13.1 and 13.2, but we also use slightly more of the total capacity than we do for Algorithm 13.2 (see Table 13.4).

This is again due to the choice of targets. Due to the targets being chosen roughly proportionate to the expected line capacities, Algorithm 13.3 gains some global information about the line capacities that Algorithm 13.2 does not have. Indeed, if we once more cut targets for Lines 2 and 3 in half (like we did to test Algorithm 13.1), then Algorithm 13.3 with weights $w_{Pr} = 0.1$ and $w_I = 0.9$ achieves slightly fewer total batches than Algorithm 13.2. The decrease is much smaller than for Algorithm 13.1 though, due to the large idle time weight $w_I = 0.9$ (Fig. 13.6).

3 Implementation

We have implemented the proposed scheduling approach for a BASF chemical plant that produces plastic additives in a batch process similar to the (simplified)

TABLE 13.4 Simulation results for Algorithm 13.3.

	Line 1	Line 2	Line 3	Line 4	Total
Average number of batches	21.88	40.25	15.74	8.79	86.66
Batches (% of expected capacity)	90.92	92.60	93.99	94.01	92.51
Targets fulfilled on average	Yes	Yes	Yes	Yes	Yes
Targets fulfilled (% of runs)	99.36	100.00	100.00	100.00	99.36
Targets fulfilled (% of start states)	93.60	100.00	100.00	100.00	93.60

model described above. In this section, we describe the challenges in applying our theoretical model to reality and how we solved them. We also describe, briefly, how the resulting application offers real-time decision support to operators.

3.1 Important steps for the implementation of our decision support tool in practice

For the production plant at hand, a full digital model as required by the approach described above, including each reactor's phase properties, the topology in which reactors are laid out, and the rules that govern how material transfer between reactors is carried out, is not available. The necessary information is, of course, available, but it is more likely to be stored in several different systems and models, as well as, sometimes, the know-how of the plant personnel. The first step, hence, is to create such a model from available technical documentation, IT systems, and expert discussions. In that course, one often needs to decide how to cope with (a) incomplete information and (b) certain characteristics of the plant that do not fully fit the model; there is a tradeoff between extending the model (at the cost of increased effort and complexity) and omitting or approximating aspects when one believes their effect to be negligible.

The resulting plant model (topology and parameters) is defined in a human-readable, easily editable file format. Thus, as the plant evolves, the plant model can be maintained without the need of changing the application source code.

When it comes to translating the current state from the PIMS, the main challenge is to recognize ongoing transfers correctly. In the PIMS, there is only one phase code used for all of the states *waiting for transfer*, *in transfer*, and potential *posttransfer operations*. Targets and origins of ongoing transfers, or the time

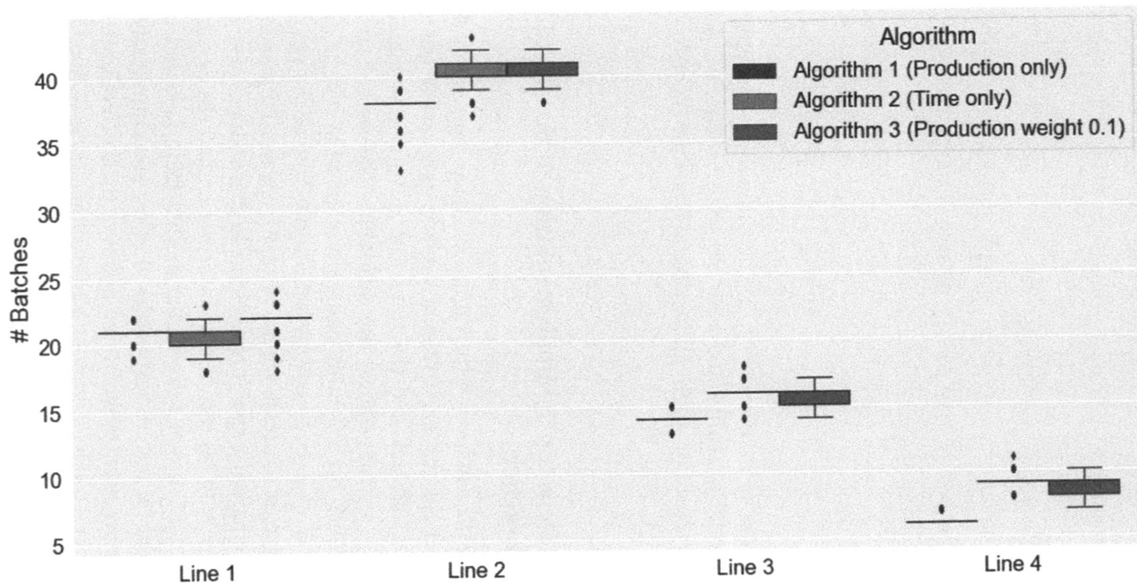

FIG. 13.6 Comparison of the batches achieved by the three analyzed algorithms as a box plot.

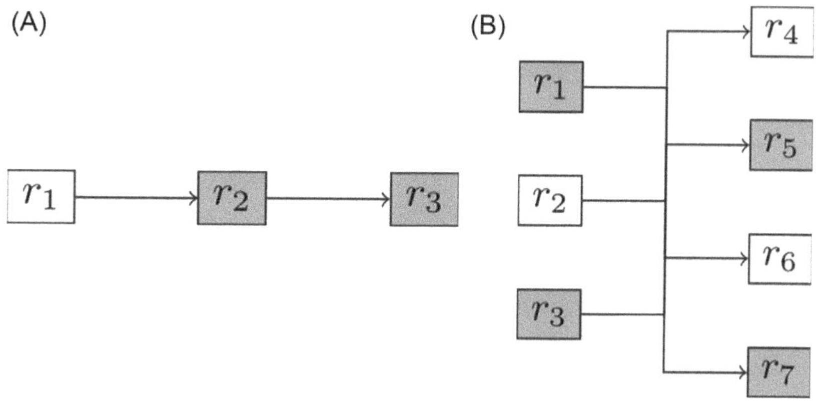

FIG. 13.7 Examples of reactor transfer situations; reactors shown in *gray* are currently in transfer. In the simple sequential case (A), the transfer must be from r_2 to r_3, hence the actual transfer time is easy to detect. In case (B), however, there are multiple options, e.g., r_1 -> r_5 or r_1 -> r_7.

since the actual transfers started, are not recorded at all. Thus, this information has to be estimated from the given data.

In the prevalent case of a single successor or predecessor, the duration of the actual transfer can be estimated by computing the intersection of matching time intervals when both participating reactors were in their respective transfer phase (see Fig. 13.7A). This does not work, however, when there are potential parallel transfers, as in Fig. 13.7B, where ambiguity about which reactors participate in the same transfer can occur. In these situations, it becomes necessary to make the best guess based on the available information, such as batch Ids, states of the involved reactors, and/or current pump activity (not all data is available for every occurrence of this problem). This can of course lead to faulty assumptions about the line state in some rare situations. Fortunately, in practice, such problems are rare and only happen in the first few minutes during a transfer.

3.2 The final application

The simulation and scheduling system described in this chapter was implemented as a desktop application including a graphical user interface (see Fig. 13.8).

When running, the application periodically (usually every 5 minutes) reads the current plant state and then runs the simulation and scheduling algorithm as described in Section 2. Based on the result, the user interface shows the expected time at which the next decision needs to be made, as well as proposals for this decision. For details on the visualization see the description of Fig. 13.8. The final decision remains in the hand of the operator, allowing him to keep control of the schedule and add any additional information to the decision-making process, which the tool may not have.

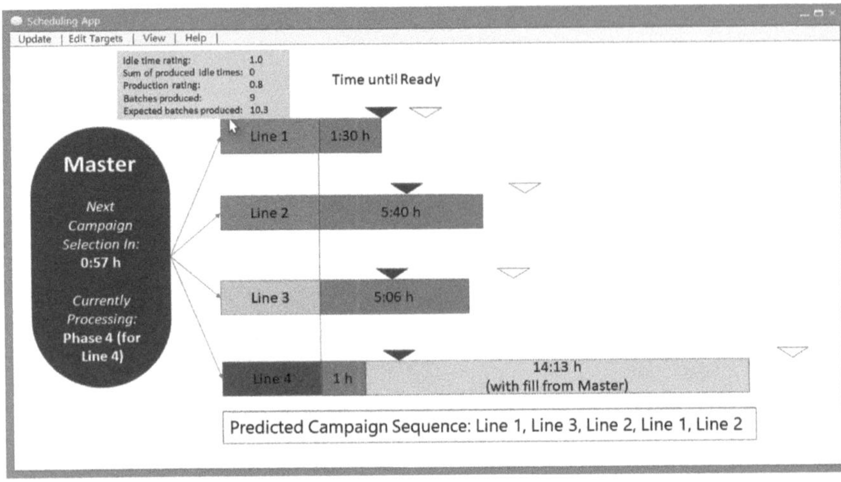

FIG. 13.8 A mock-up of our user interface. On the left side, the current state of the master reactor can be observed. Line headers are color-coded *(green to red; gray to dark gray in print version)* to visualize the line ratings computed by our tool. Further details about the line ratings can be viewed via tool tips. To the right from the line headers, the expected line empty times are displayed. Above the line empty prediction for each line there are markers that show the time Master would be ready to fill that line if chosen *(purple; dark gray in print version)* and the time when Master would afterward be ready to start a new batch *(white)*. Finally, at the bottom, there is a predicted sequence of next batches, to give an operator some lookahead to better evaluate their decision. Recall that Line 2 in our example has the shortest average bottleneck processing time, so it overtakes Line 3 at the end of the sequence.

4 Beyond real-time operative scheduling

Once our simulation and scheduling heuristics are implemented, there are more uses beyond our primary goal to set up a fully running operative scheduling/dispatching support tool at our partner's plant. In this section, we will very briefly describe two such use cases.

4.1 Use case 1: Prediction of future events and plant states

Our simulation of the plant can also be used to predict future events or partial plant states. For example, if maintenances or other manual interruptions of the production process need to be planned, it may be helpful to know when a certain reactor is likely to be in its next maintenance phase.

Given a desired partial plant state (e.g., a certain reactor in a certain phase) our simulation can read the current plant state from the PIMS and simply simulate the plant into the future until the desired plant state is reached. Of course, if the desired plant state is complex (for example involving targeted states of several different reactors), the predicted time would become less reliable, or the desired plant state may not be encountered at all.

The main work needed to enable this use case would consist of a user interface to enter the desired plant state as well as to visualize the results of the subsequent simulation in a meaningful way. All other components (simulation, scheduling, connection to the PIMS) are already available and would only need to be slightly repurposed to enable this workflow.

4.2 Use case 2: What-if analyses for plant expansion/optimization

Technical process optimization measures to, for example, increase the production capacity of the plants are continuously performed at BASF. While, of course, they already perform bottleneck analyses and similar techniques to find the parts of the production lines where optimization would yield the most benefits, a full simulation of the whole plant including scheduling of the master reactor has not been possible.

The simulation model we built for the lines can easily be extended to a larger model, which includes the master reactor and every line, so that a simulation of the whole plant can be run. In fact, we have already done so in order to test our scheduling algorithms via simulation. This larger simulation model can then easily be adapted to run what-if analyses of possible changes before implementing them. Alternative scenarios could, for example, consider different process layouts or even changes to the operating conditions leading to changes in the stochastic processing times for some production steps.

Each scenario can be simulated separately to understand the effect of each possible change would have on the total throughput of the plant, the connection via dispatching from the master reactor included. Analyzing the data from these simulations can give valuable insights concerning each proposed change and can help the engineers make better-informed decisions.

5 Conclusions and outlook

In this paper, we have described a practical scheduling problem and its solution, from mathematical modeling to actual implementation of a decision support application at our partner's plant. The core idea of our approach is to handle highly stochastic processing times and a complex plant setup using Monte-Carlo simulations to predict certain future events (such as line ready times). These predictions are then fed into an operative, online dispatching strategy which uses a construction heuristic approach to suggest the best choice for the next batch. We have shown which additional problems we encountered implementing our theoretical approach in practice, in particular when integrating with the plant's PIMS. We have also described an example for an intuitive user interface, which visualizes the results of our solution approach in an easy, understandable way. Furthermore, in the last section, we have pointed out additional benefits from the methods we implemented, which could be made usable for our industrial partners with relatively little additional work.

We strongly believe that the general idea behind our approach, predictions from Monte-Carlo simulations feeding into decision support algorithms (especially operative scheduling/dispatching), is generalizable for many other chemical plant setups and even other industries. Similar to our case study, one benefit of this approach is that once the simulation is implemented, it can be re-used for many different purposes, other than just the originally desired decision support.

Note also, that our simulation is relatively simple, needing no information about actual production recipes or chemical reactions. It only needs the asset network of the plant, together with the historic processing time data of the different production phases within the asset, in order to learn stochastic processing times. This makes the simulation model easily adjustable for other plant setups, as well as for changes within the same plant. However, despite this apparent simplicity, the Monte-Carlo simulation managed to make accurate predictions about the plant state most of the time during our tests, which was a partial result of the initial project with BASF.

To finish this chapter, we point out a few ideas for future research. We will focus on the practical method and problem, although the related problem from scheduling theory is still not well-explored either (see Section 1.2).

While we believe it is beneficial to keep the simulation model as simple as possible, further details may be needed to apply our solution approach to other plants. Additional items that may need more detailed modeling include continuous production equipment, such as filters, restrictions on connecting pipes, or special transfer settings.

With a view to scheduling, observe that our dispatching strategies are local algorithms. While this may appear natural, as only the next batch needs to be decided, it may actually be beneficial to consider predictions for more than one batch ahead. One strategy to achieve this may be to consider all batch sequences of a given length and run simulations with these to gain a larger look ahead. From initial tests, our simulation is fast enough to handle this approach still in a real-time setting for at least three and probably even five batch sequences. The open question is how to best evaluate the results of these simulations in order to derive an explainable, high-quality rating for each line to show to an operator.

Finally, as already pointed out in Section 4, further work could be done to enable additional use cases of our simulation model and scheduling approach. The two use cases described in Section 4 would be achievable with relatively little work. More work would be needed for the following interesting extension to the first use case from Section 4. Instead of simply predicting the next time a desired partial plant state appears, it would be even better to specifically schedule in such a way to achieve that partial plant state around a given time. This would need more evolved methodology to compute the most reliable way to go from the current plant state to the desired plant state in, if possible, exactly the targeted time. If such a support could be designed, it would mean invaluable help to the scheduling and preparation of maintenances and similar processes, which need the plant in a specific partial state before they can be executed.

References

[1] P. Brucker, Scheduling Algorithms, Springer-Verlag Berlin Heidelberg, 2007.

[2] L. Schwehm, Two Staged Job-Shops with Shared Machines on the First and Dedicated Machines on the Second Stage (Master Thesis), Technische Universität Kaiserslautern, Kaiserslautern, 2020.

[3] T.C.E. Cheng, B.M.T. Lin, Y. Tian, Scheduling of a two-stage differentiation flowshop to minimize weighted sum of machine completion times, Comput. Oper. Res. 36 (11) (2009) 3031–3040, https://doi.org/10.1016/j.cor.2009.02.001.

[4] F.J. Hwang, B.M.T. Lin, Survey and extensions of manufacturing models in two-stage flexible flow shops with dedicated machines, Comput. Oper. Res. 98 (2018) 103–112, https://doi.org/10.1016/j.cor.2018.05.016.

[5] A. Ouled Bedhief, N. Dridi, Minimizing makespan in a three-stage hybrid flow shop with dedicated machines, Int. J. Ind. Eng. Comput. (2019) 161–176, https://doi.org/10.5267/j.ijiec.2018.10.001.

[6] J. Yang, A new complexity proof for the two-stage hybrid flow shop scheduling problem with dedicated machines, Int. J. Prod. Res. 48 (5) (2010) 1531–1538, https://doi.org/10.1080/00207540802691374.

[7] J. Yang, Minimizing total completion time in two-stage hybrid flow shop with dedicated machines, Comput. Oper. Res. 38 (7) (2011) 1045–1053, https://doi.org/10.1016/j.cor.2010.10.009.

[8] S.S. Panwalkar, M.L. Smith, C. Koulamas, Review of the ordered and proportionate flow shop scheduling research, Nav. Res. Logist. 60 (1) (2013) 46–55, https://doi.org/10.1002/nav.21518.

[9] L.T. Biegler, I.E. Grossmann, A.W. Westerberg, Systematic Methods of Chemical Process Design, Prentice Hall PTR, 1999.

[10] S. Giritligil, S. Cesur, B. Kuryel, Overview of planning and scheduling of batch process operations, IFAC Proc. Vol. 31 (11) (1998) 455–460, https://doi.org/10.1016/S1474-6670(17)44968-8.

[11] M. Hegyhati, F. Friedler, Overview of industrial batch process scheduling, Chem. Eng. Trans. 21 (2010) 895–900, https://doi.org/10.3303/CET1021150.

[12] C.A. Méndez, J. Cerdá, I.E. Grossmann, I. Harjunkoski, M. Fahl, State-of-the-art review of optimization methods for short-term scheduling of batch processes, Comput. Chem. Eng. 30 (6–7) (2006) 913–946, https://doi.org/10.1016/j.compchemeng.2006.02.008.

[13] L.S. Dias, M.G. Ierapetritou, Integration of scheduling and control under uncertainties: review and challenges, Chem. Eng. Res. Des. 116 (2016) 98–113, https://doi.org/10.1016/j.cherd.2016.10.047.

[14] Z. Li, M. Ierapetritou, Process scheduling under uncertainty: review and challenges, Comput. Chem. Eng. 32 (4–5) (2008) 715–727, https://doi.org/10.1016/j.compchemeng.2007.03.001.

[15] I.R. Uhlmann, E.M. Frazzon, Production rescheduling review: opportunities for industrial integration and practical applications, J. Manuf. Syst. 49 (2018) 186–193, https://doi.org/10.1016/j.jmsy.2018.10.004.

[16] D. Gupta, C.T. Maravelias, J.M. Wassick, From rescheduling to online scheduling, Chem. Eng. Res. Des. 116 (2016) 83–97, https://doi.org/10.1016/j.cherd.2016.10.035.

[17] S.L. Janak, C.A. Floudas, J. Kallrath, N. Vormbrock, Production scheduling of a large-scale industrial batch plant. II. Reactive scheduling, Ind. Eng. Chem. Res. 45 (25) (2006) 8253–8269, https://doi.org/10.1021/ie0600590.

[18] M.T.M. Rodrigues, L. Gimeno, C.A.S. Passos, M.D. Campos, Reactive scheduling approach for multipurpose chemical batch plants, Comput. Chem. Eng. 20 (1996) S1215–S1220, https://doi.org/10.1016/0098-1354(96)00210-4.

[19] G.M. Kopanos, E. Capón-García, A. Espuña, L. Puigjaner, Costs for rescheduling actions: a critical issue for reducing the gap between scheduling theory and practice, Ind. Eng. Chem. Res. 47 (22) (2008) 8785–8795, https://doi.org/10.1021/ie8005676.

[20] D. Gupta, C.T. Maravelias, On the design of online production scheduling algorithms, Comput. Chem. Eng. 129 (2019) 106517, https://doi.org/10.1016/j.compchemeng.2019.106517.

[21] D. Gupta, C.T. Maravelias, Online scheduling: understanding the impact of uncertainty, IFAC-PapersOnLine 52 (1) (2019) 727–732, https://doi.org/10.1016/j.ifacol.2019.06.149.

[22] G. Joglekar, Incorporating enhanced decision-making capabilities into a hybrid simulator for scheduling of batch processes, Processes 4 (3) (2016) 30, https://doi.org/10.3390/pr4030030.

[23] G. Joglekar, Using simulation for scheduling and rescheduling of batch processes, Processes 5 (4) (2017) 66, https://doi.org/10.3390/pr5040066.

[24] K. Kuriyan, G.V. Reklaitis, G. Joglekar, Multiproduct plant scheduling studies using BOSS, Ind. Eng. Chem. Res. 26 (8) (1987) 1551–1558, https://doi.org/10.1021/ie00068a010.

[25] G.S. Fishman, Discrete-Event Simulation: Modeling, Programming, and Analysis, Springer Science & Business Media, 2013.

[26] A.A. Juan, B.B. Barrios, E. Vallada, D. Riera, J. Jorba, A simheuristic algorithm for solving the permutation flow shop problem with stochastic processing times, Simul. Model. Pract. Theory 46 (2014) 101–117, https://doi.org/10.1016/j.simpat.2014.02.005.

[27] M. van den Akker, H. Hoogeveen, Minimizing the number of late jobs in a stochastic setting using a chance constraint, J. Sched. 11 (1) (2008) 59–69, https://doi.org/10.1007/s10951-007-0034-8.

[28] G.O. Odu, Review of multi-criteria optimization methods—theory and applications, IOSR J. Eng. 3 (10) (2013) 01–14, https://doi.org/10.9790/3021-031020114.

Chapter 14

Applications of the RTN scheduling model in the chemical industry

Hector D. Perez[a], Satyajith Amaran[b], Shachit S. Iyer[b], John M. Wassick[b], and Ignacio E. Grossmann[a]
[a]*Carnegie Mellon University, Pittsburgh, PA, United States,* [b]*The Dow Chemical Company, Midland, MI, United States*

Nomenclature

Sets

$i \in \mathcal{I}$	tasks or package types
$i \in \mathcal{I}_r$	tasks involving resource r
$i \in \mathcal{I}^{PO}$	point order tasks
$i \in \mathcal{I}_r^{PO}$	point order tasks involving material r
$i \in \mathcal{I}_r^{ST}$	storage tasks involving material r
$i \in \mathcal{I}^{TR}$	changeover (transition) tasks
$k \in \mathcal{K}_i$	constraints on the multiple extent task i
$m \in \mathcal{M}_i$	extents associated with task i
$o \in O$	orders
$r \in \mathcal{R}$	resources
$r \in \mathcal{R}^{EQ}$	equipment
$r \in \mathcal{R}^{P}$	products
$r \in \mathcal{R}^{S}$	shipped products
$r \in \mathcal{R}^{ST}$	stored materials
$r \in \mathcal{R}_o$	materials requested in order o
$t \in \mathcal{T}$	time points
$x \in \mathcal{X}$	horizontal coordinate points
$x \in \mathcal{X}_i$	horizontal resources consumed by task i
$y \in \mathcal{Y}$	vertical coordinate points

Simulation and Optimization in Process Engineering. https://doi.org/10.1016/B978-0-323-85043-8.00006-4

Parameters

α_i	constant processing time ratio for task i
β_i	proportional processing time ratio for task i
$\gamma_{i,k,t}^{min}, \gamma_{i,k,t}^{max}$	upper and lower bounds on constraint k associated with the multiple extents in task i at time point t
$\delta_{i,k,m,t}$	coefficient for constraint k associated with extent m of task i at time point t
Δt	maximum number of time periods allowed for a task
$\mu_{i,r,t}$	discrete consumption/production ratio for resource r in task i at time point t
$\mu_{i,r}^C, \mu_{i,r}^P$	discrete consumption and production ratios for resource r in task i, respectively
$\mu_{i,r}^{min}, \mu_{i,r}^{max}$	discrete consumption/production ratios for the lower and upper limits of resource r in task i, respectively
$\nu_{i,r,t}$	variable consumption/production ratio for resource r in task i at time point t
$\nu_{i,m,r,t}$	variable consumption/production ratio for resource r in extent m of task i at time point t
$\nu_{i,r}^C, \nu_{i,r}^P$	variable consumption and production ratios for resource r in task i, respectively
$\nu_{i,r}^{min}, \nu_{i,r}^{max}$	variable consumption/production ratios for the lower and upper limits of resource r in task i at time point t, respectively
τ_i	duration of task i
τ_d	duration of delay d
c_r^{inv}	inventory holding cost for material r
c_r^{min}, c_r^{max}	penalty (cost) for violating lower and upper resource limits of resource r
c_r^{slack}	penalty (cost) for not matching initial and end-of-horizon inventory levels of material r
c_i^{TR}	cost for changeover (transition) task i
D_i	final due date for point order task i
D_o	final due date for order o
D_K	distance of the king ping from the front of the semitrailer (x-direction)
D_T	distance of the tandem axle from the front of the semitrailer (x-direction)
E_i	early acceptance date for point order task i
E_o	early acceptance date for order o
F_T^{max}	maximum force that can be exerted on a tandem axle
h	last time point in set $\mathcal{TT}$
H	scheduling horizon (i.e., T_h)
L_i	length (x-direction) of a box of type i

m_i weight of a box of type i
p_r sales price of product r
$p_{o,r,t}$ sales price of product r in order o that is delivered at time point t
$Q_{o,r}^{min}, Q_{o,r}^{max}$ minimum and maximum quantities of material r requested in order o, respectively
Q_o^{min}, Q_o^{max} minimum and maximum quantities of materials in order o, respectively
$\dot{Q}_{i,m}^{min}, \dot{Q}_{i,m}^{max}$ minimum and maximum production rates for extent m in task i.
V_i^{min}, V_i^{max} minimum and maximum limits on the extent for task i
$V_{i,m}^{min}, V_{i,m}^{max}$ minimum and maximum limits on extent m in task i
W_i width (y-direction) of a box of type i
W^{max} maximum weight allowed on a semitrailer

Continuous (nonnegative) variables

ϕ objective function value
$\Pi_{r,t}$ external transfer of resource r at time point t
$\Pi_{r,t}^{in}, \Pi_{r,t}^{out}$ incoming and outgoing external transfers of resource r at time point t, respectively
$\xi_{i,t}$ extent of task i at time point t
$\xi_{i,m,t}$ extent m for task i at time point t
$\xi_{i,m,t,\theta}$ extent m for task i that has been actively running for θ time periods at time point t
$\xi_{i,t,t'}$ extent of task i starting at time point t and ending at or before time point t'
F_K force exerted by the king pin on a semitrailer
F_T force exerted by the tandem axle on a semitrailer
I_r^{ave} time-averaged inventory of material r
$R_{r,t}$ resource inventory level for resource r at time point t
$R_{x,y}$ horizontal floor space resource level at location x, y
R_r^{min}, R_r^{max} minimum and maximum limits for resource r, respectively
$R_{r,t}^{min}, R_{r,t}^{max}$ minimum and maximum limits for resource r at time point t, respectively
$R_{r,t}^{min,\ slack}$ violation of the lower limit of resource r at time point t
$R_{r,t}^{max,\ slack}$ violation of the upper limit of resource r at time point t
R_r^{slack} violation of end-of-horizon inventory constraint of material r
T_t time at time point t

Discrete variables

$d_{i,t,\theta}$ indicator of a one period delay at time point t on task i that has been actively running for θ time periods
$N_{i,t}$ number of occurrences of task i starting at time point t

$N_{i,t,\theta}$ number of occurrences of task i that have been actively running for θ time periods at time point t

$N_{i,t,t'}$ occurrence of task i starting at time point t and ending at or before time point t'

$N_{i,x,y}$ indicator that the bottom left corner of a box of type i has been placed with at location x, y

1 Introduction

Over the past three decades, the Process Systems Engineering (PSE) community has pioneered the area of optimization-based process scheduling [1,2]. The essence of optimization-based scheduling is mathematical programming, and more specifically, mixed-integer programming (MIP), where a system is modeled by a set of continuous and discrete variables (e.g., batch sizes, process start times, equipment assignment) that are constrained by a series of algebraic inequalities and equalities (e.g., material balances, resource assignment constraints). An objective function is added to the resulting model, and the model is passed to a MIP solver that performs the optimization by seeking the point in the discrete-continuous feasible space with the optimum objective function value. Objective functions vary depending on what is being targeted with the model. Common examples of objective functions include profit maximization, cost minimization, production makespan minimization, and tardiness minimization. Over the years, several models have been proposed to optimize the timing and sequencing of events at a chemical plant, as well as the allocation of limited resources, such as equipment, feedstock, utilities, and operators [3,4]. At first, MIP models for scheduling were not readily accepted by the scientific community due to the claim that MIP models were too difficult to solve to render them of any practical use. However, improved model formulations [5], new algorithmic developments [6], and the evolution of computing power [7] have facilitated the adoption of these models by industry, generating value in operations and manufacturing [8].

One of the more versatile, simple, and compact models proposed for process scheduling in PSE is the Resource-Task Network (RTN). The RTN model was developed in the 1990s and has had a significant impact on the chemical industry since its inception. In the last decade, several extensions have been developed and successfully implemented in industrial processes by researchers from The Dow Chemical Company and Carnegie Mellon University. These extensions have been prompted in response to limitations that have been observed in the standard RTN formulation when implementing it in practice [9]. The strengths of the RTN paradigm have also led to its application in other areas of PSE and other industries. These include payload optimization [9], control systems [10], process network simulation models [11], supply chain transactional processes [12], and supply chain logistics [13].

2 Review of RTN model

In 1993, the group at Imperial College led by Professor Roger Sargent published a novel modeling framework for process scheduling called the State-Task Network (STN) [5,14]. This modeling paradigm regards all processes as bipartite graphs with two types of vertices: *states* and *tasks*. A state represents a material involved in the process being scheduled (e.g., raw material, intermediate, or product). A task is an operation in which incoming states are transformed into outgoing states. The STN paradigm in its original form does not account for equipment and other resources explicitly in its network representation but instead accounts for them mathematically with additional variables and constraints.

A year later, Costas Pantelides, a member of Roger Sargent's group, presented a conceptually more general and powerful modeling paradigm: the Resource-Task Network [15]. This paradigm continues representing processes as bipartite graphs but uses *resources* instead of *states*. The concept of a resource is general and includes all entities that are involved in the process steps, such as materials (e.g., raw materials, intermediates, and products), processing and storage equipment (e.g., tanks and reactors), and utilities (e.g., operators and steam). A sample diagram of an RTN is shown in Fig. 14.1, which depicts the connections between resources and task nodes, and describes process typical specifications required for process scheduling. The resulting RTN model is represented mathematically using mixed-integer programming (MIP for both discrete and continuous-time operations). Both commercial and open-source MIP solvers are available for optimizing the models. These solvers rely on techniques such as branch-and-bound [16], cutting planes

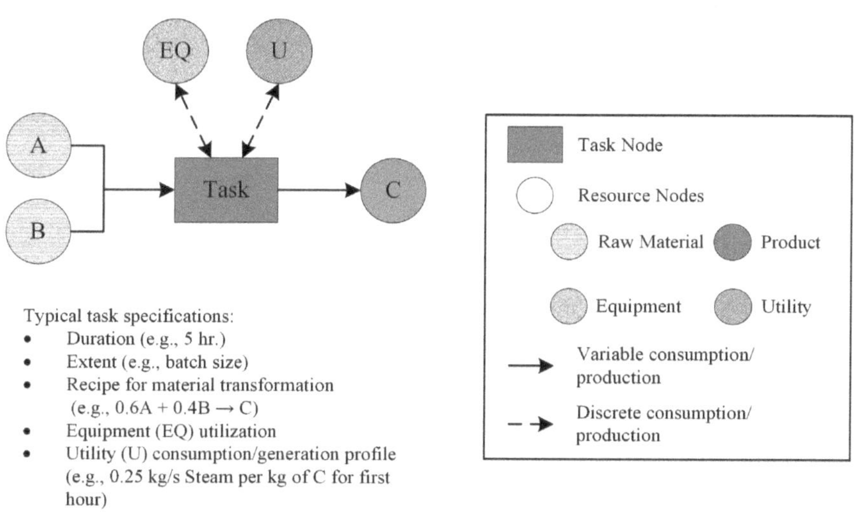

FIG. 14.1 Sample network structure of a Resource-Task Network.

[17], and branch-and-cut [18] when the models are linear. When nonlinearities are introduced in the models (e.g., when modeling inventory costs in continuous-time formulations), these may be solved using MINLP solvers that rely on branch-and-bound, Generalized Benders Decomposition [19], Outer-Approximation methods [20,21], and the Extended Cutting Plane method [22]. Another option is to linearize the MINLP models via exact linearization [23] or piecewise linear approximations [24,25].

The RTN paradigm has the advantage of producing more compact models, especially when systems have identical processing units (e.g., redundant or parallel equipment). When alternate processing units exist for the same task, a copy of the task node must be made for each alternate resource type. In its simplest form, the RTN model only requires three sets of constraints (resource balances, resource limits, and operational constraints) and three variable types (resource inventories, task extents, and task triggers). Both discrete-time and continuous-time representations exist for the RTN model.

Discrete-time representation

The original formulation of the RTN model presented by Pantelides [15] is for a discrete-time representation. The classical discrete-time formulation splits the temporal axis onto a uniform grid with time intervals of equal size as shown in Fig. 14.2. The three sets of constraints in this formulation are described below.

2.1.1 *Resource balance*

The resource level balance is given by Eq. (14.1), which indicates that the resource level for a resource r at time point t, $R_{r,t}$, is obtained by updating the resource level in the previous time point (or the initial resource value if t is the first time point) with the consumption or production of that resource and any entrance or exit of the resource into or from the system. The consumption and production of the resource is governed by the variables $N_{i,t}$ (integer)

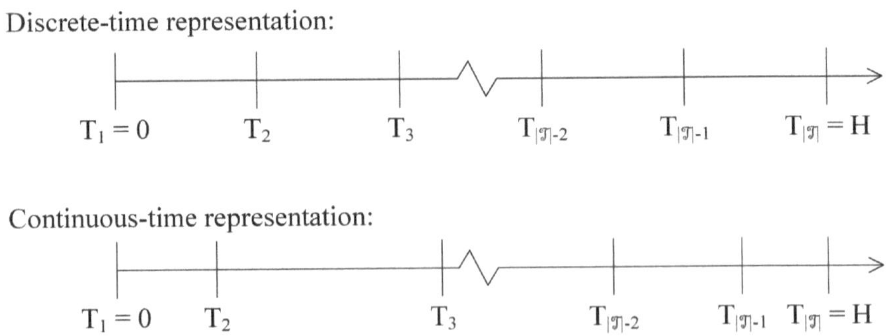

FIG. 14.2 Time points on the time axis used in discrete-time (top) and continuous-time (bottom) representations.

and $\xi_{i,t}$ (continuous), which indicate the number of occurrences of task i and the extent of the task at time point t, respectively. The parameters $\mu_{i,r,t}$ and $\nu_{i,r,t}$ indicate the consumption/production ratios relative to the number of task occurrences and the task extent, respectively. The sign on the parameter indicates if it is a consumption (negative) or production (positive) term. The parameter $\mu_{i,r,t}$ is for resources that are consumed in discrete quantities (e.g., equipment and operators), and $\nu_{i,r,t}$ is for resources that are consumed in variable quantities (e.g., materials and utilities).

$$R_{r,t} = R_{r,t-1} + \sum_{i \in \mathcal{I}_r} \sum_{\theta=0}^{\tau_i} \left(\mu_{i,r,\theta} \cdot N_{i,t-\theta} + \nu_{i,r,\theta} \cdot \xi_{i,t-\theta} \right) + \Pi_{r,t} \quad \forall r \in \mathcal{R}, t \in \mathcal{T} \tag{14.1}$$

The balance for resource r accounts for resource consumption and production in all tasks that involve the resource (set $\mathcal{I}_r$) and all-time points in the range $[t-\tau_i, t]$, where τ_i is the duration of task i, since consumption or production can occur at any time point within the task duration. For continuous processes, τ_i is the minimum duration of the continuous task, such that a continuous task can be thought of as multiple small tasks in series of τ_i duration (typically $\tau_i = 1$). The entrance and exit of resource r to the system at time t is governed by the parameter $\Pi_{r,t}$, which is positive when a resource enters (e.g., supply of raw materials), and negative when it leaves the system (e.g., delivery of a final product).

2.1.2 Resource limits

All resource inventories have lower and upper bounds as expressed in Eq. (14.2).

$$R_r^{min} \leq R_{r,t} \leq R_r^{max} \quad \forall r \in \mathcal{R}, t \in \mathcal{T} \tag{14.2}$$

2.1.3 Operational constraints

The extent of a task (e.g., batch size) is forced to zero when the task is not executed (i.e., $N_{i,t} = 0$), and is bounded between the limits V_i^{min} and V_i^{max}, when the task is performed (i.e., $N_{i,t} \in \mathbb{Z}^+$).

$$V_i^{min} \cdot N_{i,t} \leq \xi_{i,t} \leq V_i^{max} \cdot N_{i,t} \quad \forall i \in \mathcal{I}, t \in \mathcal{T} \tag{14.3}$$

2.1.4 Variable domains

Variable domains are $R_{r,t} \in \mathbb{R}^+ \ \forall r \in \mathcal{R}, t \in \mathcal{T}$ and $N_{i,t} \in \mathbb{Z}^+, \xi_{i,t} \in \mathbb{R}^+ \ \forall i \in \mathcal{I}, t \in \mathcal{T}$. It should be noted that resources that are consumed/produced in discrete quantities, such as processing equipment, may be declared as integer variables if desired. This may, in some cases, be computationally advantageous for the MIP solvers. MIP solvers may be able to further reduce model complexity during presolve by variable fixing and constraint propagation. Solvers may also more readily exploit integrality constraints when applying cutting plane methods [26,27].

2.1.5 Illustrative example

Consider a system with one reactor (RX) that transforms raw materials A and B into intermediate C, which is immediately distilled in two parallel distillation columns (DC_1 and DC_2) to produce products D and E, as shown in Fig. 14.3. The system begins with zero material inventories and has the following events: (1) a shipment of materials A (60 units) and B (40 units) enters at $t=1$, (2) the reactor runs from $t=2$ to $t=4$ (100-unit batch with a 60A/40B feed ratio), (3) the two columns separate 50-unit batches simultaneously from $t=4$ to $t=5$, and (4) all products get shipped out at $t=6$. The RTN representation for this system is shown in Fig. 14.4. Resources are consumed when a task starts and produced task completes. Fig. 14.5 shows the resulting schedule and the resource inventory levels at each time point. Note that although intermediate C is produced at $t=4$, it is immediately consumed, causing its inventory to always be zero.

2.2 Continuous-time representation

In order to overcome the limitations of having to specify the length of the time intervals in the original discrete-time RTN formulation, which gave rise to large-scale Mixed-Integer Linear Programming (MILP) models, several alternative continuous-time formulations have been proposed by various researchers. The more modern and robust version of this continuous-time formulation was developed by researchers in Portugal at the INETI and the Instituto Superior Técnico during the early 2000s [28,29]. This formulation considers time as continuous, where the system state is monitored at specified time points, the location of which is to be determined. This representation can be thought of as having a flexible nonuniform discrete-time grid as shown in Fig. 14.2, where the modeler sets the number of positions on the temporal axis to monitor, and the optimizer decides the best locations for these time points. The assumption here is that tasks can only be triggered at these time points,

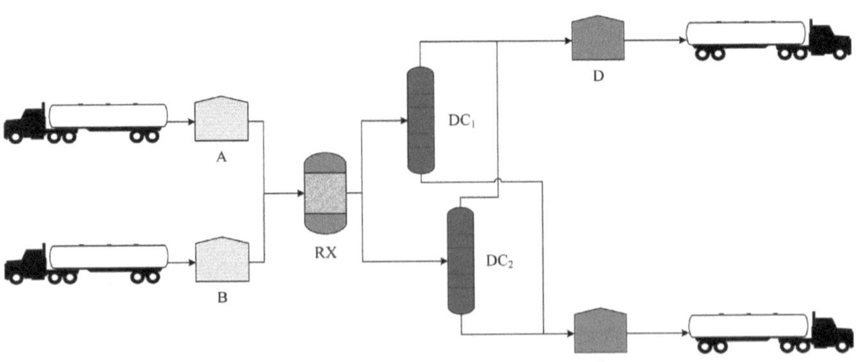

FIG. 14.3 Flow diagram for illustrative example.

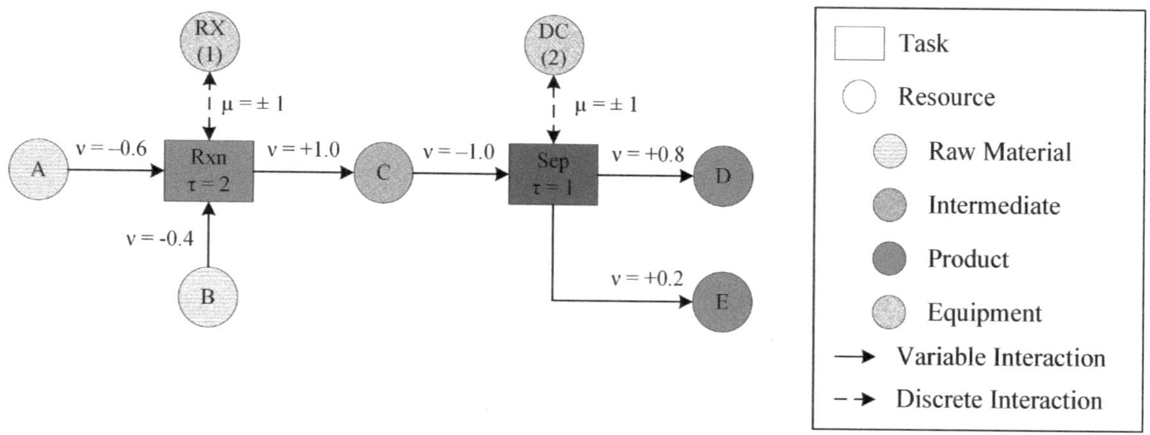

FIG. 14.4 RTN representation for illustrative example with discrete (μ) and variable (ν) consumption and production ratios (subscripts are dropped for simplicity).

Resource	Time Point 0	1	2	3	4	5	6
A	0	60	0	0	0	0	0
B	0	40	0	0	0	0	0
C	0	0	0	0	0	0	0
D	0	0	0	0	0	80	0
E	0	0	0	0	0	20	0
RX	1	1	0	0	1	1	1
DC	2	2	2	2	0	2	2

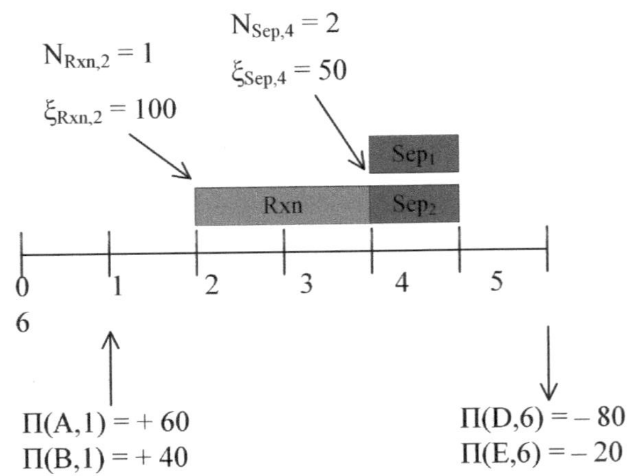

FIG. 14.5 Schedule of events and variable values in an illustrative example.

but can end at any time according to the duration of the task. A major challenge of this formulation is in determining the required number of time points. Since the optimal number of time points is not known a priori, the common practice is to successively increase the number of time points in the set $\mathcal{T}$ until no significant improvement is seen in the final objective function value. However, as pointed out by Castro et al. [29], this can become intractable and may require fixing variables to reduce model complexity and avoid long computational times. See also Lee and Maravelias [30] for an additional discussion on various strategies in this area.

2.2.1 Timing and sequencing

The difference between any two time points, $t<t'$, on the flexible nonuniform time grid is determined by the duration of all tasks starting at time point t and ending in the interval $(t'-1, t']$, as given by Eq. (14.4). The task durations can be modeled as having two parts: one composed of a fixed task duration (α_i), and the other of a variable task duration (β_i) that is proportional to the extent of the task. This constraint ensures that the time indices on the triggering and extent variables ($N_{i,t,t'}$ and $\xi_{i,t,t'}$, respectively) are mapped to the temporal axis accordingly. It also ensures that the difference between any two time points is greater than zero. Note that t and t' are not necessarily adjacent points on the time grid. The first and last time points are fixed to the beginning and end of the scheduling horizon: $T_0=0$ and $T_{|\mathcal{T}|}=H$. The feasible space defined by Eq. (14.4) can be further reduced by applying it to $t' \leq \Delta t + t$, where Δt is a parameter that represents the maximum number of time periods allowed for a task in the process.

$$T_{t'} - T_t \geq \sum_{i \in \mathcal{I}_r} \left(\alpha_i \cdot N_{i,t,t'} + \beta_i \cdot \xi_{i,t,t'} \right) \quad \forall r \in \mathcal{R}^{EQ}, t \in \mathcal{T}, t' \in \mathcal{T}, t < t' \tag{14.4}$$

2.2.2 Resource balance

The resource inventory balance is analogous to that used in the discrete-time formulation. The production and consumption terms are split into the two summations in the second term on the right-hand side of Eq. (14.5) due to the double-time point indexing on the variables. The first summation is for resource production and the second for resource consumption. An additional term is added at the end for storage tasks associated with resource r ($\mathcal{I}_r^{ST}$) to account for the consumption and release of storage resources, which is particularly relevant in continuous processes. Additional constraints for these storage tasks are presented in Castro et al. [29]. There is also no parameter for external interactions in the base formulation since the exact position of the time points is not known a priori and would require additional logic-based constraints, which increases model complexity [31].

$$
\begin{aligned}
R_{r,t} = {} & R_{r,t-1} \\
& + \sum_{i \in \mathcal{I}_r} \left[\sum_{t'<t} \left(\mu_{i,r}^{P} \cdot N_{i,t',t} + \nu_{i,r}^{P} \cdot \xi_{i,t',t} \right) + \sum_{t<t'} \left(\mu_{i,r}^{C} \cdot N_{i,t,t'} + \nu_{i,r}^{C} \cdot \xi_{i,t,t'} \right) \right] \\
& + \sum_{i \in \mathcal{I}_r^{ST}} \left(\mu_{i,r}^{P} \cdot N_{i,t-1,t} + \mu_{i,r}^{C} \cdot N_{i,t,t+1} \right) \quad \forall r \in \mathcal{R}, t \in \mathcal{T}
\end{aligned}
\tag{14.5}
$$

2.2.3 Variable domains

The variable domains are the same as in the discrete-time formulation, except $N_{i,t,t'}$ is now binary.

2.3 Discrete-time vs continuous-time

2.3.1 Representation of time

Continuous-time mathematical models are more accurate and flexible since time is a variable. On the other hand, discretization often introduces rounding error, especially for coarse time grids.

2.3.2 Model size

As the time discretization becomes more refined, discrete-time models become significantly larger. The size of the time step needs to be specified by the user to minimize round-off errors that result in low-quality schedules. This trade-off between tractability and accuracy is relatively straightforward to determine by inspecting the system time scales. On the other hand, continuous-time models typically require fewer variables. The challenge here lies in determining the number of time points to use, which requires a successive manual adaptive refinement, as mentioned previously.

2.3.3 Linear programming relaxations

Discrete-time formulations have been found to present much tighter linear programming relaxations, making them usually faster to solve than continuous-time models, despite their larger model sizes [8]. Discretization has also become less of an issue with the great advances by MILP solvers (e.g., CPLEX, Gurobi, and Xpress).

2.3.4 Objective functions

Continuous-time RTN is more amenable to makespan minimization than its discrete-time counterpart, which requires specifying the magnitude of the scheduling horizon a priori. On the other hand, discrete-time RTN handles cost minimization better than the continuous-time model, especially when inventory costs are considered, as these introduce nonlinearities in the latter because time is a variable [3]. In addition to preserving model linearity when accounting for inventory, discrete-time models monitor inventory and resources effectively because the monitoring points are fixed on the time grid.

2.3.5 *Discrete-continuous-time integration*

More recently, efforts have been made to integrate the advantages of both discrete and continuous-time scheduling algorithms in work done at the University of Wisconsin—Madison led by Professor Christos T. Maravelias, who has since moved to Princeton University [30]. Their work, which is based on the STN model, proposes solving a coarse discrete-time MIP model to fix key assignment decisions and obtain an approximate schedule. A mapping algorithm then maps the approximate solution onto units or material-specific temporal grids. A continuous-time linear programming (LP) model then refines the initial discrete-time solution and eliminates approximation errors. A similar procedure could be applied in RTN to help overcome the challenges of the different model representations and exploit the best of both worlds.

3 Industry-led developments

3.1 Extended RTN model

3.1.1 *Quality-based changeovers*

During scheduling optimization, especially batch scheduling, equipment changeovers are considered when transitioning from one family of materials to another due to safety (e.g., chemical stability), throughput (e.g., fouling), operational constraints (e.g., time to reach steady-state in semicontinuous processes), or quality reasons (e.g., product purity). Equipment cleaning tasks are included in the RTN framework via an additional changeover task. When the RTN model accounts for changeover tasks, duplicate equipment can no longer be aggregated in the same type of resource. Instead, each equipment involved in changeovers needs to be treated as its own resource to track which one is clean or needs cleaning [3].

Another type of changeover that is used in industry when a changeover is required for quality reasons is that of quality-based changeovers. The idea with this type of changeover is to have the option of forgoing an equipment cleaning step when transitioning from one material family type to another by performing several sequential batches of the second family in tandem to dilute any residual contamination from the first family. This extension was first addressed explicitly by Brunaud et al. [32]. The diagram in Fig. 14.6 details the general RTN structure for a system with a single unit (e.g., reactor) that can perform two different reactions ($Rx1$ and $Rx2$). The first reaction converts $A \rightarrow E$ and the second reaction $B \rightarrow F$. When transitioning from $Rx1$ to $Rx2$, either a cleaning step or a quality-based changeover with $N+1$ sequential repetitions of $Rx2$ can be performed. When transitioning from $Rx2$ to $Rx1$, either a cleaning step or a

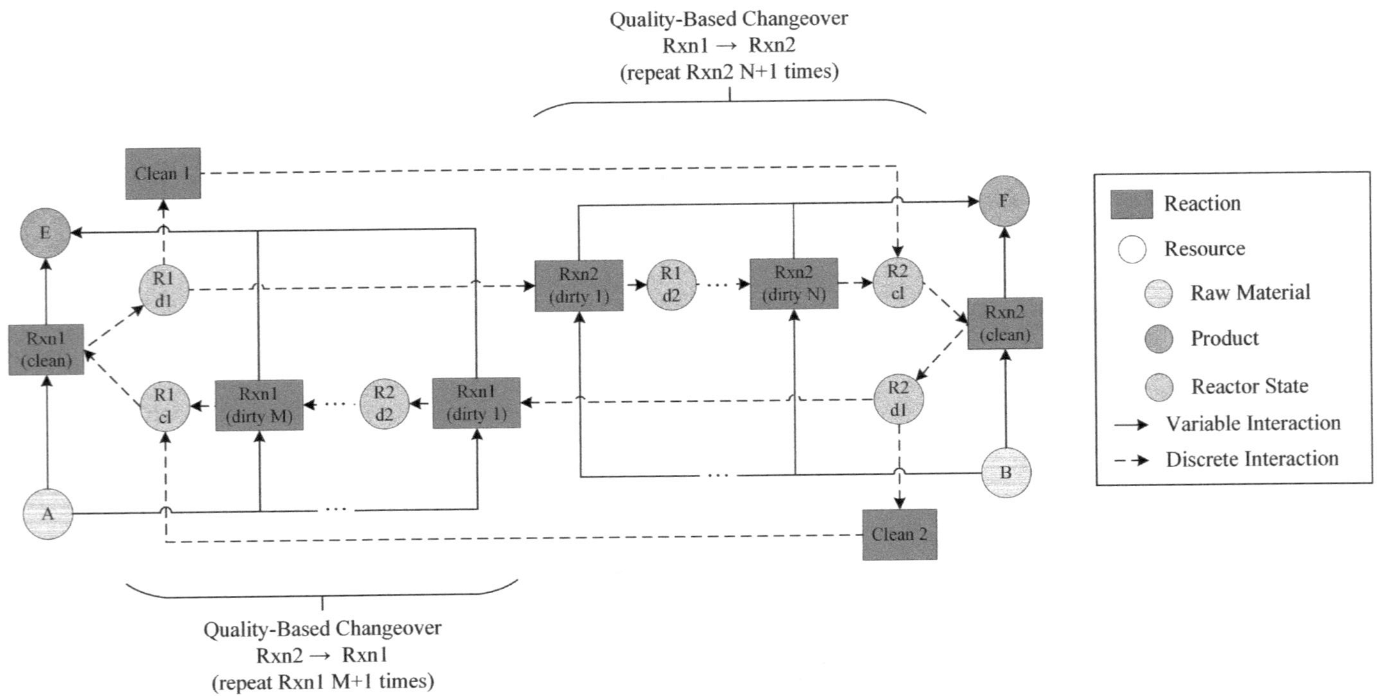

FIG. 14.6 RTN representation illustrating both quality-chased changeovers and cleaning steps to transition between Reaction 1 and Reaction 2 on a single reactor.

quality-based changeover with $M+1$ sequential repetitions of $Rx1$ can be performed.

The RTN structure has four material resources, one for each material. Instead of one resource for the reactor, which would be the case for a system neglecting changeovers, there are $N+M+2$ resources for the single reactor, representing the single reactor in its clean state ($R1cl$ and $R2cl$) or dirty state ($R1dx$ and $R2dy$ where $x \in \{1,\ldots,N\}$ and $y \in \{1,\ldots,M\}$ indicate the number of repetitions of the second task). The initial values of these resource levels should be such that only one of the reactor states has an initial value of one and the rest are zero-valued. This, along with the proper consumption and production ratios, ensures that the unit can only be in one state at a time. With a quality-based changeover, instead of two reaction tasks, there are $N+M+2$ reaction tasks. As the changeover occurs, impurities in the product tank carried over from the previous reaction are gradually diluted. Under this structure, the network complexity increases linearly with the number of task repetitions for each quality-based changeover and with the number of task transition pairs.

3.1.2 External resource transfers with time windows

Orders placed by customers or placed on suppliers can be mapped to the $\Pi_{r,t}$ term in the resource balance (Eq. 14.1) via Eqs. (14.6) and (14.7), respectively [9]. The $\Pi_{r,t}$ is separated into two terms to distinguish outgoing transfers ($\Pi_{r,t}^{out}$) from incoming transfers ($\Pi_{r,t}^{in}$). The set $\mathcal{R}_o$ contains the materials r associated with order o. The summation terms are for external material flows occurring between the early acceptance date (E_o) and final due date (D_o) of each order. It should be noted that the parameters E_o and D_o are converted from time values to time points using the discretization parameter (e.g., $D_o=5$ for a 10 AM due date on the first day when the discretization time step is 2 hours and time point set $\mathcal{T}$ is zero-indexed). In a general sense, orders may also have a minimum and a maximum quantity required for each material ($Q_{o,r}^{min}$ and $Q_{o,r}^{max}$, respectively). Slack variables can also be introduced to ensure that the model is feasible even if $Q_{o,r}^{min}$ is not met. These slacks are penalized in the objective function.

$$Q_{o,r}^{min} - Q_{o,r}^{min,slack} \leq \sum_{t=E_o}^{D_o} -\Pi_{r,t}^{out} \leq Q_{o,r}^{max} \quad o \in \mathcal{O}, r \in \mathcal{R}_o \tag{14.6}$$

$$Q_{o,r}^{min} - Q_{o,r}^{\min,slack} \leq \sum_{t=E_o}^{D_o} \Pi_{r,t}^{in} \leq Q_{o,r}^{max} \quad o \in \mathcal{O}, r \in \mathcal{R}_o \tag{14.7}$$

3.1.3 Point orders

To ensure that external deliveries or shipments are made in a single material transfer, a common requirement in make-to-order systems, a point order task can be defined [9]. The point order task duration is typically much smaller than the delivery window for the order and is often zero (instantaneous order fulfillment). Point orders are included as an additional term in the resource balance that has the same form as the standard consumption/production term. The difference here is that the summation is applied only in the allowed delivery time point window for each point order i ($E_i \le t \le D_i - \tau_i$) as shown in Eq. (14.8). The set $\mathcal{I}_r^{PO}$ is the subset of point order tasks that involve resource r. The bounds on the point order extents ($\xi_{i,t}$) are the order limits ($Q_o^{min} = \sum_{r \in \mathcal{R}_o} Q_{o,r}^{min}$ and $Q_o^{max} = \sum_{r \in \mathcal{R}_o} Q_{o,r}^{max}$) as shown in Eq. (14.9). The constraint in Eq. (14.10) ensures that each point order is triggered at most once in the allowed time window. The variable $N_{i,t}$ for the point order task i should be fixed to zero for all other time points. If multiple shipments are feasible, the right-hand side can be replaced with an integer value equal to the maximum number of shipments allowed.

$$R_{r,t} = R_{r,t-1} + \sum_{i \in \mathcal{I}_r} \sum_{\theta=0}^{\tau_i} \left(\mu_{i,r,\theta} \cdot N_{i,t-\theta} + \nu_{i,r,\theta} \cdot \xi_{i,t-\theta} \right)$$
$$+ \sum_{i \in \mathcal{I}_r^{PO}} \sum_{\theta=0}^{\tau_i} \left(\mu_{i,r,\theta} \cdot N_{i,t-\theta} + \nu_{i,r,\theta} \cdot \xi_{i,t-\theta} \right) \Big|_{E_i \le t \le D_i - \tau_i} + \Pi_{r,t} \quad \forall r \in \mathcal{R}, t \in \mathcal{T} \tag{14.8}$$

$$\left(\sum_{r \in \mathcal{R}_o} Q_{o,r}^{min} \right) \cdot N_{i,t} \le \xi_{i,t} \le \left(\sum_{r \in \mathcal{R}_o} Q_{o,r}^{max} \right) \cdot N_{i,t} \quad \forall i \in \mathcal{I}^{PO}, t \in \mathcal{T} \tag{14.9}$$

$$\sum_{t=E_i-\tau_i}^{D_i-\tau_i} N_{i,t} \le 1 \quad \forall i \in \mathcal{I}^{PO} \tag{14.10}$$

3.1.4 Manipulation of resource limits

Material balances can be written for the resource level limits as well to facilitate operations such as material transfers without having to define auxiliary tasks, making the model more scalable [9]. This is particularly useful for situations in which total material transfers from one unit to another are required (e.g., when transferring the entire contents of a tank to a truck for delivery). In the standard RTN model, this requires defining auxiliary storage tasks and storage resources to guarantee that the entire contents of a tank are transferred to the truck in a single operation. The downside of doing this is that additional integer variables for the auxiliary tasks are introduced. Instead, continuous variables can be introduced by parametrizing the resource limits (R_r^{min} and R_r^{max}) with respect to time as shown in Eq. (14.11). These new variables ($R_{r,t}^{min}, R_{r,t}^{max} \in \mathbb{R}^+$) are bounded by the static resource limit parameters as indicated in Eq. (14.12). Additional constraints, given in Eqs. (14.13) and (14.14), are then defined for these variables, analogous to the resource inventory balance in Eq. (14.1). Discrete and variable resource limit production/consumption rates are introduced for the respective resource inventory limits (i.e., $\mu^{max}_{i,r,t}$ and $\nu^{max}_{i,r,t}$ govern the discrete and continuous consumption of the limit $R^{max}_{r,t}$, respectively).

$$R^{min}_{r,t} \leq R_{r,t} \leq R^{max}_{r,t} \quad \forall r \in \mathcal{R}, t \in \mathcal{T} \tag{14.11}$$

$$R^{min}_{r} \leq R^{min}_{r,t}, R^{max}_{r,t} \leq R^{max}_{r} \quad \forall r \in \mathcal{R}, t \in \mathcal{T} \tag{14.12}$$

$$R^{max}_{r,t} = R^{max}_{r,t-1} + \sum_{i \in \mathcal{I}_r} \sum_{\theta=0}^{\tau_i} \left(\mu^{max}_{i,r,\theta} \cdot N_{i,t-\theta} + \nu^{max}_{i,r,\theta} \cdot \xi_{i,t-\theta}\right) \quad \forall r \in \mathcal{R}, t \in \mathcal{T} \tag{14.13}$$

$$R^{min}_{r,t} = R^{min}_{r,t-1} + \sum_{i \in \mathcal{I}_r} \sum_{\theta=0}^{\tau_i} \left(\mu^{min}_{i,r,\theta} \cdot N_{i,t-\theta} + \nu^{min}_{i,r,\theta} \cdot \xi_{i,t-\theta}\right) \quad \forall r \in \mathcal{R}, t \in \mathcal{T} \tag{14.14}$$

The new RTN structure for this extension is illustrated in Fig. 14.7, where the material in tank *A* is transferred to the delivery truck *B*. When the transfer task *i* is triggered, the upper limit of resource *A* is consumed by 100 units (e.g., ${\mu_{i,A,t-\tau_i}}^{max} = -100$), reducing it to zero, and thus forcing the entire resource to be transferred to resource *B*. Once the transfer task completes, the upper limit of resource *A* is restored (e.g., $\mu^{max}_{i,A,t} = +100$).

3.1.5 Resource slacks

Slack variables can be used to introduce soft bounds on resource inventories to produce schedules that would otherwise be infeasible. These include not having enough material to produce a full batch at the end of the scheduling horizon or high-demand scenarios where it may be the allowed model to violate material inventory safety stocks [9]. The slacks are penalized in the objective.

$$R^{min}_{r,t} - R^{min,slack}_{r,t} \leq R_{r,t} \leq R^{max}_{r,t} - R^{max,slack}_{r,t} \quad \forall r \in \mathcal{R}, t \in \mathcal{T} \tag{14.15}$$

3.1.6 Multiple task extents

Multiple task extents can be used when multiple operations are performed under a single governing task [9]. Doing so avoids defining binary variables

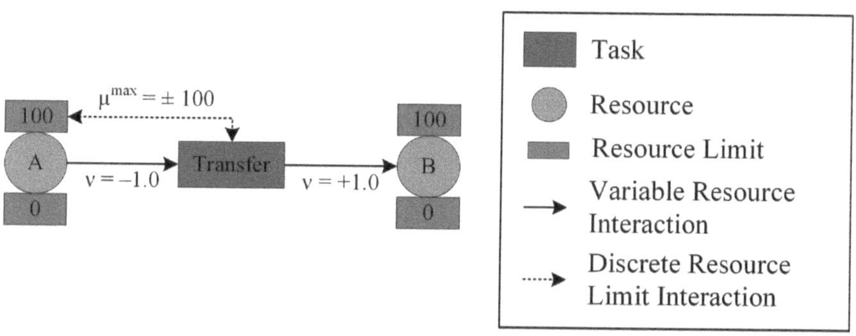

FIG. 14.7 Example of extended RTN representation with resource limit interactions.

for the triggering of each operation involved. Instead, each operation is triggered by the governing task, $N_{i,t}$, which reduces the model complexity. To implement multiple task extents, the resource inventory balance in Eq. (14.1) is modified so that the task extent and variable interaction parameters are specific to the extent number, m (i.e., $\xi_{i,m,t}$ and $\nu_{i,m,r,t}$, respectively), as shown in Eq. (14.16). The index m is also added to the task extent bounds in Eq. (14.3).

$$R_{r,t} = R_{r,t-1} + \sum_{i \in \mathcal{I}_r} \sum_{\theta=0}^{\tau_i} \left(\mu_{i,r,\theta} \cdot N_{i,t-\theta} + \sum_{m \in \mathcal{M}_i} \nu_{i,m,r,\theta} \cdot \xi_{i,m,t-\theta} \right) + \Pi_{r,t} \quad \forall r \in \mathcal{R}, t \in \mathcal{T} \tag{14.16}$$

Multiple extents can be used to represent different material sizes processed in an operation, such as when a material is being split or materials are being combined in ways that cannot be described by single extents and consumption/production ratios. These multiple extents are further constrained with inequalities of the form shown in Eq. (14.17), where the set $\mathcal{K}_i$ is the set of constraints assigned to the multiple extents in task i. These constraints can be used to enforce material balances among the multiple procedures involved in the governing task (see the bottle filling task example given in Wassick and Ferrio [9]).

$$\gamma_{i,k,t}^{min} \le \sum_{m \in \mathcal{M}_i} \delta_{i,k,m,t} \cdot \xi_{i,m,t} \le \gamma_{i,k,t}^{max} \quad \forall i \in \mathcal{I}, k \in \mathcal{K}_i, t \in \mathcal{T} \tag{14.17}$$

3.1.7 Industrial example: Multiple extents in a continuous processing plant

Wassick and Ferrio [9] show an industrial application at Dow of the extended RTN model for a continuous processing chemical plant. In the example, a 1-year production schedule for the plant is performed using the concept of multiple extents to produce three different isomers. Isomers 1 and 2 can be produced in two different grades, and Isomer 3 is produced in a single grade as shown in Fig. 14.8. Grade 1 isomers are produced in a single campaign that must last at least 14 days. Grade 2 isomers are also produced in a single campaign that lasts no less than 8 days. Campaign transitions take 3 days. Other parameters for the chemical plant are given in Table 14.1.

The network diagram for the extended RTN model for the problem is illustrated in Fig. 14.9. The diagram shown is for Grade 1 isomers. The other half of the network is the same as the one shown, where Grade 2 is used in place of Grade 1 and only two isomers are produced. A unique feature of this example is that time is modeled as a consumable resource. As a result, a nonuniform time grid is used, such that the number of time points must be specified a priori. The

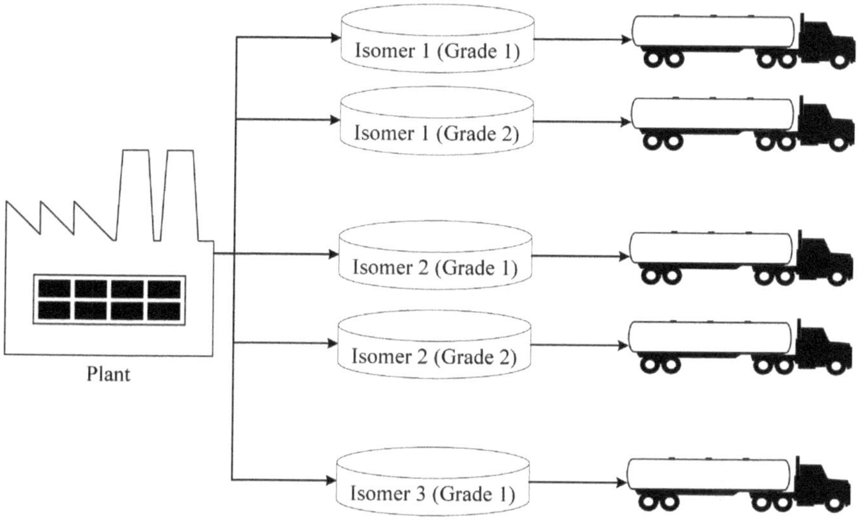

FIG. 14.8 Continuous processing plant with three products of different grades. *(Reproduced with permission from J.M. Wassick, J. Ferrio, Extending the resource task network for industrial applications, Comput. Chem. Eng. 35 (2011) 2124–2140. Copyright (2011) Elsevier, Ltd.)*

TABLE 14.1 Continuous processing plant parameters.

	Grade 1			Grade 2	
Parameter[a]	*Isomer 1*	*Isomer 2*	*Isomer 3*	*Isomer 1*	*Isomer 2*
Safety Stock	800	500	100	500	200
Inventory Capacity	5000	6049	1000	5000	2000
Daily Min. Production	75	50	6	140	20
Daily Max. Production	126	101	16	149	36
Daily Demand	95.89	76.71	12.05	21.41	4.33
Product Margin	$269.3	$361.2	$615.5	$674.0	$958.4
Inventory Cost	$27.0	$36.0	$62.0	$67.0	$96.0

[a]*All parameters are on a Mlbs basis.*
Reproduced with permission from J.M. Wassick, J. Ferrio, Extending the resource task network for industrial applications, Comput. Chem. Eng. 35 (2011) 2124–2140. Copyright (2011) Elsevier, Ltd.

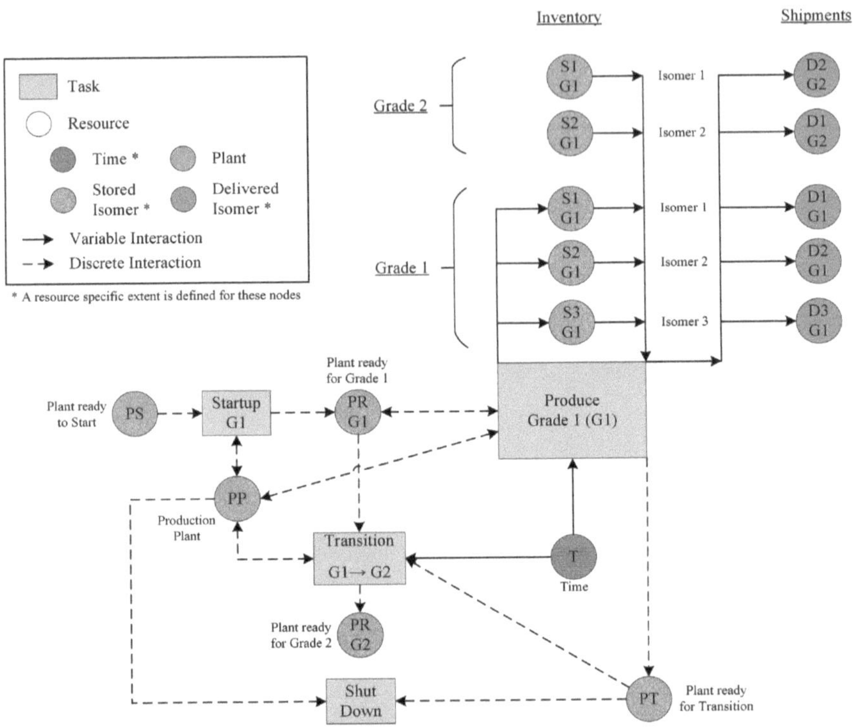

FIG. 14.9 RTN diagram for Grade 1 production and product shipment in the continuous processing plant.

duration of each task is a single time slot (i.e., $\tau_i = 1$), and the duration of each time slot is dictated by the amount of time resource consumed in each task. The time resource is initialized at 365 days and is fixed to 0 at the last time point. Determining the minimum number of time points in this discrete-time model is done offline. A few additional time points are then used with dummy tasks that consume no time to ensure that the number of time slots is sufficient and ensure that any unnecessary slots are consumed. There is a dummy task for each grade. The dummy task for Grade 1 is not shown in Fig. 14.9 but is a task that consumes the *Production Plant* resource (*PP* node) and the *Plant ready for Grade 1* resource (*PRG*1 node) at the beginning of the time slot and releases them at the end of the slot. Note that this task has zero temporal duration because the time resource is not involved.

The *Produce Grade 1* task has three types of extents: (1) campaign duration, (2) isomer production, and (3) isomer shipment. In the first extent type, the purple (gray in print version) time resource is consumed (at the end of the time slot). The second extent type has three unique extents (one for each Grade 1 isomer), in which the green (gray in print version) isomer storage resources are filled with material produced (at the end of the time slot). The last extent type has five unique extents (one for each product), in which the green (gray in print version) isomer storage resources are drained (consumed) and the blue (gray in print version) isomer shipment resources are filled (produced). Extents of the last type are indexed at the end of the time slot, allowing the model to monitor stored inventory and shipped products at the end of the campaign. Production extents are constrained to be within the production limits in Table 14.1 multiplied by the time duration extent (i.e., $\dot{Q}_{G1,m}^{min} \cdot \xi_{G1,Time,t} \leq \xi_{G1,m,t} \leq \dot{Q}_{G1,m}^{max} \cdot \xi_{G1,Time,t}\ \forall m \in \mathcal{M}_{G1,2}, t \in \mathcal{T}$ where $\mathcal{M}_{G1,2}$ are the isomer production extents [type 2] in Grade 1).

Five resources are associated with the chemical plant (orange nodes, gray in print version, in Fig. 14.9), representing the plant itself (*PP* resource) and the four states it can be in initial state before startup (*PS* resource), plant ready for production of a certain grade (*PRG*1 and *PRG*2 resources), and plant ready for a grade transition (*PT* resource). Each plant state can be either zero or one and the plant resource (*PP*) is forced to zero to ensure that the plant is always in use after startup. Isomer storage resources (green nodes, gray in print version, in Fig. 14.9) are bounded by the safety stocks and the inventory capacities given in Table 14.1. Isomer shipment resources (blue nodes, dark gray in print version, in Fig. 14.9) are bounded between zero and the annual demand for that isomer from Table 14.1.

In order to avoid unrealistic end-of-horizon effects on the produced schedule, the RTN model is constrained so that the ending inventory of each isomer matches its initial inventory: $R_{r,h} - R_r^{slack} \leq R_{r,0} \leq R_{r,h} + R_r^{slack}\ \forall r \in \mathcal{R}^P$ where $\mathcal{R}^P$ is the set of green nodes (gray in print version) in Fig. 14.9 (stored product resources), and R_r^{slack} is a slack variable that is penalized in the objective. The objective function that is maximized is shown in Eq. (14.18), which includes the annual revenue for shipped resources minus the inventory holding costs and penalties for any nonzero slacks. Since time is a variable (resource), nonlinearities in the objective are avoided by approximating the inventory cost using the average inventory for each product over the horizon: $I_r^{ave} = \sum_{t \in \mathcal{T}} R_{r,t}/H\ \forall r \in \mathcal{R}^P$. The network model consists of 7567 equations, 2955 continuous variables, and 180 binary variables, and was solved using CPLEX 10.1.0 on a 1.83 GHz CPU. The system was optimized in 51 s with an optimality gap of 0.06%. The optimized production schedule is shown in Fig. 14.10, which shows a larger production scheduled in Grade 1 campaigns. Inventory profiles for each product

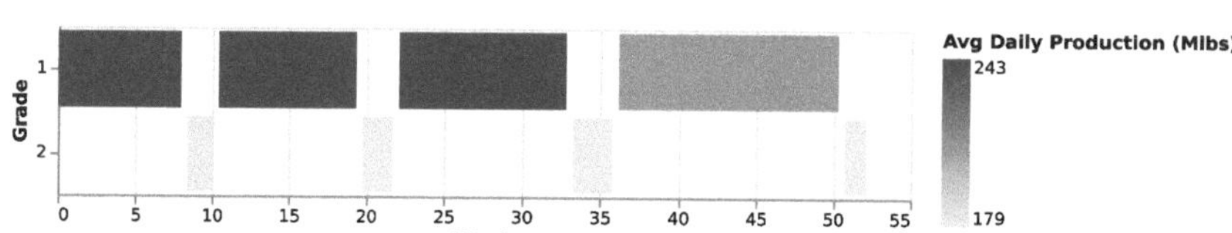

FIG. 14.10 Production schedule indicating the average daily production for each campaign.

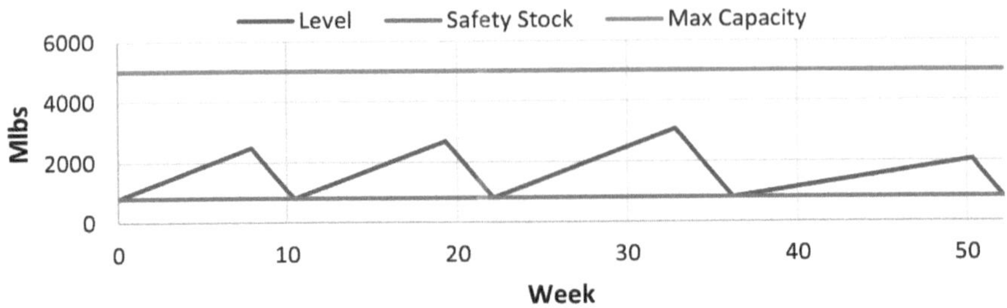

FIG. 14.11 Optimized inventory profile for Isomer 1 (Grade 1). *(Reproduced with permission from J.M. Wassick, J. Ferrio, Extending the resource task network for industrial applications, Comput. Chem. Eng. 35 (2011) 2124–2140. Copyright (2011) Elsevier, Ltd.)*

are obtained as shown in Fig. 14.11 for Grade 1 Isomer 1. The average daily product shipping rates obtained are quite stable, exhibiting minor variations throughout the year (relative standard deviations ~0.1%).

$$\max \phi = \sum_{t \in \mathcal{T}} \sum_{r \in \mathcal{R}^S} p_r \cdot R_{r,t} - \sum_{r \in \mathcal{R}^P} c_r^{inv} \cdot I_r^{ave} - \sum_{r \in \mathcal{R}^P} \sum_{t \in \mathcal{T}} c_r^{max} \cdot R_{r,t}^{max,slack} - \sum_{r \in \mathcal{R}^P} \sum_{t \in \mathcal{T}} c_r^{min} \cdot R_{r,t}^{min,slack} - \sum_{r \in \mathcal{R}^P} c_r^{slack} \cdot R_r^{slack} \tag{14.18}$$

3.2 State-space reformulation of extended RTN model

Nie et al. [10] propose a reformulation of the extended discrete-time RTN model for reactive scheduling applications with a moving horizon. The extended RTN model is reformulated to the state-space paradigm commonly used in process control. The reformulation requires lifting variables to introduce new state variables so that the system state can be fully specified at any point in time using the current state variables. The input variables in the model are the task assignments ($N_{i,t}$) and task extents, $\xi_{i,m,t}$. The state variables are the resource levels ($R_{r,t}$), resource limits ($R_{r,t}^{max}$ and $R_{r,t}^{min}$), and task history (lifted variables, $N_{i,t,\theta}$ and $\xi_{i,m,t,\theta}$). The lifted state variables ($N_{i,t,\theta}$ and $\xi_{i,m,t,\theta}$) record the past decisions in the system (i.e., the task assignment or extent for task i at time t that has been running actively for θ periods). Note that the input variables ($N_{i,t}$ and $\xi_{i,m,t}$) can be considered lifted variables for which $\theta = 0$. From a process control perspective, the external resource transfer term ($\Pi_{r,t}$) in the resource inventory balance is viewed as a system disturbance. In addition, the following dynamic state evolution equations are introduced,

$$N_{i,t,\theta} = N_{i,t-1,\theta-1} + \left(N_{i,t-1,\theta} - N_{i,t-1,\theta-1}\right) \cdot \sum_{\theta'=0}^{\theta} d_{i,t-1,\theta'} \quad \forall i \in \mathcal{I}, t \in \mathcal{T}, 1 \le \theta \le \tau_i \tag{14.19}$$

$$\xi_{i,m,t,\theta} = \xi_{i,m,t-1,\theta-1} + \left(\xi_{i,m,t-1,\theta} - \xi_{i,m,t-1,\theta-1}\right) \cdot \sum_{\theta'=0}^{\theta} d_{i,t-1,\theta'} \quad \forall i \in \mathcal{I}, m \in \mathcal{M}_i, t \in \mathcal{T}, 1 \leq \theta \leq \tau_i \tag{14.20}$$

Fixed binary disturbance parameters, $d_{i,t,\theta}$, are used to model unexpected delays in task durations, such that $d_{i,t,\theta}=1$ indicates a one period delay at time point t on task i, which has been actively running for θ periods. Consecutive disturbance parameters can be used ($d_{i,t,\theta}$, …, $d_{i,\ t+\tau_d-1,\ \theta}$) to model longer task delays of duration τ_d. Fig. 14.12 illustrates an example of the dynamic state evolution of a task with a 2-period duration that is triggered at $t=0$ (green state node, gray print version). Without disruptions, the state evolution is $N_{i,0,0} \rightarrow N_{i,1,1} \rightarrow N_{i,2,2}$. If a 1-period delay occurs at $t=1$ ($d_{i,1,1}=1$), the evolution is $N_{i,0,0} \rightarrow N_{i,1,1} \rightarrow N_{i,2,1} \rightarrow N_{i,3,2}$. If a 2-period delay occurs at $t=1$, which is equivalent to an additional 1-period delay at $t=2$ ($d_{i,2,1}=1$), the evolution is $N_{i,0,0} \rightarrow N_{i,1,1} \rightarrow N_{i,2,1} \rightarrow N_{i,3,1} \rightarrow N_{i,4,2}$.

Using a state-space formulation has important advantages. The dynamic state of the system can be fully described with the state variables at any given time. Furthermore, the future state of the system can be forecast using the input and state variables. This makes this reformulation amenable to advanced process control applications, such as model predictive control.

3.2.1 *Industrial example: Online scheduling of a mixed batch/continuous processing plant*

Nie et al. [10] present an industrial application of online RTN scheduling for the processing plant with both batch and continuous unit operations depicted in

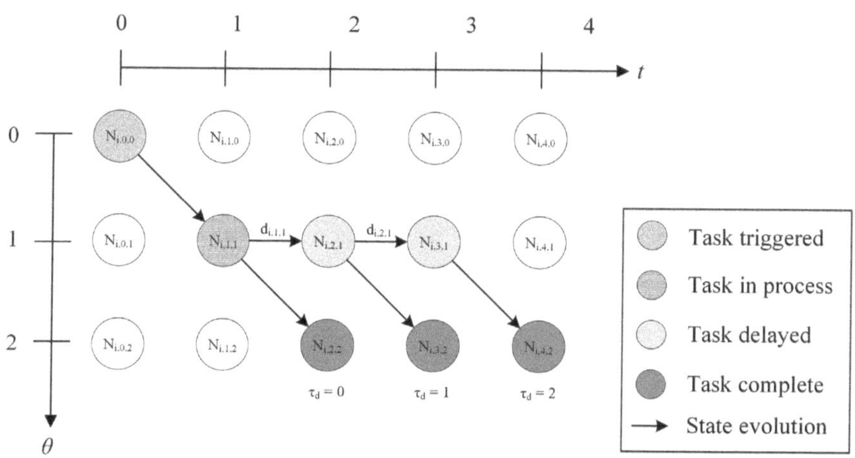

FIG. 14.12 Sample dynamic state evolution of a 2-period task with different delays. *(Reproduced with permission from Y. Nie, L.T. Biegler, J.M. Wassick, C.M. Villa, Extended discrete-time resource task network formulation for the reactive scheduling of a mixed batch/continuous process, Ind. Eng. Chem. Res. 53 (2014) 17112–17123. Copyright (2014) American Chemical Society.)*

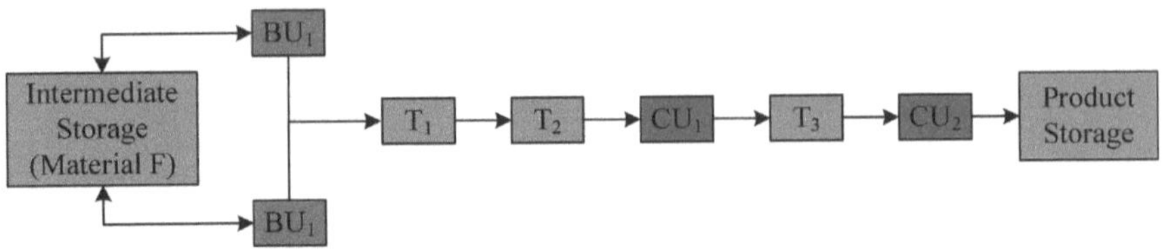

FIG. 14.13 Block flow diagram for plant with batch units (BU), buffer tanks (T), continuous units (CU), and storage tanks. *(Reproduced with permission from Y. Nie, L.T. Biegler, J.M. Wassick, C.M. Villa, Extended discrete-time resource task network formulation for the reactive scheduling of a mixed batch/continuous process, Ind. Eng. Chem. Res. 53 (2014) 17112–17123. Copyright (2014) American Chemical Society.)*

Fig. 14.13. The system consists of two parallel batch units that are followed by a serial train with two buffer tanks, a continuous processing unit, another buffer tank, and a final continuous unit, in that order. The plant produces five types of products (A–E). Products A, C, and D belong to the same product family, and B and E are their own families. Raw materials are assumed to be unlimited, except for an intermediate F that can be produced in the batch units and is consumed to produce A, B, and E. The system operation is governed by the following characteristics,

- Batch durations and sizes are fixed,
- Batch units have a zero-wait transfer policy,
- Buffer tanks have variable inlet/outlet flow rates,
- Continuous units have variable processing rates and zero residence times,
- Continuous units store no inventory,
- Product mixing is not allowed,
- Changeover tasks must be considered for different product families in the continuous units,
- The second continuous unit has a consumable resource (e.g., filter) that requires shutting down for its regeneration/replacement, and
- Materials A–F have unlimited storage capacity.

The features described above, along with the plant structure and characteristics are modeled using the extended state-space RTN model. The objective function maximizes is given in Eq. (14.21), which includes the revenue for order fulfillment within the specified time windows and penalties for product transition tasks, buffer tank inventories, and violating the safety stock levels at the end of the scheduling horizon. Sales prices for materials are indexed by t ($p_{o,r,t}$), such that higher revenue is received for early fulfillment. An important feature in the model is the small penalty on the buffer tank inventory levels. This is done to break the symmetry in the flow transfer profiles when the continuous processing units are operated under full capacity. In such instances, the model forces flow from the buffer tanks to occur as fast as possible. This avoids oscillatory behavior in the flow profiles, which is operationally favorable. See Nie et al. [10] for model parameter values.

$$\max \phi = \sum_{o \in \mathcal{O}} \sum_{r \in \mathcal{R}_o} \sum_{t=E_o}^{D_o} p_{o,r,t} \cdot (-\Pi_{r,t}) - \sum_{i \in \mathcal{I}^{TR}} \sum_{t \in \mathcal{T}} c_i^{TR} \cdot N_{i,t} - \sum_{r \in \mathcal{R}^{ST}} \sum_{t \in \mathcal{T}} c_r^{inv} \cdot R_{r,t} - \sum_{r \in \mathcal{R}^{P}} \sum_{t \in \mathcal{T}} c_r^{min} \cdot R_{r,h}^{min,slack} \quad (14.21)$$

The model is used for the deterministic scheduling of the plant with a look-ahead horizon of 72 hours (3 days). The product orders for each day are indicated in Table 14.2, where the quantity of each material ordered on a given day is 70 units. After the first 24 hours, a rescheduling is performed looking ahead for days 2–4. Three scenarios are considered: (1) no disruptions, (2) a scheduled 11-hour delay on the first batch unit (BU_1) starting on day 2, and (3) a scheduled 13-hour maintenance on the last buffer tank (T_3) at hour 60. Delays are readily introduced in the second scenario using the disturbance parameters in the dynamic state evolution equations. The maintenance event is produced by setting the upper resource limit of the buffer tank to zero for the 13 hours in the maintenance event. The model in each of the cases consists of 140,382 constraints and 143,148 variables (4380 discrete) and was solved within a 10-minute limit using Gurobi 5.5.0. After 10 minutes, the optimality gap was less than 5% in each run. The resulting schedule for the first 72 hours is shown in Fig. 14.14 with inventory profiles shown in Fig. 14.15. Similar schedules for the next three scenarios are given in Nie et al. [10]. The advantage of the state-space RTN model is in the implementation of the three scenarios for days 2–4, which are readily initialized with the relevant system history from the previous 24 hours of operation stored in the state variables. The rescheduling at $t = 24$ is smooth and does not introduce drastic operational changes. The state-space model is also amenable to responding to system disturbances.

TABLE 14.2 Product orders matrix for materials A–E (all orders are for 70 units).

Day	A	B	C	D	E
1	X				
2	X		X		X
3	X	X	X	X	
4	X			X	

Reproduced with permission from Y. Nie, L.T. Biegler, J.M. Wassick, C.M. Villa, Extended discrete-time resource task network formulation for the reactive scheduling of a mixed batch/continuous process, Ind. Eng. Chem. Res. 53 (2014) 17112–17123. Copyright (2014) American Chemical Society.

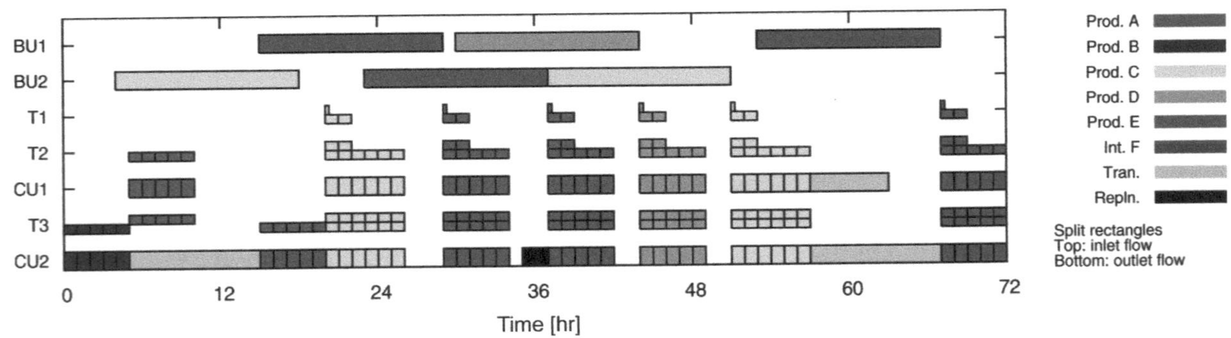

FIG. 14.14 Optimal schedule for the first 72 hours in the mixed batch/continuous plant. *(Reproduced with permission from Y. Nie, L.T. Biegler, J.M. Wassick, C.M. Villa, Extended discrete-time resource task network formulation for the reactive scheduling of a mixed batch/continuous process, Ind. Eng. Chem. Res. 53 (2014) 17112–17123. Copyright (2014) American Chemical Society.)*

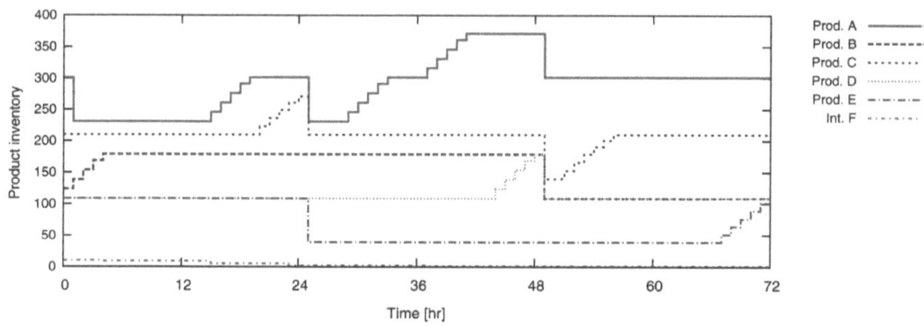

FIG. 14.15 Inventory profiles for the first 72 hours in the mixed batch/continuous plant. *(Reproduced with permission from Y. Nie, L.T. Biegler, J.M. Wassick, C.M. Villa, Extended discrete-time resource task network formulation for the reactive scheduling of a mixed batch/continuous process, Ind. Eng. Chem. Res. 53 (2014) 17112–17123. Copyright (2014) American Chemical Society.)*

3.3 RTN for spatial packing problems

Wassick and Ferrio [9] also present a new application for RTN outside of process scheduling: 1D and 2D packing problems. The RTN model for 2D strip packing optimization is time agnostic and uses a single resource: horizontal (x) floor space at location y ($R_{x,\ y}$). Thus, the number of resources is equal to the number of discrete points on the x-axis. Instead of tracking each resource over time, the resources are tracked along the vertical (y) axis. Each horizontal resource has an initial condition of unity and an upper bound of unity:

$$R_{x,-1} = 1 \quad \forall x \in \mathcal{X} \tag{14.22}$$

$$0 \leq R_{x,y} \leq 1 \quad \forall x \in \mathcal{X}, y \in \mathcal{Y} \tag{14.23}$$

A placement task i is defined as placing a rectangular package of type i with width W_i and length L_i on the floor plan. When placing a package, the (x,y)-coordinates of the bottom left corner of the rectangular box must be specified (see red X's, gray in print version, in Fig. 14.17). Thus, instead of indexing the task triggering variable with the task i and "time" y as in the standard RTN model, the triggering binary variable is also indexed by the resource x ($N_{i,x,y}$). When a rectangular box is placed at a position (x,y), it restricts the possibility of placing another box in the positions between x and $x+L_i-1$ for that same y-coordinate. Thus, triggering the task at position y, consumes the horizontal resources $R_{x',\ y}$ for $x \leq x' \leq x+L_i-1$ as indicated in Fig. 14.16. This, in turn, implies that a resource x is consumed for any boxes of type i that are placed in a position $x' \in \mathcal{X}_i$ such that $x-L_i+1 \leq x' \leq x$. The floor space inventory balance becomes:

$$R_{x,y} = R_{x,y-1} + \sum_{i \in \mathcal{I}} \sum_{x' \in \mathcal{X}_i} \sum_{\theta=0}^{W_i} \mu_{i,x,\theta} \cdot N_{i,x',y-\theta} \quad \forall x \in \mathcal{X}, y \in \mathcal{Y} \tag{14.24}$$

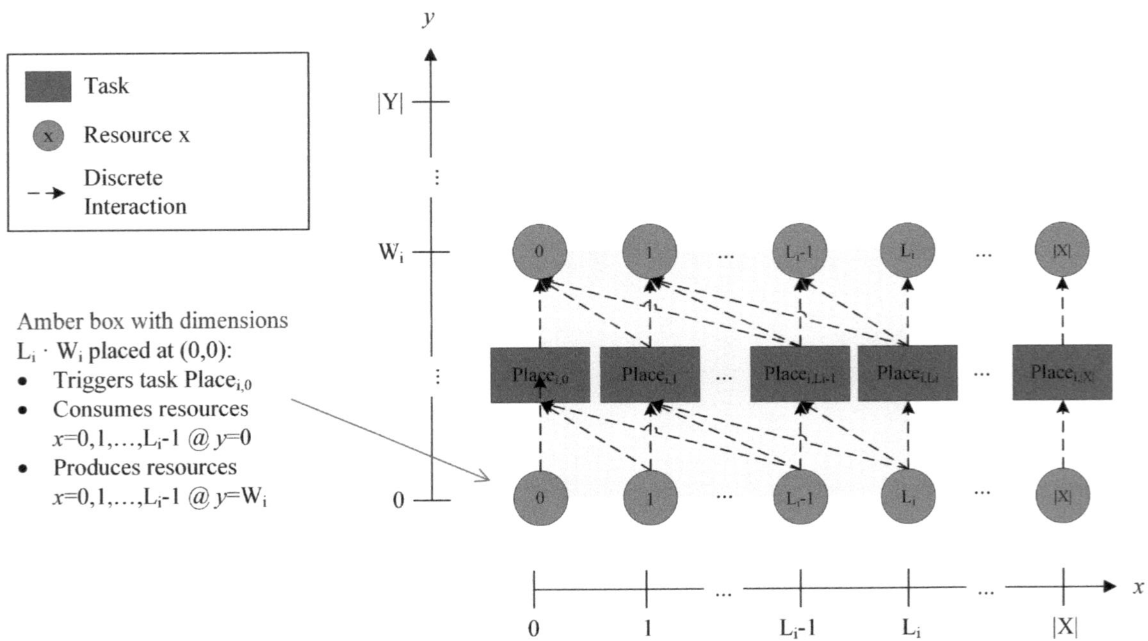

FIG. 14.16 RTN representation for a 2D strip packing problem.

Note that a resource x is consumed when the bottom left corner of a box is placed at position (x, y), and is then restored at the top left corner of the box (vertical position $y+W_i$), meaning a new box can be placed on position $(x, y+W_i)$. The width of the box can be viewed as the task "duration". The discrete consumption/production ratio μ takes the following values: -1 when $\theta=0$, $+1$ when $\theta=W_i$, and 0 otherwise. As a result, Eq. (14.24) can be expressed as:

$$R_{x,y} = R_{x,y-1} + \sum_{i \in \mathcal{I}} \sum_{x'=x-L_i+1}^{x} \left(N_{i,x',y-W_i} - N_{i,x',y} \right) \quad \forall x \in \mathcal{X}, y \in \mathcal{Y} \tag{14.25}$$

For the simpler 1D strip packing problem, the x dimension used in the 2D case can be dropped and the system is modeled as having a single resource along the vertical length (y) of the strip.

3.3.1 Industrial example: Payload loading optimization

Wassick and Ferrio [9] present an application of the spatial RTN model to load a semitrailer for payload transport. The objective is to maximize the number of packages that are loaded onto the floor space of the semitrailer. Package stacking is not allowed, and highway weight restrictions enforce a maximum total payload weight and a maximum load on the rear (tandem) axle. The weight limitations are introduced using the force balance illustrated in Fig. 14.17. The force balance for the packages (m_i), the kingpin (F_K), and the tandem (rear) axle (F_T) is represented in Eq. (14.26). The total weight constraint is given in Eq. (14.27), the moment balance in Eq. (14.28), and the rear axle weight limit in Eq. (14.29). Four different package types are allowed on the 10-unit long semitrailer. See Wassick and Ferrio [9] for model parameter values.

$$F_T + F_K = \sum_{i \in \mathcal{I}} \sum_{x \in \mathcal{X}} \sum_{y \in \mathcal{Y}} N_{i,x,y} \cdot m_i \tag{14.26}$$

$$F_T + F_K \leq W^{max} \tag{14.27}$$

$$\sum_{i \in \mathcal{I}} \sum_{x \in \mathcal{X}} \sum_{y \in \mathcal{Y}} N_{i,x,y} \cdot m_i \cdot \left(x + \frac{L_i}{2} \right) = D_K \cdot F_K + D_T \cdot F_T \tag{14.28}$$

$$F_T \leq F_T^{max} \tag{14.29}$$

The model objective is to maximize the payload weight (F_T+F_K). The resulting model (131 binary variables, 211 continuous variables, and 77 equations) was solved using CPLEX 10.1.1 at the parent node (110 iterations) without branching. The optimized loading layout is shown in Fig. 14.17, where 12 boxes that weigh a total of 44 units are loaded, placing the maximum allowable load on the rear axle (18 units). The spatial RTN model has been successfully applied at Dow to maximize payload and reduce freight costs. Additional

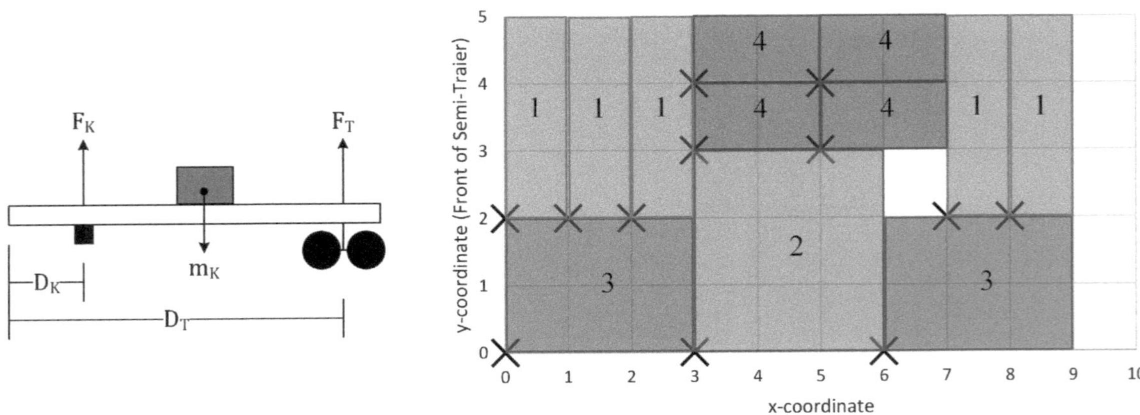

FIG. 14.17 Semitrailer force diagram (left) and optimized layout (right) for payload packing problem. *(Reproduced with permission from J.M. Wassick, J. Ferrio, Extending the resource task network for industrial applications, Comput. Chem. Eng. 35 (2011) 2124–2140. Copyright (2011) Elsevier, Ltd.)*

constraints have also been included to stay within axle load limits, account for rotating packages, and other special packaging arrangements.

3.4 RTN for transactional business process optimization

Another interesting application of the RTN model currently under development is for the online optimization of business transactional processes in a supply chain [12]. Some common business processes found in supply chains are the order-to-cash (OTC) and procure-to-pay (PTP) processes. These processes describe the end-to-end transactions that must occur on a request (e.g., actions that must be performed between when a customer places an order and payment is received for the fulfilled order). A digital twin framework has been developed to model supply chain business processes as transactional queueing networks. This concept is illustrated in Fig. 14.18. As requests (e.g., customer orders) arrive and move downstream, they go through a series of steps that are performed on the request (e.g., credit check, import/export customs forms, inventory confirmation, shipment loading, and invoice generation). The steps need not be serial, as there may be parallel transactions, as well as deviations from the "normal" flow of a request through the system (e.g., to correct errors in the order or reschedule a shipment). At each transaction, there are a limited number of resources assigned to that transaction (e.g., human or automated agents). Each transaction has a duration that is uncertain and is modeled as a continuous probability distribution, which can be specific to the request type, requestor, or agent.

A digital replica of the real processes is created from input data via a discrete event simulation that has an optimization loop as shown in Fig. 14.19. The optimization loop runs as an event in the discrete event simulation that takes as its input the current state of the business process (e.g., active requests being executed and queued requests at the different transactions in the network), along with the transactional network structure, model parameters, request details, and resource mapping. An RTN model of the business process is built with these inputs to run a scheduling optimization that maximizes the number of orders that are fulfilled on time. The resulting schedule is then translated into queue priorities and any task preemptions, which get passed to the simulation. The simulation then continues running with the new prioritized queues and preemptions until the optimization loop is triggered again. The RTN model is updated with the new system state each time the optimization event is triggered, which can occur on a periodic basis or conditional to an event occurring (e.g., system disturbance). Note that the time required to solve the optimization is critical as this dictates the duration of the optimization event in the simulation. Long optimization run times may result in queue priorities that are no longer relevant if the simulation has moved significantly from the original state passed to the RTN model.

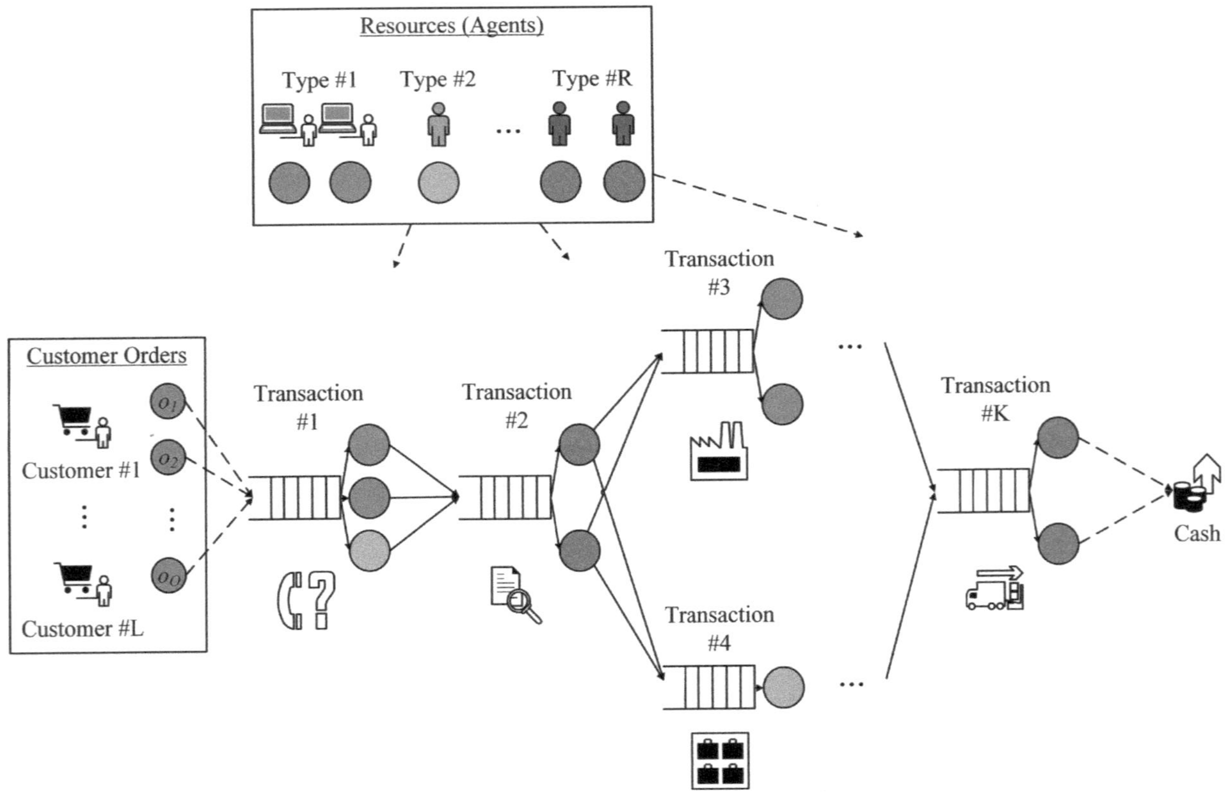

FIG. 14.18 Sample network of transaction queues in the order-to-cash business process.

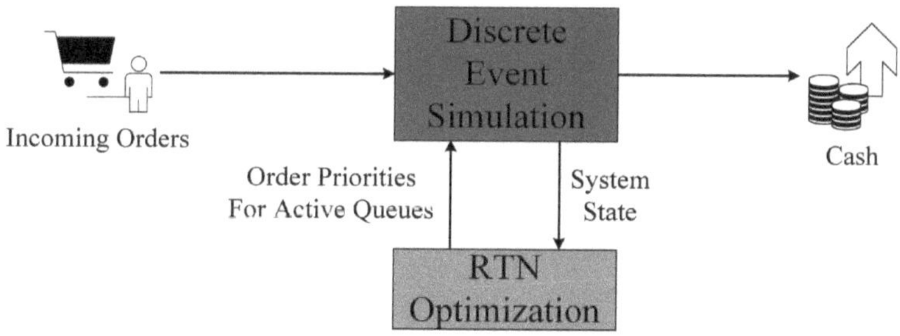

FIG. 14.19 Integrated simulation-optimization framework.

Stochasticity is introduced in the simulation via uncertainty in processing times, uncertainty in the occurrence of an exception that would require a request to take an alternate path in the network or to be recycled for reprocessing, uncertainty in request arrival dates, and uncertainty in a customer reneging. Initial case studies have shown that using an optimization loop that runs a transactional RTN has the potential to bring substantial profit gains relative to an unoptimized first-in-first-out (FIFO) method for processing incoming orders.

4 Industrial impact

The RTN framework and its extensions have had a substantial impact at Dow As an example, the online reactive scheduling using the state-space RTN has represented an annual increase in profit of over $5 million because of improved productivity and more stable operations. Since its implementation, batch wait times decreased by over 50%, operating capacity increased by 5%, and the number of manual plant interventions decreased by 75% [33]. The technological success of the application of RTN for integrated scheduling and dynamic modeling also won a Manufacturing Leadership Award by Frost & Sullivan's Manufacturing Leadership Council in 2014.

5 Conclusions

An overview of the classical RTN modeling framework in both discrete- and continuous-time has been presented in the context of process scheduling optimization, where it has proven to be a mathematically compact model, especially for problems with identical processing units and multiple pathways. Extensions to the classical model structure have been described, including (1) the network structure for quality-based changeovers, (2) constraints for dealing with external material transfers with time windows, (3) constraints to ensure the fulfillment of orders in single shipments, (4) resource limit balances to mathematically simplify material transfers, (5) constraint modifications to allow for soft constraints (e.g., minor safety stock violations), and (6) model complexity reduction via multiple task extents.

As a result of the modeling flexibility of the RTN, creativity, and expertise by the practitioner are often required when mapping existing problems onto the resource-task paradigm. For example, resources need to be properly defined, along with their interactions in the different stages of a process, which includes populating all the necessary modeling parameters and describing the system constraints. Although this may not always be as intuitive as in other modeling paradigms, once the mapping is performed, the RTN becomes a powerful tool for optimization. The computational advantages of the RTN can then be exploited to tackle many different problems, from traditional offline batch scheduling to reactive scheduling, spatial packing optimization, and business process optimization. At Dow, such novel applications of RTN have and are making a positive impact.

From an implementation standpoint, it is important to ensure that RTN model parameters can be seamlessly updated by the end-user, who is often not the original model developer, to enable its continued use in practical applications. Attempts have been made in the past to develop software packages for batch scheduling using the RTN paradigm, but these were never commercialized (e.g., gBSS software [34]). Regardless of whether such a software package is ever developed, the RTN model is expected to continue to deliver value to many industries as well as in other applications within the chemical industry.

Acknowledgments

The authors gratefully acknowledge the support and funding from The Dow Chemical Company, and the Center for Advanced Process Decision-making at Carnegie Mellon University.

References

[1] G.P. Georgiadis, A.P. Elekidis, M.C. Georgiadis, Optimization-based scheduling for the process industries: from theory to real-life industrial applications, Processes 7 (2019) 438, https://doi.org/10.3390/pr7070438.

[2] C. Maravelias, Chemical Production Scheduling: Mixed-Integer Programming Models and Methods, Cambridge University Press, Cambridge, 2021.

[3] C.A. Méndez, J. Cerdá, I.E. Grossmann, I. Harjunkoski, M. Fahl, State-of-the-art review of optimization methods for short-term scheduling of batch processes, Comput. Chem. Eng. 30 (2006) 913–946, https://doi.org/10.1016/j.compchemeng.2006.02.008.

[4] C.A. Floudas, X. Lin, Continuous-time versus discrete-time approaches for scheduling of chemical processes: a review, Comput. Chem. Eng. 28 (2004) 2109–2129, https://doi.org/10.1016/j.compchemeng.2004.05.002.

[5] N. Shah, C.C. Pantelides, R.W.H. Sargent, A general algorithm for short-term scheduling of batch operations—II. Computational issues, Comput. Chem. Eng. 17 (1993) 229–244, https://doi.org/10.1016/0098-1354(93)80016-G.

[6] R. Bixby, E. Rothberg, Progress in computational mixed integer programming—a look back from the other side of the tipping point, Ann. Oper. Res. 149 (2007) 37–41, https://doi.org/10.1007/s10479-006-0091-y.

[7] J.X. Chen, The evolution of computing: AlphaGo, Comput. Sci. Eng. 18 (2016) 4–7, https://doi.org/10.1109/MCSE.2016.74.

[8] I. Harjunkoski, C.T. Maravelias, P. Bongers, P.M. Castro, S. Engell, I.E. Grossmann, J. Hooker, C. Méndez, G. Sand, J. Wassick, Scope for industrial applications of production scheduling models and solution methods, Comput. Chem. Eng. 62 (2014) 161–193, https://doi.org/10.1016/j.compchemeng.2013.12.001.

[9] J.M. Wassick, J. Ferrio, Extending the resource task network for industrial applications, Comput. Chem. Eng. 35 (2011) 2124–2140, https://doi.org/10.1016/j.compchemeng.2011.01.010.

[10] Y. Nie, L.T. Biegler, J.M. Wassick, C.M. Villa, Extended discrete-time resource task network formulation for the reactive scheduling of a mixed batch/continuous process, Ind. Eng. Chem. Res. 53 (2014) 17112–17123, https://doi.org/10.1021/ie500363p.

[11] N. Akiya, S. Bury, J.M. Wassick, Generic framework for simulating networks using rule-based queue and resource-task network, in: S. Jain, R.R. Creasey, J. Himmelspach, K.P. White, M. Fu (Eds.), Proceedings of the 2011 Winter Simulation Conference, IEEE, Phoenix, 2011, pp. 2194–2205.

[12] H.D. Perez, S. Amaran, E. Erisen, J.M. Wassick, I.E. Grossmann, A digital twin framework for business transactional processes in supply chains, Computer Aided Chemical Engineering, vol. 50, Elsevier, 2021, pp. 1755–1760.

[13] V. Chen, Plenary: Planning Transportation Capacity at Amazon—2019 INFORMS Annual Meeting | October 20–23, 2019, 2019. http://meetings2.informs.org/wordpress/seattle2019/2019/10/23/plenary-planning-transportation-capacity-at-amazon/. (Accessed 29 January 2021).

[14] E. Kondili, C.C. Pantelides, R.W.H. Sargent, A general algorithm for short-term scheduling of batch operations—I. MILP formulation, Comput. Chem. Eng. 17 (1993) 211–227, https://doi.org/10.1016/0098-1354(93)80015-F.

[15] C.C. Pantelides, Unified frameworks for optimal process planning and scheduling, in: Proceedings on the Second Conference on Foundations of Computer Aided Operations, 1994, pp. 253–274.

[16] R.J. Dakin, A tree-search algorithm for mixed integer programming problems, Comput. J. 8 (1965) 250–255, https://doi.org/10.1093/comjnl/8.3.250.

[17] E. Balas, S. Ceria, G. Cornuéjols, A lift-and-project cutting plane algorithm for mixed 0-1 programs, Math. Program. 58 (1993) 295–324, https://doi.org/10.1007/BF01581273.

[18] E.L. Johnson, G.L. Nemhauser, M.W.P. Savelsbergh, Progress in linear programming-based algorithms for integer programming: an exposition, INFORMS J. Comput. 12 (2000) 2–23, https://doi.org/10.1287/ijoc.12.1.2.11900.

[19] A.M. Geoffrion, Generalized benders decomposition, J. Optim. Theory Appl. 10 (1972) 237–260, https://doi.org/10.1007/BF00934810.

[20] M.A. Duran, I.E. Grossmann, An outer-approximation algorithm for a class of mixed-integer nonlinear programs, Math. Program. 36 (1986) 307–339, https://doi.org/10.1007/BF02592064.

[21] L. Su, L. Tang, D.E. Bernal, I.E. Grossmann, Improved quadratic cuts for convex mixed-integer nonlinear programs, Comput. Chem. Eng. 109 (2018) 77–95, https://doi.org/10.1016/j.compchemeng.2017.10.011.

[22] T. Westerlund, F. Pettersson, An extended cutting plane method for solving convex MINLP problems, Comput. Chem. Eng. 19 (1995) 131–136, https://doi.org/10.1016/0098-1354(95)87027-X.

[23] F. Glover, Improved linear integer programming formulations of nonlinear integer problems, Manag. Sci. 22 (1975) 455–460, https://doi.org/10.1287/mnsc.22.4.455.

[24] S. Sridhar, J. Linderoth, J. Luedtke, Locally ideal formulations for piecewise linear functions with indicator variables, Oper. Res. Lett. 41 (2013) 627–632, https://doi.org/10.1016/j.orl.2013.08.010.

[25] V. Gupta, I.E. Grossmann, An efficient multiperiod MINLP model for optimal planning of offshore oil and gas field infrastructure, Ind. Eng. Chem. Res. 51 (2012) 6823–6840, https://doi.org/10.1021/ie202959w.

[26] J. Ostrowski, M.F. Anjos, A. Vannelli, Formulations for the unit commitment problem, IEEE Trans. Power Syst. 27 (2012) 39–46, https://doi.org/10.1109/TPWRS.2013.2251373.

[27] H. Marchand, A. Martin, R. Weismantel, L. Wolsey, Cutting planes in integer and mixed integer programming, Discret. Appl. Math. 123 (2002) 397–446, https://doi.org/10.1016/S0166-218X(01)00348-1.

[28] P. Castro, A.P.F.D. Barbosa-Póvoa, H. Matos, An improved RTN continuous-time formulation for the short-term scheduling of multipurpose batch plants, Ind. Eng. Chem. Res. 40 (2001) 2059–2068, https://doi.org/10.1021/ie000683r.

[29] P.M. Castro, A.P. Barbosa-Póvoa, H.A. Matos, A.Q. Novais, Simple continuous-time formulation for short-term scheduling of batch and continuous processes, Ind. Eng. Chem. Res. 43 (2004) 105–118, https://doi.org/10.1021/ie0302995.

[30] H. Lee, C.T. Maravelias, Combining the advantages of discrete- and continuous-time scheduling models: part 3. General algorithm, Comput. Chem. Eng. 139 (2020) 106848, https://doi.org/10.1016/j.compchemeng.2020.106848.

[31] H.D. Perez, S. Amaran, E. Erisen, J.M. Wassick, I.E. Grossmann, Optimization of extended business processes in digital supply chains using mathematical programming, Comput. Chem. Eng. 152 (2021) 107323.

[32] B. Brunaud, H.D. Perez, S. Amaran, S. Bury, J. Wassick, I.E. Grossmann, Batch scheduling with quality-based changeovers, Comput. Chem. Eng. 132 (2020), https://doi.org/10.1016/j.compchemeng.2019.106617.

[33] P. Hayot, Online scheduling and model-based optimization: a candid practitioner view, in: Foundations of Computer Aided Process Operations/Chemical Process Control, Tucson, 2017.

[34] N. Shah, Efficient Scheduling, Planning and Design of Multipurpose Batch Plants, Imperial College London (University of London), 1992.

Index

Note: Page numbers followed by *f* indicate figures and *t* indicate tables.

D

N

O

FSC
www.fsc.org
MIX
Papier | Fördert
gute Waldnutzung
FSC® C083411

Kolibri 360 GmbH
Ferdinand-Jühlke-Straße 7
99095 Erfurt, Deutschland
produktsicherheit@kolibri360.de